# Man-Machine System Experiments

Henry McIlvaine Parsons

*The Johns Hopkins Press*
*Baltimore and London*

Manufactured in the United States of America
The Johns Hopkins Press, Baltimore, Maryland 21218
The Johns Hopkins Press Ltd., London
Library of Congress Catalog Card Number 71-166483
ISBN 0-8018-1322-0

# Contents

# Preface

This book is the outcome of a study sponsored by the Aerospace Medical Research Laboratories of the U.S. Air Force when I was on the staff of the System Development Corporation. I wish to express my appreciation for the interest and support of Dwight E. Erlick, the Task Scientist for the project; Julien M. Christensen, Director of the Human Engineering Division; and Donald A. Topmiller, Chief of the Systems Effectiveness Branch.* Needless to say, the opinions expressed in this book are my own and should not be regarded as official views of the Air Force.

The study was undertaken to make better known a field of science that has many applications both civilian and military, and one to which many gifted researchers have devoted their efforts. It is hoped that those who work in this field in the future will be able to profit from a book which describes previous work, the methods used, and the problems encountered. A comprehensive account has not been hitherto available, and much of the research has been familiar only to those involved in particular investigations. I have been aware of the need for such an account because I myself was engaged in some of the early work.

Information for this book was acquired from a number of sources: the reports, articles, and books listed in the References; internally circulated documents of a few organizations; visits to numerous laboratories and individuals; and twelve consultants, all of whom had worked in the field. My data gathering for this book extended well into 1967.

In many instances it was necessary to visit a laboratory simply to ascertain what reports existed. In others a visit produced not only an overview valuable for comprehending individual studies but also data which had not appeared in reports. Although those who helped me are too numerous to list here, they should be assured of my gratitude. The organizations and laboratories visited included:

Applied Physics Laboratory, The Johns Hopkins University
Army Personnel Research Office
Bendix Systems Division, Bendix Corporation

*In 1969 this organization opened a new computer-based, four-terminal facility for system and component experimentation, the Human Engineering System Simulator (HESS).

Combat Development Command Experimentation Command
Decision Sciences Laboratory, Electronic Systems Division, U.S. Air Force
Disaster Research Center, Ohio State University
Human Factors Research, Incorporated
Human Performance Center, Ohio State University
Human Resources Research Office, George Washington University
Human Sciences Research, Inc.
Institute for Defense Analyses
Martin Company (Baltimore)
MITRE Corporation System Design Laboratory and other MITRE groups
National Aviation Facilities Experimental Center
Naval Research Laboratory
Navy Applied Science Laboratory
Navy Electronics Laboratory
North American Aviation (Columbus)
Office of Naval Research
Public Health Service, Division of Accident Prevention
RAND Corporation
Research Analysis Corporation
University of California, Los Angeles
Western Behavioral Sciences Institute
Willow Run Laboratories, University of Michigan

In addition, material pertaining to the extensive work done by subdivisions of the System Development Corporation (SDC) and by the RAND Corporation's Systems Research Laboratory could be acquired within SDC.

The consultants who helped assemble data and reviewed portions of this book were: Lawrence T. Alexander, William C. Biel, Alphonse Chapanis, Robert L. Chapman, Bruce L. Cusack, Harry H. Harman, William W. Haythorn, John L. Kennedy, Jerry S. Kidd, James C. McGuire, Harold Sackman, and H. Wallace Sinaiko. Their assistance has been extremely valuable. Not only did they hold a diversity of views but also they possessed imposing personal experience in the field. There should be no implication, however, that they are responsible for any of my biases or misapprehensions.

My thanks go to Dr. Chapanis and Lloyd V. Searle for helpful suggestions for improving the text.

To Mrs. Jean Fawley and Mrs. Juanita Hutchins I wish to express my appreciation for typing early versions at the System Development Corporation. I am most mindful of the perseverance of my secretary at Riverside Research Institute, Mrs. Linda Haviland, in preparing the final text. If these pages have a reasonable amount of clarity and good grammar, the reason is the assiduous editorial attention given them by my wife, Marjorie.

To few is such an opportunity given to do research on research. As the book progressed it seemed to me more than the review of a particular field. I hope that through the intensive examination of one area of application, this book also will serve its readers more broadly as a reconnaissance of the experimental method.

# Man-Machine System Experiments

# 1

# Introduction

In the last two decades a substantial amount of research has been conducted which, for want of a better term, may be called "man-machine system experiments." These have been large-scale experiments in which human subjects have interacted with machines and each other in complex system settings based to a considerable extent on simulation. Laboratory facilities, some of them elaborate, have been created for this purpose. In some cases knowledge was sought concerning the manned operation of a particular system or system component. In others the aim was to discover how human beings function in such system environments.

## NATURE OF MAN-MACHINE SYSTEM EXPERIMENTS

This experimentation has had characteristics in common with other kinds of research more familiar to behavioral scientists. It has differed not through the inclusion of any unique aspect but through a combination of aspects specified in Fig. 1. This cluster diagram shows the eight principal characteristics which converge to distinguish man-machine system experiments. Although in some experiments one or two of these may have received less emphasis than others, the pattern indicates what differentiates the studies in this book from related forms of inquiry.

Because the research has generally involved not only *multi-person situations* but also *man-machine interactions*, for example, many observers would not place it within either social psychology or human engineering, although either characterization might be legitimate. Because the human behavior it has investigated consists of tasks in *operational system settings* responding to *complex environmental stimuli*, it differs from experimental psychology as practiced in university laboratories, yet it has been laboratory-based. Because its methods have included the *manipulation, replication, and control of variables* and *objective data and quantification of results*, it seems distinguishable from games and exercises. Finally, man-machine system experiments rely extensively on *simulation*, but because they include human subjects they are distinct from simulations performed entirely on computers.

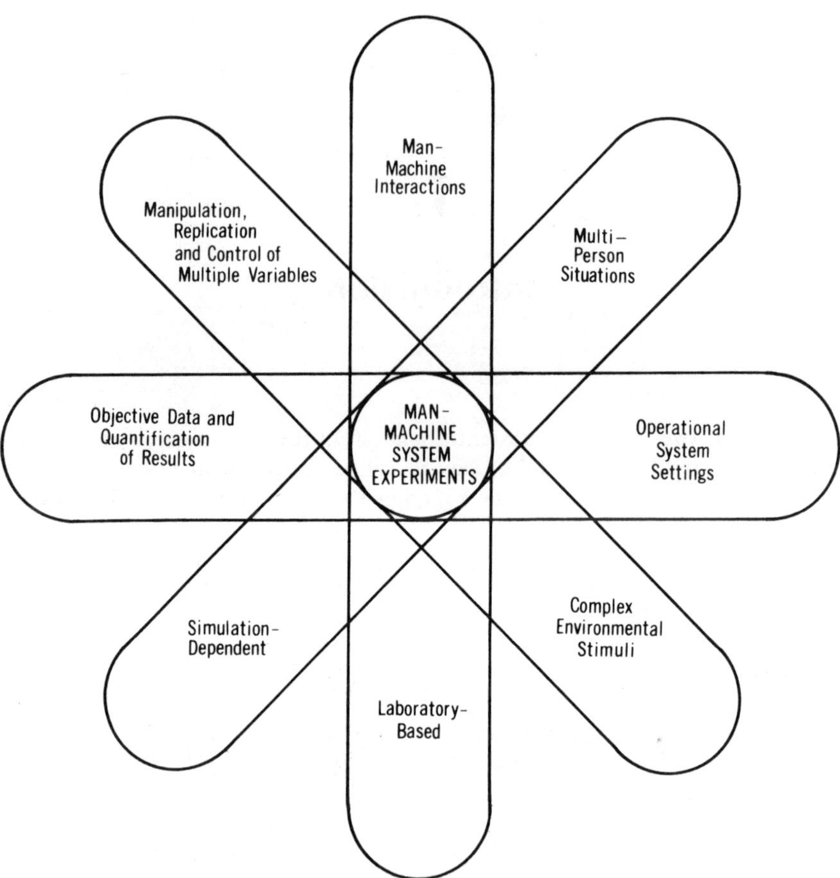


Fig. 1. Cluster Diagram Showing Characteristics of Man-Machine System Experiments.

As a hybrid type of research, man-machine system experimentation has neither been claimed by the more traditional types of researchers as their own nor become widely known through publication in journals or presentations at professional meetings. It might be characterized as part of the field of "human factors," the study of man-machine relations; but past experimental programs are unfamiliar to many in this field who are fully conversant with other kinds of system research and man-machine experimentation. Although many investigations have contributed to system building, the scientific community at large has remained unaware of this research. What is anomalous about such an undeserved fate is that this experimentation has been both pioneering and costly; greater recognition might have been expected due to either characteristic. Some sets of experiments have cost a million dollars or more. Many complex problems have been faced concerning management, design, simulation, subjects, and measurement. To be sure, not all experiments have been so expensive or extensive and not all have been productive. Nonetheless this has been research transcending in scale and challenge much experimentation better known to behavioral science.

Most of the reports of man-machine system experiments have remained relatively inaccessible except to those directly involved in the system concerning which the research was conducted. Few reports have reached the open literature in books or journals. Many reports have been classified and their downgrading has occurred only recently. Perhaps the greatest difficulty in gaining access to this research has been the lack of any published account which fully describes it or indicates what studies it embraces.

There have been several brief reviews of the work in this field (Licklider 1962; Sinaiko 1962; Singleton 1964) but none has been comprehensive or provided a bibliography. Other authors (Chapman 1965; Davis and Behan 1962; Haythorn 1963*a*; and Kidd 1962) have touched on method and role, but none has attempted to give a full picture. As a result, efforts to learn about this body of experimentation have been piecemeal, usually occurring only when researchers have started to do something similar.

It hardly seems fair that those who have worked so hard in a significant scientific enterprise should see their studies consigned to semi-oblivion. What are all these studies? Table 1 gives an overview. It tries to characterize approximately thirty major programs and more than a dozen others according to the kind of system involved, objectives, and simulation, as well as identifying the laboratory and sponsor and indicating the approximate dates. The programs have differed greatly in their extent, and "program" has been used somewhat loosely for purposes of aggregation. Table 1 also indicates the number of experiments in each program and the number described in this book, about two-thirds of them substantially, the rest briefly. It has been difficult on occasion to say what was an experiment. Those tabulated have varied greatly in size; as will be seen later from the text, in a few programs experimental method has been less rigorously applied than in others; and some programs have included considerable associated research which was also experimental.

Although a number of experimental projects during World War II must be regarded as the forebears of this research, the first major laboratory experiment came in 1951. Virtually all of the laboratories have been situated in government agencies, not-for-profit organizations, and universities. The sponsors have been primarily the Navy, Air Force, Army, Office of the Director of Defense Research and Engineering, Federal Aviation Agency, and National Aeronautics and Space Administration. Some studies were self-sponsored.

By and large, the experiments resulted from technological developments that placed new requirements on men to work together in military or civilian equipment aggregates which grew out of the new technology. Because in many cases these aggregates were developed as distinct entities for definite purposes, they came to be called systems.

Here the phrase "work together" may be interpreted in several ways. In one sense it means that the outputs of one individual or set of individuals are the inputs to another, and vice versa. In another, it means that individuals operate in some co-ordinated fashion on the same general task, sharing the load. In a third sense, there may be a nodal position where information is received and actions

**Table I. MAN-MACHINE SYSTEM EXPERIMENTS AND LABORATORIES**

(*Note*: Dates are approximate. Experiments in totals are the "man-machine system" type, liberally interpreted. Totals of experiments described in text are in parentheses; such descriptions vary from extensive to cursory. Many programs included other, associated experiments.)

| Organization, time period, sponsor, number of experiments | System and operations | Objectives or major variables | Principal simulation, subjects |
|---|---|---|---|
| Project Cadillac, New York University, 1948–55, Navy, 6(6) | AEW&C aircraft for fleet air defense: surveillance and interception | Communication procedures, capacity, team composition | Equipment mock-up and simulated radar, Navy officers and men |
| Psychological Research Associates, 1950–57, Army, 6 (5) | Infantry rifle squads, small arms fire | Evaluative tests, training methods, team composition | Instrumented field sites, Army officers and men |
| Technical Development Center, 1950–59, Civil Aeronautics Administration, approx. 25 (1) | Terminal and approach air traffic control (civil) | Air lanes, airports, displays, procedures, configurations | Equipment mock-up and simulated radar, civilian professionals |
| Willow Run Research Center, University of Michigan, 1951–54, Air Force, 1 (1) | Land-based air defense: weapons assignment | Operator capacity, crew size, displays | Equipment mock-up and computer paper tape, civilians |
| Systems Research Laboratory, RAND Corporation, 1952–54, self-sponsored and Air Force, 4 (4) | Land-based air defense: surveillance | Organizational behavior, system adaptation and training | Equipment mock-up and simulated radar, students, Air Force officers and men |
| Electronics Research Laboratories, Columbia University, 1952–54, Air Force, 4 (4) | Land-based air defense: interception | Manual vs. semiautomatic systems, operator capacity, console design | Equipment mock-up and field site, simulated and real radar, Air Force officers |
| Naval Research Laboratory and Chesapeake Bay Annex, 1952–56, Navy, 9 (8) | Shipboard CIC for fleet air defense: surveillance and interception | Displays, data transfer methods and devices, capacity, CIC operations | Equipment mock-up and simulated radar, Navy officers and men |
| Laboratory of Aviation Psychology, Ohio State University, 1952–61, Air Force, 19 (19) | Radar-aided terminal air traffic control (military) | Displays, procedures, task distribution, training | Equipment mock-up and simulated radar, Air Force officers, students |

**Table I. (Continued)**

| Organization, time period, sponsor, number of experiments | System and operations | Objectives or major variables | Principal simulation, subjects |
|---|---|---|---|
| Lincoln Laboratory, 1953–55, Tri-Service (Air Force, Army, Navy), 2 (1) | Land-based air defense: surveillance | Task distribution, display devices | Equipment mock-up and simulated radar, Air Force men |
| Research Division, New York University, 1954–57, Army, approx. 2 (0) | Land-based air defense: surveillance | Task distribution, electronic countermeasures | Simulated and real radar |
| Project Michigan, University of Michigan, 1954–58, Army, 3 (2) | Battlefield surveillance | Data collation, tracking hostile units | Automated TTY and display overlays, Army personnel, civilians |
| Road Research Laboratory, England, 1955–56, 1962–63, Department of Scientific and Industrial Research, 3 (3) | Vehicular roadways, traffic | Traffic flow, vehicle capacity, car-following | Special roadways and regular vehicles, civilians |
| Operational Applications Laboratories at Shaw AFB, 1956, Air Force, 1 (1) | Tactical air missions: interdiction, return-to-base, interception | Manual vs. semiautomatic systems, capacities | Field site, simulated and real radar, Air Force officers and men |
| System Development Corporation, 1957–58, Air Force, 4 (4) | Land-based air defense: surveillance, interception | System training methods | Field sites and simulated radar, Air Force officers and men |
| Coordinated Science Laboratory, University of Illinois, 1957–59, Tri-Service, 2 (2) | Shipboard CIC for fleet air defense: surveillance, interception, weapons assignment | Levels of automaticity, capacity | Digital computer and simulated radar, Air Force officers and men |
| Combat Development Experimentation Center, 1957 to 1966, Army, approx. 25 (3) | Army field operations, tactics, organization, devices, equipment | Comparisons, evaluations of effectiveness | Instrumented terrain, Army officers and men |
| Systems and Human Factors Laboratories, System Development Corporation, 1958–60, 4 (4) | Land-based air defense: surveillance | Training, crew turnover, debriefing, decision-making | Abstracted radar displays, civilians, students, girls |

**Table I. (Continued)**

| Organization, time period, sponsor, number of experiments | System and operations | Objectives or major variables | Principal simulation, subjects |
|---|---|---|---|
| System Development Corporation, 1958–63, Air Force, 12 (12) | SAGE air defense: surveillance, interception, weapons assignment | System training, team training, manual support, computer programs | SAGE and manual field sites, simulated radar, Air Force officers and men |
| Logistics Systems Laboratory, RAND Corporation, 1958–64, Air Force 4 (4) | Logistics, aircraft maintenance, ICBM squadrons, bases, organizations | System comparisons, policies, arrangement, organization | Digital computer and simulated paper, Air Force officers and men, civilian specialists, RAND staff |
| Second IDC, System Development Corporation, 1959, Weapons Systems Evaluation Group, 2 (2) | Land-based air defense: surveillance, interception | Effects of electronic countermeasures | Equipment mock-up and field sites, simulated radar, Air Force officers and men |
| Navy Electronics Laboratory, 1959, Navy, 1 (1) | Shipboard CIC for fleet air defense: surveillance | Data transfer methods | Equipment mock-up and simulated radar, Navy men |
| Operational Applications Laboratory, 1959–61, Air Force, 3 (3) | Land-based air defense: weapons assignment | Decision-making | Outdated prototype and processed tracks, Air Force officers |
| National Aviation Facilities Experimental Center, 1959 to 1966, Federal Aviation Agency, approx. 54 (13) | Terminal, approach, and en route air traffic control (civil) | Traffic flow, procedures, equipment arrangements, airport location | Equipment mock-up and simulated radar, civilian professionals |
| MITRE Corporation 1961–62, Air Force, 3 (2) | Air Force Hdq. Command System (473L) | System exercising, problem-solving, procedures | Current system with digital computer, Air Force officers and men |
| Boston ATC Test Bed, MITRE Corporation, 1961–63, Federal Aviation Agency and Air Force, 6 (6) | En route air traffic control | Methods of computer support | Never-used SAGE equipment and digital computer, civilian professionals |
| Information Processing and Control Facility, Ohio State University, 1961–63, Air Force, 3 (3) | Hypothetical center for processing aerial reconnaissance data | Team-composition, feedback, capacity | Paper reports, card inputs, computer, students |

**Table I. (Continued)**

| Organization, time period, sponsor, number of experiments | System and operations | Objectives or major variables | Principal simulation, subjects |
|---|---|---|---|
| Human Resources Research Office, 1961-64, Army, 3 (3) | Tank platoons, infantry | Training methods, stress | Miniatures and terrain model, field sites, Army officers and men |
| Simulation Facility, System Development Corporation, 1962, Air Force, 1 (1) | Strategic Air Command Control System (465L) | Information requirements, displays | Projection displays, Air Force officers |
| System Simulation Research Laboratory, System Development Corporation, 1962-64, self-sponsored, 3 (3) | Hypothetical terminal air traffic and intelligence systems | Laboratory shakedown, procedures, organizational functioning | Special consoles and digital computer, students |
| Applied Physics Laboratory, Johns Hopkins University, 1962-65, self-sponsored, 5 (5) | Hypothetical shipboard CIC for fleet air defense: weapons assignment | Decision-making, collaboration with computer | Equipment mock-up and human simulation, Navy officers |
| Command Research Laboratory, System Development Corporation, 1963-64, Advanced Research Projects Agency, 3 (3) | Hypothetical center for assessing nuclear strike effects | Decision-making: Bayesian processing | Paper-and-pencil, digital computer, display, students |
| Institute for Defense Analyses, 1963-65, Department of Defense, 21 (many brief) (21) | Multiperson communications, conferencing | Use of teletype and telephone networks | Regular office spaces and phones, civilians, high-level military officers |
| Grumman Engineering Corporation, 1963-65, National Aeronautics and Space Administration 3 (3) | LEM lunar landing, orbital docking, LEM/CSM rendezvous | Performance, procedures | Equipment mock-up, civilian and military pilots |
| Command-Control Simulation Facility, Ohio State University, 1963-66, Air Force, 12 (6) | Hypothetical center for threat evaluation, data assembling | Decision-making: Bayesian processing | Paper reports and digital computer, students |

**Table I. (Continued)**

| Organization, time period, sponsor, number of experiments | System and operations | Objectives or major variables | Principal simulation, subjects |
|---|---|---|---|
| Command Research Laboratory, System Development Corporation, 1964, Advanced Research Projects Agency, 5 (5) | Hypothetical command center for nuclear war | Group vs. individual displays, decision-making | Special consoles, projection display, digital computer, students |
| MITRE Corporation, 1964–65, Air Force, series (1) | North American Air Defense Command Combat Operations Center (425L) | System testing, procedures, effectiveness evaluation | Prototype or actual equipment, consoles, computer, Air Force officers and men |
| Submarine Tactics Analysis and Gaming Facility, Electric Boat Division, 1964–65, Navy, 2 (2) | Hypothetical tactical units in mutual opposition | Decision-making: various vulnerabilities and probabilities | Abstracted display, civilians |
| Group and Environment Design Laboratory, Princeton University, 1964–65, Navy, 3 (3) | Hypothetical competing groups of tri-service commanders | Decision-making: individual differences in integrating data | Verbal data, students |
| Disaster Research Center, Ohio State University, 1964–65, Air Force, 1 (1) | Columbus, Ohio, police radio dispatcher room | Effects of intracity airliner crash on police operations | Equipment mock-up and human simulation, police officers |
| Martin Co. (Baltimore), 1964–65, National Aeronautics and Space Administration, 3 (3) | 7-day Apollo lunar mission | Performance over extended period | Equipment mock-up, test pilots |
| Emergency Operations Research Center, System Development Corporation, 1965–66, Office of Civil Defense, 2 (2) | Hypothetical city, nuclear attack | Types of displays | Displays and manual inputs, municipal officials |
| System Design Laboratory MITRE Corporation, 1965–66, self-sponsored, 2 (2) | Tactical air control center: interdiction, mission planning | Manual vs. computer-aided system (AESOP) | Equipment mock-up and digital computer, civilians |
| Army Personnel Research Office (later BSRL), 1966-Army, (0) | Image interpretation | Multiperson team effectiveness | Digital computer and photography, Army personnel |

such as assignments are ordered. All three types of working together may occur in the same system. The point is that the situation is not simply that of a single individual interacting with a machine, but one involving a number of people and machines, among whom the actions of any one man-machine combination may influence and be influenced by those of others.

The technological development which had the greatest impact in fostering this kind of man-machine research seems to have been radar, in conjunction with the production of high-performance military and commercial aircraft in large numbers. Other developments have been increasing dependence on communications and displays and the use of computers in military systems. One should not be misled into believing that these are the only technological developments responsible for man-machine experiments. However, it is worth asking why they have been dominant.

Radar has had the function of extending man's sensing capabilities, and thereby has generated large amounts of data for a number of man-machine combinations to process and respond to in a co-ordinated and purposeful fashion. Developments in communications and displays not only contribute to data-processing load but entail their own data-interchange demands on man-machine performance. Computers also contribute to the information processing load through their incorporation of large data bases in their memories and ability to process vast amounts of data at great speed; and they, too, entail their own demands on man-machine performance.

In view of this context, it is hardly surprising that those who have conducted man-machine system experiments have come primarily from two categories. Predominant have been psychologists trained as experimentalists and interested in engineered systems. The other category consists of engineers, operations analysts, and others occupationally involved in the development of some particular system, or intrigued with the development of simulation. Although the partnership has sometimes been uneasy, the two types have generally worked together in an effective manner to join together man and machine in the laboratory.

As stated at the outset, some experiments—most of them, in fact—have sought knowledge about a particular system. This might concern some piece of equipment, a training technique, procedures, or certain conditions affecting performance. Other experiments have tried to acquire generalizable knowledge about the way human beings perform in system settings. How do operators and managers make decisions? How do they develop their procedures? How do they communicate with each other? It is important to recognize the heterogeneity of objectives before going further into this book.

It is possible that some objectives may have been more easily achieved than others. The reader may come to his own conclusions as he proceeds. He will have a chance to compare them with the author's before he is finished.

## ORGANIZATION OF THE BOOK

The main purpose of this book, stemming from the study which gave rise to it, is to help future experimenters conduct man-machine system experiments. No guidebook exists. Hence, in the descriptions of experimental programs prime emphasis is placed on how researchers have gone about their tasks. To set the stage for these descriptions, the next chapter deals at length with problems of method. Readers interested only in the factual account of the experiments may prefer to skip it. Others may want to refer back to it later as the historical data make the generalities in the chapter more meaningful. Appendix II is also directed at method; it contains what little systematic advice on method has previously been published.

The next twenty chapters describe the programs and their experiments. As mentioned earlier, "program" has been used somewhat loosely. Sometimes it means no more than a set of studies at the same location on the same general theme. In any case it should not be inferred that "programs" have been planned as such in systematic fashion. More often later experiments in a set have developed from earlier ones in an unpremeditated manner or from unforeseen problems facing the researcher. The "program" approach was selected to give structure to the work, to make description more economical, and to indicate how experiments actually were related to each other. This approach added considerably to the toil of description, since few experimenters have capped their work with a report fully summarizing and interrelating the individual studies.

Not all experiments in all programs are described in detail. When a program has included numerous similar experiments, descriptions are given only for a sample that is believed to be representative. Summary information is provided about the others. In a few other cases descriptions are brief because it was impossible to get enough information. Otherwise, experiments are described in some detail. Such coverage demonstrates the scope of the work, which is difficult to grasp for those who have not taken part in it. More important, it has been necessary to provide enough detail to describe the system involved, the methods used by the experimenters, and major results. It might have been preferable to give more austere accounts if original sources were readily accessible to readers; as noted earlier, most of them are not, although they are listed in the References. The reference data from which the material in the twenty chapters was reduced came to more than five million words.

The number of experiments in a program varies. In some instances only a single experiment was completed, although more may have been planned. Some chapters consist of a single program or set of experiments; in others there are a number. They are aggregated in various ways; according to sponsoring agency, research organization, or topic. Thus the organization of chapters is not standardized and is somewhat arbitrary, due largely to the heterogeneity of the subject matter and varying availability of information. As a consequence, the order of chapters is only approximately chronological.

The criteria for inclusion of a study are those set forth at the start for defining a "man-machine system experiment." The boundaries were established according to what seemed best to fit the pattern of this type of research. No doubt some readers will object to certain inclusions or exclusions. To each his own pattern perception! Another criterion was the extent to which an experiment involved the kinds of problems and methods characteristic of this research. There seemed to be little point in including those experiments whose methodology has been well established and publicized.

Since a number of fields are closely related to man-machine system experimentation, it seemed useful to describe these. Readers might like to know what was *not* being included under the main theme. Chapter 23 gives overviews of these: system testing, small group studies, gaming, and all-computer simulation. Another supplement to the accounts of experiments is Appendix I, which describes plans for facilities that were never built, or if built, never saw an experiment.

The last two chapters approach man-machine system experiments from a more general viewpoint. Chapter 24 looks at their place in man-machine system research as a number of researchers have viewed it. Highlights of their commentary have been reproduced. Chapter 25 examines the strategy of this kind of research. It deals with objectives and accomplishments. It covers relationships between programs and facilities, cost and benefit. In addition, Appendix III describes some of the contributions of man-machine system experiments to general knowledge about systems. Chapter 25 may aid those responsible for man-machine system research and development to determine the extent to which this kind of experimentation should be encouraged in the future and the directions in which it should go.

# 2

# Problems of Method

As the previous chapter indicated, methodology is a major consideration of this book. One of the book's purposes is to help future practitioners conduct man-machine system experiments. This chapter will outline some of the problems which such practitioners will face.

To a considerable extent the problems of method in this kind of research are those of experimentation in general. There is no need to describe here how to perform experiments, since other texts (e.g., Chapanis 1956, 1959; Kerlinger 1965; McGuigan 1960; Sidman 1960; Townsend 1953; Underwood 1957) have provided such descriptions for psychological research; still other writers have concentrated on the use of statistics. However, there are methodological problems in man-machine system experimentation which do call for attention. Although for the most part they exist also in more conventional experiments, in some instances their resolution has seemed so straightforward in such experiments that little has been written about them. In other instances they have received more analysis. In either case, the scale and complexity of man-machine system experiments accentuate their significance, so much so that some of these problems may appear to be specific to this research.

These experiments have been characterized by elaborate facilities and apparatus and by large numbers of people. Some of the apparatus may represent part of the system being investigated. Instrumentation for collecting data may be extensive. Other equipment presents the complex stimuli that evoke performance. Communications connect subjects with each other and with experimenters. A computer is likely to be part of the apparatus.

The people involved may include a sizable staff to conduct data-taking sessions, large numbers of subjects, engineers and computer programmers, advisers, and those who collect information about the system under scrutiny or analyze the data from the experiment. Still others design and produce the simulation inputs—the stimuli that evoke performance.

The methodological problems of man-machine system experiments will be discussed under the headings of management, design, simulation, subjects, and measurement. The diversity and extent of people and equipment suggest why management deserves special consideration in this kind of research. Design must cope with a particularly wide range of variables and with accompanying requirements for control. Simulation inputs of considerable volume and intricacy must be designed, produced, and presented; much of the simulation has no parallel in other types of psychological experimentation. Subjects, who may have to be trained in advance, often work as teams at specific tasks, along with quasi subjects; their selection and supervision can challenge the researchers. Great quantities of data are collected and reduced; multiple measures must be selected because of the many parts and purposes of the system studied.

How can researchers acquire an understanding of the methodology of these elaborate experiments beyond that derived from their experience with other experimentation? There are three ways. The best way is to take part in one as an experimenter. The next best way is to become familiar with a large number of such experiments vicariously by reading about how they were done. The chapters which follow try to provide such an opportunity. The third way is to read a guide, like this chapter.

As a guide it dwells as much on problems as on their solutions. It is assumed that the reader will profit just from warnings to "worry about this" and "you can get into trouble with that." Although much positive guidance is included, and Appendix II includes checklists which have appeared in the literature, it is impossible to present a simple handbook suitable for all experiments. They are too diverse. The systems they examine differ widely. Individual experiments are usually embedded in larger programs with varying facilities, objectives, and costs. As Chapter 25 points out, the objectives of man-machine system experiments vary from seeking particular or ad hoc knowledge to seeking general knowledge, and from exploration to verification. The methods or tactics of an experiment will depend on the over-all strategy chosen. Methods have to be tailored to the individual experiment.

The guidelines and problems set forth in this chapter are not traced to the situations in which they became apparent; there is no wish to imply criticism of specific researchers and laboratories. But the reader can assume there is good reason for including each guideline and problem. Perhaps the problem was solved badly, perhaps the guideline was not followed. On occasion the solution required much innovation or effort. This chapter should alert readers to significant facts brought out about experiments in the chapters which follow. But the material in this chapter is also drawn from unpublished experiences and views of many experimenters, including the author and the consultants who assisted him.

## MANAGEMENT

Management functions include planning for all the phases of this research, the acquisition and administration of resources, and the organization and coordination of operations in each phase.

It is convenient to separate the life of a man-machine system experiment into three phases, with subphases within each. Indeed, all experiments can be viewed as multiphasic. The laboratory occupation phase, during which the data are gathered, is the middle phase. Despite its key role in the experiment, it can be the shortest, although it may continue for weeks or even months in larger experiments and call on the participation of scores of individuals. It can include preliminary shakedown sessions for the equipment, for computer programs, and for experimental staff, as well as indoctrination and training sessions for the subjects. It may also incorporate exploratory sessions or even subsidiary experiments to develop values of independent variables, performance measures, and methods for presenting inputs and collecting data. The actual data-taking sessions are the core of the experiment, as in other experiments.

The first phase covers preparation. It may take several times longer than the second and be costly. The system being investigated must be thoroughly studied if it is one presently in use, or carefully created if it is a future or hypothetical one; much information about it must be assembled. The simulation inputs must be designed, produced, and checked out; and if part of the system is to be represented within a computer, each model must be programmed and also checked out. The formal design of the presentation of independent variables must be created. Suitable performance criteria and accompanying measures have to be selected, and techniques must be developed during this phase to collect, reduce, and analyze data. Further computer programming may be required for the data-reduction methods. The subjects must be assembled. For large-scale experiments, all sorts of manuals must be written detailing the procedures to be followed during the data-taking sessions. The acquisition of a facility, of apparatus, and of staff can also be regarded as part of the preparation phase, if the experiment is the first in a program or if changes are needed in these resources.

In the third phase, the performance data are reduced and analyzed, an extended process even with computer support. The researchers write reports and otherwise disseminate the results. The reports may include recommendations. The third phase may also last several times longer than the second. Except in duration and in methods of dissemination, the third phase in man-machine system experiments resembles that in psychology experiments in general.

### *Composition*

The composition of the management for this kind of enterprise is critical. Those who exert authority must understand the system and the methodology of the research to be able to provide guidance. Arbitrary control by those who have little professional interest in the work can create discord and inhibit the effectiveness of the staff.

One approach is to assemble a partnership of gifted and hardworking scientists. Both creativity and industry are essential. Man-machine system research is no place for the indolent. Some management structures, such as a rigid hierarchy, should not be installed automatically simply because they have been effective in other circumstances. Further, the role of ideas and the value of experience in this kind of research cannot be overemphasized. Much time may be required to build up a competent management and to acquire a competent staff. When further experimentation is projected, the retaining of trained personnel may be essential to its success.

The research makes so many different demands that some diversity in technical expertise, interests, and personality is valuable. Although it may be necessary to have a formal leader of the management team, its members need to solve problems in common. Different members may have to assume technical leadership during different phases of the experiment and for different aspects of its management. Although such an arrangement may generate debate at times among the members, it is possible to establish a consensus and take action on that basis. (Please note this is not advocacy of decisions by majority vote.)

The management should acquire and maintain constant communication and close liaison with other groups and disciplines whose interests do not lie primarily in the experimental research but whose knowledge and support are critical. These include engineering and computer programming groups and experts familiar with the system investigated in the experiment. A management committee which meets frequently–daily during the experimental runs–should include engineering and programming representatives. The committee should have representation from all the major personnel groups concerned with the experiment (except subjects).

*Planning*

The planning of any experiment obviously involves the objectives discussed in Chapter 25 and the place of the experiment in any over-all program. The point to be made here about objectives is that the goal should be clearly established. Management policy, particularly that coming from levels above the research managers, should remain consistent. Otherwise the direction of effort will keep changing, preventing the extended, systematic follow-through of ideas. At the same time, the research management must have the flexibility to redirect the experimentation as a result of its own experience in the laboratory.

Planning should be consistent with resources, including funding. Too great a divergence between aspirations and accomplishment may indicate ingenuousness on the part of the researchers and possibly a greater talent for dreaming than for doing. In addition, plans which are grandiose but lack sufficient substance about the knowledge to be sought are less likely to be funded. This is especially true of proposals for expensive facilities–although such proposals have sometimes succeeded, with results not always to the credit of the proposers.

Unless they are simply window-dressing, written plans may be helpful in the advance structuring and specification of effort. Plans can include schedules, budgets, and descriptions of all the tasks which have to be accomplished in each phase of the experiment. Scheduling should be realistic and should incorporate contingency planning. Items of equipment are often delayed, and computer programs usually take longer to complete than forecast. The unexpected is certain to occur, especially in field experiments. Not only is weather unreliable, but equipment malfunctions seem inevitable. These may occur in equipment which seems tangential, such as air conditioning or power supply, but which is essential. For a variety of reasons, both equipment and subjects may become unexpectedly unavailable.

It may be well to have a general plan, supplemented by detailed plans. The general plan shows the major goals, resource requirements, over-all arrangements, and milestones. The detailed plans incorporate the specifications of apparatus, programming and computer models, computer usage, simulation, data collection, data analysis, subjects and their participation, staffing, experimental design, and procedures for conducting the data-taking sessions. These procedures may be set forth in handbooks or manuals for the experimenters, for the subjects, and for personnel who function as quasi subjects. Checklists are useful as well. The manuals state how to present simulation inputs and collect data; they also provide training information to guide the subjects during the data-taking sessions. Other manuals describe the system being investigated and indicate how it is represented in the experiment with respect to equipment, procedures, personnel, and configuration. Plans and manuals should be updated as necessary.

It is important to keep records or logs of what has happened during each phase of the experiment, especially during the phase of data-taking sessions. Sometimes there are deviations from the detailed plans and schedules. During the second phase the logs should be kept daily and weekly, and they should pinpoint problems. Planning conferences should be scheduled during the first phase, and should occur daily during the shakedown or rehearsal periods of the second phase.

Although they share the same kinds of problems, planning for a man-machine system experiment is far more difficult than planning for more conventional psychology experiments. One of the major challenges is the selection of independent variables or values of variables (e.g., input "load") and the selection of dependent variables or measures. Little may exist in the way of prior research or theory to guide the experimenter. As in other kinds of experiments he may conduct pilot studies, which he may or may not call experiments. These may be few or many, limited or full-scale exploration experiments. Or he may search less systematically by manipulating independent variables and measures during the shakedown subphase of his experiment. If his experiment's objective is essentially discovery rather than verification, he may even redefine his independent variables and measures during the experiment itself—usually with a resultant loss of certainty about his results. Then he is "planning" during the experiment as well as beforehand. More will be said later about this process of "sequential planning."

The planner also faces a quandary in projecting the time requirements for the shakedown subphase, regardless of any preliminary study. How long will it take to familiarize the experimental staff with procedures; to remove "bugs" from procedures, equipment, simulation inputs, and computer programs; and to adapt both staff and subjects to the simulated "culture" and to such features as compressed time? There may be an unfortunate tendency to minimize the time needed. How long will it take to train the subjects to some desired level of performance, such as a "steady state," before the experiment begins? In the ordinary university laboratory the research psychologist faces the same questions, but generally not on a scale involving a large staff, many subjects, perhaps extensive computer time, and the occupancy of a large facility with many support personnel.

The experimental design may also call for the achievement of a steady state of performance (constant performance level, such as a baseline) during the experiment, to permit comparison with another steady state after some variable has been altered. To develop time schedules, the planner may have to predict how long it will take to reach each steady state. Even approximate forecasts are difficult. This is just part of the scheduling problem. Should the experiment consist of relatively few long sessions or many short ones? One argument in favor of short sessions is that a malfunction or error requiring a rerun has less grievous an impact. This is one way to plan for contingencies.

### *Acquisition and Administration of Resources*

The resources which the management must acquire and administer consist of a facility, apparatus, frequently a computer, and staff. In each case the requirements tend to be greater than in most psychological experimentation. One advantage of planning man-machine system experiments in terms of sets of experiments, or programs, is to spread the cost of resources among a number of studies.

*Facility.* Depending on the particular situation, the facility may be built for the experimentation or it can be an existing location which may be converted to serve the needs of research. Such locations may be indoors or outdoors. Indoor sites range from offices to which relatively simple equipment is added temporarily (e.g., communications), to operational centers, such as one on a ship or a functioning air defense unit temporarily taken over for the experiment. Those outdoors may be areas of terrain, large or small, highly instrumented acreage permanently assigned to experimentation or a small area equipped simply with stakes through which soldiers in the experiment must pass.

When an indoors facility is created for the research there is a risk that it will be proposed and built before the research is planned. This unfortunate inversion of needs will be discussed further in Chapter 25. On the other hand, designing a facility to meet the needs of a particular experiment may inhibit the flexibility, versatility, reliability, and expandability that such facilities should hopefully possess (Ernst 1959).

Flexibility and versatility may be needed when it is difficult to predict the size and number of teams that will work in particular spaces. Movable and removable partitions can help solve this problem, along with space for cables under removable floors. Locations will be needed for session administration, for activities of quasi subjects (described later), and for subjects' presession and postsession briefings and discussions.

One of the factors which can limit flexibility is the installation of permanent one-way viewing windows permitting the experimenters to keep an eye on the subjects. These may be desirable, however, when it is necessary to keep subjects under surveillance to make sure they are not violating instructions. In addition, some performance measures may be acquired through visual monitoring, although the importance of such measures may be exaggerated. (In any case, the subjects and their displays may be too far away to be seen by the monitors' unaided eye.) One reason for including extended one-way viewing windows in the facility design is to enable visitors to observe the experiment, but this rationale may simply support the notion that the facility is primarily a showpiece. Alternatives to permanent, extended one-way windows are small, movable, one-way viewing ports and closed circuit television. The concept of raised observation decks with one-way windows should be carefully examined before these are designed into a facility.

*Apparatus.* The apparatus that the management must acquire and maintain includes simulation devices which produce inputs to the subjects, calling on them to perform in certain ways; displays which present those inputs; devices through which subjects make their responses; telephonic and other communications; instrumentation to record the responses; and a timing system to coordinate all operations. Except for the first and last two categories, the apparatus may be equipment that system operators regularly use in their operations—a radar display, for example, a console, a tank, a radio mouthpiece and earphones. The apparatus may be used by the subjects at the operational scene or brought to the laboratory. Or the equipment may be simulated in some fashion in the laboratory—by a television display, a control panel, a miniature tank, a telephone terminal.

Simulation of operational equipment raises questions of fidelity or realism. If it is prototype equipment it is likely to require continuing maintenance, both remedial and preventive. Management must be equally, or even more, concerned about the maintenance of the simulation input devices and the recording devices than about their cost. Subsequent attention will be given in this chapter to the fidelity and reliability of the simulation inputs and the appropriateness and reliability of the recording instruments.

*Computer.* In some experiments the device for introducing simulation inputs may be as simple as a script; that for collecting performance data may be no more complex than a tape recorder. But many will need access to a computer both for transmitting inputs to display surfaces and collecting performance data. The computer may be used for producing the inputs in the first place and for reducing and analyzing the performance data. As noted earlier, a computer may also model or simulate parts of the system being investigated. To generate inputs to which the subjects must respond, the computer may in fact first receive and process within one of its models the outputs from the subjects or from another model. And if the system itself is computer-based or computer-aided, the computer in the experiment can represent the one in the system.

Another computer function can be the immediate assessment of subjects' responses, as in umpiring their decisions. It can select subsequent simulation inputs according to the assessment made. In this sense it is "reactive," an aspect of simulation discussed later.

To participate in such ways during data-taking sessions of the experiment, the computer must operate in a real time mode or there must be buffer computing equipment which does this. Even with a time-shared computer the buffering equipment may be necessary to deal with delays that might otherwise follow subjects' responses; the buffer equipment can also temporarily store responses and generate displays, such as printouts.

If the computer is the one in the system investigated and also can produce the simulation inputs, hold them in storage, and then transmit them for display, the temptation is great to exploit this linkage. But there is a risk. The computer will have produced for presentation only that information which people in the system encounter after it has been computer processed, not the information as it reaches the computer from various sensors and communication channels. There is also a risk in the use of the system computer for performance recording and data reduction. What gets reduced may be only what the computer senses as a result of subjects' switch activations, such as pushing buttons. More will be said about these problems subsequently. What must be stressed here is the significance to management of these limitations.

What exists outside the computer's domain may constitute critical simulation inputs for an experiment or critical performance data. But their critical nature may not, and probably will not, be recognized by those whose activities lie entirely within the computer domain—programmers or data system analysts. This is perfectly natural. The neglect is likely to occur if computer programmers dominate the research and development organization within which the man-machine system experiment is conducted, or if they guide the experimental effort itself.

Several other problems can arise in the exploitation of a computer for man-machine system experiments. One is analogous to that cited concerning the building of a facility. If the computer is acquired before the research it will support is planned, the inversion of needs may inhibit the development of a superior research program. This inversion can also contribute to another problem—the estimation of required computer capacity. This can be very difficult, and underestimation can be unfortunate. A third problem is the amount of time needed to install the computer, check it out, and develop its software (programs). Underestimation is likely here, too.

*Staff.* Large-scale experiments require large staffs. In view of the various phases and subphases through which an experiment progresses, the staff supporting the management must be flexible. Flexibility can be achieved in part through using the same personnel for different functions within the laboratory organization. It may also be advisable to borrow personnel from other organizations. For peak loads, as during data-taking sessions, it may be necessary to employ part-time personnel or contract personnel, but since staff must be trained, it can be advantageous to maintain an in-house supply for an on-going research program. Contingency plans must be prepared to make substitutions when key personnel are absent. It is essential that laboratory supervisors be experienced and well informed.

Job analyses may be useful for ascertaining the manning requirements for experiments. Staff functions in data-taking sessions include supervision of the entire operation, recording of difficulties and other events, monitoring and controlling the activities of subjects so that they follow experimenters' instructions, participation in presenting the simulation inputs, and participating in the recording of the subjects' performance. It is important to assign enough individuals to perform all these functions properly.

Those who take part in presenting simulation inputs may operate devices which produce simulated radar signals of aircraft on display scopes. The individuals follow a script in manipulating switches and knobs on the device, or they respond to commands from subjects. In other words, they simulate pilots. In other situations, individuals represent organizations with which the experimental subjects communicate, such as interacting agencies or superior echelons of command. In still other cases they may take the roles of hostile troops in a field experiment. In this book all of these types who simulate or represent people but are not called "subjects" are designated as "quasi subjects." They are part of the staff, and in many cases they also perform the functions of making the subjects follow instructions and of recording performance data. In other cases these functions are carried out by individuals who have no simulation roles.

Clearly the acquisition and administration of the staff, including quasi subjects, are management responsibilities which are not found in other psychological research, at least not to the same extent. There are also support personnel, e.g., advisers or consultants; engineers; and, when a computer is being used, programmers. The advisers may come from the organization which operates or will operate the system being investigated. They can furnish useful information about procedures, problems, personnel, and the operating environment. They can also help disseminate the results of the experiment. Often the advisers are also subjects.

Engineering personnel should definitely be either present or on immediate call during the data-taking sessions to deal with equipment malfunctions; so should programmers to cope with computer program emergencies. In either case the management should be skeptical about estimates of the time needed to restore normal operations. One approach is to multiply estimates by a factor of three.

Programmers and programming are related to the research management in other significant ways. For example, the design and production of the programs to accomplish the various computer functions already described are highly technical activities. They are also very time-consuming and expensive. The research management must make certain that what goes into the programs meets the research goals and that the programming does not exceed time and money limits. Although it is essential to have a sufficient programming staff and a realistic programming schedule, wishful thinking tends to produce the underestimation mentioned earlier. In addition, a programmer may go to unnecessary lengths to make his program a thing of beauty as well as utility.

Another management problem is the estimation of how much computer time should be reserved. Program development may require five to ten times as much computer time as the actual data-taking sessions will need, and four or five times as much computer time as will be needed for data reduction and analysis. The management cannot afford to take on faith the programmers' estimates of necessary computer time, of needed programming time, or of the number of programmers required. Some of the experimental staff must be sufficiently knowledgeable about programming to assess cost, time, and programming capability. These people should be able to communicate with the programming staff.

In addition, it should not be assumed that the programmers will become interested in fields alien to their own on their own initiative. Rather, the management should attempt to bring them into the planning of the experiment and to indoctrinate them in experimental techniques and human factors considerations.

*Organization and Co-ordination of Operations*

Although management responsibilities for organization and co-ordination of operations during the three phases of a man-machine system experiment are in many ways obvious, certain of them will be emphasized. The various necessary activities during the phases have been previously outlined.

*First Phase.* In the preparations phase, a major undertaking can be the creation of a simulation production capability. This means a capability to produce simulation inputs on a massive scale. The nature of such production is discussed later. Such experiments consume vast amounts of performance-evoking stimuli, which usually must vary considerably in an experiment. These stimuli represent the complex events which impact on the system. They vary because the experimenters want to know how loads which differ quantitatively and qualitatively affect the operators' and the system's performance. They may also vary so operators do not encounter the same inputs in succeeding sessions. Much automation and specialized technology can be required in the mass production of inputs which must be thoroughly checked out before being used in the experiment.

Another aspect of simulation is the modeling of parts of the system or organization which will be represented within the computer. Although the models must be programmed, their design consists of assembling considerable data-base information and preparing extensive specifications of operational relationships among many organizational and equipment entities. To prevent incompatibilities between models, it is wise to plan and construct them as a single package, not piecemeal. Also, it is best to begin with aggregated models and flow charts and work toward greater detail rather than in the other direction. Model designers should be informed as to what kinds of performance data will be required so they can allow for the accumulation of such data at advantageous points. The modeling process can be exploited to integrate much of the preparation for the experiment. There should be frequent meetings to review work being done, with publication of minutes, agreements, assignments, and pending questions.

The construction of the data base for the models, or for other simulation in the experiment, can present serious problems to the management. If the experiment includes a depiction of a current system, it is necessary to find out how that system operates. This may not be easy. Those in charge of the system may be willing to provide the information, but it may not be written down anywhere. The researchers may have to go and inspect the system at first hand. Then they describe it in a handbook, along with the embedding environment. The field study should come in the first phase; it should not be an after-the-fact inquiry.

The programs for the models must be thoroughly checked out during the first phase, along with the programs for input presentation, data collection, and data reduction. Unless the data collection and reduction programs are completed in this phase it may not be possible to acquire and reduce all the desired data automatically, and extensive, laborious manual intervention may become necessary. Adequate time should be allowed for debugging all computer programs, and the work of the programmers should be fully documented with flow charts, notes, and descriptive material as it progresses through the construction of programs and their debugging.

*Second Phase.* The need for shakedown or rehearsal sessions has already been indicated. These train the staff, help them improve their procedures, and further check out the apparatus, inputs, and computer programs. It has also been noted that such sessions may be exploited to search for appropriate levels of load and other independent variables and to examine methods of collecting data and the apparent usefulness of various measures. During these shakedown sessions staff members must play the roles of subjects. Since various mishaps are likely to occur, the actual subjects should not be allowed to participate lest they be disillusioned; they must have their own training or indoctrination sessions. Both the shakedown and training sessions may be numerous, together lasting even longer than the data-taking sessions.

The data-taking sessions themselves require much co-ordination between the individuals and devices making simulation inputs and the individuals and devices recording data. Clock-dependent precision is called for. Co-ordination is particularly difficult to achieve in field experiments, partly because communications are less reliable and participating elements, such as aircraft, may not show up. One of the management problems is to determine when an absence, a mistiming, or a malfunction should require the session to be interrupted or terminated and when it should be disregarded. Although sometimes equipment malfunctions can be simply regarded as a normal aspect of operations, they can also ulcerate the management.

As mentioned earlier, it is advisable to keep a continuing record of data-taking sessions, logging all deviations from expected operations and ground rules. The management may also establish a quality assurance function. An individual or small staff then becomes responsible for continuously inspecting all experimental operations, eliciting comments and criticisms from the staff and support personnel, acquiring discrepancy reports, and uncovering incipient difficulties.

*Third Phase.* A major management responsibility is the dissemination of information about the experiment and its results. If no report is forthcoming, or its distribution is seriously curtailed, it is reasonable to suspect that the experiment did not turn out the way someone wanted it to who had a stake in the outcome. If the report is sketchy, the reason may be poor experimental design, resulting in the researchers not having enough assurance concerning the results to want to publish much about them. It is also true that many researchers dislike writing reports; one way to find out whether this was a reason for inadequate reporting is to examine how much was put down on paper before the experiment was conducted.

Reports can be well written or poorly written, with a clear or confusing format. Management should consider itself responsible for producing a lucid, readable report, with effective presentation of figures and tables. When a program has been completed, a summary report should cover the entire program. Often two types of reports are advisable, one for the "customer," setting forth the results and any recommendations, another describing just how the experiment was performed, for professional inspection.

Management responsibility for dissemination of information does not end with the issuing of a report; briefings should be given as well. Laboratory demonstrations can also be held, and follow-up discussions should be offered for those who could profit from the results. If the report is classified, unclassified portions may be published separately, as may subsidiary findings of interest to special audiences. If the experiment had some ad hoc objective concerning a particular system, any generalizable results should be identified as such and disseminated.

## DESIGN

Experimental design may be defined as the specified arrangement of the conditions that produce data. To elaborate on this phrase it will be necessary to explore the nature of design in experiments in general, but this will be done to introduce those features which constitute particular problems in man-machine system experiments. These arise from the cost of such experiments, the multifaceted character of the situations they investigate, and the variety in the over-all objectives of this research.

(For present purposes no attention is given to so-called "ex post facto" experiments or "data analysis designs," in which the independent and dependent variables are selected after the circumstances in which they are embedded take place. The use of correlational techniques for investigating man-machine systems is discussed in Chapter 25.)

Experimental design consists of the following steps:

1. Stating an experiment's purpose.
2. Selecting the independent variables and their states (values, levels) which will be manipulated (varied) or held constant, and identifying variables which cannot or will not be manipulated or held constant.
3. Selecting the dependent variables, the criteria for the measures to be applied to them, and the measures.
4. Taking steps to prevent other variables, or the independent variables themselves, from having effects which would diminish confidence in the results.
5. Determining the number of replications (repetitions) of unique conditions (combinations of states of independent variables).
6. Segmenting the entire operation of data-taking into parts, specifying their durations and numbers, and scheduling them.

An optional seventh step is to select the kinds of statistics to apply to the results for testing their statistical significance. Such statistics can help the researcher make confidence judgments about the results. The other six steps are necessary whether or not significance statistics are used. The applicability of such statistics may depend on how these steps are carried out. But it is best to regard statistics as an aspect of design, rather than design as a consequence of using statistics. Campbell and Stanley (1963, 1966) have summarized the relationship thus:

> Good experimental design is separable from the use of statistical tests of significance. It is the art of achieving interpretable comparisons and as such would be required even if the end product were to be graphed percentages, parallel prose case studies, photographs of groups in action, etc. . . . Use of significance tests presumes but does not prove or supply the comparability of the comparison groups or the interpretability of the differences found.

The dependent variables, their measures, the criteria for these measures, and the significance statistics which may be exploited are discussed later in this chapter. At this point we shall concentrate on the other design steps.

*Purposes*

Frequently there is a temptation to oversimplify the general reasons for conducting experiments. For example, it may be stated that experiments test hypotheses; some do. Perhaps every experiment that relates differences in results to differences in independent variables could be interpreted as doing so when accompanying statistical treatment tries to disprove the null hypothesis that no difference has resulted. But the wide range of purposes that in fact exists should be acknowledged if the reader is to understand the variety of man-machine system experiments, as well as others.

*Verification.* Possibly this is the most common purpose, and for that reason it may be mistaken as universal. The researcher not only asks whether a difference in states of an independent variable results in a difference in the measured dependent variable, he wants to be confident about this cause-and-effect relationship. His quest is for certainty.

*Exploration.* The exploration-verification distinction, as mentioned earlier, will be discussed further in Chapter 25. When exploration is the dominant purpose, the researcher is more interested in discovery than in certainty. He is looking for variables or states of variables which for some reason are important, and the experiment reveals them.

*Generalization.* The researcher performs the experiment so he can generalize from the results. The experiment may test a theory, or the researcher may select independent and dependent variables in a way which will permit him to extend the results to many other situations.

*Ad Hoc Answers.* Only a particular situation is of interest. The relationship between generalization and ad hoc objectives is also discussed in Chapter 25. It is fair to say that "applied" research deals with ad hoc objectives.

*Description.* Here the emphasis rests on measuring the response to a situation or stimulus rather than on the differences between responses resulting from the differences between situations or stimuli. In some examples the effective stimulus is specifiable, as in psychophysics. But in other contexts the origin of the dependent measure, e.g., error frequency, often is not identified.

*Functional Analysis.* This is also descriptive, but such analysis aims at showing a systematic relationship between at least three values of an independent variable and some dependent measure, their co-variation in some pattern. Included are parametric experiments which reveal the effects of two or more related independent variables—both parameters of some phenomenon—and the relationships between the sets of effects.

*Evaluation.* As in description, the emphasis lies on measuring performance rather than in tracing effects to causes. The performance may be matched against some standard, which defines acceptability. Or the purpose may be to ascertain peak (maximum) performance (capability).

*Correlation.* This objective and such techniques as regression analysis may or may not be admitted into the company of experiments, depending on a researcher's personal biases. Although correlation is often used to measure co-variation among performance variables, it can also be applied to co-variation between a dependent variable and a system or input variable not manipulated by the experimenter.

*Comparison.* Two entities, such as two systems, are compared with each other. They react to the same requirements, and their performances are subjected to the same kinds of measurement.

*Diagnosis.* The paramount concern is to find out what led or might lead to some result. It is a kind of trouble-shooting. As in the case of verification, it depends on the manipulation of independent variables to find out how this variation affects the dependent measures. As in the case of exploration, the researcher may also be trying to find out which independent variables to manipulate.

*Methodology.* Rather than directly seeking substantive knowledge, a researcher investigates the effectiveness of some laboratory procedure, simulation technique, measurement instrument, or other methodological feature to use later in seeking substantive knowledge.

More than one of these objectives can characterize an experiment. They are not all mutually exclusive. To some extent, all have been found in man-machine system experimentation. They impose differing requirements on experimental design.

The variety of purposes for doing an experiment has not been emphasized by most authors of books on research and design. An exception is Kaplan (1964). He has pointed out that "the usual discussion of experiment in the philosophy of science focuses on only one type, which we may call the nomological experiment." This has the purpose of verification. According to Kaplan, it "aims at establishing a law, at proving or disproving some hypothesis or other." The most familiar type is the "crucial experiment."

Kaplan has called those experiments with the purpose of exploration "heuristic." They are undertaken "to generate ideas, to provide leads for further inquiry or to open up new lines of investigation." Some simply are aimed "to see what would happen if. . . ." Kaplan used the term "fact-finding" for experiments with the purpose of description. He also described some experiments as a "boundary" type—"to fix the range of application of the laws."

### *Independent Variables*

The knowledge sought through an experiment is defined by its independent variable or variables, of which man-machine system experiments usually have more than one. Perhaps more than one primary question is being asked. Perhaps a researcher wants to know how the answer to a primary question depends on various circumstances (other variables).

Something (the independent variable) receives different magnitudes or forms (states, levels, values) in an experiment to see how the difference(s) affect(s) something else (the dependent variable). Most of the general objectives just outlined call for the researcher to select the independent variables that will vary and the states each will have. He will be able to manipulate some variables by assigning states at will. In other cases he will have to accept the states available or select among them. The determination of independent variables and states is likely to be particularly difficult in man-machine system experiments because of the large repertoire from which to choose. The researcher may also have fewer sources of inspiration. What have been some of the determinants of choice in this and other kinds of experimental research?

1. *Theory.* The experimenter deduces that a certain variable should have a particular effect. He sets out to prove his hypothesis.

2. *Hunch.* This is also called "insight." It comes out of the blue, or stems from "armchair speculation" or some degree of analysis.

3. *Prior experiments.* These produce questions, or indicate some lead worth following. In particular, analysis may show that an independent variable in a prior experiment consists actually of two or more variables. These must be manipulated individually to find out which was responsible for the earlier results.

4. *Pilot studies.* Some are brief and often somewhat unsystematic or even disorderly experiments. But, as mentioned earlier, they may also be careful and fairly extensive explorations. They can be particularly effective for determining the states of an independent variable to put in the main experiment; they may investigate a number of levels chosen from the entire range.

5. *Preliminary sessions.* Before the experiment proper begins, the researchers may conduct preliminary shakedown or rehearsal sessions. In addition to trying out data collection methods, performance measures, simulation technology, and laboratory procedures these sessions may examine one or more of the independent variables to settle on the most appropriate states. This is most likely to occur in the case of input load. Then the preliminary sessions take on the same role as pilot studies.

6. *Computer modeling.* A large number of independent variables and levels of these can be put into an all-computer simulation. Those which lead to important differences in results can be regarded as candidates for the experiment. To

select the appropriate ones, the researcher must consider what assumptions were made in the computer simulation.

7. *Careful and comprehensive examination of the "real world" situation related to what the experiment will investigate.* In the case of man-machine system experiments, this is the system, whether in existence or being planned. Such an examination can help a shrewd analyst pinpoint key variables and their states.

8. *Correlation studies.* Quantitative data from the "real world" or from a system test or exercise are obtained about two or more variables. Correlation analysis may then show that they vary together. If a number vary together, factor analysis may yield explanatory factors. Correlations and factors may suggest experimental treatment of the variables to show functional relationships.

9. *Surveys.* Individuals in the real world provide opinions and other reactions in questionnaires or interviews. Such subjective data may help indicate what objective experimentation should investigate.

*Categories.* One approach to the selection of independent variables and their levels is to review a checklist of the categories and subcategories of such variables. Underwood (1957) set forth four categories for psychological experiments—environmental, task, instructional, and subject. The major categories of interest in man-machine system experiments total a dozen or more, although perhaps a majority could be fitted into Underwood's set:

Procedures and policies
Personnel requirements
Design, including communications and automation
Training techniques
Organization
Decision-making methods
Input loads, including "noise"
Input selections (with equivalent loads)
Ambient conditions
Subjects, or teams of subjects
System resources
Tasks
Instructions from the experimenters
Embedding and interacting organizations (including "enemy")
Feedback from performance
Repetitions of conditions
Time factors
Order of encounter of experimental conditions

These are categories of variables which directly affect human performance. There are other classes of variables that affect system and subsystem performance but influence human performance only indirectly, if at all. These are not discussed in the present context.

Input load has special importance for man-machine system experiments. Since it consists of most of the information given to the subjects, as a consequence of which they must act to perform their tasks, it forces their performance. Because its variation can show how the system operates under varying requirements, it may be an important independent variable even when another is the one of primary concern. When systems are developed to cope with heavy or special inputs, it is a variable of great importance indeed.

*Alternatives to Selection.* The researcher has several alternatives to selecting every state before the experiment begins. He can simply defer some of these to a later experiment, if another is expected. He can relegate one or more variables to one or more relatively brief supplementary experiments which are then planned to follow the main study. He can plan to change the variable states during the course of the experiment, depending on what happens; and he can even arrange the experiment in two parts, the second to incorporate new elements resulting from what is learned in the first. In a sense this is planning a later experiment. Finally, he can presume that new variables will emerge during the experiment, that is, make their importance apparent. He may then present and vary these later in the experiment.

But changing variables during the experiment or introducing new ones is not feasible in experiments which explicitly investigate processes of change in performance, such as learning and training. In others such a recourse can cast doubts on what brought about the results. This problem, which also goes to the heart of the difference between verification and exploration objectives, will be discussed shortly in connection with other matters that reduce confidence in an experiment's cause-and-effect findings.

*Nonmanipulated Variables.* In addition to those which are selected for systematic variation, the researcher must deal with two other types of variables. In one case he can select some level or value of the variable and hold it constant during the experiment. There are two reasons for doing this. The experimenter should be able to specify the conditions under which the results have been obtained. He should identify all of these constants. The other reason, discussed subsequently, is that if the variable is not purposely held constant, it may vary in some fashion which will distort the results.

The other type of variable is the kind which does not permit the experimenter either to vary it or to select some level at which to hold it constant. Then the experimenter can simply try to hold it constant at the level at which it is available, if this can be identified. If he cannot, then he may try to pick instances of the variable in a random manner as a way of eliminating or reducing bias. This approach is more effective when there are many instances. If the experimenter cannot hold the variable constant or if he purposely randomizes it, he should try to record the states or levels it assumes during the experiment. Then he at least can state the conditions under which he got his results.

Campbell and Stanley (1963, 1966) have offered a somewhat different classification of variables in addition to "manipulated" ones which the experimenter can assign "at will": "potentially manipulable aspects . . . that the experimenter might assign in some random way . . . but rarely does"; "relatively fixed aspects of the environment . . . not under the direct control of the experimenter but serving as explicit bases for stratification in the experiment"; " 'organismic' characteristics" which the experimenter can measure but not alter; and "response characteristics" which "usually appear as covariates or dependent variates."

In general, the experimenter must ask to what extent the states of a variable are manipulable, measurable or definable, and stabilizable. As will be seen shortly, he must also examine stability across repetitions in terms of identity, assured or assumed equivalence, and fluctuations.

*Selection of Manipulated Variables.* How does the researcher go about selecting the variables he will manipulate? His sources of inspiration have already been listed, but how he exploits such sources is another question. Much depends on his judgment, especially in assigning priorities.

Obviously, the variables have to be linked to particular aims of the experiment, some more apparent than others. For example, if the primary aim is to determine how the system will operate under various loads, the characteristics of the input must be varied to satisfy this aim. If the primary aim is to determine the benefits of computer automation by comparing two systems, the alternative systems constitute the two states of the key independent variable. But the parameter of input load, as noted earlier, should be varied as another independent variable to show how the systems differ under different loads. Conceivably one system might be superior with low loads, the other with high loads.

There are certain risks. One is the risk of including a variable because it is easy to define, present, or manipulate. It may have apparent validity because it is indeed specifiable and variable, but in fact it may be trivial. The other is the risk of exclusion. Some variables may be difficult to simulate or measure. They get a low priority and are deferred to a later experiment, which never takes place. Because the large number of potential independent variables in man-machine system experiments includes some which can be trivial and others which can be difficult to handle, these risks are especially worrisome.

A major consideration is the size of the experiment. How many independent variables should be included, with how many states for each variable? The number of states per variable will probably differ among variables, and often the number is set by the nature of the variables. For example, if two communication procedures are being compared, the variable is communication procedure and it will have two states. If there are five crews as subjects common to all conditions, the crew variable will have five states. Simply investigating the effects of an innovation, such as a new equipment feature, crew position, or environmental influence, should actually mean introducing two states of the variable the innovation represents: its presence and its absence. Its absence can be regarded as its "zero state." The variable which most often comes without a preordained number of values or levels is the input load. It has been noted that to select states to meet the experiment's aims, researchers may have to conduct a pilot study or preexperiment sessions.

It is likely to be more difficult to establish how many variables to introduce than their states, but the experimenter's options are bounded by a number of factors, especially resources and costs. Because of the substantial investment required in simulation, equipment, programming, subjects, and staff, it is desirable to get as much out of an experiment as possible. This means investigating a number of variables. The complexity of man-machine systems also makes this desirable; there is much to investigate. In addition, if two or more variables are manipulated within an experiment the researcher can ascertain the interactions between them, that is, the extent to which results are affected by combinations of states. He may wish, for example, to find out which is the most effective combination. But the cost also puts an upper limit on the size of the experiment; the bigger it is the more it costs. Further, the availability of subjects may limit its duration. And, of course, the longer an experiment lasts the greater is the risk that something will malfunction: equipment may fail or a subject may become ill.

The size or duration of an experiment's data-taking phase depends on the number of independent variables, the number of states in each, the ways in which these are combined for presentation, the number of repetitions (replications) of each combination, and the duration of data-taking for each such combination. Man-machine system experiments require the researcher to examine the trade-offs among these. For example, for a given size, the more replications there are the fewer variables there can be, and vice versa. Similarly, more variables or states of variables can be handled if some orthogonality is sacrificed, that is, if they are combined in ways which are to some degree confounded.

*Types of Design.* Man-machine system experiments generally include more than one independent variable and often more than two states for a single independent variable. Hence the most suitable design for them is a multivariate one, which is also susceptible to the analysis of variance method for testing the statistical significance of the results.

In some instances the design can be a complete factorial, in which every state of every variable is combined with every state of every other variable and every combination of the other states so the total of different combinations (unique conditions) is the product of the states. By incorporating a number of variables and their states, factorial design shows how differences in one variable are affected by differences in others—their "interactions." Man-machine system analysis may want to disclose these. For example, one training technique may be shown experimentally to be better than another under high loads but not under low, or for some subjects but not for others.

In other instances the number of unique conditions in a complete factorial design would be so large that, as suggested above, it would be necessary to omit some variables or variable states. Instead, the researcher may resort to incomplete factorial designs, such as Latin squares, Graeco-Latin squares, repetitions or blocks of these, or "nested" designs resembling the "mixed" designs described by Lindquist (1953). When the total number of states in an experiment is very large, because there are many variables, incomplete factorial designs can take the form of fractional factorials. In these each state must be introduced as many times as any other, but the experiment can contain only one-half, one-quarter, or even one-eighth as many combinations of states of variables as the complete factorial design would involve.

Latin square and Graeco-Latin square designs make it easy to design the order of presentation of variables as a semiorthogonal, independent variable into an experiment to provide complex counterbalancing. Such squares are often used when all subjects or teams of subjects experience each of the other variables and their states. They cannot serve in certain experimental situations–in comparisons of training techniques, for example. Each subject would have to learn the same task with each technique–a manifestly impossible situation; this is the nonreversibility problem. The very economy these designs provide in reducing the number of unique conditions results in partial confounding. All of the interactions are not represented in the design. That is, if there are three variables, each state of any one is not combined with all the possible combinations of the states of the other two. Fractional factorials admit even more confounding. An advantage of a complete factorial design is that it shows *all* the effects of states on each other.

Still another kind of experimental design which can be used to some extent in man-machine system experiments is what has been called "time series" by Campbell and Stanley (1963, 1966) and "steady state" by Sidman (1960). The state of an independent variable is changed, or a new variable is introduced, after performance has achieved a constant or uniform level (or constant rate of change). This can be done many times and any consequent alterations in performance noted. This design lends itself to sequential designing during the experiment. The researcher does not have to specify all of his variables and their values before the experiment starts, since he can introduce one whenever performance has reached a steady state. Among the drawbacks of this design in man-machine system experimentation are (1) limits on the situations in which it can be used; (2) the need to have many replications of unique conditions to demonstrate a "steady state"; (3) the difficulty of defining such a state; and (4) some uncertainty as to what really caused a change in performance. It may be necessary, for example, to reintroduce a previous condition as a way of determining whether simply changing conditions–change in itself–brought about a difference between steady states.

The time-series design has had proponents and opponents. Because of the flexibility it permits, it may seem particularly suited to the objective of exploration and discovery. But as will be brought out again in Chapter 25, discovery requires finding enough reason to believe the results of exploration are worth further inquiry in the form of verification. The researcher must have some assurance about the cause-and-effect relationship between the variable or variable state he introduced and the consequent performance, tending to rule out any alternative reason for that performance. This means extensive replication of conditions before and after manipulation. In man-machine system experiments such replications are time-consuming and costly.

Perhaps unfortunately, the literature on experimental method has concentrated on design for verification to the neglect of design for exploration. This may be due to the origin of much of the literature in the application of statistics. But even if significance statistics are applied, some of the rules or standards might be different for exploration experiments–for example, the so-called confidence levels expressed in percentages (generally .01 and .05). As Davis and Behan (1962) have suggested with regard to pilot studies, the requirement might be made more liberal, such as .10. It has occurred to others that the researcher might simply present the level reached instead of stating whether it satisfied some criterion. Davis and Behan (1962) have also indicated that in pilot experiments the experiment should be "more concerned about making a so-called Type II error than making a Type I error. Technically speaking, a Type II error is made when the null hypothesis is erroneously accepted. The experimenter, in other words, accepts the hypothesis that a variable is not significant when it actually is." Such thoughts about statistical confidence can be extended generally to the role of confidence in the outcomes of man-machine system experiments which have exploration as their prime objective.

It is realized that to many readers unfamiliar with the terminology, the foregoing treatment of particular experimental designs will seem abstruse, while to others it will have been too cursory or superficial. Each type of reader may want to look further into the literature, including the authors cited.

*System Interactions.* The very systems they investigate often complicate the selection and presentation of independent variables and their states in man-machine system experiments. Systems are characterized by interactions. These occur between systems, between subsystems within any system, between functions, and between individuals. In this context "interaction" means that what one entity does influences what the other does, and vice versa. During an experiment, as in the real world, each shares in determining what inputs are received by the other, what procedures are employed to cope with them, what learning takes place, and what motivates the people. The experimenter cannot be sure that the states of the variables he put into the design will be the ones impacting on particular system elements; and he may find some new variables intruding that he did not expect. His input load variable is likely to confound others because he may be unable to reproduce this variable's states in successive replications of experimental conditions.

In one situation, two or more systems are represented in the same experiment. They may compete with one another, even to the extent of simulated hostilities. Both may be aggressors, or one the aggressor, the other the defender. Or they may co-operate. In any case, if the actions of one are communicated to the other, the latter's performance will almost certainly be affected. All sorts of errors can occur in an experiment, especially in rapidly paced situations, due to communication delays or outages and mistakes in reports. For example, in two-sided combat a weapon actually put out of action may continue to fire and cause "false kills." Targets may be destroyed twice or the wrong target designated as destroyed.

What can the experimenter do about this, other than forego putting both systems into the experiment—a solution he may very well select? One recourse is simply to limit intersystem communication. Each system registers its reaction to the other's action only by telling the experimenter what this reaction would be or what it would expect. The other's action will have resulted entirely from a preestablished input. This approach limits each system's reactivity to the other to a set of single moves or episodes.

Another approach is to represent one of the systems by some of the experimental staff or by a computer. This agent will have been carefully instructed or programmed to react in ways which are always similar or equivalent, and thus reproducible, in response to each type of action by the system represented by experimental subjects. This approach greatly reduces the amount of unpredictable reactivity in the experiment, but the demands on staff or computer can be extensive. A relatively simple example is the elimination from subjects' displays of simulated radar signals from hostile aircraft which their system shot down. Even this may not be too easy to do.

Unfortunately it is not always feasible to come up with standard reactions to subjects' actions when these can vary widely. If such actions can be sufficiently constrained by experimenters' instructions, then instructions can also reduce reactivity when both sides are played by subjects. But the subjects may be unable to play their roles realistically.

Still another way to try to cope with intersystem reactivity is to lay stress in the design on independent variables which are not likely to be differentially affected by varying inputs which are the outputs of an adversary. These would be variables that were well established before the experiment began and thus would not be altered during its course. They might be the knowledge each side had beforehand, the personalities of the commanders being represented, the size and configuration of each organization, or the channels of communication.

In a second situation, two or more subsystems interacting with each other within a system produce the serial processing problem. As Davis and Behan (1962) have pointed out, subsystems may operate in series or in parallel. When they operate in series, as soon as an input furnished by the experimenter goes through the first subsystem he can no longer control and often cannot even describe the input entering the next subsystem. It consists of the output from the first, which has processed it in unpredictable ways.

This means that the subsystems later in the processing series may receive input loads which are too small or too heavy, too perfect or too degraded, to demonstrate how those subsequent subsystems would perform under other circumstances. Furthermore, as in the case of intersystem interaction, the inputs to the subsequent subsystems will vary unpredictably and cannot be reliably reproduced. The system as a whole would be properly diagnosed only if it were certain (an unlikely circumstance) that the processing by the early subsystem—such as one performing surveillance in air defense—would always remain the same. There might be no comprehensive way to diagnose difficulties in a later subsystem such as interceptor control.

Probably the only solution to this problem is to examine the later subsystem by itself. Davis and Behan (1962) have suggested that rather than examining the system as a whole, it may be preferable to "break systems into subsystems, to study the subsystems independently and then to put the subsystems together in a selective fashion." When the boundaries of the system domain are shrunk, so that the experiment examines only a "downstream" subsystem, special simulation problems arise. The inputs to that subsystem have to be designed to represent just what might have come out of the antecedent subsystem under a variety of circumstances. Such simulation is entirely feasible when both subsystems handle computer-processed data. But difficulties develop if the output of the first subsystem includes much unwanted content—"noise"—the characteristics of which are hard to synthesize.

As for interactions between functions and between individuals, these occur with great frequency in man-machine systems. The experimenter faces many design questions. How many functions can he afford to omit, either to simplify the simulation or to create a more abstracted and controllable situation? Similarly, to what extent should the actions and interactions of individuals be limited, through instructions, to fixed procedures? Should procedural flexibility be allowed, or even encouraged, to see how new procedures might be evolved by the subjects during the experiment? Which is better, to present clear options as alternatives, or let the subjects evolve them in the hope they will evolve some that had not occurred to the experimenters? Will subjects spontaneously detect alternative courses of action so that they can make choices and decisions, or should these alternatives be forced on them as differing states of an independent variable?

One of the roles of simulation in man-machine system experiments is to evoke performance as it presents some of the independent variables and states of variables. How explicitly it evokes and limits performance is a problem of method the researcher must resolve. Simulation also can help prevent uncertainty about the results by contributing to control—the next topic.

*Assurance Methods, Internal*

In those experiments in which independent variables are manipulated to see what differences in results come from differences between states of a variable, it is not enough just to vary the variable. It is important to know how much confidence can be placed in the apparent cause-and-effect relationship. It is convenient to regard confidence in two ways. One is to look at the assurance that can be placed in the results, considering factors only within the framework of the experiment. The other is to ask how confidently similar results can be attributed to analogous variables and their states in the "real world"; in man-machine system experiments this is the system being investigated. Campbell and Stanley (1963, 1966) and Campbell (1969) have thus distinguished between "internal validity" and "external validity."

What can threaten either type of validity is the possibility that some other variable is responsible, in large extent or small, for the results. It can do this in a number of ways:

1. It cancels out the effects of a difference between two states of the focal variable, the one being examined.
2. It subtracts from the effects.
3. It adds to the effects.
4. It is entirely responsible for the effects; the focal variable is not.

All four of these effects can be called "confounding," in those circumstances where it is not determined what the contribution of the other variable is. If a confounding variable affects the difference in results associated with the difference between states of the focal variable, it threatens internal validity. It threatens external validity even if it exerts no differential effects.

Another frequently used term is "contamination." Here this term will refer to assurance-threatening factors involved in the conduct of data-taking sessions, particularly the processes of data collection and measurement, which are treated later in this chapter. Both confounding and contamination have been called "bias."

Obviously it is important to hold confounding to a minimum. Assurance that a particular variable is responsible for the results depends on making certain that some other variable is not. Most of the methods for achieving internal validity have been called "experimental control." In man-machine system experiments such control is likely to be more difficult to accomplish than in simpler kinds of experimentation. An effort will be made to specify some of the threats to internal validity in man-machine system experiments, to suggest some of the tactics for counteraction, and to indicate some of the difficulties in implementing such tactics.

Before considering man-machine system experiments in particular, it will be useful to subject the phenomenon of confounding to some analysis. What is the confounding variable, a manipulated variable or a nonmanipulated variable?

In a multivariate experiment it can be a manipulated variable. That is, one of the independent variables confounds the effects associated with another independent variable in the experiment. Confounding is related, of course, to interaction. In each case the effects associated with the states of one variable may differ because of the states of another. If the experimental design makes this influence ascertainable, the relationship between the variables is interaction; if not, it is confounding. When this influence cannot be fully specified, as when all the states of each variable are not combined with all the states of the others, confounding is "partial."

On the other hand, in either univariate or multivariate experiments the confounding variable can belong to one of the nonmanipulated varieties already described. These differ according to the experimenter's ability to identify them, to keep them in a constant state, and to select that state. With all it implies, identification of variables is not as simple a process as it may seem. Depending on the counteractions to be taken, it may require definition of a particular variable or simply definition of a class of variables. If it is not certain to be present, the experimenter must estimate the probability of its occurrence. He must also try to judge how seriously its confounding with the independent variables would distort the cause-and-effect relationship; some confounding variables have relatively little additive or subtractive effect. In the identification process, the experimenter has to rely on many of the same sources of inspiration he uses in selecting the independent variables to manipulate. Confounding variables have to be plausible, just as do independent variables. Their identification is not all-or-nothing.

Nevertheless, for convenience in the present discussion nonmanipulated variables will be divided into those which can be identified and those which cannot. These two categories plus the category of manipulated variables give us three kinds of confounding variables.

*Counteractions.* To counteract each kind there are various confidence tactics. For the first category, manipulated variables, the counteractions are orthogonality, equivalence, counterbalancing, and replication. For identifiable nonmanipulated variables they are replication, constancy, preclusion, randomizing, contrast, and refinement. For unidentifiable nonmanipulated variables they are replication, preclusion, and contrast. What do these terms mean?

Although orthogonality may be defined in statistical terms, here it is used simply to indicate that one independent variable in a multivariate design is not dependent on another. All variables affect each other equally. Each state of a variable is presented in combination with each state of every other. As indicated in the preceding discussion of confounding by an independent variable, state *A* of variable I must occur (equally often) with both state *A* and state *B* of variable II. Otherwise states *A* and *B* of II might have different effects simply because state *A* of I accompanied one of them and not the other. Complete orthogonality occurs in a complete factorial design, where each state is presented in combination with each possible combination of other states. Orthogonality varies in degree. There is less of it in incomplete or fractional factorial designs, in which some of the combinations of states are omitted and consequently some of the interactions between variables cannot be ascertained.

By equivalence is meant that when some state of a manipulated variable is repeated, its characteristics are equivalent or even identical, from repetition to repetition. Often they cannot be identical.

Counterbalancing has various usages, but in this book it means the prevention of confounding that could arise from order of appearance. To counterbalance, the experimenter sequences the presentation of states of variables in such a way that one follows and precedes another equally often. In an experiment on a two-state variable, the order would be *ABBA BAAB* etc. In more complex designs the order of presentation is itself treated as an independent variable (e.g., as in Latin squares, mentioned earlier), orthogonal to the others. Without counterbalancing one state might confound another if the effects of one could carry over to the other but the same opportunity was not afforded to other states, or if the reciprocal opportunity were not provided.

Replication can mean the repetition of the states of a variable either by simply repeating a state or a combination of states or by repeating a state to produce a number of different combinations. In both cases it helps average out fluctuations in the state's level or value. The repetition of combinations also provides an index of random variance, that is, variation resulting from unidentifiable differences within variables or from measurement operations. In the analysis of variance test for statistical significance, the ratio between this chance variation and the variation apparently resulting from explicit differences between states indicates the degree to which the latter variation may be attributed to chance.

In some cases the constancy of a nonmanipulated variable and the equivalence of a state of a manipulated variable during successive replications may reduce the random variance in performance. As has just been pointed out, one function of replication is to furnish an index of this variance, which should be kept at a low level for testing statistical significance. If a variable state fluctuates, performance may fluctuate more than it might otherwise, and the variance may thereby increase. This probably will occur when the state is the team of subjects, but it does not necessarily occur in reaction to fluctuations in other variables, such as certain kinds of input. Empirical investigation of this relationship has been limited in man-machine system research.

The term "constancy" serves as the label for one of the most used counteractions against confounding: keeping a variable in the same state throughout the experiment. The variable then can have no first-order differential effects by unintentionally co-varying with the manipulated states. The state of a constant may be chosen by the experimenter if the variable happens to be alterable; if it is not, what is available remains the same, or so the experimenter hopes. A constant may have some unattributable fluctuation in its replications, contributing to the random variance just discussed. Sometimes a constant is called a "parameter." Constancy of a nonmanipulated variable resembles equivalence among repetitions of a state of a manipulated one.

Although most counteraction tactics deal with the effects of variables, manipulated or nonmanipulated, one type seeks to prevent a variable from having any effect at all. Preclusion, or if you will, exclusion, arranges a laboratory situation so a great many possible "impacters" which would evoke some kind of performance simply do not occur. This counteraction has also been called "removal" or "screening" (Townsend 1953). Simulation, as such, restricts the stimuli which are presented to subjects. Preclusion also limits the kinds of performance permitted. This is done through instructions and by specifying the methods and devices for responding. Likewise, preclusion can restrict feedback information and motivational consequences, either positively reinforcing or aversive. The claim could be made that preclusion actually holds a confounding variable constant in a zero or nonexistent state and thus is a special case of constancy.

Randomizing means co-presenting a nonmanipulated variable in a chance or random manner with the states of an independent variable or combinations of states of two or more such variables. Frequently the experimenter does this when he suspects or knows that the nonmanipulated variable will assume different states, but he cannot manipulate these or select one to hold constant—or he does not want to.

Randomizing inhibits systematic associations of which the experimenter may be unaware. It also is viewed as important for its role in testing for the statistical significance of results. One manifestation of the nonmanipulated variable is as likely as another to be combined with any state of a manipulated variable. For this rationale to work, the nonmanipulated variable must be combined a substantial number of times with each of these states so there will be a random distribution associated with each. This supports the concept that uncontrolled variation originates randomly. Randomizing, which may be complete or can be restricted in various ways, may be applied to such variables as order of presentation and subjects when these are not manipulated and when randomizing is feasible.

The term "experimental control" is often taken to signify just one of the counteractions against confounding—the use of a comparison group of subjects. Not only is this counteraction tactic just one of many but it should be viewed more broadly. If only one state of an independent variable is presented in an experiment, it is possible that the same effects would occur when the variable was not introduced at all. Then the results would be due to some other variable.

This possibility can be investigated by including the zero state in the experiment. For convenience "zero state" defines the condition in which either a quantitative or a qualitative variable is absent. If the results from the zero state are different from the nonzero or focal state, there are grounds for assurance that the focal state was responsible for the results that followed it. In the present discussion this is called the tactic of contrast.

The contrast tactic has been particularly popular with qualitative variables which have only two states, present or absent. But this tactic comes into play whenever two or more states of an independent variable figure in an experiment, even though they may not include the zero state. As long as confounding by nonmanipulated variables is prevented through other counteractions, a difference in results is attributable to the difference between states. However, the experimenter may want to include the zero state to see what the difference is between the effect of some state and the consequence of omitting the variable altogether.

It should be clear that this counteraction must be accompanied by others. For example, if the variable is a training technique, its introduction and its absence must be tested on different groups of subjects. The absence group, usually called the comparison or control group, must resemble the introduction group as a result of randomizing or constancy. A number of counteractions may often be found working together in a well-designed experiment.

The counteraction of refinement, sometimes called reduction, is an extension of contrast. Refinement proceeds thus. In one experiment a variable state consists of a number of qualitatively different elements, for example, a whole set of specified procedures. This set is compared with the absence of all its members. Results indicate the set's superiority, but there may be confounding among the elements; one element may be entirely responsible, but all get equal credit. In further experiments the elements themselves and combinations among them become states of the independent variable. In each case one of the procedures is the contrast (comparison) state; if its results are negative, it is omitted in the next experiment. Finally the element (or elements) responsible for the superior results is isolated. In short, states of a variable (or variables in a class) are progressively refined.

Especially in man-machine system experiments, complete success in preventing all confounding is more an ideal than reality. The final tactic, which is not really a counteraction, is to report in the account of the experiment all cases where the experimenter had reason to believe some confounding occurred. He should also report the various counteractions he took. His reporting will help the reader determine how much confidence to place in the results.

*Origins of Confounding.* The chances of confounding in man-machine system experiments are greater than in most because (1) there are more variables which have to be dealt with, and (2) many factors make the various counteractions more difficult to apply. Some problems can be foreseen, some cannot. The latter include circumstances that are expected but do not occur and those unforeseen that do occur, such as mishaps of many varieties.

The many categories of potential independent variables previously listed illustrate the wide range from which confounding can come; the sources of independent variables can also be the sources of confounding. The need to conduct experiments of a multivariate type, discussed earlier, increases the likelihood of the confounding of one manipulated variable by another. Confounding by nonmanipulated variables is made more likely by the nature of the material with which man-machine system experiments deal: machines and men.

As Chapanis (1959) has pointed out, machines which are designed to operate consistently or alike do not necessarily do so. These can be the system machines as well as those which help collect data. (Contamination by data-collection and measurement apparatus is discussed later.) The machines can be the simulation devices which generate the inputs. Their operators also make mistakes. Machines are subject to malfunctions, breakdowns, and differing accuracies, all of which can differentially affect—co-vary with—states of an independent variable in any experiment. For example, if a data-taking session has to stop and restart due to a computer program difficulty, the subjects will have received extra practice in the particular experimental conditions of that session. But mishaps are not the only problem. When operational equipments are compared as the experimental objective, laboratory circumstances may distort the comparison. For example, a lag in presentation time may accompany one display but not another.

Equipment difficulties and computer program problems are analogous to the "institutional events" which Campbell and Stanley (1963, 1966) have cited as threats to internal validity under the heading of "history" (preceding or during one measurement period but not another). In field situations there are more, for example, weather, illumination, availability of resources.

Human beings are more variable than machines. Subjects differ from one another in their performance in an experiment, but the causes of the difference are difficult to assess beforehand. Individual performance fluctuates under the same conditions and varies nonlinearly. Ernst (1959) has pointed out that if any operator shares his time among many tasks, the division of his attention may lead to variable performance in any one task. Further, many operators may share the same task, and their relative contributions may vary. These are some of the interactions between functions and between operators mentioned earlier. Nonlinearity in performance may stem not only from acquiring skill in the task's procedures but also from adopting or inventing new procedures. Because "the behavior of subjects is relatively unrestricted" in man-machine system experiments, "instead of having to select among a small number of predetermined actions, a subject may do something which the experimenter has in no way provided for or expected" (Davis and Behan 1962). These within-subject changes, which are a function of time, fit that Campbell and Stanley category of psychological and biological processes (under the label of "maturation") that also threatens internal validity by occurring before or in one experimental condition and not another.

Teams of operators develop new procedures for working together or alter old ones, as Davis and Behan (1962) have noted. This can be viewed as a kind of team "maturation" process that can confound other variables. Instructions, SOPs (standing operating procedures), and surveillance by the experimental staff can limit the extent of such "procedurization," if that is desired, but in a complex team operation this process cannot be prevented entirely.

Some attention has already been given to the problems of exerting experimental control over situations where sets of subjects, as well as individual subjects, react to each other. There is the serial processing problem, where subsystems process inputs in series rather than in parallel, and also the problem of competing or co-operating systems. In each instance the outputs of one entity become the inputs to the other. The inputs vary in ways which prevent orthogonality, equivalence, constancy, and above all replication. Intersystem competition, including conflict, characterizes games and gaming. Inevitably these tend to permit much more confounding than do experiments, and although they may serve educational and explorational objectives well, their outcomes are characterized by limited certainty.

If implicit or explicit competition between two systems in an experiment exists without the interchange of inputs, motivational reactions among the subjects may confound either state of the system variable. Even in noncompetitive situations, military subjects (or their superiors) may think they are being individually tested. Subjects may like one system more than the other. Intersystem rivalry, the eventual emergence of a "winner," introduces a new and uncontrolled variable. To be sure, this may be inserted in certain situations deliberately to bring about the very complexity, variety, and uncertainty that may enhance innovative performance. But then the burden lies on the researchers to introduce the states of this variable in a nonconfounding manner.

In general, motivational variables, including those defined by special cues and reinforcing and aversive consequences, have to be handled with great care. So do situations where motivations can produce subject behavior that competes with performance on behalf of the system. Kinkade et al. (1963) provided some illuminating illustrations of what happened in an experiment on air traffic control.

Certain potential sources of bias can affect the results of an experiment when precautionary restrictions are not imposed. For example, in a real-time simulation evaluation study, some visitors "dropped in" to see a new system development in operation. These people were not particularly interested in observing the control conditions; they wanted to see the test of the experimental condition. Some of these visitors were high status men and their presence during the test of the experimental condition probably biased the results. This bias could have been eliminated by restricting visitors during the actual test run. Demonstration runs could have been scheduled and these runs could have been conducted in such a way that they would not have influenced the results of the evaluation study. In another simulation study where female "pilots" were employed, it was found that controllers tried to make dates with the girls while they were operating in the test conditions. Measures of frequency and duration of communications were definitely affected by this behavior and other measures could have been differentially affected. That is, if one test condition required very little of the controller's time and attention, he might have devoted his extra time and attention to unnecessary conversation. Thus, a real difference between the test conditions might not have been demonstrated in the results. Again, this source of bias could have been removed by restricting the behavior possibilities.

A recurring problem in man-machine system research is getting enough subjects. It is accentuated by the need for subjects to be representative, a matter to be discussed more fully. There are seldom enough individuals available to carry out all the desirable counteraction tactics against confounding. This is true whether the individuals are actual operators of the system being examined or drawn from some other source. Paid individuals in large numbers are expensive. A man-machine system experiment is likely to need many persons, for long durations. Skilled system operators may be unavailable in sufficient numbers, even when the research has strong management support. In addition, it is often difficult to keep military personnel consistently on hand during data-taking sessions; they have competing duties, go on preestablished leaves, or are reassigned.

This resource problem is compounded by the fact that in most of the experiments individuals comprise teams. In many cases these are fairly large. In the experimental design, each team is one subject. If each team consists of a dozen persons, five subjects would require sixty individuals. There is still another difficulty. It is often desirable that each subject remain as much as possible the same while other variables are being varied. This means that the composition of the team should stay constant, especially since particular skills are required for each position and for interactions between positions. But this is not feasible if the individuals keep changing.

As Davis and Behan (1962) have pointed out, teams as subjects are rarely (if ever) numerous enough to randomize, that is, to assign a number of them randomly to each state of a variable. There may not even be enough to establish any control groups in implementing the confidence tactic of contrast.

When it is possible to establish teams as control subjects, that is, subjects in the control condition, they may be limited to one or two, just as those in the experimental or focal condition may total only one or two. This means there are not enough subjects in the contrasting conditions to make it likely that differences between the subject teams in one condition and the teams in another condition are much the same as those between teams in general. Apparent differences in results between conditions may really come from differences between teams. Although differences between teams can occur because one team happens to have better performers on the average than the other, they also can result from differences between two or a few individuals. Even in a large team, some of the operators fill key positions. If the individual in a certain key function in the control state is much more (or less) competent than his counterpart in the focal state, team performances may differ regardless of the states.

Confounding also can result from the so-called "Hawthorne effect," named for the location of an industrial establishment where this well-known but poorly defined phenomenon was first systematically observed. The performance of a set of operators improves just because a change is made in its circumstances in an experimental setting, regardless of the nature of the change. What has been poorly defined is the origin (or origins) in the complex of factors which make up a change of circumstances in an experimental setting. The origin has been viewed as the attention given to the subjects or as other contact with the experimenters, such as inadvertent reward or other positive reinforcement, perhaps in conjunction with the very fact of change.* Among others, Davis and Behan (1962) have

*See Roethlisberger, F. J.; Dickson, W. J.; and Wright, H. A. *Management and the worker* (Cambridge, Mass.: Harvard University Press, 1939). This account indicates there were many confounding variables.

noted the possibility of the Hawthorne effect in man-machine system experiments. It could confound a particular variable state. It could also make the experimental conditions and results unrepresentative of the real world by adding an influential variable not found there—a possibility discussed under "external validity."

Even when an experiment investigates something other than training or learning as such, subjects may still learn during the experiment. As already noted, this is one of the nonlinearity factors that can lead to confounding, if more learning is associated with one state of a variable than another. In man-machine system experiments this problem may have a special impact due to the dual nature of learning in such systems. As mentioned earlier, operators may not only acquire skills but they may also change or adopt procedures. The process of procedurization when it occurs spontaneously can be as confounding as the acquisition of skill or knowledge.

In the acquisition of skill or knowledge a problem may arise that is particularly severe in such experiments. This is the memorization of a simulation input so it is easier to handle if it is repeated. Some inputs, such as those simulating air attacks, have complex patterns and properties requiring quick judgments or precise visual discriminations. On subsequent presentations of an identical input, the operator "knows it by heart." He does better, but the experimenter may be misled into thinking the improved performance results from another variable state associated with the subsequent presentation.

Still another difficulty occurs when a man-machine system experiment examines some proposed improvement in a system that is actually a constellation of new features. If the change does produce better performance, the researcher still does not know which feature did it. Very possibly the system could be improved faster and at less cost if he did.

*Obstacles to Using Counteraction Tactics.* Such are some illustrative situations that can cause confounding. Another way to look at the difficulties of man-machine system research is to review the various tactics described earlier and the obstacles which can keep experimenters from exploiting them fully.

*Orthogonality.* This may have to be incomplete because of the large number of variables and variable states to be examined in a single experiment.

*Equivalence.* Mishaps and errors of many kinds in large-scale and intricate experimentation can change a particular state of a variable from what was planned.

*Counterbalancing.* It may be too costly or awkward to change the equipment configuration frequently. In addition, sessions may have to be conducted out of order, due to inadequate or lost resources or other untoward events.

*Replication.* Since repetitions of experimental conditions extend an experiment's duration and cost, there should be only as many as necessary. Although some statistical techniques may help, it is not easy to determine how many there should be.

*Preclusion.* Simulation and instructions accomplish only so much. The complications introduced by interactions between systems and between subsystems have already been described.

*Constancy.* The same kinds of factors that lead to inequivalence affect constancy, and there are others. The variability of both men and machines has been noted earlier; too often the causes are unpredictable.

*Randomizing.* The unavailability of enough teams as subjects to randomize them has already been pointed out. The same is true of sets of equipment. Inputs usually cannot be randomized because the researcher wants either to present a realistic simulation or to vary the load systematically.

*Contrast.* Again there is the question of enough subjects. On a broader basis, the experimenter must decide whether to invest his resources in contrast or in replication. This is a difficult trade-off problem.

*Refinement.* An experiment cannot have an infinite number of variables and variable states. The number involves more than a trade-off with replication and with orthogonality. How far should the researcher refine his variables within an experiment? What should he defer to a subsequent experiment?

*Some Solutions.* As said at the outset, this chapter poses at least as many problems as it answers. But there are some ways to cope with confounding.

For example, take a situation where there are not enough teams to randomize them or assign different teams to contrast conditions. Let us presume none of the independent variables embodies an irreversible process which changes the operators, such as learning in an experiment on techniques of training. To deal with this situation, the researcher manipulates a variable called "teams." Each team becomes a state of this variable and is associated with the other states of other variables. These may include a "zero."

The design may be a "complete factorial" or some version of an "incomplete factorial" with the latter's fewer experimental conditions (combinations of states of variables) and partial confounding. Kidd and Michels (1959) favored the Latin square version mentioned earlier, which admits three variables, all with the same number of states. A Graeco-Latin square admits four variables. Davis and Behan (1962) advocated the "mixed" designs in which the same subjects are used for some comparisons of variable states and different subjects are used for other comparisons.

The economy of the square must be considered with reference to the number of subject teams that are desired to increase assurance about their representativeness. This is a matter of *external* validity. Since there must be as many teams (treated as a variable) as there are states of other variables, some multiplicity of teams is inherent in the square design. If external validity were not a primary consideration or if it could be assured in some other fashion, the researcher could turn to a complete factorial design incorporating a single team. This would be more economical than a square. If a complete factorial design and a square design had the same number of teams, the square would be the more economical.

When only a few teams can be composed from the number of available individuals, there are several ways of assignment and matching to maximize the number of teams or their equivalence. These are discussed later in the section on *Subjects.*

To preclude learning or practice from confounding variable states (in a non-training experiment), all the subjects should be trained before the experiment begins. They may need classroom indoctrination and individual practice as well as team training. All should get the same information in briefings. Before he terminates the training, the experimenter should try to find out whether each team has reached a steady state of performance. He should get quantitative data, even though he may have to use judgment rather than significance statistics to conclude that performance has stabilized. He may well want to establish some criterion of stability in advance. Another thing he can do is design the experiment so he can find out how much learning occurred during the data-taking sessions, if any. This may require successive blocks of sessions, each block containing all conditions in the same arrangement.

The experimenter can also try to equalize the effects of practice during the experiment by making sure that each state of each variable occurs about equally often during each stage of the experiment and precedes and follows each other state with the same frequency during each stage. Accordingly, he introduces the counteraction tactic of counterbalancing by treating the order of presentation of variable states as itself a variable, with different orders constituting different states. Alternatively he may use *ABBA BAAB* patterns within parts of the experiment; these lend themselves to more flexible design. Still another approach is to randomize the order.

The special practice effects which can result from memorizing a simulation input can be prevented, at least in large part, by techniques described later in this chapter. Essentially, the inputs are varied in ways which keep them equivalent or constant in load but conceal their resemblance to previous inputs. The experimenter should, if possible, try to find out whether the different versions of the inputs within any load level had differential results. To do this he can design the different versions into the experiment as states of an independent variable. In any case, he should refrain from changing the inputs arbitrarily during the experimental sessions, even though he may be tempted to do so if he finds they lack challenge or, instead, offer too much. Rather, he should test the inputs in preliminary sessions and adjust them before the experiment proper begins.

Simulation in which the inputs are preestablished is a potent method of applying the confidence tactics of preclusion, equivalence, and constancy by keeping out extraneous factors and reproducing the same inputs. But given the variability of human involvement in simulation and even of electronic processing, not to mention that of surrounding circumstances, a concurrency approach is desirable when it is feasible. Concurrency makes simulation even more consistent. By concurrency is meant the presentation of the same inputs to different systems or under different procedural conditions at the same time. The comparison situations—the different systems, for example—would have to be operated in the experiment in parallel. Such a arrangement also saves time. Concurrency may be even more effective when inputs are real rather than simulated, since it is so difficult to repeat the same real inputs consistently.

Next to simulation the most effective preclusion agent is probably instructions to the subjects, as in any experiment with human subjects. Yet these require even more care in preparation and delivery than they do in other research, due to the complexity of the experimental environment and the number of alternative courses subjects may adopt. Instructions must be fully written down and thoroughly understood by the subjects. The experimenters must make clear, at least to themselves, where the subjects must be constrained in their procedures and where they have latitude. All instructions should be reported in the description of the experiment. Members of the experimental staff must monitor the data-taking sessions to make sure that instructions are being followed, and they should record deviations.

Preclusion must also be practiced by the experimental staff, including its managers. Precautions should be adopted to keep the staff from giving subjects inadvertent cues or feedback that could distort results. The large number of individuals in a subject team, the variety of on-going situations, and the number of experimental staff members combine to make this a bigger requirement than in other kinds of experimentation. Quasi subjects should be monitored to make sure they behave the same way with all subjects and in all conditions. Visitors should be excluded, or at least the subjects should be kept unaware of their presence. One way to handle them is to hold demonstration sessions after the experiment is finished, or during the experiment with the staff as subjects; note the foregoing quotation from Kinkade et al. (1963). It is unlikely that visitors can be manipulated so that their presence can be made common to all conditions.

As in other kinds of experiments, the formal design of a man-machine system experiment must be planned and recorded to include the schedule of sessions and intrasession conditions; and a log of what actually occurred must be maintained. Possibly the most challenging tasks are determining in advance how many replications of a condition are desirable (or judging during the experiment when there have been sufficient), and planning for contingencies (or dealing with them as they arise). There are no simple prescriptions for these tasks.

There are significant advantages in establishing and maintaining a rigorous design. Another approach is to attempt a kind of flexible modular design, in which certain successive blocks (within sessions or day-by-day) have the same combinations of experimental conditions, so that when disaster strikes there is less damage to the design, and the experimenters can adjust the number of replications as they go along. But the dangers of confounding are greater, and less confidence can be placed in the results. More will be said about this in Chapter 25 in comparing the objective of certainty with that of discovery.

A somewhat different approach to flexibility favors dividing an experiment into many short sessions instead of a few long ones. As suggested early in this chapter, less of an investment is lost if a malfunction or error occurs late in a relatively short session and it has to be aborted. It is also possible that if conditions remain the same throughout, subjects' interest will last throughout the session when it is short. Since shorter sessions mean more intermissions, there will be more opportunities to make any necessary changes to equipment or computer programs or to give instructions to subjects when incipient difficulties come to light.

Certainly session length should not be governed by some factor of convenience or tradition, as in the psychoanalytic hour or those university experiments with student subjects held during class periods. In some situations length is a matter of external validity. For example, it may be best to match a session's duration to that of the system situation being simulated. If the effect of duration itself is of interest in a study of space flight, the session should last as long as the actual flight. In other situations time may be compressed, a technique discussed shortly under *Simulation.*

In many man-machine system experiments it is important to avoid any "end effect," that is, some effect on performance because the subjects know the session is coming to an end. One way to prevent this is to have sessions of irregular duration. Another is to add some time, perhaps of varying length, to that part of the session in which data are gathered.

In this discussion the term "session" has meant the continuous period during which subjects are performing in a laboratory. But the terms "session," "run," and "trial" have not always been clearly distinguished in the research literature (as the following chapters indicate), any more than "condition," "problem." "set," "period," or "phase." There might be less confusion if "condition" were understood to be each combination of states of variables; "set" were to include all conditions, and "subset" some portion of them; and "run" were to mean each run-through or replication of a set. The term "problem" is often taken to designate some simulation input coinciding in length with a session.

It is clear that a single session can include a single condition, some subset of conditions, or even all of them. Possibly a single condition could extend over a series of sessions. In addition, data might be recorded only during certain periods within a session.

### *Assurance Methods, External*

To this point the treatment of experimental design has centered on how confidence in the results depends on features strictly within the framework of the experiment itself. As noted initially, another aspect of confidence is the degree of certainty that the outcomes match what would occur in the world at large, outside the laboratory. As with internal validity, some tactics to increase external validity are associated with independent variables, others with dependent variables or measurement. The latter will be discussed in the fifth section of this chapter. The former are briefly described here, along with attendant difficulties.

1. Perhaps most important is the representativeness of each state of a variable, whether a constant or one of some number of states. This means each piece of equipment, task, subject, procedure, environmental influence, and simulation input. Does each typify what it represents? This implies two further questions. Are the most important values or states of the variables in the real world represented? Is the range or variety of these represented?

Although these questions of sampling are in essence no different for man-machine system experiments than for other kinds, it can be much more difficult to meet sampling requirements or to know whether they have been met. Special problems are associated with simulation and subjects. For example, if a system is being developed so it can handle very heavy peak loads of traffic—for example, an air defense system—the simulation should incorporate such peak loads. Yet they may not be representative in the sense of being normal. There should be as many subject teams as possible, to make the sample of teams more likely to reflect the actual population of teams. Yet it has already been pointed out how difficult it can be to assemble a number of teams. The questions still remain, how many suffice and how should they be chosen to make the sample sufficiently representative? More will be said about subjects further on in this chapter.

It may be assumed, and usually is, that a single set of equipment can represent all the sets in the same class. Yet, as has been observed, operational equipment is variable to some degree, like other equipment, and is subject to biases and malfunctions. The assumption may be false. Further, equipment may be operated in an experiment to represent the hardware in a future system. How that hardware will actually be designed can only be a best estimate at the time of the experiment. Or equipment in the experiment may represent what is operating in a current system being compared with a future one. Yet by the time the experiment is finished and the results published, the current or "benchmark" system may have so changed that the comparison is no longer valid. The same may be said about computer programs and team procedures.

The representativeness of procedures is a difficult matter in man-machine systems experiments for another reason. Partly because in teams people interact with each other, it is not possible to make them follow prescribed procedures precisely; there is always some deviation. This is one reason why interactions lead to variable performance, as we have seen. Furthermore, it can be difficult to prescribe precisely, in advance, the procedures which team members should follow. In fact, in some experiments the experimenters may want to give them latitude to see what procedures they evolve or how they change what has been specified. If procedures in an experiment are so variable, how can the researcher be assured they properly represent those in the real system? How can he know whether the options to change and evolve procedures match the team's options in the real system?

2. Another, related method of increasing confidence in an experiment's external validity is to include the multiplicity of variables which are known to influence performance in the real world. Here, as in other ways, the effort to achieve external validity competes with that for internal validity. It has been noted that the greater the number of variables with varying states, the more time-consuming and laborious it is to implement such tactics as orthogonality and replication.

But important variables which might be introduced as constants may be disregarded for a different reason. They are omitted because they seem to be too difficult to incorporate. These are such aspects as the adversary and his actions (including his deceptions and errors), or actual combat, or special environmental circumstances like weather and zero gravity. Although the difficulty may sometimes be insuperable, the researcher may simply follow the line of least inconvenience. More will be said about these things in the section on *Simulation*. Researchers may also omit significant events or contingencies because they are rare. It is not clear how to replicate rare events to achieve internal validity without jeopardizing external validity by repeating them.

3. In the real world performance is frequently influenced by developments which cannot be specified in advance. To enhance internal validity, the experimenter employs the counteraction tactic of preclusion to forestall such effects. Since they are not allowed to intrude or emerge in the experiment, the experimental situation may be regarded as losing some of its external validity. This is the old story of increasing confidence in what an experiment yields by limiting its field of view. The variables that are manipulated hopefully remain uninfluenced by confounding from unpredicted or unspecified circumstances. Man-machine system experiments are sometimes criticized for "artificial" restrictions on the situations they embody, because the complexity of actual systems encourages critics to favor a more naturalistic approach. What needs to be asked, however, is what is actually precluded in an experiment, if anything? To what extent does the preclusion make the effects of variable states that *are* included unrepresentative of the real world?

4. One form preclusion takes is to limit the performance that subjects are permitted. This is done by specifying the procedures to be followed and the points at which decisions must be made. Yet as was noted a short while back in discussing the representativeness of procedures in the real system, operators may have considerable latitude in what they adopt. They may also come to decisions at other times than those which might be specified. If this is so, similar latitude ought to be extended to the experiment to support external validity. Yet by limiting the amount of preclusion, internal validity could suffer.

5. If preclusion takes away something, experimentation as such may add something to threaten external validity. It may influence the behaviors of subjects during the experiment. This can occur in ways other than by taking data and applying measurements, described later, although possibly these are the most usual occasions. Pretesting as an experimental device, for example, may sensitize subjects so they subsequently perform differently than they would in real life. Although counterbalancing cancels out the effect of order from the viewpoint of internal validity, the effects of initial conditions in the experiment on later performance may cause that performance to differ from performance in the actual world. Clearly, practice, either early in the experiment or before it starts, may give operators skill levels uncharacteristic of real operators. Contact with the experimenters may supply cues and consequences (reinforcements) not encountered in reality. The "Hawthorne effect" has already been noted. It can easily be a greater threat to external validity than to internal validity by affecting performance in all conditions rather than affecting it differentially. Safeguards against some of the effects of experimentation per se, including the Hawthorne effect, will be discussed in the section on *Subjects*.

6. As noted previously, the state of an independent variable in an experiment is often a composite of two or more states which might be manipulated separately. If one composite is found superior to another, the experimenter still does not know which component makes it better. Although it may be more practicable to manipulate the composites at first, in subsequent experiments the components can be examined separately in a process of successive refinement. Such refinement is needed to support external validity. While, for example, a combination of training techniques *abc* might surpass *xyz* both in an experiment and in

real life, *xbz* might do so also. To make an experimental finding more applicable, the reason for the effect should be narrowed down by eliminating irrelevant factors, such as *a* and *c*. Then the finding can be related more precisely to real-world situations.

7. Valid extrapolation to the real world also depends on all the experimental circumstances under which the finding was reached and the extent to which such circumstances are deemed representative. Hence the experimenter should specify in his report not only the states of all independent variables but also the values of all variables held constant as well those which varied autonomously but were measured.

8. Confidence in the applicability of a laboratory experiment's results can be increased if a field test is conducted afterward and similar results are obtained. The field test need not be as comprehensive as the experiment, and significance testing of the relationship between two sets of results is not essential. The comparison may have to be made on the basis of judgment.

## SIMULATION

Simulation covers a great deal of territory, and an entire book could be written about it even if it were confined to man-machine system experiments. To many it is the most challenging aspect of such research, with the result that it often overshadows other problems of method which also need attention. One hazard is that simulation may come to be viewed as the primary goal, so experiments are conducted to make use of it, rather than the other way around.

Views of simulation's role in man-machine system research on the part of numerous authors are set forth in Chapter 24. All-computer simulation, for example, is discussed there. Anticipating that chapter slightly to give the reader an overview of methodological problems, this section will look at what is simulated and what does the simulation—the objects and the agents; uses and advantages of simulation, and some criteria; the production of simulation, and its presentation; the fidelity of simulation, including noise and reactivity; and time compression and expansion.

### *Objects and Agents of Simulation*

If simulation is taken to mean the representation of something by something else, everything of interest in a man-machine system experiment may be simulated. This includes the system being investigated and its subsystems, the events and situations which lead the system to react, the system's environment, other systems and organizations, the system's operations and its components. Physical components include machines or equipment: vehicles, such as ships, tanks, automobiles, and aircraft; radars; missiles; facilities, including displays, consoles, pushbutton panels, keyboards, and computers; and communications, for example, telephone, radio, and pieces of paper. Components also include people—the operators of the machines and the system managers. Components may be interpreted further to embrace the processes and procedures through which the system functions.

The environment to be simulated may be land, sea, or air, and various objects or phenomena in it—vehicles, or geographical features, or weather. The other systems and organizations range from one in which the system being examined is a part to those with which it has dealings. A process can be a computer program.

Those real-world features which tend to figure most prominently in simulation are the events and situations which lead the system to perform. An air defense system reacts to hostile bombers and to its own interceptors. A logistics system reacts to malfunctions in equipment and to supplies or shortages of spare parts. A police dispatching central reacts to an urban disaster and to police and ambulance units. Conceivably all these originating events and objects could be simulated directly, and sometimes they are. But usually the system itself deals with them less directly—through radar returns from aircraft, through messages or records on paper about malfunctions and spare parts, and through radio or telephone calls from police units. Accordingly, it is likely to be the radar echoes, the pieces of paper, and the radio calls that are simulated, not that from which they emanate. In the simulation of inputs then, that which is simulated may well be some signal or message instead of the original article. The system's outputs may also be signals or messages for reception and processing elsewhere; these too are simulated.

The agents of simulation are more various than is generally realized. Among them are equipment, environments, and people. Actual equipment may simply be assigned to a different role, as when a SAC (Strategic Air Command) bomber simulates a Soviet bomber. Ordinary office telephones may function as a simulated system's phones, or an intercom system may represent a radio network. Actual equipment may be modified or reduced, as in the case of an immobilized automobile or an aircraft cockpit. Actual equipment may be closely copied, as in a mock-up. Equipment may be used that is thought to resemble the real equipment: displays, consoles, keyboards and panels, the computer itself. The displays may be computer-generated on a cathode ray tube or projections of slides prepared manually. Some lamps may represent the lamps in the real system. An actual environment occasionally simulates the environment in which the equipment functions. Some instrumented terrain is the scene of a simulated battlefield. A special road is traversed by an experimental automobile.

This kind of simulation has both advantages and disadvantages. Aircraft and consoles already exist and are the real thing, or almost so. But their availability can be uncertain, and hidden costs can be considerable. If large numbers of aircraft are required at frequent intervals, availability may be out of the question. Smaller numbers may be difficult to control as planned. Individual aircraft fail to appear where and when they are expected. Communications are available, but may fail or become overloaded with other requirements, such as those for operational use. Actual terrain seems readily available as an experimental setting for maneuvering soldiers, tanks, or low flying airplanes, but adverse weather may prevent its use.

An alternative is to resort to miniature terrains (scale models) associated with miniature tanks or with cockpit simulators engaging in low level flight. Miniature cars may be guided along miniature roadways. For certain objectives, notably experimenting on the performance of controlling the vehicle, it is not possible to shift too far from the real objects. But where other kinds of performance are of primary interest, the environment may be represented otherwise, by maps, photographs, cathode ray tubes (e.g., radar scope simulators), and map-like or diagrammatic (e.g., geometric) surfaces on which markers, signals, or tokens represent moving objects.

Words, numbers, and graphical materials also can and do represent, in an experiment, the physical real world directly. Diagrams may simulate all system elements, information flow, and actions taken. More familiarly, words and numbers (and sometimes voltages) simulate elements, processes, inputs, and outputs. The words and numbers occur in scenarios and scripts, including scripts of inputs for human simulators; in handbooks and descriptions of system operations and environments for system operators and experimental staff; in instructions to subjects concerning the procedures to follow; and in computer programs. Such programs may contain models of: (1) the external events and situations or the inputs which generate system responses; (2) the system elements and processes which operate on such inputs; (3) the system outputs; and (4) even the human operators and their performance. All-computer simulations consist entirely of such symbolic program models; man-machine system experiments may incorporate some of them, along with simulation by such agents as people, equipment, environments, and symbolic and graphical materials in other media.

What about people as simulation agents, rather than as objects of simulation? Obviously, people are most often used to simulate people. But as quasi subjects one or a few individuals also may simulate (1) an entire organization, such as a superior echelon or command or an information source, e.g., a weather bureau; (2) a man-machine subsystem, such as an aircraft; or even (3) a computer, with the human simulator processing inputs as though he were a program and composing verbal outputs as though he were a computer-driven display.

Clearly some simulation agents can portray a variety of objects, and objects can be represented by a variety of agents. There can be many different combinations. The relationship between object and agent can find itself at various places along a reality continuum. Representation may be "first-order," where the agent simulates the object itself, or "second-order," where it represents some signal or message emanating from the object and in that sense representing it.

Together with the data-gathering agents in an experiment, the simulation agents constitute what some (e.g., Davis and Behan 1962) have called the "meta-system." Connections between simulation agents and data-gathering agents (discussed subsequently) may have to be close and continuous. The same individuals or the same computer may perform both functions. A record should be kept of the simulated inputs as well as the outputs to verify what actually went into the system to make it perform.

### *Uses and Advantages of Simulation*

Simulation has other uses in man-machine systems besides experimentation. It can support training, design, communication between disciplines, and non-experimental evaluation or testing, such as physical testing of components. For example, radar simulation might be used to train operator teams, to experiment on operator procedures, and to test the capacity of a radar scope (e.g., a plan position indicator) to display signals. This does not mean, however, that the same simulation equipment is suitable for each purpose, although this is not always understood. Unless it was expressly designed for multiple use, it is likely to be appropriate only for the purpose for which it was designed. For example, the kinds of signals needed to experiment on operator procedures may not have to be as diverse and precise as those for testing the equipment. More precision and reliability may be needed for experimentation than for training. Similarly, the same simulation inputs or problems may not be optimal or even suitable for two purposes, such as training and system evaluation. The former may call for a range of inputs, the latter for greater realism. On the other hand, a production technology for creating inputs for training may very well be turned to creating them for evaluation.

For experimentation, probably the greatest advantage of simulation is the control it provides over inputs and other system features. As noted earlier, it contributes to preclusion, the tactic of preventing unwanted variables from confounding those in the experiment. It also facilitates replication, the repetition of experimental conditions, together with the constancy or the equivalence which such replication requires. It is much easier to repeat a simulated input coming either from a script or a computer tape than to rerun an exercise with actual aircraft or other objects.

Even on the first run-through, simulation is likely to be more reliable than real-world inputs. The availability of simulated elements is greater than that of elements subject to malfunctions, constraints, and interference. Simulation can also project future circumstances, incorporate infrequent events, present input loads which are not otherwise feasible, and vary these under fairly precise control. It eliminates hazards to people and equipment and may cost far less than operating the objects that are simulated. However, simulation is by no means inexpensive. Its realism may leave much to be desired. Some things are very difficult to simulate properly, such as weather and the effects of actual hazards on performance.

What are some of the criteria by which to judge the effectiveness of some proposed simulation device or approach? Chapter 24 responds to this question further. Here, the following criteria are proposed for man-machine system experiments. Researchers may have to make trade-offs between them, or between them and other requirements; they should at least try to determine what the trade-offs might be.

*Cost.* This consideration should include the possibility of initial design for mulitple use, as well as the cost of a versatile capability for producing simulation inputs.

*Precision and Reliability.* These concerns used to be even more significant when analog computers and electromechanical devices predominated. They are less pertinent to digital computers, which are more precise and reliable.

*Ease of Production and Availability.* This theme is discussed at some length below. Input content may depend on how easy it is to produce.

*Ease of Presentation and Manipulability.* This theme is also discussed below. Presentation is closely tied to production.

*Fidelity and Verisimilitude.* Other terms include completeness, validity, realism, and generalizability. Although relatively little can be said with certainty about how to handle this criterion, the theme receives considerable attention further on in this section. There is often a risk that fidelity may be sacrified to ease of production or presentation.

*Level of Detail.* Clearly, the level of detail of simulation relates to its fidelity and the simulation's purpose. If the experiment is designed to answer highly specific questions, it must contain a vast amount of detail and represent many functions and entities. But this makes it more costly and analysis more difficult. Rauner and Steger (1961*a*, 1962) observed: "The simulation must be abstract enough to permit manipulation of a reasonable number of variables, but must not be so abstract as to cast doubt on the validity of results." The level actually chosen in an experiment which these authors described was one which "seemed to feel right to both the Laboratory staff and the participants," but, these writers said, some rationale should be indicated for the selection. They further suggested that in the future it might be possible to vary the level of detail among the applications of models.

Rauner and Steger (1961*b*) also wrote that one function of increasing the level of detail was to help validate an experiment's results. They added that "By providing a detailed, tangible representation of reality, game simulations make it easier for planners to understand the overall model than if it is more abstractly drawn." In addition to "understanding," they said, a proper level of detail would provide to a planner "greater confidence" in the results of the experiment and "greater ease" in applying them. But here, too, was a dilemma. The more the detail, the greater the loss in "manipulative power." Because of the greater cost and effort required, the fewer must be the runs and the less the confidence in the reliability of results.

*Production of Inputs*

Input simulation must first be produced, then presented in the experiment. It is information on magnetic tape, film, audio tape, paper, or some other medium. Production consists of assembling the information and putting it on the medium.

The simplest assembly method is to copy inputs which have entered the real system. These might be actual sonar signals recorded in a destroyer or submarine. An experiment on image interpretation might use photographs copied from those which were interpreted in a real surveillance system—or even the same photographs. Recordings of voice communications—rerecorded or not—might be replayed. Teletype messages from actual operations might be reproduced. The experimenter might even incorporate historical data about weather and battles.

Production of this kind of simulation input is relatively easy but it is constrained. Although the researcher may reduce the constraint somewhat if he himself can control the environment and system operations giving rise to the recordings, all there is on hand for simulation is a replica of what happened. This may be less than, or different from, what is desired.

A variant of this method is the regenerative recording technique for computer-based systems (Sackman 1967). The computer in the system records the inputs which the system received from diverse sources as well as the actions taken by operators through manipulating switch-associated controls (switch-actions). This replay technique serves primarily for re-examinations of the data.

Another assembly method is to combine recorded or previously experienced inputs to create a composite different from any that occurred earlier but composed of actual events. For example, a sample of flights for an experiment on air traffic control might be created by combining a number of actual flights and their patterns for a particular airport. Similarly, actual commercial flights might be copied and combined to provide background traffic for an experiment in air defense. Although the experimenter still depends on what has occurred, he can generate a new composite.

The third and most usual method is to create inputs rather than reproduce them. It gives the versatility and completeness that otherwise are hard to achieve. The design task is a major effort. The research staff must extensively examine the system and environment to be simulated and collect a vast amount of data, so that the material created will resemble reality. For example, the staff of an experiment on logistics must ascertain all of the hundreds of different malfunctions that might require spare parts, the many situations that might cause the malfunctions, and the probable frequency of each malfunction. For a large-scale experiment in air defense, scores of bomber tracks must be projected on the geography of radar coverage in a manner reflecting the speed and altitude capabilities of the aircraft, their probable tactics and targets, and the temporal and spatial relationships between them.

In addition, of course, the experimental staff would have to collect a great deal of information about the operations and capabilities of the real system, so that the simulated system would reasonably match it. Such information in an air defense experiment would include data on radar coverage, on interceptor performance, on communications, and on computer processing.

To design each session's input from scratch for a large number of experimental sessions would be an almost overwhelming task. Accordingly, a number of methods have been devised to facilitate the synthesis of inputs. These include (1) designing "libraries" of specific, indexed inputs to select from; (2) standardized techniques of altering a small number of such inputs; (3) programmed templates of typical inputs which can be aggregated in a great variety of combinations; (4) retrofitting, whereby new elements are added to a total input or some are deleted that are no longer appropriate; (5) modularization, which permits combining elements in various ways to create new total patterns; and (6) rapid design of new inputs for unforeseen situations.

These methods and much of the design process can be semiautomated by means of a computer. This is possible when inputs are symbolic, and relatively easy when the symbols are standardized, such as radar returns, and their positions can be expressed in mathematical terms. When the symbolic material takes the form of words, expressing information or calling for action, the extent of automation in design depends on how stereotyped and classifiable are the messages and other verbal materials which have to be manipulated. The design of much of the material cannot be automated. Even so, verbal inputs designed and produced by hand are much easier to create than physical objects; it has already been pointed out that symbolic materials of all kinds lend themselves to simulation.

The fabrication as well as the design of input materials is a major effort, and if it could not be largely automated it would be most laborious and expensive. A computer can easily generate magnetic tape and printouts of verbal material. Special devices can be constructed to transfer information from magnetic tape to film.

*Load.* The importance of input design and production lies partly in the frequent need, already noted, to vary the input load as an independent variable, giving it a number of values or states. Since the input is the major factor in evoking system performance, such performance may be expected to vary according to the input load's quantitative and qualitative differences. Effective methods of design and production permit systematic variation in load.

In some experiments the system's response to differences in load may be the principal objective, to evaluate its ability to deal with an enemy threat, for example, or to handle different amounts or kinds of traffic. As pointed out earlier, in other experiments load is varied to learn whether the states of some other variable, such as procedures, achieve relative superiority according to the input loads imposed on the system. A complete picture of system performance is gained only if it is tested over a range of loads. Some systems are built to replace earlier ones specifically in order to handle heavy input loads. To determine whether these systems accomplish their mission, it is essential to test them experimentally with the loads for which they were designed. In most cases only simulation can do this.

At an early stage in system development, input load can be varied through simulation to obtain estimates of how well operators at consoles can cope with different loads. Such experimentation can indicate not only the proper design of consoles but also the number of consoles (and operators) for peak loads. (Unfortunately, this has not always been done.)

Input loads can be compared with each other quantitatively by tallying the number of similar events or situations to which the system, subsystem, or operator must respond. The times covered by the tallies must be the same. The number of events per unit of time can be regarded as the load's time density, a measure of comparison. An even better time density measure is the ratio between time available and number of action-evoking events.

The difficulty system operators have in responding to events is a function not only of time density but also of the nature of particular events. Some require more time than others. The time relationships between them can be determined empirically. By assigning weights, different kinds of events can be combined in the time density measure. In any case, inputs that are qualitatively different from each other should be designed into the load. Some may have such inherent difficulty that system performance may be severely degraded.

It was observed in the previous section of this chapter that frequently an experiment requires two or more sets of inputs that have constant or equivalent loads but are different. They are different so the operator will not be able to handle an input more easily simply because he has memorized it. Equivalence or constancy means that the two different loads place the same demands on operator processing.

The researcher can adopt several approaches to achieving both difference and equivalence/constancy. One is to maintain the same elements but disguise them. For example, he may start the same aircraft tracks from different locations, by creating a mirror image or rotating all track origins 90 degrees. He may reorder the starting times of events. He may assume load equivalence among all input events of the same type, such as an aircraft's change in speed. He may characterize inputs according to such equivalence properties as variety, redundancy, distribution, uncertainty, distraction, and intensity. He may determine weighting factors empirically, as just noted for establishing time density.

Finally, as suggested earlier, he may design the experiment with "equivalent" inputs as an independent variable to find out whether the different sets actually impose the same difficulty. This would be indicated by measures of time and error in processing. Subjects can also be queried at the end of an experiment as to whether they recognized the equivalence of inputs.

*Presentation of Inputs*

Some simulation inputs can be acted on by subjects directly, whereas some require a transducer to convert them into forms which operators or other equipment can process. Messages or reports on paper can be provided directly to a subject. Voice tapes need only a playback device connected to a phone line, and slides need only a projector. Magnetic tapes are transduced by a computer and peripheral equipment so that the information on them appears in printouts or on a computer-driven cathode ray tube display. Films containing optical radar signals need some kind of transducer to convert the marks on the film into electrical signals at a radar. Scripts need human transducers who speak their contents into telephones, tell them to subjects on a simulated battlefield, or convert them into manipulations of knobs and switches on devices. The devices in turn convert them into moving signals that appear on radar displays or into data pulses that are further converted by a computer into electronic signals on a display. (Some simulation inputs processed by a computer are outputs from the subjects who have produced them by processing other outputs from the computer.)

Much of the history of transducers has been one of early unreliability of electromechanical and analog devices, lack of precision in the analog devices, and increasing improvement in transducer design. Preventive and corrective maintenance of transducers and constant checking on their accuracy are important requirements; such checking applies equally to human transducers, the quasi subjects described in another section of this chapter.

*Location of Transducer and Inputs.* The location of a transducer in the simulated system determines the nature of the simulation input that enters it, just as the location of direct inputs to a subject determines their nature. Where in the system should the transducer be located, or where should the information be directly presented?

In a surveillance system, an observer in a helicopter sees some action on the ground and composes a radio message about it. The content of that message is not, let us say, a complete or accurate representation of the ground situation, which is difficult to discern, complex, and partly obscured. The ground situation has been transformed. The radio message is sent to a communication center, received, and recorded, possibly exactly as it was sent, possibly not. In accordance with procedures and as a result of other information, its contents may be supplemented or otherwise altered, and it is reformatted for further transmission to a computer center. There it may be recopied and reformatted for entry (possibly without error) and the computer, through its program, may filter it, supplement it, or otherwise alter it in combining it with other information. This is a third conversion. The final contents do not represent just what occurred back there on the ground.

In another surveillance system, a radar "sees" an aircraft and produces electronic signals which are registered for visual inspection on a plan position indicator (PPI) scope. The information is "noisy." Misleading, irrelevant, or missing data characterize the signals, misrepresenting the aircraft or its position. The signals are automatically transmitted to a device which converts them into binary pulses; what comes out of the converter is not the same content that went in. The pulses go to a computer. The product of its processing as seen by an operator on a display does not have the same characteristics that went into it, nor the same characteristics that could be seen at the PPI. Again, after this third conversion, what is viewed may not adequately represent the object in the air.

Now suppose the first system is to be simulated in an experiment. The situation on the ground conceivably could be simulated so it was seen directly by the simulated observer in the helicopter. On the other hand, the observer's message might be the thing simulated (thus including the effects of one conversion), or it might be the communication center's transmission (with the effects of two conversions), or the computer's output (three conversions). Correspondingly, in a simulation of the second system, actual U.S. aircraft might fly through a radar's coverage in an experiment to simulate a Russian bomber. Or a transducing device at the radar might put simulated radar signals into the radar's receiver. Another transducer might enter simulated signals somewhat different in form and content into the converter. Finally, magnetic tape with still different content and format might be transduced by the computer.

In either system the proper format and content of the simulation input must differ at each point of entry, because of the changes which occur during each conversion. Those changes must be reflected in the design of the input if it is to represent that in the real system. Yet these changes may become progressively difficult to specify as more conversions occur, partly because changes in a later conversion compound those in an earlier. The difficulty increases with the extent of "noise" in the original transmission—from the ground to the helicopter observer or from the aircraft to the radar; the nature of noise is discussed subsequently.

One way to achieve the proper content and format is to place the transducer and the simulation inputs as far forward as possible in each linkage. The inputs can be designed with greater validity if they enter the system after a single conversion. Another solution is to give in to the temptation of by-passing the problem and neglect the conversions entirely.

This temptation may be greatest in a computer-based system. In such a system it is relatively easy to produce the simulation inputs for that system stage which follows computer processing, the display stage. The system's own computer, in fact, can be used for this production. Yet it is exceedingly difficult, perhaps impossible, to produce inputs for this stage that take into consideration the changes that would have come from various pre-computer conversions. As a result, in experiments on computer-based systems, information may be presented to subjects in a fashion that misrepresents operational reality by disregarding the effects of conversions.

This by-passing of proper input design occurs in part because of the ease of production of inputs for entry into the system at the display point. It is also a question of system boundaries, discussed in Chapter 25. If the simulated system includes the locations at which the conversions take place, transducers and inputs can be placed at those locations. In any case, if experimenters introduce inputs directly into the simulated system's computer, they should state in their reports the extent to which the outcome might have differed if the simulation had included conversions in the information flow.

Similarly, Sackman (1967) has cautioned that in experiments on computer-based systems, researchers should obtain data about those manual operations of operators that are not registered by the computer. This requirement can mean such an extensive data collection effort that the researchers might prefer to forego it.

*Developments in Simulation Technology*. Simulation technology in computer-based systems has become very sophisticated in recent years. One development deals in part with the problem that has just been discussed, that of serial processing information through subsystems. However, it is concerned with serial processing of data after the computer has already received the data from other sources, rather than with the sequence between original source and input into the computer. As suggested in the section on *Design*, when the serial processing problem involves only interactions between operators and the computer, it is more amenable to solution, probably because the problem is focused on content.

It was pointed out earlier that since the content on arrival at one subsystem depends on the antecedent processing by other operators it is difficult to achieve experimental control. The researcher cannot guarantee either the same input to the subsystem from replication to replication or the input he desires. As observed further, one solution is to operate the subsystem by itself. Then it is necessary to simulate the effects of the prior subsystem on the inputs. These effects came about as a result of switch actions by operators interacting with the computer. Simulation of switch actions is easier than simulation of other kinds of conversions and has been a technological advance.

Switch actions can be simulated as having been "perfect" or as possessing whatever lags and errors the researcher wishes to design into the input. With this technique the same input can be repeated reliably and with realism with respect to the actions taken within the preceding subsystem. Operators in that subsystem need not participate in the experiment.

Another technique is concerned with tagging each input item distinctively—for example, giving it a number. Then the computer keeps track of what happens to it. It can determine whether the item actually enters the system and at what point it changes. This helps the computer itself collect data about its own performance. It permits an experimental staff member at a console to change an input element during the experiment—for example, in reaction to some adversary action. It also enables the staff to monitor the progress of elements throughout the experiment as they appear with their tags on a display.

In a computer-based system human intervention may become necessary when automatic processing requirements exceed certain programmed limits. A feature of simulation technology can force such manual intervention. First the limits of automation have to be determined by test. Then inputs are introduced which exceed the limits. The computer can be programmed to record all instances where the desired inputs actually were introduced and the manual actions which operators took in consequence; it also classifies and summarizes them. This technique is useful in experimentation which emphasizes human intervention.

When the system represented in an experiment is computer-based, a computer—even the system's own—not only can produce the simulation inputs but also can present them to the subjects, record their performance (and its own), and make judgments about their performance (and its own). It has been pointed out that the linkage between production and presentation is simpler when the same computer can do both. The warning must be repeated, however, that this technological accomplishment may lead to a serious problem. The simulation of the system may be limited to those inputs which enter the computer and those outputs which it produces. Yet some inputs may be handled and some outputs produced by people and other machines; these will not be considered in the simulation. This deficiency is not the same as the failure of the simulated system to deal with the various conversions in information before it enters the computer, but both inadequacies may occur together.

All the simulation inputs produced by a computer do not have to be generated before an experiment starts. Indeed, immediately following an experimental session the models in the computer, representing some part of the system or some related system, may produce inputs or printouts, tape or cards, for the next session, in reaction to what the subjects and system have just done. These reactions can be generated even during a session, the computer then presenting inputs to subjects on cathode ray tube displays.

In a technology which is demanding, fascinating, and continually being improved by man's ingenuity, it is hardly surprising that presentation and production of simulation inputs by computers absorb much time, effort, money, and attention. Nor should it be surprising that the ease and availability of computerized production and presentation could lead both to a neglect of other types of simulation and to engaging in simulation for simulation's sake.

*Fidelity*

Simulation fidelity is clearly important to assuring an experiment's external validity. Yet little experimental evidence has been gathered about requirements for simulation fidelity or realism in man-machine system experiments. Experimental research on simulation fidelity has centered on that needed for training aircraft pilots; various degrees of realism have been compared for their effectiveness in these studies. In the field of air traffic control, two kinds of graphical or diagrammatic simulation have been compared with equipment (radar simulators and consoles), for experimentation and training. These studies will be covered later in this book. As previously noted, there is considerable further discussion of simulation, including simulation fidelity, in Chapter 24; much of that is of a general nature. Here it seems best to review questions which may more immediately perplex prospective experimenters.

An analysis of fidelity of simulation in man-machine system experiments raises a number of questions. Is objective realism necessary or does subjective realism suffice? In any case, when is fidelity desirable and when is it undesirable? When may it be traded off in favor of other needs? What features have proved most difficult to simulate validly and what factors have stood in the way of simulation fidelity? Kidd (1962) commented that "the whole issue of fidelity in simulation is still unsettled" and "it would be very helpful to have more than an intuitive feeling for the establishment of fidelity values for all the attributes of a simulation facility."

*Verisimilitude.* Subjective realism means that the simulation evokes the same behavior from experimental subjects as the real objects, events, or inputs would, regardless of the extent of similarity. The simulation is real enough to the subjects, even though objective scrutiny shows differences, perhaps large ones. The illusion of reality is created, one might say. Perhaps the best term is verisimilitude.

How can the experimenters know whether subjects are experiencing subjective reality? Probably the best test is whether the subjects respond to the simulation as they would to the reality. As Kidd (1962) put it, there is some indication from research that "the essential factor for achieving valid results is the nature of the response requirements imposed on the human operator."

In the absence of empirical evidence, experimenters can try to "look through the eyes of the subjects" by putting themselves in their roles, or by listening to their comments. They can interview the subjects and make them place simulated items on a reality scale. Subjects are likely to comment candidly if verisimilitude is absent. Subjects may "fight" the simulation, blame it for their difficulties in performance, or even fail to respond to it. However, a researcher can often avert such behavior and strengthen verisimilitude if he explains to the subjects initially what is unrealistic and why it was necessary to create the simulation in that way.

When they are unsurprised and given an advance explanation, subjects may exhibit far less antagonism to low-fidelity simulation.

Subjects are more tolerant about simulations of sensor data where technical problems account for diminished realism than about computer models of operations with which the subjects are familiar. A subject tends to insist on realistic performance characteristics of an aircraft which he flies himself. He also worries about loss of realism apparently due to the experimenter's lack of sophistication. The experimenter must give no evidence that he is ignorant about the operator's system. Subjects should also be warned to expect malfunctions in operational equipment, in transducers, and in computer programs. As previously noted, such malfunctions, which can bedevil experiments, should be held to a minimum through preventive maintenance, and engineers and programmers in sufficient number should be standing by to take quick remedial action.

If objective reality is desired instead of verisimilitude, the researcher faces a different question. How does his symbolic simulation relate to what the system's sensors would actually be sensing? One way to proceed is to arrange and measure the objects of the sensing. For example, if the simulation is to consist of aerial photographs, the developer of the simulation first creates the ground objects which will later be photographed–if they do not already exist–and measures them at the scene. Thereby he establishes "ground truth." If in the experiment an image interpreter fails to identify an object in the photograph, it is definitely known that the object was there and was photographed.

The same approach might be extended to radar and sonar, through the concepts of "air truth" and "sea truth." Recordings of radar signals and electronic countermeasures, for example, would be made during flights and emissions of actual aircraft. They would be used for calibration of synthetic material or possibly for simulation directly.

*Reasons for Nonreality*. One reason to abjure too comprehensive a striving for reality is that such an effort can be self-defeating. The experiment becomes too unwieldy or costly. Rather, realism should be sought in the critical aspects of the experiment. But the researcher must be careful that simplification in what he regards as peripheral does not, in fact, affect what he regards as central. For example, it might be felt that simulated inputs for a surveillance system had to be simplified to make it possible to study procedural changes in processing. But in the actual system the processing procedures might be influenced by features of the inputs that were lost in the simplification.

On occasion, the level of fidelity is simply a matter of resources. Consider simulation of radar. It is possible to produce and control with relative ease a great many radar signals of many aircraft as "canned" (stored) inputs, if subtle characteristics of the signals and diverse characteristics of noise are omitted. To present these subtle and diverse characteristics is relatively easy if it is necessary to generate signals of only one or a few aircraft with accompanying noise in alterable form. But to produce and manipulate a great many signals with subtle characteristics accompanied by extensive diversity of noise would be extremely expensive and technically difficult. (Noise in this context has very complex forms and patterns rather than some stochastic distribution, which, of course, is easy enough to produce.)

Time compression, discussed subsequently, may be regarded as a way of compromising reality to conserve resources. It increases the amount of system performance per unit of time. Space compression is not practiced so widely; instances of it include miniaturization of automobiles and highways, tanks and terrain. Organization compression, however, has been frequent. In an experiment a single individual, a subject or quasi subject, plays the role of an entire organization. He not only enacts the main person through whom the organization communicates but also provides the information and makes the decisions that would come from elsewhere in the organization.

Ernst (1959) described the difficulty of achieving complete realism thus:

> It is necessary to decide first upon the degree of realism that is required, knowing that simulation will yield relative rather than absolute results. Complete realism is probably impossible and certainly undesirable, because over-complication, with its inevitable inaccuracies, would mask the basic issues and unduly confound the results. However, the representation must be realistic enough to embrace all significant factors, and these must certainly be kept under experimental control. Given adequate control, undesirable interactions between such factors need not be thought of as confounding the experiment, for their delineation may become the principal objective of the experiment.

Too narrow a concern about realism may conflict with objectives in experimental design. An experimenter may wish to vary inputs and environmental circumstances systematically to find out how the system operates under a range of conditions. In a military system, however, operational personnel may regard only one or a few of these conditions as normal or probable and thus realistic. They may prefer to restrict the conditions to what they regard as the threat, whereas the researcher wants to stress the system, perhaps even to breakdown.

In justification, the researcher may point out that the threat is a matter of probability. It can take many forms, and not enough may be known about these. Even if substantial knowledge does exist about the threat, it may not be available to the researchers; but it is more likely that whatever knowledge exists is fallible. Researchers cannot have complete confidence in the data they get even about their own forces, much less the adversary's. Data about both may be colored by wishful thinking. Critical information should not be taken on faith. Predictions about what the adversary would do may be based largely on what, under similar circumstances, one's own side would do.

Researchers may wish to forego some kinds of realism because they believe that by presenting abstract inputs they can generalize more widely from the results—or at least just as widely as they could with concrete inputs that might be more difficult to produce and present. They may use checkerboards or cell-like designs instead of geographically detailed maps, markers instead of vehicles, or dots instead of electronically generated radar signals.

Experimenters may contrive an entirely imaginary system, with displays that are not intended to be identical with those in a current or planned system. The hypothetical system may have functions resembling those in current systems, but the operational procedures are invented, not borrowed. Such simulations, discussed in Chapter 25, are found in experiments seeking general knowledge, not knowledge about a particular system.

Researchers may prefer to make use of a hypothetical system if they are investigating new procedures or equipment for a particular system with personnel from that system as subjects. Those personnel may be biased against the new procedures or equipment because of familiarity with the old; or they may be biased in favor of them. It may be possible to eliminate or modify that bias by putting the procedures or equipment into a system with which they are not familiar. Such a system may have to be designed in great detail.

Under some circumstances it may be necessary to use experienced subjects, although the experimenter would like to know how less experienced ones would perform. Since too familiar a setting might make the task too easy, the experimenter may resort to a composite of settings.

One of the questions researchers face in devising hypothetical systems, as well as in simulating real systems, is the level of detail to incorporate. This has already been touched on with regard to simulation criteria. What facilitates generalizability? If the answer is abstraction, then this has to be defined in terms both of level of detail and relatedness to a real system.

Finally, it may be necessary to compromise with reality for ethical reasons. There are widely held views that subjects should not have to encounter certain kinds and degrees of stress. For example, in an experiment on vehicular traffic, should it be possible for the subject driving a simulated automobile to knock down a simulated pedestrian?

*Obstacles to Fidelity.* Certain matters seem especially difficult to simulate. These include acceleration or deceleration in air or land vehicles as well as vehicle movement in general, and actual combat. Compromises with regard to movement include putting a stationary automobile on rollers with external "passing" scenes, or giving pitch, roll, and yaw movements to a stationary cockpit, also with views of terrain as though this were being overflown. In the simulation of a battlefield, mines and other explosives may be detonated. Al though it is not feasible to simulate death and destruction fully, soldiers may be instructed to drop in their tracks. The signals of aircraft and ships may disappear from radar scopes, and simulated reports may say that they were destroyed. Civil defense officials may see external devastation, simulated by projections of doctored photographs, but the emotional impact cannot be the same as the real thing.

In other cases poor fidelity may be attributable to a combination of difficulty in simulation and inconvenience, or simple disregard, as noted in the discussion of external validity. In air defense studies there may be a tendency to give low priority to simulating wind, weather, altitude, degradation in radar returns (e.g., fading, low blip-scan ratio), reactive course changes by enemy aircraft, feints, and enemy mistakes. In other studies, too, sensor inputs may be insufficiently degraded. Peripheral activities may be omitted.

A major obstacle to fidelity in simulation is the difficulty in projecting the future, a topic also commented on in connection with external validity. The researcher may have to predict some future input load, such as the frequency of aircraft arrivals and departures at an airport. Or, he may be simulating a command and control system which is being developed. In facing the updating problem, he must foresee changes in computers, in programs, in consoles, in displays, in personnel, and in procedures. If his simulation must be designed while the system is still in the conceptual stage, his problem is even more difficult. Chapman (1960*b*) observed:

If one wishes to predict real-life system behavior from simulation studies, one must use representative stimulus conditions and system components–both men and machines. Unfortunately, most fancy systems are developed to meet anticipated rather than existing conditions. I don't know how one is sure that he has drawn an adequate sample from a population of undefined conditions that do not exist.

The problem of fidelity in getting people to simulate people is discussed later. When people are represented by parts of computer program models, there exists a tendency to make simplifying assumptions about their behavior. It becomes necessary to consider motivational factors and their effects as well as probable error rates and times required to complete tasks. Not enough is usually known about error rates to make the simulation valid. It is possible to make estimates of time required, but these have to be distributed in accordance with the distribution of competence among actual system operators. Data on such distributions are not generally available. Although error and time data can be collected by the researchers from the system being simulated, this may not be easy to do. Computer modeling of human performance in systems is further covered in Chapters 23 and 24.

If it is difficult to model detailed human reactions to pre-established inputs, it is more difficult to model a total simulation in which the inputs change as a result of the human reactions. The inputs resulting from the reactions must be planned in addition to the reactions themselves. Then another step follows in which the reactions to the new inputs must be predicted, then the inputs resulting from these, and so on. Very quickly in the sequence of steps it becomes impossible to set forth the inputs unless in each step the range of possible human reactions is very limited indeed and the inputs resulting from each can be readily predicted.

This is the problem of introducing reactivity or responsiveness into an experiment. Both reactivity and noise deserve extra attention as problem areas in simulation fidelity.

*Reactivity*. The simplest case of reactivity is the one in which inputs are stored on some medium, like magnetic tape, for presentation through a transducer. Although such inputs are generally unalterable except by re-entering the production process, there are a few ways to give them some flexibility.

The easiest technique to implement is deletion. If elements in the simulation package are tagged with identifiers, e.g., numbers, which the computer can recognize, the elements can be removed during the experimental session. For example, simulated radar signals from hostile bombers that are shot down can be eliminated from the displays used by surveillance operators and interception directors. A member of the experimental staff watches the action at his own console. When he sees a "kill," he manipulates switches to transmit to the computer the tag number of that simulated bomber and the instruction to eliminate it. Its radar signals do not reappear during the session. No additional inputs have to be placed on the tape as a consequence of the reaction resulting in a kill.

In contrast, consider a different sequence of events in real life. When the bomber pilot detects the approach of an interceptor attempting a kill, he changes course, or speed, or altitude, and perhaps thus avoids being shot down. To simulate this evasive action would mean placing many new position inputs on the magnetic tape, reflecting the new course, speed, or altitude in the bomber's simulated track. It would also mean the elimination of those originally there. This has not been feasible in simulations of large-scale air battles, although new on-line programming techniques raise the possibility of making such substitutions on a modest scale. Deletion, on the other hand, does not require adding inputs resulting from new characteristics.

Along the same line the technique of deletion can be used to alter inputs from session to session, possibly as the result of what occurred in the preceding session. This is done before an experimental session starts by inserting supplementary instructions via a keyboard or some IBM cards to remove certain programmed inputs. For example, the tape might contain thirty simulated bomber tracks. An instruction to the computer might withhold fifteen of them from one session, while another instruction could withdraw a different set of fifteen from a second session, thus producing two different sets of inputs with the same tape.

More complex is the technique of producing and storing a number of alternative inputs which might be the consequences of an operator reaction. These might be stored on a magnetic tape or in a script which directs a human simulator. Also stored is some kind of guidance for making the choice among the alternatives. This guidance may state which characteristics of the human reaction should result in one alternative, which in another. There is more flexibility in the situation where a human simulator is following a script, since he can interpret the operator reaction and the guidance instructions together to select the alternative.

With human simulators who are not forced to follow a script even more flexibility is possible. The human simulator can concoct a new input instead of having to choose among stored alternatives. Consider the simulation of voice messages from an observer in a helicopter. The contents of a voice tape are unalterable; at most a device or a human being might make selections among them. But a human simulator could create and transmit new messages. He would not be constrained by choosing among alternatives and by guidance rules for doing so. The human simulator, perhaps with some device, could also fabricate the inputs which were the consequence of his innovation. For example, were he simulating a bomber pilot, he might not only take evasive action but also create the subsequent track.

Such flexibility typifies free-play simulation, in contrast to scenario simulation. These can be combined, just as input simulation may be a mix of computer and human sources. But generally in man-machine system experiments there has been relatively little free play. The human simulators, which this book calls quasi subjects, are constrained by rules and procedures. The roles they play are assigned to individuals associated with the experimental staff rather than actual subjects for the very purpose of standardizing their reactions and making sure they follow instructions.

Reactivity becomes far more complex in an experiment when the inputs to an individual, team, or system are the outputs of another individual, team, or system. It is difficult to predict the particular input at each step because it proceeds from one of a number of possible human reactions. The input which brought that reaction in turn resulted from a number of possible reactions, and so on. When one side in a two-sided game is modeled in a computer, it is possible to model explicitly only a relatively few steps, and then only if the alternative reactions at each step are few in number. Tic-tac-toe can be modeled for its entire course, but chess can be modeled only for a limited number of moves unless some rules or tactics are incorporated into the program to guide the selections among alternatives. The difficulty of modeling one side of an interteam competition or conflict in an experiment can be imagined. It usually seems better to let actual subjects engage in interaction with other subjects.

The difficulties that such interteam competition poses for experimental control have been well covered in the section of *Design*. All that needs to be said here is a word about fidelity aspects.

If motivational effects of competition can confound variable states, they can also be desirable because they resemble what happens in the real world. Competition may also make the environment more complex for each team, the events more various, and the uncertainty and challenge greater. If real-world competition does not exist in the experiment, the subjects may tend to adopt procedures and strategies that merely seem adequate—they "satisfice." Reality is distorted.

A very different effect can come from implicit competition, the kind that may arise when two similar systems are being compared in an experiment without interaction between them. The subjects may compete with each other, exerting extra effort, if they know they are in a comparison situation. If they are informed about their performance they are likely to do this even more. As previously suggested, confounding may result. Reality may be distorted, too, when competition occurs in an experiment but not in the real world, and each system may perform better than it would in actuality.

*Noise*. Any system which processes information must cope with noise. Simulation in man-machine system experiments must include the simulation of noise to be valid. Noise can be a critical aspect of the input.

What is noise in this context? The term here is meant not to stand just for a particular auditory stimulus or for electrical phenomena above the level of which a signal must rise to be detected. Rather, it is used in a more general sense, with borrowings from the unwanted nature of noisy sound and other electrical phenomena which accompany signals. It means factors in the input to system operators and equipment that interfere in some way with the correct discrimination of information being transmitted. The information may be a typed message, for example, a radio transmission, or a series of radar signals.

Noise can take effect in several ways (Parsons and Perry 1965). It can simply blot out the information, like an ink blot covering a written word. For example, radar signals from an aircraft may be lost in a myriad of other radar returns from clouds or mountains; the returns are so numerous and close together that the aircraft signals are indistinguishable. Noise may distort the information by eliminating or adding some feature, or by doing both to different features at the same time. An instance would be a typographical error which dropped "not" out of a sentence or an error which changed "not" to "now." Radar jamming keeps the scope operator from knowing the range of the jamming aircraft, although the jamming still indicates the aircraft's direction. Camouflaging a ship hampers the discernment of its outline. Finally, noise may consist of false information which is a counterfeit of the real. A hostile radio may transmit misleading messages as though they came from a friendly source. One frequent form of this type of noise is the production of many apparent signals, so that the observer cannot tell which is the one he wants.

Noise is particularly effective when two or all three of these types of noise occur together. For example, if due to noise some radar signals from an aircraft fail to appear on a display, their track may be already difficult to distinguish. If some false targets or counterfeits are added, the track may be utterly obfuscated.

Noise may be categorized as coming from nature, from a system's or operator's own actions, or from an adversary. Noise can be clerical errors in transcribing radio messages, heavy vegetation concealing troops from view, feints by an attacking submarine being tracked by sonar, multiple sources of data resulting in some overestimate of resources, or radar-detected "chaff" surrounding the path of an incoming missile. Because it can be so effective in confusing the other side, hostile forces employ noise against each other. In radar systems the attempts to employ it are called electronic countermeasures (ECM), those to cope with it electronic counter-countermeasures (ECCM).

As noted previously, noise, especially ECM from hostile sources, is difficult to simulate in all its diversity because the more effective it is the more difficult it is to describe and predict. If description and prediction were easy, so would be the ECCM. As has been pointed out, this kind of noise is not random. Nor does the need to simulate it spring from any requirement to present "perceptual realism" as such. Rather, noise is important because of the actions or nonactions that result from it. It leads operators to do things they should not do and fail to do things they should; and their discriminative behavior then influences the actions of other operators. The test of effective simulation of noise is the discriminative behavior that results.

In a few man-machine system experiments, noise and its effects have been the focus of inquiry, yet in others it has received insufficient attention in the simulation. Why? The difficulty of creating it has been noted; its simulation takes time, money, and skill. Its complexity and subtlety also make it difficult for experimenters to understand. Security constraints add to its mystery. Noise often leads to an "ostrich syndrome." Although the head is not actually hidden in the sand, the matter is disagreeable enough to be avoided.

The simulation of noise may simply be postponed to a later date, which never arrives. At times it is argued that state-of-the-art improvements in the system will shortly eliminate the noise. In addition to this wishful thinking, it may be urged that the system should be evaluated first under optimal conditions. If it fails under these, then it might as well be discarded. Unfortunately, if it succeeds under these, it may be accepted without a subsequent evaluation under less ideal conditions. This risk is a real one in man-machine experimentation oriented to system evaluation.

Still another reason why noise is often disregarded in simulation comes from an emphasis on a system's processing or decision-making stage. This may be somewhat removed in time and space from the system's "front end" where the noise enters and where information conversions occur. The significance of these conversions has already been pointed out. Noise and conversions often interact in distorting information (Parsons and Perry 1965). In reality, decisions, e.g., evaluations of threat, are greatly influenced by noise and by how the system handles it, but the commanders who make the decisions may not fully understand this. They may leave the noise problem to others. Experimenters interested in decision-making may do likewise. Whether they are designing it, operating it, or evaluating it, those involved in a computer-based system tend to disregard the system's front end. It does not seem to be where the action is. But it is certainly where most of the noise enters.

*Time*

The time during which events and human performance take place in a real system may be variously simulated in the laboratory. It may be the same, compressed, or expanded. Compression ratios in man-machine system experiments have been 3:1, 7:1, 48:1, and others. To help design and organize the inputs the researchers may even resort to a double compression. First, they assume that all the events which might normally occur in, say, thirty days in a real system actually occur in ten days. Then they compress the ten days into ten hours, with one hour of laboratory time representing one day—but actually three days in the real world.

Compressed time need not flow smoothly. That is, within a laboratory session there may be a series of "clock jumps," with the result that segments of the session may incorporate events occurring at or close to their rate in the real-world system, but the total session will represent a fraction of its real-world counterpart because time has been skipped by means of the clock jumps. Uneven progression may derive also from "event-stepping," where time is skipped between certain events so that events rather than the clock regulate the simulation. This might be called "event compression."

There are a number of reasons to compress time in an experiment. One is that without compression long-duration processes simply could not be simulated, as Kidd (1962) has pointed out:

> Some real systems have very long feedback delays. Most productive businesses, for example, require decisions about which the consequences may not be denotable for months or years. Even with extensive support, researchers cannot afford to represent such conditions in the laboratory. Foreshortening of the time intervals between certain processes, then, becomes a highly desirable expedient. Still another facet of this same problem centers on certain "critical-incident" phenomena in system development. For example, under ordinary circumstances, the growth of an organization is a lengthy process. The primary expedient here is

usually the introduction of artificially high levels of environmental stress in the hopes of forcing the pace.

By shortening laboratory tenancy, time compression saves money even in situations where it might be feasible to reproduce real time. It also can increase the interest and attention of the subjects, because matters with which they have to deal occur more frequently. By eliminating minor real-world activities, simulation sometimes creates voids. These are reduced by time compression.

On the other hand, time compression may cause problems because of its lack of simulation fidelity. Compressed performance may be distorted performance. For example, too much compression may exaggerate the importance of co-operation and co-ordination within an organization; people may have to work together more closely than they would in the real system to get a job done in the time allotted. The faster rates of performance needed to accomplish the task may result in more errors. Time compression may discourage innovation, there being less time in the simulated situation available for contemplation or discussion. Human learning might be unrealistically facilitated. By making subjects more attentive, time compression may prevent the occurrence of an inattention that characterizes the real system on occasion and that leads to poorer performance.

Such are some of the factors to consider in selecting a compression ratio. The tempo of events in a system and its parts is another factor. The durations between important events in a real system may vary among its subsystems and in a real organization among its various functions, so any effects of a particular ratio of compression could vary among the parts of a simulated system or organization. Presumably one should determine which is the most important subsystem within the context of the experiment and tailor to it the experiment's time compression ratio. It can be particularly important to make certain that thereby some desired number of events or operational cycles is included in the experiment.

Should the same ratio occur throughout an experiment? Differing ratios may be used provided the experiment is designed to prevent changes in ratio from biasing results. (One objective might even be to try to find out whether differing ratios have differential effects.) Variation in compression ratio might be one way to tailor compression optimally for different subsystems, some being optimized in one part of the experiment, others elsewhere.

Subjects become oriented to compressed time as though it were real time, although no data exist on how long this process takes and how complete the orientation becomes. As a matter of fact, we regularly experience compressed time in verbal simulation of the world in much of what we read, not simply in history books but in a novel where a page perused in a minute covers an hour or a day of activity. Another illustration is the comic strip, which may be viewed in a fraction of the time taken by the events it represents.

Expanded time has not been introduced as such into man-machine system experiments, but has figured in one method of simulation associated with some of them. In graphical simulation, a system's information flow is represented in a series of diagrams on which an individual can trace what happens to an input after it enters the system. He may take much more time to do this than the real system and its operators would. Essentially this kind of expanded time is self-paced. It might be exploited for fine-grain examinations of behavior, or to find out what would happen if the rate were slowed down in the real system.

Another temporal consideration is "empty time." This has occurred only too frequently in man-machine system experiments when a piece of equipment has malfunctioned. The equipment can be an operational unit, a simulation transducer, a data recorder, or a computer performing one or more of these functions. The most important thing is to have an engineer or programmer on hand to make repairs. But the researcher should also make contingency plans for filling this empty time.

A discussion of time in experiments would not be complete without some mention of computer processing time. If system operations are simulated symbolically in a computer model, they occur in much faster time than they do in the real system. The compression ratio is very high. This is attributable not only to the speed of the computer but also the symbolic nature of the simulation and the aggregating of system details which symbols make possible.

Although computer processing time is so fast, a computer in an actual system may also be made to operate in real time, or in "realtime," as Sackman (1967) has phrased it, to match the system's real time. This means that the computer is receiving data from the front end of the system as they enter, not according to the speed with which all the data could be processed; and the computer is regulating actions elsewhere in the system in accordance with the tempo imposed on those actions by the other parts of the system or environment. Usually in such a situation the computer is said to be "on line." Sackman (1967) has discussed "real time," "realtime," and "nonrealtime" processing at some length.

In a man-machine system experiment, a computer may function in both real and fast time. It may operate "on line" to present inputs to operators at consoles, for example, at a real time rate. Similarly it may record the operators' outputs as these occur. However, in reducing the data and performing statistical analyses it operates in fast time, as it also does in input production.

## SUBJECTS

In this book the people in experiments who simulate people in systems are called either "subjects" or "quasi subjects." Experimenters have also called subjects "participants" and "players," or they may be labeled "actors" or "simulators." It would be interesting to know whether the label they bear in a particular research program affects their performance.

### *Selection*

Undoubtedly the selection of subjects is one of the principal problems faced by man-machine system experimenters. Some of the reasons have already been indicated in the section on *Design*. The importance of selection arises partly from the wide range of individual differences that has come to light in the experiments this book describes. Those who have played the roles of operators, commanders, and decision-makers have differed widely among themselves. Although such individual differences have not been systematically investigated in most real systems, it can be presumed that they also exist there. The significance of individual difference has been noted by Davis and Behan (1962).

Individual differences in team situations can result in differences in team performances, even though teams sometimes resort to compensatory procedures to counteract the inferior performance of an individual. An operator processing information in series may be fast or slow, careful or error-prone. Someone in a nodal position may be well-organized or readily confused in assigning tasks and monitoring them. Commanders vary in aggressiveness and degree of risk-taking. In each instance the team output can be substantially affected by individual performance, even when the team is fairly large.

The great differences in capability which have been uncovered in man-machine system experiments have on occasion constituted the major or only difference in experimental results. The discovery has usually been serendipitous, since personnel selection has rarely been a focus of the experiment. It is possible that great differences in motivation have contributed to the differences in performance.

Variability among potential subjects and teams makes it difficult to achieve either external or internal validity. External validity calls for a representative sample of subjects or teams. Such a sample can consist of individuals clustering around a performance mean, or it can require a distribution of capabilities. A biasing of subjects toward very effective or very ineffective performers is likely to distort the evaluation of a system, a set of equipment, or a set of procedures. It can also distort experimental comparisons. One set of equipment, for example, may be superior to another when operators are highly talented, but it may be equivalent or even inferior if they are inept. Some range of competence among subjects can demonstrate this interaction between equipment and operator performance levels.

On the other hand, some kinds of differences in equipment design can lead to a consistent difference in performance whether the subjects are inept or talented. To be sure, if the researcher wants to predict absolute rather than relative performance—the "evaluation" previously mentioned—he should select subjects with competence or aptitude levels similar to those of future operators.

As Sinaiko and Buckley (1961) and Morgan et al. (1963) have pointed out, design engineers do not qualify as representative subjects in experiments which investigate the equipment they have designed. Not only are they too familiar with its operation, they have aptitudes above those of the eventual system operators. They also have an inevitable partiality toward their creation that will affect their performance in a comparison study.

To achieve internal validity the experimenter must try to maintain constancy in his subjects and teams if these experience all the conditions of the experiment. If different subjects and teams have to be assigned to different conditions, he has to seek equivalence among them. As we have seen, randomization of teams is rarely feasible in man-machine system experiments. Some kind of matching may be attempted to make teams equivalent. This will be discussed shortly in connection with methods of team composition. Nonequivalent selection of teams constitutes one of the threats to internal validity described by Campbell (1969) as the differential recruitment of the comparison groups. Another threat is the differential rate of autonomous change resulting from selection—what Campbell called "selection-maturation interaction." Confounding can result from either threat.

*Sources.* The representativeness of subjects depends in part on their source. Sources for subjects in man-machine system experiments have included military officers and enlisted men, air traffic controllers, test pilots, municipal officials, policemen, company employees, university students, and women and girls who were none of the foregoing. The military personnel have often come from the system being investigated, or from a current system related to one to be examined. The same may be said of air traffic controllers. Test pilots have simulated astronauts who, to be sure, have been drawn from the same ranks.

But representativeness can compete with other considerations. Students and company employees may be readily available and their continuing participation more reliable. Women and girls have the same trade-off advantages and may be regarded as especially "manageable." If less representative subjects take longer to train before the experiment begins, this may be an asset; it means they lack competing habits, predispositions, or biases already acquired on the job. They are "naïve." If two systems are being compared, they are not predisposed toward or against one of them due to past experience.

The factor of professional inexperience may be discounted to a considerable extent if the experiment undertakes a comparative evaluation. The comparison study looks for differences in results rather than absolute values. The differences obtained with students may approximate those achievable with professionals, because lack of professional background affects each state of the focal variable equally. If an experiment seeks generalizable rather than ad hoc information, results based on amateurs may seem just as generalizable as those obtained with professionals, or even more so.

Thus there is no simple method of determining the appropriate source of subjects. The purpose of the experiment must be considered, as well as trade-off factors. User experience saves training time. In intricate ad hoc experiments it may be essential, especially if it is important to know the actual effects of each state of some variable. As a further advantage, user subjects may transfer some of the learning they acquire in the experiment to the real-life situation. If they are high-ranking officers or officials, they can acquire a deeper understanding of the experiment and its purposes than they could otherwise. In consequence the experimental outcome may have a greater impact.

In team situations experimenters may want to make use of teams that already exist instead of creating new ones. Presumably in existing teams the members are already accustomed to working with one another. Informal procedures have been developed. Incompatible members may have been filtered out. Such a team is more representative of real life than one which is artificially created for the experiment.

*Number*. As explained earlier, each team is a single subject in the experimental design. It is often difficult to obtain more than a few teams or crews, due to cost and unavailability of people. Yet the more obtained, the greater the chances that the experiment will have a distribution of teams approaching that existing or to be expected in the real world. The relationship inevitably remains fortuitous, since stratified or proportionate stratified sampling is out of the question. Not only are there too few teams but the actual characteristics of teams in the real world are usually undetermined.

It was pointed out in the discussion on *Design* that the needs of internal validity and external validity may conflict in establishing the number of teams. For example, a single team experiencing all states of independent variables can be the best arrangement for assuring internal validity by reducing confounding. It "serves as its own control." The experimenter may have just so much time available before the experiment starts. He can give a single team intensive training so its various procedures are well established and its performance highly consistent. Then the random variance in its performance is lessened and differences in the effects of states of variables are more apt to be statistically significant. In contrast, the experimenter could use the same amount of time to assemble a number of teams, training each to only a limited extent in following procedures. These teams are also more apt to be representative. But at the same time the procedures, being poorly followed, will not validly reflect those that might be used in the real system, and the random variance (error term) will be large. Accordingly, the experimenter may choose the first course as that which best reconciles internal and external validity, especially if he thinks that differences in procedures may have more impact on the system than differences in teams.

As will be seen in the discussion of team composition, it is possible to create a number of teams out of a set of subjects smaller than the total of team positions multiplied by the number of teams created. This can be done, for example, by putting a number of different subjects in the key position, or positions, keeping the subjects in other positions the same. This arrangement is more apt to achieve external validity than relying on a single subject in an experiment in a critical position. How much can be generalized about decision-making from an experiment which incorporates only one individual as the decision-making subject?

*Screening*. One approach to acquiring either a representative sample or equivalence among subjects is through examining and screening candidates. Even if it is not possible to eliminate many because there are too few available, the examination process may tell something useful about those who take part in the experiment. This information should be included in the experiment's report.

Of particular importance is each individual's prior applicable training. If all individuals in teams cannot be associated with all states of variables, the level of experience of those associated with one should be equated with the level of those associated with others. Another way of trying to achieve this equivalence, of course, is by training all subjects to some established level before the experiment starts; this will be discussed shortly. Achieving equivalent experience in comparison teams without such pre-experiment training is possible only if the subjects have had an opportunity to encounter all the various variable states in their working lives—or if they have had no opportunity to encounter any of them.

In the case of military personnel and civilian professionals who are system operators, some information can be obtained about their over-all competence in their regular jobs. This may be exploited either for estimating the distribution of competence in the experimental tasks or for trying to compose equivalent crews. The information can be sought from their commanding officers or supervisors, if only in the form of ratings along various dimensions or percentile rankings. Biographical information can also assist in making inferences. Biographical data include age, academic grades, and jobs held. Academic grades may suggest the level of verbal ability, and jobs the amount and kind of experience in leadership. Neither supervisors' ratings nor biographical data are likely to be of much help, however, in examining candidates for such complex tasks as problem solving.

Alternatively, or in addition, the researchers may give the potential subjects tests for intelligence, aptitude, personality, or performance (achievement). Probably the chief advantage of intelligence, aptitude, and personality tests is to eliminate extreme cases at either end of the scale. Otherwise the tests may have insufficient predictive power for the researchers to gauge how any one individual will perform in particular tasks in the experiment. In preference to these tests, potential subjects should be tested on the tasks themselves, similar tasks, or components of the tasks, such as reading or simple discriminations. In fact, the researchers may be able to conduct some preliminary sessions of the experiment before assigning subjects to positions, to determine who should do what. In addition these sessions may provide some indicators of leadership ability and interperson compatibility, to compare with results of psychometric assessment.

Even if considerable information can be assembled about individual competence, it is not certain that its exploitation will indicate the relative competence of teams or make them equivalent. Not enough is known as to how individual abilities are related to team abilities. It can be useful to keep a record of individual as well as team performance in an experiment to cast more light on this area. In addition, interperson activities which can be examined only in a team context need to be separately recorded and studied. These have not been as pointedly investigated in man-machine system experiments as they deserve. Such activities include load-sharing, task-assignment, and communication behaviors.

One further requirement in screening subjects may develop in certain military situations. When a military organization faces anything it may regard as a test, including a man-machine system experiment, it is likely to assemble its best performers to compose the team or crew to be tested. In some quarters this is known as the "tiger team." The researchers may have to take steps to prevent this if they want to find out how normally composed crews will perform.

*Crew Integrity*. If subjects drop out of an experiment, their loss may have differential effects on the outcome. Campbell (1969) has called this origin of confounding "experimental mortality." Turnover can occur in man-machine system experiments when subjects leave a team by reason of illness, other duties, vacation or leave, or for some other reason. Substitutions have to be made, and the team loses some of its constancy. This misfortune is most likely to occur in operational settings, where there are competing and often overriding demands on personnel.

*Team Composition.* Teams can be composed in a number of ways:

1. Randomly assigning individuals to each team.
2. Matching individuals to each other and then assigning those within each match to different teams; this assignment can be random.
3. Assigning individuals so teams perform equally well under the same conditions, according to some measure; the teams are matched, not necessarily the individuals.
4. Assigning individuals to positions according to some arbitary rule, applied consistently, such as alternating assignment to teams according to subjects' alphabetical order.
5. Accepting crews as they already exist.
6. Rotating individuals through crew positions.
7. Assigning some individuals to fill the same positions in all crews while other individuals vary between crews. Those common to all crews handle "routine" functions, whereas those in different crews handle "key" functions. The latter may be permanently assigned to functions or may rotate through them.
8. Matching individuals in different teams to each other after the experiment through the statistical operation called analysis of co-variance.

In turn, teams can be associated with states of an independent variable or variables in several ways. To repeat what has been said earlier, these are:

1. Randomly assigning teams to states, a number of teams per state. This is not usually feasible in man-machine system experiments because there are seldom enough teams to justify randomization.
2. Making teams equivalent to each other in some fashion and assigning a different team or teams to each state. It may be difficult to achieve equivalence. A multiplicity of teams for each state may be needed to average out differences.
3. Associating each team with each state, as in a factorial or incomplete factorial design (e.g., the Latin square). The teams then themselves constitute states of an independent variable.

Note that in the foregoing summaries teams are not the same as groups of subjects used in experiments on individual behavior. Teams have team behaviors and are measured for team performance. As stated earlier in this chapter, from the point of view of the experimental design a team is a subject. Also, certain parallels exist between assignment of subjects to teams and assignment of teams to states. In particular these are the operations of randomizing and matching. Methods of team composition and methods of team assignment also may interact. Some methods of team assignment may be helped, or hurt, by some methods of composition. This point is worth examining further.

Might random assignment of subjects to teams help the random assignment of teams to states by reducing the number of teams needed? Could it lessen the biases among teams? If so, it might be argued that the fewer or smaller such biases, the fewer the teams that would be needed in randomization. But such double randomization might work in just the opposite direction. Random assignment of individuals to teams could result in major differences among the teams if by chance persons with widely different capabilities were placed in corresponding key roles in different teams. The double randomization question merits more attention from experts in experimental design and statistical testing.

Matching of individuals to make teams equivalent has already been discussed in connection with screening. It was pointed out that screening for matching could be based on biographical data, tests of various kinds, and the extent of experience in performing the system's tasks. Matching of individuals is only as good as the basis is for predicting performance in the experiment, and the predictor may not be valid. Nevertheless, matching has been widely used in psychology experiments on individual subjects. It has been advocated by Chapanis (1959) but not by Campbell and Stanley (1963, 1966), who favored its use only as an adjunct to randomization. However, their objections concern undifferentiated groups of individual subjects in education experiments. The composition of teams in man-machine system experiments involves other considerations.

The method of matching individuals to compose teams might be combined with another method, that which distinguishes between filling critical positions and filling routine positions. Only those individuals filling each of the critical positions would be matched, the others being the same in all teams. If potential subjects were in short supply, this limited matching would have an advantage over matching for all positions. Although for good matching there should always be more individuals to choose from than positions to fill, matching among a few positions in a team requires a smaller pool of potential subjects than matching among all of them.

Matching individuals through the method of co-variance seems more applicable to individuals in undifferentiated groups than to those in teams performing tasks which differ among the individuals. The method of matching teams through some measure of team performance must also be considered. Not only does it share with individual matching the problem of finding a suitable basis for matching, it encounters a further difficulty. Unlike individuals, a large number of teams whose performance may be assessed and from which matches may be chosen may not already be on hand. It may be best to compose teams by matching individuals and then to compare team performances as well.

Rotation of individuals through positions in a team is a method of composition that can be combined with randomization. Individuals can be assigned randomly to sets of subjects and then rotated within each set to compose a number of different teams. Even within a single set, consisting of as many subjects as there are in a team, rotation can produce a multiplicity of teams—as many as there are positions within a single team. Although rotation can yield a substantial number of different teams, it has a major drawback. The continuing alteration of roles may weaken the interpersonal relationships that can be important in team performance.

Like randomizing and matching, rotation is a way to constitute a different team for each of the states of a variable. Because of the transferable learning that may occur regardless of the position filled, rotation may be less desirable than the other methods in experiments which investigate nonreversible processes, such as training. All three methods can be used as well to compose teams in experiments in which each team encounters all the variables' states. Although their value in association with this way of assigning teams may be considerable, it may seem less obvious because other methods are also available. These are the methods of composing teams according to some arbitrary rule, and of accepting them as they exist.

In this discussion so far, methods of composing teams have been examined with respect to their effects on internal validity—the avoidance of confounding. Choice of method may also depend on the desire for external validity. As noted earlier, the greater the number of teams in an experiment the higher the chances are that they will represent the teams in the real world. Methods of composition should thus be evaluated according to how many teams they can create with a given number of subjects.

Matching runs into some difficulty in this regard because effective matching requires more potential subjects to be available than subjects chosen. If the pool of potential subjects is small or the size of the team is large in relation to that supply, rotation of subjects through positions becomes an attractive method of increasing the size of the team sample. Since the kinds of positions in the team and their requirements can vary so much in man-machine system experiments, one of the split methods of composition may seem preferable. Some of the subjects are common, some rotated. The other split method, in which the non-common subjects occupy the same positions throughout the experiment, requires more subjects for the same number of teams but not as many as are needed if each team is composed in its entirety of different subjects.

### *Training*

If the subjects in a man-machine system experiment start out with more proficiency in dealing with one state of a variable than another, clearly any difference in results can come from the difference between proficiencies rather than the difference between states. Proficiency, then, is a confounding variable. An appropriate counteraction is to transform it into a constant. As indicated in the discussion of screening, one way to do this is to try to select individuals or teams possessing the same proficiency in dealing with any state. The other is to try to bring the individuals or team to the same level of proficiency with each state by training them.

If there are individuals who have dealt previously with all the states, it may be possible, as suggested earlier, to use biographical data or tests to find those with equivalent capability. However, one or more of the states may be entirely new to all subjects. In that case it may be advisable to select subjects with no experience with any of the states; their *lack* of proficiency is the constant. But it should be realized that they will acquire proficiency during the experiment and that this may happen faster in relation to one state than to another. Then the experiment should be continued to the point where proficiency has leveled off for each state. Otherwise it is possible that any difference in results associated with states is really attributable to faster learning with one than another.

Starting the experiment with untrained subjects has other advantages. Subjects will not have acquired performance habits with one state that impede, through "habit interference," performance with another. The use of "naive" subjects can also indicate, within limits, how the level of performance is related to the interaction between amount of training and the states of the focal variables.

Alternatively, researchers often go about training subjects to equal proficiency on all states of variables because the available or desirable subjects already have considerable proficiency with one of them. When this approach is taken, the experiment's results cannot be generalized to unpracticed operators. This limitation is unimportant if, as is so frequently the case, the researcher is concerned only with system performance when the system has been operational for some time.

The training is conducted shortly before the experiment begins. One of its benefits is that random variance in a subject's performance may be lessened if he is highly practiced. The amount and kind of training are discussed shortly. Although it may be assumed that training, as such, will match subjects by bringing them to equal proficiency with all states, the researcher may also go through a matching procedure. On the assumption that it will remain at the measured amount, he measures the proficiency of the subjects and assigns those with equal levels to corresponding positions on different teams. Thus the individuals undergo the performance or achievement testing mentioned earlier. The measurement of performance must be independent of the states so that it indicates equal proficiency across them.

Even after extensive pre-experiment training the performance of subjects may continue to improve during the experiment. The experimenter would like to know whether this has occurred and, especially, whether it has occurred more in conjunction with one variable state than another. He can design the experiment to determine this. If two or more run-throughs of all experimental conditions follow one another, differences in results between run-throughs will indicate that the teams were still acquiring proficiency in the first run-through. The data should indicate whether the change is larger for one variable state than another.

*Amount of Training*. The criterion of equal proficiency across states of the focal variable or variables is some constant level of performance with each of them. To rely on equal performance scores across states would not take into consideration the differential effects of the states themselves, the very question the experiment was investigating. Fixed durations of training would not eliminate the differential effects of earlier experience. Instead, researchers should train the subjects until their performance reaches a steady state in each of the major experimental conditions. If the focal variable of the experiment consists of sets of equipment, subjects should practice with each set. Once performance has leveled off, the researcher may assume that the subjects are performing on each set with the same proficiency they would have acquired in the operating system.

But how does the experimenter know that performance has truly reached a steady state? He faces two problems. One is to determine when enough data have been acquired to give a convincing appearance of stability. The data must be continuously reduced and the performance measures displayed. The determination that performance is close enough to a stable level will be a matter of judgment. Of particular importance is the requirement, during the planning of the experiment, to leave enough time for the training that may be needed. Underestimation is likely.

The other problem is the possibility that a steady state does not really indicate the end of improvement. Teams, like individuals, develop new procedures which can cause an abrupt shift. As Chapman (1960*b*) stated the problem, "the learning function is not a smooth continuous one but a series of jumps that are the product of problem solving and procedural innovation." The researcher can try to prevent such innovation through instructions and other constraints, unless he is actually hopeful it will come about.

*Kinds of Training.* Preexperiment training of subjects can take several forms. The primary one is practice in the system operations which the experiment will investigate. This is the training just discussed. The simulated inputs should be different from those in the experiment so no specific learning or memorization of inputs can carry over to the experiment. The training sessions should be different from the pilot or exploratory sessions which the experimenters also conduct before the experiment starts. As already noted, the subjects should not take part in the exploratory sessions, which are intended to check out staff, procedures, inputs, computer programs, equipment, and so forth.

Another kind of training is training in component tasks or training of an ancillary nature. The component task may be touch-typing for an experiment involving computer-associated keyboards, map reading for one in which troops maneuver, talking on the radio for one in air traffic control. Interactive tasks may also be taught, such as the distribution of multiple assignments and transfer of responsibilities between operators. Component tasks, whether individual or interactive, are likely to be better taught through training focused directly on them than through training in team situations. Training of an ancillary nature could include physical conditioning, as in the case of subjects simulating astronauts.

The third kind of training can be summarized as indoctrination. Through lectures, handbooks, and discussion, the subjects are familiarized with the system being simulated, its operations and environment, the experiment and experimental operations, and the constraints on what they can do as subjects. All these instructions are much more far-reaching than those usually given in experiments concerned with behavior. Some of them will be discussed further in connection with *Motivation* and *Control.*

### *Motivation*

When motivation is discussed in connection with man-machine system experiments, those confusions which normally accompany the term are augmented by a duality of reference. "Motivation" may mean the extent to which the motivation of subjects realistically resembles that of the people whom they are simulating. Or it may mean the motivation of the subjects to play the roles given them. The distinction is important.

*Realism of the Motivation.* Presumably subjects, to act like the people in the real system, should be affected by the same motivational factors. These factors come in two related sets. One consists of what might be called "establishing circumstances," the other of the consequences of performance that cope with these circumstances. For example, some circumstance involving the real system might establish a threat or stress, and the system operator would take action resulting in its termination or future avoidance. In the experiment the subject should act in a similarly motivated fashion as a result both of the simulated threat and his performance in dealing with it.

Although the establishing circumstances are important, it can be extremely difficult to invest them with the same effectiveness in the experiment that they have in the real system. An impending collision between aircraft on a simulated radar display may not have the same impact on an air traffic controller that it might if it appeared on a real radar display. On the other hand, the successful prevention of the collision, as such, conceivably could influence subsequent performance as much in a simulation environment as in a control tower.

When the consequences of performance are made known to the subject, we have feedback. It has both motivational and informational effects on the subject's subsequent behavior. In a man-machine system experiment it is important to arrange the consequences of performance and their feedback to the subject so they closely resemble actuality. Although this may be easier to do than arranging the establishing circumstances, it is not always a simple matter. Consider, for example, an experiment which simulates a control center in a ship. If the ship is destroyed by an enemy, one realistic consequence would be that the experimental session would immediately terminate, but the experimental design may call for continuing it for its pre-determined duration to collect more data.

Complete fidelity cannot be achieved in simulating establishing circumstances such as combat and catastrophe. Even when they are signified by messages and signals the subject knows they have not occurred. But the realism of the message or signal per se as the conveyor of the circumstances may help bring the motivation of the subject close to that of the real system operator, who would be also dealing only with messages or signals. In addition, extensive indoctrination on the circumstances through briefings and instructions may augment their impact. Finally, it may be beneficial to use as subjects those already familiar with such circumstances and their gravity; on this basis, experienced military personnel would be more motivated than college students in an experiment on nuclear warfare.

An obvious way to give the simulated establishing circumstances full motivating power is to convince the subjects that the circumstances are real. This may be done in certain situations simply by not letting the subjects know that the circumstances are simulated. However, experiments which thus employ concealment are open to charges of unethical professional conduct, even if the subjects consent to take part.

In any case, the experimenter will want to know how effective the simulated circumstances were from the point of view of motivation, and how effective the simulated consequences of performance were. He can get some indication by probing subjects to learn their subjective reactions. They can register concerns, intentions, expectations, desires, and feelings both by verbalizing them and by rating them on scales. Probing should link queries about such reactions closely to the establishing circumstances and performance consequences. Even so, there will not necessarily be one-to-one correspondence between subjective reactions and the motivating factors.

*Motivation to Play the Role.* To motivate subjects to be good simulators of other people it is necessary to establish circumstances to which the consequences of good role-playing can be related, and to bring about those consequences. For example, the researcher can ask the subjects to participate enthusiastically and when they do so he can reward them. He can also tell them how important their participation will be because of the significance of the research, and then when they play their roles well he can indicate they are supplying important data because they are doing such good work as simulators. Financial rewards may be given, not only for perfect attendance and punctuality in arriving at experimental sessions but also for effective role-playing according to ratings by experimenters of particular performances. This process of rewarding for effective simulation performance may be accompanied by punishment for poor simulation performance, and it should be applied only on an individual basis, along with identification of the good or poor role playing. Unless, and perhaps even if, it is well handled, the researcher runs the risk of rewarding or punishing the individual's performance as a system operator rather than as an actor. This risk of thus contaminating the experiment's results may be the reason why little if any effort has been made in man-machine system experiments to strengthen role-playing in this fashion.

The researcher should also seek to prevent subjects from *mis*playing their roles. Subjects may leave their roles temporarily to chat with the experimental staff about nonsystem matters. They may try to cheat to achieve a better personal showing in some component task. In fact, military subjects may be inappropriately motivated in their system performance because they think they are being personally evaluated in the experiment; and their commanding officers may be especially concerned and try to influence experimental design or data collection to avoid the risk of a bad showing. Subjects may also engage in by-play with quasi subjects. They may resort to unauthorized activities during slack periods to avert boredom. Researchers can try to prevent such misplaying of roles by giving careful and repeated instructions, keeping subjects under surveillance, explaining the purpose of the experiment, and giving punishments (e.g., fines). This brings us to the whole question of *Control.*

### *Control*

Those who conduct man-machine system experiments have to exert control over their human subjects to a greater extent than do most behavioral scientists. The subjects are more numerous and occupy the laboratory for longer durations.

Instructions are lengthier and more difficult for the subjects to follow. More occasions present themselves for deviating from the constraints the experimenter has set. The purpose of control over subjects is to prevent factors from intruding that would make the experimental situation unlike the system operations being simulated or would confound the effects of independent variables.

*Indoctrination.* Questions of control begin with the introduction of the subjects to the experiment. Their indoctrination should be done by the experimenters themselves to assure an adequate understanding. Not only should briefings be given but handbooks of instructions should be prepared covering experimental operations and rules for subject behavior. The researchers may wish to give tests to make sure the handbooks have been read and memorized.

How much understanding about the experiment is it necessary to impart? The researchers may want to hold back some information if its release would jeopardize achieving valid results, even though subjects might prefer to know everything. For example, the details or even the over-all nature of scenarios and simulation inputs might be withheld to forestall advance preparations by the subjects. (In field experiments other kinds of advance preparations can be thwarted by holding no-notice sessions.) Some of the variables being investigated, such as load, may remain unspecified, to limit unrealistic motivations and expectations. Unless some purpose is stated, the subjects will try to figure one out. In some situations the purpose can be explained without injuring the experiment. In others it may have to be disguised. It could be unwise in an experiment investigating spontaneous team behavior, such as procedurization, to divulge to the subjects the purpose. Their procedurization would no longer be spontaneous.

Concealment of some of the methods of simulation may be as warranted as concealing the simulation inputs prior to their presentation. No harm is done to a subject by pretending his outputs are being processed by a computer when actually a human being is simulating the computer. But the researchers should examine each instance of deception to make sure it will not have any unfortunate long-duration effects.

*Supervision during Sessions.* Continuous and meticulous surveillance and supervision of subjects are essential during experimental sessions. Members of the experimental staff should be designated for this function and instructed how to carry it out. It is important that enough personnel be allocated; researchers can easily underestimate the need. The number required will increase with the degree of activity imposed on the subjects by the inputs. As the load rises, some kind of help may be needed from automation, or the monitors, by instruction, may resort to sampling. In any case, the subjects must remain under visual and auditory surveillance through closed-circuit television or one-way windows and through connections to telephone lines or microphones picking up conversations.

The purpose of such supervision is to make sure that the subjects follow the instructions and rules given them. Many of the instructions are part of the simulation. Rules specify what subjects are permitted or forbidden to do as subjects. For example, they may make it clear that subjects should telephone each other rather than shout across the room if in the real system they would be miles apart. They state in what activities subjects may engage during idle periods, and in what they should not. They indicate to whom subjects may address questions among the experimenters, with whom they should not talk, and other aspects of contact between subjects and experimenters or quasi subjects. Such contact must be carefully regularized. Socializing during sessions should be prevented. The rules may establish certain off-limits locations in the experimental area, such as places where the experimental staff works.

Subjects become forgetful, and competing interests lead to inappropriate behavior. But the fact of being under observation and being corrected can minimize such deviations, hopefully without distorting the realism of the simulation. If deviations do occur, the experimental staff or quasi subjects should record them, or the subjects themselves may be induced to do so. Often the experimenters can detect behavior which forewarns of some emerging problem, and this can be prevented before it becomes serious.

*Information Exchange among Subjects.* Sometimes unsupervised discussion among the subjects within a crew may be systematically arranged in post-session meetings. The subjects can discuss system operations or problems about the experiment itself. This is a way of helping the subjects develop new system procedures or of finding out, by recording the discussion, what they are feeling and saying about the experiment. Possibly the process also develops cohesiveness, makes the subjects more involved in the simulation, and develops independence from the experimenters. Unless the experiment is focused on procedurization itself, the adoption of such post-session discussions should depend on an analysis of what the discussions can usefully contribute and how they may add an uncontrolled factor to the experiment.

*Information Exchange between Teams.* If one team develops new procedures on its own, it becomes important to pass these along to other teams in the same experiment to keep procedures equivalent among the teams. On the hand, precautions should be taken to prevent exchange of information about simulation inputs and other matters a crew should not know about beforehand. Different sequencing of input material and scenarios is one solution, competition between crews another, although the latter may breed its own difficulties.

*Information Exchange between Subjects and Experimenters.* In addition to instructions concerning how to behave as subjects, information going from experimenters to subjects can include the results of their performance. The outcome of a team's performance may be summarized after each session. This procedure may make the experiment more interesting for the subjects and foster competition. If knowledge of results would be similarly provided in the real system, realism is enhanced. If not, the teams may perform in a way uncharacteristic of real life. In any case knowledge of results, if furnished, should be given in the same regular and specified manner to each team, with the same type of content. This can include both team and individual scores, including errors. Informal, casual information about experimental results and performance should be scrupulously avoided; there is too great a danger that one team or one experimental condition may receive more of this kind of reinforcement than another. All indications of praise or blame must be similarly excluded.

Information going from subjects to experimenters may include (1) suggestions about experimental operations; (2) attitudes, expectancies, and self-instruction on the part of the subjects; and (3) evaluations of some of the states of variables in the experiment, such as equipment designs, training techniques, and procedures. Such evaluations will be discussed under *Measurement*. Attitudes, expectancies, and self-instructions may be sought from the subjects through questionnaires, interviews, and rating scales. Scales and questionnaires can be administered impersonally through a computer terminal at the end of a data-taking session.

*Hawthorne Effect.* The phenomenon of enhanced performance because the subjects are taking part in an experiment has been discussed earlier as one that can threaten either external or internal validity. According to Davis and Behan (1962), when "motivation has been artificially increased by manipulating the social (laboratory) climate, it becomes extremely difficult to generalize experimental results to everyday field operations," but such manipulation is justified if one wants to find out "what kind of performance can be expected with optimum motivation." It has been conjectured that the "laboratory climate" can be manipulated to minimize its effect on subjects as well as to strengthen it. Among the steps which may minimize it are the isolation of subjects from experimenters to reduce contact between them and shut out comments by experimenters about subjects' performances, and subjects' self-management of their affairs.

### *Quasi Subjects*

As we have seen, man-machine system experiments have frequently included sets of people who represent personnel associated with the system being investigated, but who are not regarded as genuine subjects in the experiment. These "quasi subjects" simulate pilots, embedding organizations, hostile forces, or a commander's assistants who handle communications or displays. Their roles are important but peripheral or are simply responsive to the subjects. Quasi subjects may be selected from the experimental staff or its support elements, military personnel, or personnel hired for the job. Women and enlisted personnel who have never flown a plane may simulate pilots; psychologists or clerks may pretend to be generals or members of their staffs. They provide contact between the subjects and the external world by giving inputs, receiving outputs, enforcing and clarifying procedures, and setting limits and constraints. At times they collect data.

But no data have customarily been collected about the performance of quasi subjects. Although experimenters may have presumed that this performance did not affect the experimental results, undoubtedly it has. In some experiments the performance of subjects has had to be further processed by quasi subjects before it was measured. In others the quasi subjects affected subjects less directly; they were simply the subjects' principal human contact in the laboratory. Though generally they have not been called on to engage in dynamic interplay with subjects but rather to maintain *non*adaptive performance, it is not certain how consistent that performance has been. Finally, in occasional experiments quasi subjects have been given equipment or tasks which were a major focus of the experiment.

Research reports have failed to specify how quasi subjects were selected and what training they received. Although it has been assumed that they remained constant quantities during an experiment, they may have learned to perform their tasks better, just as subjects do unless they are thoroughly pretrained or were fully trained before becoming subjects. The need for pre-experiment training or selection on the basis of prior experience applies to quasi subjects as it does to subjects.

If social interactions develop between quasi subjects and subjects, the behavior of both can be affected to the experiment's detriment. The likelihood of such interaction increases when the quasi subject is an attractive clerk or secretary and the subject is a military officer. Quasi subjects should be cautioned as much as subjects against any social contact between these groups, within the laboratory or on the outside.

To play their roles effectively, quasi subjects should become completely familiar with the people or organizations they are simulating. They should know procedures, jargon, and geography. They should be so knowledgeable and practiced that when the quasi subject makes errors they are those of the professional rather than the amateur.

In real systems, people make errors. Pilots make errors. Generals make errors. Communicators make errors. Display keepers make errors. Unless the quasi subjects make errors they will be playing their roles unrealistically, and the inputs and feedback they furnish to the subjects will misrepresent what happens in the real system environment. But the circumstances of their participation in the experiment limit the likelihood of error. Presumably they should be required, through instructions, to make occasional, typical mistakes, programmed by the experimental design into the input.

Data should be collected about the errors that quasi subjects make, whether these are purposeful or accidental. Data should be gathered also about other aspects of their performance. In representing embedding organizations, the quasi subjects should be consistent, following rules assembled in handbooks before the experiment begins. Although rules may also have to be developed at times as an experiment progresses because all contingencies cannot be foreseen, they must be recorded and standardized so that all the quasi subjects respond similarly to similar contingencies. When new problems develop, it may be necessary to refer them to laboratory supervisors for resolution.

A number of other guidelines may be suggested. In talking with subjects, the quasi subjects should be serious and should emphasize the importance of the experiment and its goals. They can help make the subjects goal-oriented, and help them to understand the simulated system and to believe in the simulation. Quasi subjects can participate to a major extent in exerting the control over subjects discussed earlier, by making certain that the subjects' procedures remain within established limits. Thus, quasi subjects can function as control staff—and data collectors—as well as simulators.

Researchers should establish criteria for selecting quasi subjects and publish these criteria in their reports, along with the data about their performance and information about their training. It is also necessary to assign quasi subjects to experimental conditions and subjects in a fashion that will prevent or minimize confounding. If each team or variable state is associated with a unique quasi subject, it will not be possible to distinguish the effects of the quasi subject from those of the team or state. Preferably, quasi subjects should be rotated through teams and states, or each should be common to all of them. The latter approach is not feasible, however, when two teams or systems are performing concurrently; and it can overload the quasi subject when subjects being handled in common impose substantial requirements on quasi-subject performance.

When a quasi subject, or two or three of them, represents an entire organization, the simulation involves what earlier was called "organizational compression." The experimenter may assume that because the quasi subject is merely the mouthpiece for an aggregate, his personal characteristics and style of expression are not especially important. But this viewpoint may oversimplify matters. How closely should the characteristics and style resemble those of the contact in the real world? If that person is a communicator, he may still reflect some of the manner and reaction of the organization's head or commander. If he is the head man himself, there is all the more reason to want to represent him realistically.

As a technique, organizational compression merits more analysis. Researchers should determine the conditions under which quasi subjects can properly represent aggregates of people.

## MEASUREMENT

Finally we come to the entire process of measurement. It includes the collecting of data, through both instrumentation and human means; the determination of criteria; the selection of measures in accordance with various specified requirements; and the analysis of the data together with the confidence testing of the results. These subprocesses are highly interrelated. The measures selected must satisfy the criteria and identify the kinds of data, yet the feasibility of obtaining data exerts some effect on the selection of measures and criteria. Preselected methods of analysis influence what data are collected.

Each of the subprocesses calls for considerable description, because each makes special demands in man-machine system experiments. A vast amount of data gets collected, through a multiplicity of methods. Care must be taken lest this aspect of measurement itself contaminate the results. Criterion selection is related to the objectives of the system and the experiment. System objectives are likely to be numerous, sometimes conflicting, occasionally disguised. Many different measures are obtained because of the diversity in system performance. Some are more useful than others in fulfilling the specified requirements. Data reduction is a big undertaking and usually requires a computer, which may also be helpful in data gathering. Analysis and confidence testing are major tasks because of the many measures and the multiplicity of independent variables and their states that are generally found in the experiment's design.

### *Data*

The collection of data in an experiment should be planned well ahead. The methods should be set forth in planning documents and tried out in the check-out sessions preceding the data-taking sessions. To the extent feasible, planning should embrace the selection and amount of data to be collected, the sources of data and agencies of collection, and provisions to prevent contamination.

*Selection of Data.* What are some guidelines for selecting the data? The most obvious one would seem to be relevance. The data should be pertinent to the measures chosen for stating the experiment's results. Data by themselves are simply records, which may or may not be expressed in quantitative or summarized terms.

According to one view, the measures should be entirely established before the experiment starts. But from a different viewpoint it is possible and necessary to predetermine all the measures and thus all the kinds of data only in those experiments which have a verification objective. For experiments which aim at exploration and discovery, the value of some of the potential measures may become evident only after the data have been gathered. Selective planning of all data collection is not feasible. In fact, a discovery-oriented experiment is partially defined by the fact that some of its measures can originate from the data.

Other criteria for data selection have been set forth by Meister and Rabideau (1965) as objectivity, quantifiability, validity, reliability, automation, and economy. Some of these seem more applicable to measures then to data, others fit both. An additional criterion is the need for redundancy (back-up) data.

Not all data are quantifiable. Records of particular problem situations or critical incidents, for example, are qualitative and can be expressed only in words. The same may be said about responses to open-ended queries in questionnaires and statements of opinion. Such statements may become quantitative only to the extent that they are contained in rating scales or rankings. Nonquantifiable data can furnish useful insights and are often desirable in exploration experiments. By their nature they do not lend themselves to summarization. Hence they are likely to be disregarded in reports because they take up inordinate space.

In general, data related to time are easier to collect than data concerned with errors. Yet accuracy can be more important to system success. The difficulty in getting error data comes from problems in recording as well as the rarity of some errors, including crucial ones. Frequency data, e.g., of communications, are often so easily obtainable that they are collected without regard to their value—which may be limited.

Indeed, a criterion for data selection which should *not* carry undue weight is that of data availability. It can be difficult to acquire performance data in some situations. For example, either careful human monitoring or complex instrumentation is required to record human head and hand-signal movement on the part of vehicle drivers in traffic simulations; eye movement is still more of a problem. On the other hand, movements of foot pedals and steering wheel can be recorded much more easily. But that does not mean that hand, head, and eye movements should be neglected. Along the same line it has been pointed out earlier in this chapter that switch actions at computer-connected consoles can be readily recorded by the computer itself, whereas numerous other operator activities, some very significant, cannot be; but that does not mean they should be neglected.

The criterion of economy is related to the amount of data collected and cost and effort of processing. Should more data be collected than the researcher knows or suspects will be processed? In the case of data that are computer-recorded and can be easily computer processed, this question is not so significant. But where considerable manual processing is needed it has been a major dilemma in man-machine system experiments.

There are a number of reasons for collecting more data than will ever be reduced. Some surplus data will be collected simply to back up primary data from another source; this other source may at times malfunction, or ambiguities may arise. A hunch may dictate the collection of data, and it may or may not need follow-up. Still other data may serve during the experiment only to help the experimenters understand what is happening. In discovery experiments, large amounts of data may be collected just to see what they yield. Although this procedure may seem expensive, the extra cost may be small indeed compared with that of rerunning the experiment to collect data which the researcher had neglected to obtain. Still another reason for acquiring data which will not be used to any great extent is the experimental design which requires that the system being examined reach a steady state before and then after a change in an independent variable; the data collected during periods of change may not be pertinent. In other situations the data from the transition periods are important and those from the steady states are relevant only by indicating that these were reached.

Yet researchers should guard against waste that can be foreseen. For example, motion picture cameras and films are expensive, and experimenters should be wary of using them. Data cannot be extracted easily from film and reduced; coding and tabulating human performance recorded photographically are even more difficult than coding and tabulating voice communications recorded on voice tapes.

The transcription of such tapes can be a long and laborious effort. It is preferable to try to code and tabulate directly from the tape to bypass the need for transcription. In any case, recordings do not always have to cover an entire experimental session. Samples may be taken, provided the subjects do not know this; the sampling should be disguised. If the entire session is recorded, the data transcribed and reduced may come from samples of the recordings.

Although waste should be avoided, how and when to avoid it depend on the circumstances. For example, if the recorded data are essential to the analysis of results but cannot be reduced because of the cost and effort involved, the experiment has been wasteful. Careful attention should be given to ways whereby manually collected data can be converted for computer processing. If vast amounts of data are collected that are never used, probably either the data should have been reduced and analyzed or more were gathered than were needed or could be used.

*Agencies of Collection.* As agencies of collection it is conventional to distinguish between instrumentation, which is mechanical or automatic, and human observers, whose operations are manual. In turn, manual collection can be either direct, through observation of on-going activities, or indirect, through examination of records or recollections (Meister and Rabideau 1965). Automatic collection can be differentiated into the kind characterized by recording devices and the kind where a computer collects data on operations performed on itself.

Automatic and manual data collection may be alternatives between which the researcher must choose, or they may occur together. One may back up the other or data collection may include both mechanical and manual elements. A motion picutre camera may back up human observers (or vice versa). A human being operates a stop watch or a tape recorder. A human operator may have to measure photographs or transcribe and categorize data from voice tapes. The connection between data recording and data reduction must be considered in grading the automaticity of the collection agency.

Facility of processing the collected data is also one of the criteria for evaluating the agent and making a choice between automatic and manual methods. Other criteria include accuracy–frequency and amount of error; precision–the sensitivity of the method; reliability–the likelihood of malfunction; variability–variance among devices or people; capacity–the volume and speed of registration; flexibility–adaptability to different situations; complexity–ability to discriminate and record patterns of situations or events; neutrality–lack of bias or collusion; selectivity–ability to screen out or interpret data; dependence–need for maintenance of equipment and training of observers; and cost–dollar expenditures for acquisition, operation, and upkeep. Many of these criteria have been listed by Meister and Rabideau (1965) as trade-off characteristics.

Inaccuracy, bias, and variance in human observers have been the major factors which have led researchers to emphasize automatic recording in man-machine system experiments. Davis and Behan (1962), in pointing out that monitors "with their various aids for data collection constitute another kind of system–in effect, a metasystem," have warned of the results of overload:

> The same variables which influence the behavior of the data-processing system of interest also influence the behavior of the data-collection system which monitors the system of interest. Since one of the most pervasive variables affecting all system performance is input load, the experimenter must always be sure that the individuals collecting data are trained to recognize and collect the data required, and that in the process of collecting this data they do not become overloaded. When data collectors become overloaded, there is an interaction between the system that is being observed and the system doing the monitoring. In such a situation, the response measures being collected are no longer related solely to the system under observation, but reflect instead the behavior patterns of both systems.

Although instrumentation undoubtedly assures greater objectivity in data collection, Meister and Rabideau (1965) noted that objectivity and subjectivity represent "a continuum, not a dichotomy." They argue that in system testing "no clear-cut advantage *in general* for either type of methodology is apparent." They also note:

In observation the data reported are already partially evaluated, because the human cannot observe without in some way organizing or interpreting his observations. The very data collection categories he used tend to include a built-in evaluation, which cuts down on the amount of data gathered, because the categories are inherently *data selective*.

The observer's very freedom to select, however, is largely responsible for the errors that he introduces into his data. He often responds to irrelevant stimuli or fails to respond at all; or his biases may distort the data.

*Contamination*. In the discussion of *Design* it was mentioned that factors related to measurement could "contaminate" the results of a man-machine system experiment and thereby diminish the confidence to be placed in them. Contamination associated with dependent variables is perhaps as important to consider as the confounding of independent variables. A dependent variable is contaminated if any of its states—its measured values—is affected by the process of measurement so that the independent variable is not exclusively responsible for those values. Either internal or external validity can be threatened, depending on whether the contamination affects the outcomes differentially or jointly.

Contamination can develop out of a number of aspects of data collection. Campbell (1969) listed instrumentation, testing, and instability. The instrumentation factor can affect results due to changes in the measuring instrument, whether this is human or automatic. Human observers, Campbell noted, vary in recording skill and efficiency as a result of learning and fatigue, among other reasons. As mentioned earlier Chapanis (1959) has emphasized that mechanical devices vary as well as humans.

It must be presumed that learning and fatigue affect human data collectors in man-machine system experiments, although researchers have not tried to determine to what extent. Another kind of potential human contamination factor is the experimenter's expectation of what the results will be, including his expectation that a difference between states of an independent variable will result in differences in a dependent variable. It is well recognized that such expectancies can contaminate experiments on behavior in the course of data collection. Social interactions between experimenters and subjects can influence data-collection procedures in complex experiments as well as simpler ones. In fact, in some man-machine system experiments such interaction has taken the form of collusion between subjects and data gatherers.

By the testing factor Campbell meant that taking one test could affect a subject's scores on a second test, or that publication of a set of measurements could influence a second set of measurements of the same individuals with access to such publication. In man-machine system experiments this factor would not only include feedback of results to subjects during the experiment but also the effects of data-taking during one session on performance in a subsequent session. Subjects might channel their performance in some fashion if they discovered what kinds of data were being gathered. Data collection should be as unobtrusive as possible.

By "instability" Campbell meant sampling fluctuations, the variation in repeated or equivalent measurements. He noted that this threat to internal validity was "the only threat to which statistical tests of significance are relevant." Both mechanical and human data-collection methods are subject to such instability, but mechanical devices presumably have less of it. Another way to look at this factor is in terms of precision; variance is inversely related to the precision of measurement as well as its reliability. Objective data collection can increase internal validity in particular if it increases precision, reliability, and accuracy.

Good experimental design can help prevent contamination. When preclusion and other counteractions are brought to bear against confounding, the events resulting from different states of an independent variable are more predictable. Such is notably the case in man-machine system experiments in which simulation tends to limit the number of possible events to be observed. When events fall within the realm of those expected, it is easier to observe and record them. The data that are gathered are more accurate, reliable, and precise.

Contamination related to data collection can also result from the way inputs are organized, and it can be prevented through better organization. It should be realized that performance lags behind input. Performance resulting from inputs introduced toward the end of a session could occur only after the session is terminated. If the input load has to build up early in the session, performance at a point near the start could be superior to later performance simply because it has not yet dealt with the inputs introduced at that point. In other words, data collection is not concurrent with input introduction. Experimenters must take this asynchronism into account in arranging states of independent variables within experimental sessions.

*Instrumentation.* In man-machine system experiments in laboratories and many field situations the principal recording devices are tape recorders, cameras, event recorders, keyboards, stop watches, and devices which register switch actions. Tape recorders pick up face-to-face, telephone, or radio communications. Cameras may register what people are doing but they are more likely to record events on displays, such as large maps or simulated radar displays. Observers may use keyboards to record various events in some code; event recorders do this automatically. A switch action, such as pressing a button on a console, can also be tallied by an event recorder, but switch actions are mainly the inputs to computers; switch actions are the human behavior that digital computers can record. (Analog computers can record other behavior that can be transduced into voltages, such as movements and pressures.)

Some kinds of experiments in field situations call for even more instrumentation, because actual people and objects are being observed rather than signals and messages. Special instruments may be needed to determine where troops and vehicles are located as they move across some terrain in poor or even in good visibility. Other instrumentation is needed to indicate automatically the outcomes of action, such as firing. Such instrumentation may be needed because human observation alone would be too fallible. However, it can also be extremely elaborate and expensive and endow an experiment with such an aura of objectivity that other experimental requirements are neglected.

In selecting instrumentation for data collection, researchers should be concerned about its reliability and maintainability as well as its cost. As already indicated, they should also consider how readily the data can be taken from the recordings and processed for analysis. Manual intervention in making such transfers can be improved through practice and through coding the data from the medium directly on IBM cards or magnetic tape.

Although digital computers can directly register only switch actions, there can be a considerable variety of switches and a much greater variety of switch "meanings," that is, the function of a particular switch. One console button can be associated not only with different sets of functions but also different elements of symbolic displays. Light pens and light guns, which are also switch-action devices, can select any item in alphanumeric or pictorial displays on a cathode ray tube. Nonetheless, there are other human actions which must be recorded manually or through devices such as a tape recorder and a camera, subject to the same limitations as in a noncomputer system.

The operator switch actions which have been recorded can be replayed by the system's own computer, along with the inputs received by the system and the computer's own responses. Such replay through "regenerative" techniques (Sackman 1967) mentioned earlier permits re-examination of the data.

Not only can the system computer record switch actions, inputs, and computer responses, it can be programmed to reduce these data for analysis. In fact, the data can be processed during an experimental session. As has been pointed out before, the linkage between data collection and reduction is an intimate and economical one—so much so that only the perceptive researcher will seek out other kinds of data that are less accessible. He will not imitate the drunk who leaves the dark alley where he lost his keys to search for them under the street lamp. Astute researchers will also understand the space limitations that may be encountered in computer or buffer storage and in computer-generated evaluations. The formulas programmed into the computer for refereeing the outcome of some operator action, for example, may be highly sophisticated, but still may not match the intricacy of human umpiring.

*Human Data Collection.* There are four kinds of persons who may gather data in a man-machine system experiment: members of the experimental staff, quasi subjects, operational personnel, and the subjects themselves. Operational personnel, who are sometimes recruited for the purpose in field experiments, have been found to be highly unreliable. Subjects can provide data along several different lines. They may either furnish their own views as to what the experiment is demonstrating or evaluate their own success or failure. They may evaluate the states of an independent variable, such as equipment design. They may make suggestions about how to improve the matter being investigated, such as a display, or what is needed in procedure or policy. The variance which characterizes their opinions and the divergence between their opinions and objectively derived data will be discussed shortly.

Human data collectors gather data directly from performance during experimental sessions and from subjects' comments during those sessions, from their comments in post-session discussions and debriefings, from interviews and subjects' progress critiques, from pre-experiment and post-experiment questionnaires filled out by the subjects, and occasionally from workshops in which the subjects participate. Researchers may profit from trying to optimize these channels. For example, discussion leaders can be counseled and trained, and procedures can be developed to encourage participation.

As gathered by all four kinds of human data collectors, data may be "raw," categorized, or evaluated; as Meister and Rabideau (1965) have noted, these tend to be mixed. Categorization means defining units of performance, such as components of an air traffic controller's verbalizations, so the units can be tallied and their durations measured. As in all content analysis, defining units can be a tantalizing task. Categorization may be established in advance so the units can be recorded directly or after the raw data have been recorded. Evaluation takes advantage of the human's pattern-recognition capabilities. For example, in his recording the data collector may relate a unit of performance to its goal or to its effect on other subjects; how much of this to do should be made clear to the data gatherer.

The data collector is often called on to make judgments of the success or failure of a subject's performance, such as intercept control or firing at a hostile tank. Although human umpiring of this kind relies extensively on interpretation, it may be assisted by providing the referee with clearly defined criteria and instructions about contingencies. Judgments of an number of umpires can be combined. Umpires can be required to indicate in their analysis why the particular outcome occurred.

Visual observations and estimations ("eyeballing") may be recorded by experimental staff, quasi subjects, and even subjects, on recording sheets. Auditory data may be similarly gathered. Again, the data may be raw, categorized, or evaluated. They may be samples rather than a continuous recording. The same sources can provide descriptive material, such as descriptions of critical incidents and ratings of subjects' performances.

Thus, subjects may rate themselves. Ratings can also be regarded as data about discrete units of performance or other matters. For example, subjects may rate the difficulty of their simulated jobs or tasks. They may rate their satisfaction with them. Opinions and evaluations of all kinds, including umpire judgments, can be expressed in ratings and rankings. Ratings of the importance of the performance judged, such as a decision, can accompany the ratings of its quality. In combination they provide a score which reflects importance as well as excellence. Ratings or rankings may be applied by subjects or staff to the simulated inputs with which the subjects have to deal, along various dimensions. These can include hazard, difficulty, or trouble; then scores derived from these ratings or rankings when the system saturates or breaks down measure the effectiveness of the particular state of the independent variable being examined at the time. Subjects may provide ratings of their subjective feelings, moods, motivations, and emotions. In short, ratings are a crude way to quantify qualitative phenomena, in which man-machine system experiments abound.

*Subjective Opinion.* When data from subjects take the form of opinion, the wide range of opinions invariably encountered bears testimony to the risk of relying on any of them. The opinions may be related to procedures or equipment, and the subjects may be experts, such as pilots or air traffic controllers. Even two such experts are likely to have diverging opinions.

The danger of designing equipment or procedures according to the views of "users" is further demonstrated when those views, obtained from subjects, are compared with measurements of data collected through instrumentation. An occasional experiment has compared such measures with subjects' impressions of their own effectiveness in relation to different aspects of inputs; the objectively derived measures contradicted the impressions. In more experiments a contrast has been drawn between a subject's evaluation of, and the actual performance data derived from, some display or display feature, console or console configuration, or other equipment which could be subjected to human engineering analysis. The subjective opinions and objective data have disagreed.

*Improvement of Human Data Collection.* As Meister and Rabideau (1965) have observed, the techniques of human observation and data recording deserve more study than they have received. These authors commented that the observational process could be improved by training observers in cue discrimination and data recording, in specifying the cues in advance, and in simplifying the observational task. It has been demonstrated that skill in data collection increases greatly with practice, both in accuracy and capacity. The criteria of what to watch for and how to evaluate should be clearly set forth in advance and given to the observer. Subjects' actions should be specified according to well-defined, discriminable segments. Observers can be helped by good design of recording sheets. Cues and categories can be already printed on them, and their formats should facilitate the time-ordering of data and provide space corresponding to the amount of data to be recorded.

Recording forms can either explicitly reject or call for diagnostic data as well as performance data. Observers can record "difficulty" situations in addition to checking those on a list. They can record apparent causes of each difficulty in addition to marking a checklist of possible causes. Observers may note inferred relationships between events. Particular procedures for recording evaluative data can make such data more useful. For example, observers should be cautioned not to use such high-generality phrases as "too much time taken" or categories which are too broad. Explanations of why they were made may be required to accompany judgments; so may the data on which they were based. These may be spoken into a tape recorder at the time of judgment. In general, such dictation can be helpful in stimulating subsequent recollection.

Data from a number of observers can be pooled to obtain aggregates or collective opinions. The observers can be checked and tested to determine how well they are performing in the experiment. For this purpose some kind of mechanical recording may be needed in parallel. Such recording may be desirable in any case during the experiment, for double checking and back-up.

### *Criteria*

Criteria may be regarded in man-machine system experiments as general terms or statements describing dependent variables. Although criteria can be viewed as qualitative as well as quantitative, they must be rendered into quantitative form to be generally useful in experiments. Measures do just that. They are more specific than criteria, which they define operationally. A number of different measures can pertain to a single criterion (for example, system safety), and a number of criteria can figure in an experiment (for example, safety, effectiveness, and resource consumption); thus, the experiment may include a number of dependent variables.

In verification experiments in particular it is desirable to specify both the criteria and the measures in the experiment's plan. The criteria have to be selected first, because they play a major role in determining the measures. Other reasons for selecting measures will be taken up shortly. What are the criteria for selecting criteria?

A criterion should reflect system objectives, assist in making system diagnoses, or be pertinent to the independent variables in the experiment. Some criteria may satisfy more than one of these considerations. As a matter of fact this is generally the case in ad hoc man-machine system experiments, because independent variables are manipulated either to show how the system performs under varied circumstances or to find out what parts or features of the system are responsible for the level of performance. In the general knowledge type of experiment the researcher is entirely concerned with linking his dependent variables to his independent variables, although to do this he may make use of system performance criteria which he assumes reflect system objectives. He does this to give his results reality and generality. For example, in an experiment on training techniques the results indicating relative effectiveness may be expressed in system outputs.

Experimental criteria are associated with performance, not static characteristics. In man-machine system experiments the performance of interest is a combination of human performance and machine performance, with varying emphasis on the two elements. How the system performs depends on both the men and machines which comprise it, and the same is true of its subsystems. Even at the component level, performance is what is done by a man and a console, a man and a radio, a man and a rifle. Machine-only functioning and man only behavior do occur in man-machine systems, of course, and occur in experiments about them, but they are not the dependent variables in man-machine system experiments.

Although criteria are associated with performance, they can consist of what performance involves rather than what performance is. For example, the criterion may be cost. If two policies result in equally good performance, how much more does one cost than the other? If two systems are loaded until they break down, what is the difference in loads at the point of collapse? In neither case does the criterion assess performance directly.

Still other criteria are used in evaluating systems but cannot be introduced into experiments themselves. These include maintainability (unless the experiment happens to be investigating system maintenance), survivability, adapability to change, and ability to be integrated with other systems. An experiment on training methods based on simulation would not examine the number of personnel needed to install the method or the outlays needed for producing the simulation inputs, although these points would have to be carefully analyzed before any decision was made to adopt the method which the experiment showed was superior.

When criteria are based on the objectives of the real system being simulated, researchers can run into a number of difficulties. Real systems usually have multiple objectives (some of which are more explicitly acknowledged than others), but relative weights among objectives are seldom specified. For example, a real air defense system may exist to protect both population and industries and to prevent the destruction of one's own offensive weapons (and incidentally to benefit in one way or another those who designed it, produced it, and operate it). Even if some general distinction of relative value could be made between protecting populations and protecting industries, how could the criterion of protecting populations be quantified? Communities differ in composition and size. How much is a human life worth? The problem of incommensurable criteria extends to other kinds of systems. Some allocate resources over extended periods of time. What would be the criteria of their effectiveness–criteria that could actually be helpful? The same question can be asked about broad questions of policy and strategy.

In general, it can be urged that criteria for man-machine system experiments should be comprehensive, relevant, and quantifiable. They may have to be multiple, and when they are the researcher may try to combine them through weightings or other techniques. For some purposes system output provides a common criterion. Uhlaner and Drucker (1964) pointed out that a common metric based on system output could permit comparisons among the contributions of training, selection, and equipment design to system effectiveness.

*Measures*

Like independent variables, dependent variables in man-machine system experiments come in many varieties, and the researcher faces the task of choosing among them. Just as he is likely to select a number of independent variables, he will find it profitable to make use of a number of measures to get the most out of an experiment. A man-machine system experiment is characterized by a multiplicity of measures reflecting the multiplicity of criteria they define. Each measure will say something about the performance of the system, subsystem, components, or individuals in it. Some measures will be more effective than others in demonstrating differences generated by the states of the independent variables.

Although several guidelines will be presented here for selecting measures, researchers may wish to follow the precedent of some of their predecessors by incorporating into a research program an experiment to investigate the usefulness of measures to be used in the other experiments. Such inquiry may also come in pilot studies or pre-experiment sessions (if data reduction can be rapid enough). In any case, since the kinds of measures chosen depend in part on the nature of the system and its operations, the researchers should become as knowledgeable as possible about that system and its parts. They should consult system operators and get help from analysts who have been developing measures of the same system for other purposes. Researchers should describe in the reports they write about an experiment the rationales they used in selecting measures.

*Guidelines*. Various requirements have been stated by Kidd (1962), Sinaiko and Buckley (1961), Morgan et al. (1963), and Meister and Rabideau (1965). There is general agreement that measures should be reliable and valid. Reliability means that the value found in a particular measure recurs in repeated measurements under the same circumstances. Before rejecting a measure as unreliable, the researcher must make sure that variability is due to instability in measurement rather than poor control over independent variables and constants. One objection to using subjective judgments as measures is that they tend to be unreliable. Lack of reliability threatens internal validity, as noted in the discussion of data; sampling fluctuations in the data come to light in the instability of the measures derived from them.

The validity of measures means their pertinence to the matter being investigated, their contribution to the external validity of the results. There are several ways to interpret this. One is to relate the measure to the criterion. If the measure represents the criterion, then it is the criterion that must be valid—by being the same in the experiment as it would be in the real system, for example. However, if the same measures are used in the experiment that are used in the system, validity can be inferred without involving the criterion; this is possible only in ad hoc experiments.

Validity can be construed as "relevance," "meaningfulness," and "criticality," in the sense that some measures are more valid than others. Too much precision may be irrelevant. For example, measures of reaction time averaging around one-half second at a console might have little relevance to system operations if response latencies up to two seconds had no differential effects on these. Tabulations of observed behavior, whether communication frequencies or other tallies of performance, should not be undertaken unless they provide information to which some criterion gives meaning. Criticality may be more difficult to assess; perhaps it simply means to forego trivial measures.

Other requirements include comprehensiveness, sensitivity, and availability. The number of measures may be fewer in experiments which focus on the effects of independent variables than in those whose aim is to evaluate the system; they are chosen to indicate the nature of the relationships. If they are not sensitive in revealing relationships, they are discarded. This does not mean that multiple measures should be foregone completely. Numerous measures involving a variety of dependent performances can add generality to general-knowledge experiments. In evaluative studies that place emphasis on the system, "the practice is usually to collect as many measures as can be shown to vary with the independent variables used. This is a wasteful procedure, but not an entirely unreasonable one" (Davis and Behan 1962). It is reasonable because unless the measures associated with all important criteria are incorporated, incomplete conclusions may be drawn about system effectiveness. Productivity measures should not be neglected, for example, while speed and accuracy measures are included.

Availability is synonymous with feasibility. Some measures may be infeasible because the events which produce the data that are measured do not occur frequently enough. This is one of the problems of measuring certain kinds of error. Cost of instrumentation may also limit feasibility. There is also another side to availability. A measure should not be sought simply because it is available. Measurement is not a justifiable end in itself.

*Types of Measures.* Measures have been categorized in various ways. Some concern total system performance, some components. Some are final or end measures, some intermediate. Some relate to outcomes, some to processes. Some are evaluative, some diagnostic. Actually all of these binary groupings have much in common. Probably the main purpose of component, intermediate, and process measures is diagnostic, for system trouble-shooting (Parsons 1962). They can reveal the locations where poor performance degrades the entire system. End measures, sometimes called "single-payoff" measures, cannot do this in systems in which inputs are processed serially. They conceal the problems that arise earlier in the processing.

Another purpose of component measures is to predict end measures when these cannot be readily obtained. Here the question of serial processing again enters. If a subsystem early in the chain does poorly, later subsystems are not likely to help matters. Then the performance of the early subsystem does predict system performance. But if it does well, the measure of its performance cannot predict system performance because a subsequent subsystem may do poorly. A measure of performance of the last subsystem in the chain may constitute one part of a total system measure and thereby help predict it. The other part consists of the inputs which the system originally receives.

Systems which are not characterized by serial processing are not subject to this analysis. For these, attempts may be made to show how well component measures predict total measures by determining correlations between them. Component measures from experiments may also be put into computer models in which system performance is simulated. The computer generates the end or system measurements.

A type of component measure which has been inadequately exploited in man-machine system experiments is the measure of interaction between components, between individuals, and between subsystems. More analysis is needed to set forth the various interface measures which might be put to use. Such analysis would have to start with taxonomies of interaction. These might profit from studies of critical incidents in system functioning. Critical incidents are a class of dependent behaviors which cannot be measured, but they can supply insights that might lead to new measures.

Again let it be noted that most measures in man-machine system experiments describe the joint performance of men and machines (or computer programs), not that of one or the other alone. Although there are machine (or program) contributions to more of their experiments than psychologists in other fields generally concede, in man-machine system research the duality of source is unquestionable. For diagnostic purposes it may be desirable to sort out the respective origins of errors, for example. But frequently the operator and his equipment (or program) constitute a performing unit in which the relative contributions can be partialled out only through component experiments in which operator factors and machine aspects are separately manipulated as independent variables. The composite nature of performance in man-machine systems limits the generality of the results of experiments on them, since the machines (or programs) may differ widely from one system or kind of system to another.

Many classifications of measures emphasize four: those based on time (including latencies and durations); those concerned with accuracy and error; those associated with amount accomplished, or production; and those which indicate the resources consumed. But in some man-machine system experiments measures have been evolved which at best seem only partially related to these. For example, some air traffic control experiments have embodied "hazard" scores and "user convenience" scores. These were derived through ratings and rankings by experts.

Ratings and rankings are convenient ways to create composite measures. Humans combine subjective or even objective measures simply by ranking one state of an independent variable ahead of another. Composite measures may also be developed by assigning weights to the component measures and merging them in a single synthetic measure. But the determination of weights is not easy, nor is it always clear how two measures that might be combined are related to each other. With regard to air traffic control, for example, Kidd (1962) wrote:

> Safety and delay en route are both meaningful criteria of system performance. The dual facets appear to be related in reciprocal fashion but the correlation is far from perfect. The difficulty comes from the fact that any simulated system which yielded enough mid-air collisions to make statistical data processing significant would be a very poor system. The comparison of such a system with operational systems which have near perfect safety records is meaningless.

It is desirable at times to try to achieve comparability of different measures. As in other kinds of experiments and measurement, one way to do this is to change particular measures into ratios or proportions, such as the ratio of inputs to outputs. These can be expressed as percentages. Another approach is to determine the location of a measurement in a distribution common to many measures, such as a normal or Gaussian distribution—the bell-shaped curve—and to transform the particular measure into the generalized one for such a distribution.

### *Analysis*

Since the analysis of measurements of data in man-machine system experiments conforms to analysis in experiments in general, it will receive no extended treatment here. Readers are referred to standard texts on statistical analysis. A chapter on statistical methods by Chapanis (1959) is recommended for its clarity and its use of examples from human engineering, which brings it closer than most treatments to the context of man-machine system experiments.

For readers unfamiliar with statistical analysis, it might be explained that descriptive analysis consists primarily of aggregating individual measurements, or demonstrating relationships between them, and showing relationships between aggregates. Individual outcome measurements associated with a state of an independent variable are aggregated into single measures which indicate central tendencies such as means and medians. The extent of dispersion of the individual measurements so aggregated is expressed as a variance or standard deviation; their relationships to each other can be graphed or tabulated in a frequency distribution. The relationship between individual measurements in two different sets can be graphed in correlation diagrams, and the closeness of the co-variation can be expressed as a correlation coefficient. The measurements in the two sets may derive from two different measures of the same state of a variable, measures from different states, or measures from different variables.

Differences between means (or medians) are calculated to show the effects of the differences between the states of an independent variable. In an experiment incorporating more than one independent variable, means can be calculated for each state of one variable in combination with each state of the other variables and across all their states. Differences between means can be illustrated in graphs or tables which show the values of the means; the spatial or arithmetic divergences indicate the differences. When there are more than two independent variables, presentation of the means related to the states in all of them in a single graph or table tends to become unwieldy. Thus, all the results even for one of the measures in a large, multivariate man-machine system experiment are not usually shown in the same display.

The means associated with different variables in a multivariate experiment can also be related to each other, two at a time, through a correlational technique called partial correlation. The effects of the other variables are eliminated. The correlation coefficient indicates the extent of co-variation. Multiple correlation shows the effects of groups of variables.

This brief summary should suffice to suggest how measured data in a man-machine system experiment are summarized and contrasted. Because so much in the way of data is collected and measured, clearly the processes of summarization and contrast can become formidable. Fortunately computer programs exist to do the kind of processing required. As indicated earlier, if the subjects' responses in an experiment result directly in switch actions which can be transmitted to the computer, processing can occur during the experimental session itself. If the responses can be quickly transduced into computer-accessible form, processing can at least be accomplished between sessions.

Necessarily most of the data from an experiment must be measured, summarized, and contrasted in the publication of results if the results are to be understood by readers. Otherwise there would be too much to print and too much to digest. The communication process depends on quantification and summarization. When results are not quantified, as in descriptions of incidents, lengthy narratives or series of narratives are required, and these must include many complex details of system operations if the readers are to understand them.

In man-machine system experiments the key elements in the summarized quantifications of results are means and differences between means. The means describe system performance and component performance under varying circumstances. The size of differences between means is an index of the degree to which differences between states of an independent variable are important.

Variances or standard deviations have less importance but do deserve brief comment in two respects. A substantial dispersion among the measurements of a task may betoken some major errors on the part of the performer. Thus, a large variance may be a cue to this aspect of performance; if it is, the researcher should so indicate in his report. Variance which cannot be accounted for through differences between states is regarded as random, as noted earlier in this chapter; it is used in ratios between it and the variance attributable to differences between states for calculations of statistical significance, to be discussed next.

*Confidence Testing.* Another part of analysis is the testing of the differences between means for their statistical significance. The need for such testing has been debated at length among behavioral scientists, and man-machine system experiments have not been exempted from the debate. Without question significance testing can contribute to confidence in the internal validity of results in many such experiments, so it has considerable value. It provides a quantified index of confidence instead of leaving the degree of assurance to individual judgment. For verification experiments, where the demand for certainty is high, it has particular appeal; it is less crucial for discovery experiments.

It should be understood, as suggested in the section on *Design*, that significance testing does not by itself demonstrate whether confidence should be placed in the results. Since it cannot be justifiably undertaken unless certain counteractions against confounding are embodied in the experimental design, it does reinforce their use. But it does not prevent or eliminate all confounding, nor does it assure external validity. Unless these limitations are well understood, the design may be inadequate and statistical testing may lead to false confidence.

As explained previously, tests for statistical significance seek to determine whether a difference (variance) between means might occur through chance, rather than because of the difference between states of an independent variable. A significance test hypothesizes that a difference did occur by chance and seeks to disprove this null hypothesis. An index of variation in measurements which *cannot* be attributed to the variable's states is called an index of chance variance and is contrasted, in a ratio, with the variance between means. The size of the ratio indicates (in tables) the probability that the variance between means has occurred by chance rather than because of the states of the independent variable.

The size of the ratio and hence the probability figure is influenced by three factors: the size of the difference between means, the number of data (observations) of which they are composed, and the extent of chance variance. The greater the chance variance the smaller will be the ratio and the larger the probability figure for the null hypothesis. The larger the difference between means or the larger the quantity of data, the larger will be the ratio and the smaller the probability figure. The smaller the probability figure the higher the confidence should be that the difference between means results from the difference between the states of the variable. By convention, experimenters have depended on two cut-off figures as separating confidence from no confidence. These have been 5% (.05) and 1% (.01), the latter being the more stringent.

*Problems.* This thumbnail sketch serves to introduce a number of problems which can characterize all experiments in behavioral science and, perhaps, man-machine system experiments in particular. Allusions to some of them have been made earlier.

In the first place, the chance or random variance should be kept as small as possible. This means, as we have seen, taking steps to limit fluctuations in measures. The researcher tries to assure the stability of instrumentation and observers. It also means taking steps to limit fluctuations within a state of an independent variable in repeated occurrences. The researcher tries to maintain constancy and assure equivalence. In man-machine system experiments these efforts are especially demanding.

Second, the experimenter must get enough data through replications of each experimental condition. Each state may be repeated with the same subjects (teams), but repetitions with equivalent or randomized subjects can also count as replications. But how many replications are required? The magnitude of a man-machine system experiment calls for keeping the number to a minimum because of cost and the need to include other, time-demanding counteractions.

Considerable understanding of statistical analysis may be required to form the best estimate of the required number of replications. The experimenter may wish to get advice from a statistician. In some designs the number must be determined in advance. In others, which thereby have more flexibility, a modular design can enable the experimenter to iterate the experimental conditions. Then he must have rapid data processing at his disposal to determine, through "sequential analysis," whether enough data have been gathered for significance testing. In any case, it is clear that the planning of the experiment should incorporate the requirements for this type of statistical analysis; they cannot be an afterthought.

In some experiments with less rigorous designs, changes in states of an experimental variable are preceded and followed by extended steady-state periods of performance to enable the researchers to judge reliable differences occasioned by the change. Steady or stable states also require many replications to yield considerable amounts of data. The reason is essentially the same as in significance testing, although statistical treatment usually is foregone. Instead, experimenters tend to rely on visual presentations of results (e.g., graphs) to make judgments that enough data have been gathered to warrant comparisons. Although some statistical procedures exist for testing the significance of differences between means in these time-series designs, here too experimenters often rely on judgment. This point might be restated by saying that researchers may make judgments of statistical significance as well as estimates of data sufficiency, without benefit of statistical testing. If a difference between means is a substantial one, as shown, for instance, in graphical form, and if the researcher knows he has collected a great deal of data for each state of the independent variable, he may conclude that the difference between means would yield a very low chance figure if he did go the the effort of making a statistical test. He judges "by eye," so to speak. His confidence is enhanced if he has some basis for knowing as well that random variance has been small. A graphical presentation of a steady state level of performance may furnish such a basis.

A third problem in significance testing in man-machine system experiments is related to the foregoing question of amount of data. In such experiments there are likely to be many occasions where data should be discounted or discarded for one reason or another. Equipment malfunctions are only one reason. In addition, there will be occasions where data are not recorded. Advance planning should cover these contingencies so the remedies applied do not violate premises on which tests for statistical significance are based.

One of the assumptions of conventional significance tests is the normal or Gaussian distribution of measurements. Yet in man-machine system experiments a frequently used measure is reaction time (latency). Zero time marks one limit of the distribution and the mean is close to it; as a result, distributions of reaction time are highly "skewed" rather than normal. Statistical texts or experts should be consulted for methods of dealing with this problem.

The fifth problem concerns the difference between practical significance and statistical significance. The practical difference between two means in a man-machine system experiment is the difference that really concerns the designer of some new equipment, the developer of a new training technique, or the innovators of a new procedure. If the difference amounts to only a few percentage points, is the innovation worth worrying about? The difference may be significant statistically yet be very small from a practical viewpoint. Clearly, practical significance should be discussed by the researcher in his report; he should not infer that one alternative should be adopted rather than another simply because the difference between their performances proved to be greater than chance. Yet the researcher should also consider whether a difference in means of a few percentage points might not be important. What if it meant, in an air traffic control system, the savings of lives because one or two fewer air collisions would occur? What if it meant, in an air defense system, the saving of a city because two or three more hostile bombers were shot down?

Finally, questions must be raised about the probability levels used for making the confidence-no-confidence judgment about differences between means. Are the familiar .01 and .05 levels appropriate to all man-machine system experiments? In the discussion of *Design* earlier in this chapter, the point was made that two types of error can result from abiding by one of these levels. On the one hand, the difference between means might really have resulted from the differences between the states of the independent variable, even though the probability level found for the null hypothesis in the significance test was higher than .05 (if this was the cut-off chosen). This would be a Type II error. On the other hand, if the probability level that was found was .05 or less, the difference between means might nonetheless have been really due to chance. This would be the Type I error.

As Chapanis (1959) and Davis and Behan (1962) have pointed out, researchers may want to set probability values at levels higher than .05 to avoid the Type II error. One situation could be a discovery-type experiment. Another would be an experiment in which alternative system designs were being compared prior to system development. If all cost about the same amount, the Type II error would be the one about which to be most concerned. If results showed that designs differed by a fair amount in apparent effectiveness, the researcher might be willing to be a little less confident about the statistical significance of that difference. His viewpoint might be further influenced by the kind of impact resulting from a genuine difference.

In such situations the experimenter would be required to set a new cut-off for the probability of the null hypothesis, such as .10, .20, or even as high as .40. Where should he set it? One solution would be to set none at all. Rather, he would simply state the probability level found by the significance test, describe the circumstances in his report, and let readers come to their own judgments of confidence. Whether this approach might be followed more widely in all experimentation is beyond the bounds of this discussion.

# 3

# Forebears

The forebears of complex man-machine system experimentation were programs sponsored by the National Defense Research Committee (NDRC) of the Office of Scientific Research and Development (OSRD) during World War II. Particularly deserving recognition were (a) the studies in the control of gunfire conducted for the Applied Psychology Panel of the National Defense Research Committee and (b) the Systems Research Laboratory of Harvard University, which investigated naval combat information centers. Other kinds of pioneering "human factors" research in World War II were sponsored by the National Defense Research Committee, the Navy, and the Army Air Force (Fitts 1947).

## GUNNERY STUDIES

The history of the Applied Psychology Panel and its predecessor, the Committee on Service Personnel, has been recorded by C. W. Bray (1948), executive secretary of the committee and later chief of the panel, succeeding W. S. Hunter. W. E. Kappauf (1947) has also recorded some of the work, to which many outstanding experimental psychologists lent their talents. One of the projects started in 1942 was a study of Army antiaircraft artillery, directed by L. C. Mead and W. C. Biel, with Tufts College as contractor. These researchers established the validity of a synthetic tracking trainer developed at Tufts (Hudson and Searle 1944) by comparing training on this device with training on the gun director which it simulated. They also created proficiency tests for antiaircraft trackers and investigated their value in training. Further, as Bray has noted, they were involved "in acceptance tests of new equipment and alternative operating procedures."

The multifaceted nature of the project illustrates a trend in the evolution of this early research. Although interests were initially focused on personnel classification, they broadened during 1942–43 to include proficiency testing, manuals, and training. For example, J. L. Kennedy, technical aide to the panel, undertook a project to improve the training of Navy rangefinder personnel. He drew on experience acquired in a research effort on selection and training of Army heightfinder operators headed by C. H. Graham and W. J. Brogden, with

Brown University as the contractor and Kappauf as the principal experimenter. The methodology and success of this earlier effort contributed greatly to the progress made in later projects. The Navy research to which it was extended resulted in a University of Wisconsin project, directed by Brogden and D. G. Ellson, at a new Navy school for rangefinder and fire-control radar operators at Fort Lauderdale, Florida.

Before long the involvement of applied psychologists widened still further to include studying the design of equipment. (The term "human engineering" was not current at the time.) In 1944 a joint project of the panel and the Armored Medical Research Laboratory investigated the sources of errors in Army field artillery and, among other things, "developed several pilot models of new gunsight scales which were designed to eliminate many of the errors made at the guns" (Bray 1948). Also in 1944 the Applied Psychology Panel established a large field laboratory at Laredo Army Air Field, in Texas, to conduct research leading to improvements in the design of gunsights in the flexible gunnery equipment of B-29 aircraft.

By and large, these early investigations did not incorporate the experimental complexity of the studies with which this book is principally concerned. For example, although two-man or three-man gunnery crews were sometimes the focus, there was only one "Panel investigation of methods for training groups of men to operate as teams," according to Bray (1948). In that study teams of four, five, or more men worked together at high speed on the Navy Gun Director, Mark 37.

However, some of the problems characterizing more complex experiments were explored in the early research, notably measurement and simulation. It seemed essential to find a way to determine how large an error a gunner might make in tracking aircraft. Doing this with gun-camera photographs of the reticle and field of view not only proved unreliable and arduous but also seriously delayed giving knowledge of results to the gunner. A second technique, qualitative observation by instructors, lacked accuracy and reliability. Accordingly, W. C. Biel and his colleagues adapted a third measurement technique, using a supplementary telescope aligned with the gunner's sight, to score tracking performance for training. A reticle already incorporated into this checksight enabled an observer with a stopclock to register time on target. This checksight technique was widely copied. It spread from the Army to the Navy and eventually was followed by a photo-tube-and-amplifier device which could be signaled by a searchlight on the target aircraft. The Applied Psychology Panel evaluated and helped improve this device.

Another innovation in measurement achieved less success. L. C. Mead, L. V. Searle, and K. S. Wagoner developed a system for the remote recording of aiming errors in field artillery. Although guns had to be more firmly emplaced than was usual in field artillery practice, this technique seemed suitable for further experimentation on field artillery errors but, according to Bray (1948), the Army showed no interest. The panel sold the materials for salvage.

In addition to the Tufts tracking trainer, mentioned earlier, and certain other training simulators, a major simulation project was undertaken for experimentation on B-29 and other airplane gunner equipment. One of the B-29 gun stations was simulated by a mock-up and a target image was projected on a screen. Since a leading objective was to evolve a method for rapidly measuring gunner performance, scores were recorded automatically and remotely in several ways. This equipment helped remove the dependence on airborne testing, a dependence which then, as later, was both constraining and frustrating due to problems of co-ordination, weather, and unexpected unavailability of aircraft. An interesting further innovation placed target simulation in the airborne environment. D. G. Ellson developed a remotely controlled system to simulate targets and record performance for test and training within the B-29 while aloft. One product was a polygraph recorder for registering separate motions. The ground mock-up also led to a synthetic ground trainer.

## COMBAT INFORMATION CENTER STUDIES

There was an extensive Applied Psychology Panel program in the selection and training of radar operators, as well as in voice and Morse code communications. But the radar-associated investigations of greatest interest for the present account were those of the Systems Research Laboratory of Harvard University, initiated by L. Beranek and S. S. Stevens. Shipboard studies, supported by the Bureau of Ships, showed the need for a laboratory in which experimentation could be conducted under more controlled conditions. With NDRC-OSRD funding, such a laboratory was established for simulation experiments at Beavertail Point, Jamestown, Rhode Island. The last eight months of 1945 saw the publication of fifteen reports, some of them covering the studies afloat.

The principal figures in the project were Beranek and C. T. Morgan, the directors, with R. C. Morton in charge of laboratory outfitting and engineering. The staff included psychologists (e.g., W. R. Garner), physicists, and time-and-motion engineers, with representatives from radar design and operation, sonar equipment and training gear, communications engineering, gunnery training, and operations research. Contract assistance in time-and-motion work came from Purdue and New York Universities. Harvard, Massachusetts Institute of Technology, and Beavertail itself developed radar and display simulators.

On January 1, 1946, responsibility for the Beavertail facility was shifted to The Johns Hopkins University (to which Morgan and Garner had transferred), with support from the Navy's Office of Research and Inventions, later the Office of Naval Research, under the cognizance of the U.S.N. Special Devices Center. Research was also undertaken in the Electrical Engineering Laboratory of Johns Hopkins under F. Hamburger, Jr., and then in the Psychological Laboratory at Johns Hopkins. The Systems Research Field Laboratory at Beavertail moved to Baltimore in 1948 as the Systems Research Laboratory in the Institute for Cooperative Research at Johns Hopkins. In 1952 cognizance over the Johns Hopkins research was assumed by the Systems Coordination Division of the Naval Research Laboratory. In 1953 sponsorship of the Johns Hopkins research program was shifted to the Psychological Sciences Division of the Office of Naval Research, where it remained until the contract terminated in 1959. The work of the Johns Hopkins investigators is described at the end of this chapter.

The Beavertail facility appears to have been the first laboratory to receive the "system" label. It was put into operation with remarkable speed, thanks to the Navy's desire to improve combat information center operations on its fighting ships as a consequence of experience in the Pacific, particularly in the Okinawa campaign. Combat information centers (CICs) were the man-machine complexes where radar and other information was viewed on various display scopes, evaluated, and distributed for weapons and battle direction. The display scopes were mostly "plan and position indicators"—usually named without the "and" or abbreviated as "PPIs"—which presented radar echoes in distance (range) and direction (azimuth, true bearing) from the radar in a map-like fashion. CICs were a product of the rapid development of radar during World War II. Since they had not existed as such prior to radar, their arrangement, associated communications, procedures, components, and interfaces with the other older combat elements of the ship reflected the haste of development. They constituted a ready and rich domain for applying what would later be called "human factors" study and improvement.

*Harvard Laboratory Studies*

In the Beavertail laboratory were placed a number of operational equipments—PPI displays, voice communications, and large displays for recording (plotting) paths of aircraft—to simulate two shipboard combat information centers. In the six experiments conducted and reported by the Harvard University Systems Laboratory, the objectives of inquiry and methods used were as follows:

(1) Amount of information—range and bearing data—that could be communicated from a CIC to antiaircraft gun directors with gunnery liaison officers reading from a horizontal display (air plot) of aircraft paths (tracks) over shipboard-type phones to antiaircraft directors (Systems Research Laboratory Staff 1945*g*). Experimental variations: the number of gunnery liaison officers—one or two; and when there were two, a single phone circuit was shared or there were separate circuits.

(2) Performance of the plotter at a horizontal display (air plot) in making fixes (plots) to locate hostile raids upon receiving range and bearing data by shipboard-type, sound-powered phone (Systems Research Laboratory Staff 1945*j*). The speaker read from prepared material (on maneuvering board paper). Experimental variation: the time-density of information transmitted to the plotter; that is, the raids ranged from one to eight coinciding in time, and readings per minute were one, two, or three. In this experiment the subjects were well-practiced Harvard students. In other experiments they were former Navy enlisted personnel hired as a group for the experimentation.

(3) Performance of the plotter at the air plot in plotting fixes told over a shipboard-type phone from prepared material consisting of four, five, or six raids, with one transmission per raid per minute (Systems Research Laboratory Staff 1945*k*). Experimental variation: standard Navy plotting method vs. an improved method using colored range circles and a simple speed and course indicator. Analysis included micromotion measurement from motion pictures.

(4) Value of plotting boards at command stations (Systems Research Laboratory Staff 1945*l*).

(5) Evaluation of a new item of equipment for manual plotting, the automatic target positioner, an adjunct of the dead reckoning tracer (Systems Research Laboratory Staff 1945*i*).

(6) Evaluation of PPI display equipment showing simulated radar targets, with the operator reporting each appearance of a radar echo (blip) in range and bearing to a recorder over a shipboard-type phone (Systems Research Laboratory Staff 1945*e*). The load during eleven tests varied from a single stationary target to five moving targets presented simultaneously. Experimental variation: two different types of PPI consoles, each with or without a particular additional component which might aid in determining range and bearing. Each of the four methods was tested with four different operators, who had received six weeks of CIC training but no concentrated practice on this equipment. Measures included accuracy, number of reports per minute, and time delays in reporting.

Only in this last study were radar echoes simulated. When the laboratory was established, it was planned to depend on actual aircraft for aircraft radar targets and to use, initially, a piece of training equipment which could simulate slow-moving surface targets. In this study, by altering the simulated rotation rate of the radar, the trainer was made to simulate automatically some targets moving at 80 knots; by manual settings, operators simulated target speeds of 400 knots. Actual aircraft flew in some of the laboratory experiments.

A new radar simulator for surface targets, finally installed and operable in 1946, was adapted from a radar trainer built by the University of California, Division of War Research at San Diego, the CIC Problem Generator and Display System Model 1 UCDWR. The new equipment could project a geographical plot of the motion of as many as six ships and two torpedoes on a screen, while simultaneously displaying upon a PPI scope a relative plot of the same problem centered on any one of three ships (Systems Research Laboratory Staff 1945*n*, *o*). For more than four months five men, on the average, were engaged in development and construction work on the electronic circuits to ready this device for Beavertail use, and considerable additional work was required to put into working order the electromechanical drive units which had been manufactured by a company in Los Angeles. The difficulties in getting the electromechanical components to simulate high-speed aircraft portended the problems encountered with other radar simulators in the years to come.

The laboratory might well have gone on to more complex CIC experiments if the war had not ended. This is not to say that no studies in a more complex context were ever conducted. Indeed they had been—before the laboratory became operational—on shipboard in exercises with actual aircraft simulating hostile target aircraft. In fact, the project's energies and reporting were largely devoted to the shipboard studies, which were concerned, as were those in the laboratory, primarily with the role of the CIC in antiaircraft target designation. The laboratory experiments were actually executed to complement the previous studies done at sea.

### *Harvard Shipboard Studies*

One report, "Summary of Afloat and Ashore Studies of AA Target Designation Systems," summarized the shipboard and laboratory work (Systems Research Laboratory Staff 1945*m*). Six reports (Systems Research Laboratory Staff 1945*a, b, c, d, f, h*) described separately the collection of data on five cruisers, a battleship, and a destroyer, with resulting recommendations. In addition to data collected from regular training exercises, four special exercises were organized particularly for this research. Although the aircraft in all instances were real, the antiaircraft fire was simulated. Operational personnel functioned at their duty stations as they would in actual combat operations.

The first study reported took place on a heavy cruiser of the Baltimore class, the USS Boston. There were three exercises, the first for practice, while the others each included about thirty air attacks directed from the ship to simulate Japanese kamikaze, i.e., suicide, missions. Two procedures were compared for "coaching" (i.e., passing radar target information to) Mk 37 directors for 5-inch guns; two were compared for 40 MM. directors; and three compared for Mk 57 and Mk 63 directors for machine guns. In one exercise antiaircraft defense was combined with shore bombardment, and in another with surface engagement. A fourth exercise combining AA defense with fighter direction was canceled because of a change in the ship's orders.

According to the study report, "fourteen Scientific and Naval personnel participated in the experiments," and "Continuous recordings were made of the sound power circuits used in voice coaching of the gun directors; photographs were taken of activities in CIC; and written records were made in the directors of the designation range, ranges and bearings, acquisition range, and acquisition time of each raid designated to a director." ("Designation range" meant distance to the hostile aircraft when the gun director was told to acquire it; "acquisition" meant the director saw it.) The procedures were compared in such terms as percentages of targets acquired or taken under fire and average acquisition range or time. Results also compared performances of different types of directors; they related acquisition time to coaching rate and coaching rate to the number of directors being coached simultaneously; and they showed how various factors differentially affected coaching rate and accuracy. In addition, as a by-product, it was found that land echoes interfered with both surveillance and fire-control radars.

The special exercises on the USS Boston were conducted, as were all the shipboard studies, off the West Coast. They appear to have been the only ones to introduce experimental comparisons systematically. Special exercises were also held on the USS Tucson and a destroyer, the former engaging separately in AA target designation, surface engagement, fighter direction, and shore bombardment, the latter in AA target designation and surface engagement with air attack. Tactical situations and loads were varied and some effectiveness data were obtained, resulting in recommendations of more training in CIC procedures and more exercises to obtain evaluative data.

Possibly such recommendations represent what one can expect from system or subsystem performance data gathered in "exercises" into which systematic comparisons have not been designed. However, other types of measures were obtained which led to specific improvements. In the USS Tucson as well as the USS Boston exercises, and in training exercises on three other cruisers (USS

Miami, USS Nashville, and USS Louisville), time-and-motion data were recorded from the movements of CIC crew members and, in one ship, from the performance of radar scope operators and plotters as well as from communications. Rearrangements of equipment resulted from many of the extensive recommendations based on these data, and subsequent activity measurements validated these arrangements.

In the summary report on the shipboard and laboratory target designation studies, a number of comments dealt with experimental methodology. "The personnel involved in the experiments were all trained Navy or civilian CIC crews," the report noted. "Most of the experiments were repeated using different equipment operators so that the result would not depend on the unusual abilities of any one man. The equipment used was standard service gear, and the testing methods were devised to provide adequate and reliable scores for each operational (laboratory or shipboard) test." However, the report noted that performance did not occur during battle or after the crews had been at general quarters for an extended duration, so that results "will be optimistic in most cases if compared to results obtained during the stress and confusion of combat."

It is possible, the report said, to maintain to some extent in shipboard research the constancy of operating conditions and crew efficiency when comparative evaluations of equipment and methods are conducted, but not as well as in the laboratory, where more rigorous testing can eliminate some equipment and methods from consideration for shipboard use. So, "operational tests, when possible, should follow the laboratory tests. It is unfortunate that because of the pressure of the war and the lack of a suitable testing laboratory some very inadequate target designation methods and equipment were pressed into shipboard use that would have been discarded or improved had they been appraised first in a laboratory."

The Harvard CIC investigations were truly a pioneering effort. Partly because the reports have been classified until relatively recently, their story has not been publicly presented before. The discerning reader will detect even in this brief account many innovative features and perennial problems. The studies arose from an acute operational need. They were oriented toward finding solutions to particular problems. They introduced experimental methodology into system testing, even to some extent in the operational environment–on board ship. They tried to integrate laboratory and on-site inquiry. They combined various investigative techniques and exploited a variety of measurements (omitting significance statistics). They introduced simulation of the external environment in conjunction with performance by trained personnel on operational equipment.

The reports themselves attempted to present their contents and conclusions in a manner that would help get their message across to those who might profit thereby. They were characterized by clear and simple English, lucid summaries, short sentences, and paragraphs and sections with suitable headings, with many explanatory photos and drawings, intelligible graphs to show quantitative results, illuminating tables, effective internal arrangement, and brevity. As noted earlier, some recommendations were adopted, but the extent to which those who could profit from these studies and their reports actually did so cannot be deduced from the reports themselves. Probably no one knows. Operational application, so far as is known, never became the subject of additional investigation and report.

### *Johns Hopkins Studies*

When the Beavertail laboratory was taken over by The Johns Hopkins University at the start of 1946, a research program continued there until 1948 before moving to the university itself, where similar research had already begun. The Johns Hopkins program was indeed prolific, with more than 170 reports published by the end of 1953. It took a form somewhat different from the Harvard CIC studies. It was entirely laboratory-centered, with emphasis on variables which could be investigated through the performance of individual operators working with some piece of equipment; and eventually the research dealt increasingly with visual, auditory, motor, and more complex processes where experimental findings had basic applicability. From the start the "Systems" in "Systems Research Laboratory" had reference to research on elements within man-machine systems, rather than research directed at these systems or subsystems in their entirety. As an indicator of the trend, the Psychological Laboratory eventually became the sole locus of research, most of which was published in psychology journals.

Much of the research in the earlier period was concerned with characteristics of radar displays, particularly plan position indicators, and with the performance of radar operators. Some studies were evaluations of particular equipment; others were concerned with effects on human discrimination of parameters of the PPI's cathode ray tube and the signals which it could produce. Aids for obtaining information from these displays, e.g., range and bearing information, also received attention, and work was done on coding, plotting boards, scale reading and dials, and controls. Experiments were conducted using simulated radar inputs, and considerable effort was devoted to developing adequate simulation equipment for this purpose. There were important by-products, for example, "Applied Experimental Psychology" (Chapanis, Garner, and Morgan 1949), as well as major contributions to "Human Factors in Undersea Warfare" (Panel on Psychology and Physiology 1949). The Johns Hopkins research program had far-reaching effects in establishing sound human engineering practice in equipment design, not only in radar-associated equipment but in many other types.

Illustrative of the Johns Hopkins work at Beavertail were such reports as "A study of factors affecting operation of the VG remote PPI" (Garner 1946); "The relative efficiency of a bearing counter and bearing dial for use with PPI presentations" (Chapanis 1947); "Brightness of grease pencil marks on a vertical plotting board" (Gebhard and Newton 1947); "Some experiments with the VF aided tracking equipment" (Gebhard 1948); and "Accuracy of visual interpolation between scale markers as a function of the number assigned to the scale interval" (Chapanis and Leyzorek 1950). In the case of the last three, the data were collected at Beavertail and processed at Johns Hopkins. The output of the Systems Research Laboratory and Psychological Laboratory at the university included "Visibility on cathode-ray tube screens: Intensity and color of ambient illumination" (Williams and Hanes 1949); "The relative discriminability of several geometric forms" (Sleight 1952); and "Some design factors affecting the speed of identification of range rings on polar coordinate displays" (Garner, Saltzman, and Saltzman 1949).

An interesting question is whether the returns from this component-oriented work and from the years of research which followed were greater than those which would have accrued from an equivalent investment in more complex system experiments with their attendant large costs arising from simulation, subjects, and experimental staff. As with most questions about alternative research strategies, probably no answer would find 100% agreement, even by advocating a combination of strategies.

# 4

# Project Cadillac

The first big program of complex simulation-based man-machine system experiments was called Project Cadillac. It was conducted by New York University for the Navy at the latter's Special Devices Center in Port Washington, New York, from January 1948 through February 1955. Experimental work began in earnest only in 1951 and included six large-scale, multioperator studies and seven individual operator studies, some of them with considerable complexity of load variables.

The project derived its name from a code term given at the end of World War II to a potential technique for obtaining radar information about suicide (Kamikaze) raids against the fleet at Okinawa. By flying their aircraft at low altitudes, Japanese pilots were able to escape early detection by the radars in the United States ships. They were protected by the curvature of the earth because radars have line-of-sight "vision." The immediate solution was to station radar-equipped picket ships some distance from the fleet. After the war had ended another technique was implemented: placing radars in high-flying aircraft to effect airborne early warning (AEW). The term "Cadillac" continued to be applied to Navy AEW aircraft, including the WV-2 (Lockheed) Super Constellation, which was outfitted not only with search and heightfinding radars but also with a number of consoles enabling officers and enlisted personnel to conduct both surveillance and intercept operations. This aircraft itself was unarmed and it was designed to function as an airborne combat information center similar to the shipboard CICs described in Chapters 3 and 5, but with more limited functions. It had one radar which provided information about the distance and direction of hostile aircraft and another which indicated their altitude.

While the WV-2 as a flying CIC was still under development, a remarkable innovation occurred. A mock-up of its CIC portion was built at the Special Devices Center as a locus of analysis and experimentation, with prototype equipment and special electronic circuits and facilities. Project Cadillac conducted a number of human engineering analyses and extensively and experimentally investigated procedure and load variables, first in the surveillance function and subsequently in interception control. Six Navy officers with CIC experience and several enlisted personnel were attached to the project as advisers and experimental subjects; the composition of this contingent varied over the years. R. L.

Chapman was the project's technical director in 1950–51; later this position was held by S. Veniar. A summary report prepared by J. J. Regan, who followed V. J. Sharkey as technical monitor, has described the project's ventures and publications (New York University Cadillac Staff 1956). Most but not all the publications have been declassified.

The mock-up, constructed in the years 1949–50, incorporated a layout which had been judged superior to a number of others in a questionnaire survey and analysis (Adiletta and Chapman 1951). It included five plan position indicator (PPI) consoles for surveillance and/or interception, one heightfinding console, a status board, and stations for the commander and radio communications. The facility did not include the distance/direction radar itself, from which, in real life, the surveillance or interception consoles would receive their signals, nor the altitude radar for the heightfinding console. Rather, in the laboratory outside the mock-up were thirty 15–AM–1 target generators, twenty-four to simulate radar returns from aircraft, the remainder to simulate surface targets. Each could produce a moving radar echo (blip) at all of the PPI scopes.

The 15–AM–1 target generator was the forerunner of a similar device, the 15–J–1c, which was also developed at the Special Devices Center, primarily for training but widely used also for experimentation (see Chapters 5, 6, and 7). Dynamic target (blip) information was generated by a mechanical computer assembly employing ball and disc integration and driving a set of potentiometers to obtain electrical signals proportional to the range and bearing of each simulated target. The actual position information was generated in rectangular coordinates. Electronic comparator and time-coincidence circuits allowed these signals to be resolved into video components for display on a PPI.

Human simulator operators, working from scripts, inserted track origins, courses, and speeds at these devices, and a supervisor at a scope monitored their output for accuracy (although exact measurement was not attempted). The inputs into the mock-up consoles were recorded (tracked) on another display and observed by the experimenter who was managing the session. Voice communications by intercom among the subjects inside the mock-up and over a radio link to a presumed fleet commander were recorded on a set of tape recorders. PPI scopes inside the mock-up were photographed during some of the studies.

## PRELIMINARY STUDIES

The first experiments in Project Cadillac were individual operator studies of surveillance activities. The most elaborate (Sinaiko, Lefford, and Taubman 1951) explored methods to be used in later studies for systematically varying stimulus inputs. Events of several classes—new targets, target disappearances (fades), reappearances, course changes—occurred with five "time densities"—one, two, three, four, or five events per minute. Probability of detection, it was found, decreased with increasing time density for each class and was lower in some classes (course changes and fades) than in others; "in a gross sense," the experimenters reported, "a hierarchy of kinds of discrete events has been established in terms of the difficulty for radar operator to detect them." Each of six subjects went through five 25-minute runs, in each of which time density increased every 3 minutes. Subjects had to perform all the surveillance tasks in addition to detection, such as plotting tracks—that is, target paths—and reporting track characteristics. Each run contained fifteen tracks with differing origins, courses, and speeds.

In a prior study (Lefford and Taubman 1950) four subjects, in four runs per subject, simply detected, tracked, and reported targets and their characteristics without any fades, reappearances, or course changes, the number of new tracks progressively increasing to fifteen during 30 minutes. Detection latency depended on the number of targets already on the scope face. Data were also obtained on operator accuracy and productivity in the nondetection tasks. One point of interest in these two experiments was the requirement for the subject to maintain his whole range of task behavior in a situation where the component task of detection was the principal dependent variable being measured. A brief study (Lefford 1949), at another location before the laboratory was available, examined operator accuracy in making unaided visual estimations of target bearing from the AEW aircraft with off-centered PPI presentations; less than 4% of the estimations had more than three degrees of error.

The real goal of the project was to conduct large-scale system experiments in surveillance and intercept operations, not these individual operator studies. The system and simulation equipment was ready by mid-1950. Practice sessions were held. But how to approach such system experimentation presented something of a problem. There were no precedents to serve as models. (The Harvard studies described in Chapter 3 were either virtually unknown to the researchers or regarded as inapplicable; they are not mentioned in any of the Project Cadillac reports.) Eventually, at the end of 1950, something was undertaken which today might be considered as the obvious first step, a listing of system variables. These were classified under three general headings (Chapman 1951): behavioral or procedural variables, personnel or environmental variables, and design variables. Approximately twenty were spelled out in the first category, with a varying amount of itemization for each variable, and ten in the second; design variables did not receive much attention, because in this program it was difficult to make major hardware changes within short time spans. This specification was essentially a description of what might be manipulated experimentally in a way that system or subsystem performance might be differentially affected. Conferences and discussions followed, and the revised listing served as the source not only of independent variables selected for the first experiments but also of requirements for experimental control.

Possibly one reason why it had been difficult to take this step earlier was the relative unfamiliarity—largely unavoidable—of virtually all the research people, at all levels, with man-machine systems, AEW operations, and multivariable experiments. In addition, it became clear that staffing and leadership of this kind of complex, large-scale research inevitably present challenges. Inexperience and lack of well-established guidelines seem unlikely to promote humility or harmony, especially if the professional linkage is tenuous between research managers and researchers, when they come from diverse disciplines with differing approaches to the task. Interpersonal discord within the project was accompanied by a considerable turnover of research personnel, who, when they achieved some degree of sophistication, did not necessarily remain long on the scene.

## FIRST MAJOR EXPERIMENT

The first two multioperator experiments were closely related. Both were concerned with the manner in which surveillance data were reported from the airborne CIC to the OTC, the officer in tactical command of the fleet which the AEW aircraft was helping to defend. The procedural variable was the communication channeling of reports. In an indirect procedure, each of the surveillance operators could report his detection and tracking data on an internal telephone network (intercom) to a status board keeper in the aircraft. The latter wrote the data on the display and a "talker" transmitted the data by simulated radio to the simulated OTC. Or in a direct procedure, each of the operators could report his data directly by simulated radio to the OTC. In this case the status board keeper monitored the simulated radio and wrote the reported data on his display.

In the first experiment, conducted by L. S. Rubin and H. M. Parsons in January, 1951, and reported later (Rubin 1954), another communications procedure variable was explored along with each of these procedures. In a "round robin" procedure all the surveillance operators monitored the communication channel and reported in turn; in a "selective switching" procedure, they were cued as to when to make their reports by signals from the status board keeper. Each of four surveillance operators searched for, tracked, and reported targets in a 90-degree sector (quadrant) of the radar coverage of 360 degrees. The status board keeper and talker performed as already indicated. A heightfinder operator provided target altitude information. A commander supervised operations and provided general tactical evaluations. An assistant commander at a scope monitored the entire radar coverage and the communications, and he arranged transfers of target responsibility from one surveillance operator to another when targets crossed sector boundaries. Thus, nine subjects worked together in the experiment. Measurements of individual operator performance were not isolated in the analysis of results; rather, the output of interest was the crew's.

Training which preceded the experiment included two hours of practice on each of the four communications procedures, and the subjects had had about forty-five hours of other kinds of practice with the equipment. There were sixteen experimental sessions, two per day, each lasting about 80 minutes; the duration varied somewhat so the subjects could not predict the termination. Without the subjects' knowledge, data were taken only from a 20-minute sample period which varied in its temporal position within the session; this restriction on data collection facilitated data processing. The variation, which had four values, constituted another independent variable to create a 4 X 4 experimental design, arranged in two Graeco-Latin squares to determine order of presentation of each value of each of the independent variables.

The organization of inputs was relatively intricate. Twelve simulated attacking bombers (targets) were introduced during the 20-minute sample period, together with six course changes and eight sector crossings; eight tracks were carried over from the earlier part of the session but faded as new targets entered so that the load remained at about twelve tracks. Thus, although variation in courses and other characteristics assured differences in the over-all pattern, difficulty of search and reporting remained approximately the same. This was true also of the 10-minute periods which preceded the sample period. The post-sample periods had much greater heterogeneity, hopefully to provide camouflage and because they could not affect performance during the sample period. The construction of the scripts to obtain a combination of variety and equivalence seemed a challenging task at the time, since the only precedents for guidance came from the two earlier individual operator experiments.

Among the measurements taken were the totals of all operator reports, totals of reports of various types, latencies of reports (intervals between events and the OTC's reception of the pertinent data), and accuracy of target position reports. In this and subsequent experiments, most of the data for analysis came from the voice tapes of intercom and radio reports. These tapes had to be transcribed and the contents tabulated, a lengthy but necessary process. It was not found feasible to process such data during an experimental run. In addition, latency data required comparisons between the times of recorded reports and actual times of events; and for accuracy data, comparisons had to be made between recorded reports and actual target positions, courses, and speeds.

Operators were able to produce more reports of target speed in the indirect, selective-switching condition than in the others; estimating speeds was believed to be more difficult than estimating courses or plotting target positions. For most of the measures, however, analyses of variance showed no statistically significant effects from the independent variables.

## SECOND MAJOR EXPERIMENT

The second experiment in surveillance data reporting, undertaken by H. M. Parsons in April 1951 and reported by Cusack and Parsons (1953), was more ambitious and also produced more statistically significant results, accompanied by a number of recommendations of importance to operational effectiveness. It will be described at some length, not only because of its complexity but also because the experiments which followed incorporated many of its techniques. Also, the best way to set forth the scope and complications of this kind of experimentation is to describe an experiment's features, even if these do not entirely engross the more casual reader.

There were seven independent variables with two, three, four, or six values per variable: communication procedure, number of surveillance operators, composition of the team, amount of input load, successive sessions, time during the experimental session, and input script. Teams of seven to nine subjects each, performing the same functions as in the preceding study, conducted surveillance operations during six sessions of 215–225 minutes each, one session per day.

As already indicated, the reporting or communication procedure varied between the direct and indirect methods investigated in the first experiment; each incorporated the "selective switching" procedure. Two, three, or four surveillance scopes were manned. The six available officers were randomly assigned to the commander, assistant commander, and surveillance operator positions to form three different teams, the composition of each differing from the others. The same three enlisted personnel always filled the other positions. Four input loads were designed, each for a 10-minute segment of time, varying in equal increments of presumed difficulty (see below). Each of the six sessions was divided into three blocks of 70 minutes each. The script variable was included, not to investigate differential effects of scripts but rather to present differing though equivalent inputs to the crews, in order to minimize any special learning effects. There were three input scripts, each containing the variations in load. After the experiment, statistical analysis showed that, as planned, they had no differential consequences.

Although there were only six sessions, each included twelve time segments from which data were obtained, since each of the three time blocks per session had four successive 10-minute segments (in which load varied from segment to segment, always in a three, one, four, two order). The four segments occupied the last 40 minutes of each time block. The time-block, team-composition, number-of-surveillance-operators, and script variables were related to each other through two Graeco-Latin squares, one for each communication procedure. One of these procedures was followed in sessions one, two, and five, the other in sessions three, four, and six; the team-composition and sessions variables were not completely orthogonal with each other. With such a design, the only interactions which could be tested statistically were those between load and number of operators, between load and procedures, and between number of operators and procedures. However, it was possible to pack a $2 \times 3 \times 3 \times 4 \times 3 \times 6$ complex into only six experimental sessions.

The method of constituting equivalent input loads differed somewhat from earlier techniques. The basic units of load for the surveillance operators were "target-minutes" and "events." A "target-minute" was defined as the movement of a target across the face of the radar scope for one minute, regardless of the target's speed or course. Two targets, appearing at the same time but fading after 4 and 9 minutes, respectively, would contribute 13 target-minutes to load; since appearance and fade were events, together the two targets contributed four events to load. The four load levels in the experiment, for target minutes and events, were 40 and six; 80 and twelve; 120 and eighteen; and 160 and twenty-four. Proportionality had to be maintained also among types of events. The first 30 minutes of each block, a "pre-sample" period, were divided into 10-minute segments which contained differing loads, but the loads were equivalent among the pre-sample periods in different blocks. The session always lasted 5 to 15 minutes after the last data-taking block. The subjects did not know that data were being taken only during certain portions of the session, nor was the nature of the input revealed to them beforehand.

As usual in this kind of experimentation, the advance preparation of the input scripts consumed much time and effort. The input was designed to convey "operational realism," in that within any session most of the targets appeared from the direction of expected attack rather than from many different directions. The latter pattern, which would have loaded the surveillance operators more evenly, had characterized the prior experiment and the individual operator studies. But the more realistic situation was selected so results could be applied to the operational situation actually faced by airborne CICs; the Navy officers indicated afterward that the aim was achieved. Although the same stimulus situation appeared repeatedly, this duplication was concealed from the subjects (it was hoped) by (1) rotating the attack pattern so it appeared from a different direction from one session to the next; (2) varying the order of the scripts from session to session; and (3) varying the inputs in the pre-sample and post-sample periods.

Each surveillance operator searched, tracked, and reported on targets either in a quadrant, a third, or a half of the total surveillance area, depending on whether the number of manned consoles was four, three, or two, respectively. In addition to the requirements on these operators, the assistant commander had to perform co-ordinating and supplementary tasks and the commander had to send tactical evaluations to the fleet, some of them in response to scripted requests.

### *Experimental Findings*

Surveillance report data were measured for production totals, latency, and accuracy as in the preceding experiment, and also for rate of reporting. The first three kinds of measurement were applied also to the commander's report. A substantial number of analyses of variance resulted in various statistically significant conclusions and accompanying recommendations.

The two communication procedures, it was shown, did not differ over-all according to production, latency, or accuracy measures, but the indirect procedure was more productive with heavy loads. With this procedure the same amount of information was reported faster between the aircraft and the OTC, leaving more clear-channel time. Accordingly, the indirect procedure was recommended for adoption. Nevertheless, the status board keeping involved in it constituted a bottleneck, so the experimenters urged that the display and the methods for using it be improved.

It made no difference in production, latency, or accuracy whether there were four, three, or two surveillance operators; and the amount of input load had no interaction effects. Accordingly, it was concluded the surveillance would not suffer if one or two of the consoles were used for interception control or became inoperable due to malfunction.

Surveillance capacity was saturated between the third and fourth levels of input load; report production increased as the load level changed from one to two and from two to three, but not when it changed from three to four. Further, as load density increased, surveillance operators produced proportionately the same number of reports of target courses and speeds and proportionately less information of other types. Another interesting finding was that although the percentage of detections decreased as load increased, detection latencies remained the same among the three higher load levels. As in the prior experiment, operators produced somewhat fewer speed reports than course reports, and the speed reports had longer latencies.

There was no decrement in the performance of the surveillance operators over the three and one-half hours of continuous operations, whether performance was judged according to production, latency, or accuracy, and regardless of the number of operators, the load level, or the communication procedure. Accordingly, it was concluded that the duty period of AEW crews could be at least three and one-half hours.

Marked differences between performances of crews indicated individual differences among surveillance operators but not among commanders. A program of personnel selection was advocated for airborne combat information center officers directed toward surveillance rather than command capabilities.

## FURTHER SURVEILLANCE EXPERIMENTS

Just prior to this second multioperator experiment, another study of individual operators (Veniar 1953) examined the surveillance performance of four of the Navy officers as subjects during variable-length sessions, averaging about four and one-half hours for five successive days. The subjects performed at the same time at the scopes in the mock-up, each receiving the same input; there was no operator interaction or action as a crew. An operator had to report on signal at 3-minute intervals during 20-minute sample periods consisting of the last two-thirds of each of nine 30-minute units during the session; the operator was signaled also at irregular intervals but the data then provided were not analyzed. Nine new targets were introduced; six were already on the scope and these faded during the period, so targets at any one time varied between six and eleven; events varied between three and six during any 3-minute segment. Production of reports during a session dropped about 1% in each successive half-hour. This decrement was traceable to course and speed reports, particularly the latter; there was no significant decrement in detection or tracking performance. There was no systematic change from day to day.

A third multioperator experiment (Rubin and Connolly 1954) was conducted considerably later to investigate the effects of different approaches to the heightfinding function on system performance. In the AEW CIC an enlisted man at a console manipulated the aircraft's heightfinding radar to determine the altitude of other aircraft, since the surveillance radar provided only azimuth and range data to the surveillance consoles. The operators of the latter had to get altitude data from the heightfinding operator to include in their reports of target information. They could do so in any one of four ways. Two were manual, in that surveillance operators got altitude data over the intercom. Either they could initiate requests, or the heightfinding operator could simply keep telling them what he had. Two others were electronic, involving the placement of a marker (tag) over the target blip at each console scope to designate its position. The electronic methods differed according to the manner in which priorities were assigned by the assistant commander to operator requests.

In addition to examining these four methods, this experiment varied the number of surveillance operators among four, three, two, and one; it varied input load among three levels—six, twelve, and eighteen targets; and it varied the individuals who acted as heightfinding operators among four enlisted men. There were forty-eight sessions of 8 1/3 minutes each. Data were collected concerning the number of altitude reports, course reports, speed reports, and position reports. This experiment extended the investigation of the number-of-operators variable in the second experiment by including a condition in which there was a single operator. It also took a different approach to session arrangement and simplified the data gathering and analysis tasks. In brief, the results favored the electronic technique for obtaining altitude information, with a preset, automatic method of determining priorities. A single surveillance operator was relatively ineffective for all four categories of reports, and two such operators produced fewer course and speed reports than the three-man and four-man teams. The analyses of variance also showed significant differences among the individual performances of heightfinding operators.

A final multioperator experiment concerned with surveillance was conducted still later in the Project Cadillac program (Schapiro and Guastella 1955). Since the report has remained classified, none of the results will be described here. There were five independent variables. The number of surveillance operators was varied among four, three, two, and one. Two radio channels were compared with one. Status displays were either an improved common display or an individual display for each surveillance operator. Input load varied among four levels, and there were three different "executive teams." These variables were organized through a factorial design to yield 192 combinations and permit the analysis of the effects of all interactions between variables. Four 80-minute sessions were held on each of twelve days, with each session divided into four 20-minute segments. Data were collected from the 7th through the 16th minute of each segment. Loads consisted of various combinations of target-minutes and events, as in the second multioperator study. The types of measurement were also much the same, except for some additional measures of the performance of the "executive team." The total crew was somewhat larger than the crews in the previous multioperator studies, with new manning requirements. In order to form three distinct executive teams of three officers each, two Air Force officers participated along with six Navy officers and a chief petty officer. Two Air Force enlisted men joined seven from the Navy in manning the positions of status board keeper, talker, heightfinding operator, and plotter. Each executive team served for four consecutive experimental days.

## INTERCEPTION-CONTROL EXPERIMENTS

In the interception-control phase of the Project Cadillac program, three individual operator experiments were conducted to determine the effects of load parameters on the performance of an interception controller and to generate a general measure of task difficulty based on these parameters. Attention will be given to this work because it made possible the design of the inputs for a multioperator experiment which followed. A fourth individual operator experiment examined the usefulness of two manual devices for aiding the derivation of intercept vectors, that is, courses to fly to make an interception. The last experiment—a multioperator type—combined the surveillance with the interception function. Otherwise, in all the interception control studies it was assumed that the surveillance operations had already been accomplished, and there was no prior surveillance activity similar to what had been investigated earlier in the program; the interception function was examined as an isolated subsystem.

In every individual operator study one Navy officer served at a time as controller, and referees judged whether or not his attempted interception was successful according to certain criteria. Data were obtained also on the time between the hostile aircraft's detection and its interception (penetration time) and on communications. The first experiment (Fox and Connolly 1953), with six subjects each of whom underwent thirty trials, had four independent variables: the subjects, the number of targets (attacking aircraft), their speeds, and a temporal load condition. Target totals were one, two, three, four, five, or six. When there was more than one, their approaches to a "protected line" were simultaneous. There were two possible speeds. The temporal factor consisted of the time during which a target could be usefully intercepted, that is, between the earliest possible point of interception and the time its track would cross the protected line. This available time was two, four, six, eight or ten minutes. The controller had to direct his interceptor aircraft from a simulated airborne position to make a cut-off intercept, one interceptor per target. That meant that he had to track both the targets and the interceptors on his PPI display, make visual estimates of the proper intercept vectors, and transmit these by simulated voice radio to a simulator operator, who would insert them as headings to fly into the equipment by which radar returns from the interceptors were simulated. The penetration distances of attackers significantly increased and the percentage of successful interceptions significantly decreased as a function of the increasing number of multiple interceptions required and the decreasing amount of time available.

These two load parameters were combined in a ratio—time available divided by number of interceptions—to form an index of interception difficulty. In the next experiment (Connolly and Page 1953), which had three parts or subexperiments, this index was explored further. In the first part, two other load parameters were investigated to determine whether they had any differential influence over the effects of time available. The first parameter was tested in nine problems of four interceptions each, four subjects encountering each problem twice; the nine problems resulted from combining three of the parameter values with three times-available. The second parameter was examined at the same time by random assignment within problems. The second part of the experiment investigated whether different values within the same ratio would produce different degrees of control success; they did not. Seven sets of values included seven multiple-interception loads, from two to eight, each presented three times to three subjects. In the third part both ratios and values within ratios were varied in seventeen combinations, and the subjects of the second part faced each of these twice; other aspects remained the same. The results of the second part were confirmed, in that differing values within the same ratio had the same effects. Further, differing ratios, as might be expected, did have differential effects, generating monotonic, curvilinear functions for the criteria of both penetration time and attackers destroyed (kills). This result made available, the experimenters noted, "a carefully determined and practical scale of difficulty by which intercept 'stimuli' can be drawn up for future experimentation upon the airborne CIC system, in its parts, or as a whole."

There remained one further experimental check. What would happen in multiple interceptions if the target inputs were not simultaneous but simply overlapped in time, that is, were temporally displaced? In a third individual operator experiment (Fox and Page 1954), performance of three controllers was studied at four levels of overlap, in thirty-six trials per controller. Multiple interceptions totaled three, five, and seven, and time available was 2, 4, and 6 minutes. In this, as in the second experiment, the interceptions were not the simple cutoff type; at a distance no more than four miles behind it, with certain limits on relative heading and bearing of the target, the controller had to turn the simulated interceptor to approach the target from behind (rear hemisphere attack). It was found that percentage of overlap did not significantly affect the success of interception control.

### *Multioperator Experiment*

The multioperator experiment on interception control followed (Page and Connolly 1954). Its primary purpose resembled that of some of the multioperator surveillance experiments, namely, to determine how many operators (and consoles) were needed under various levels of load; in this case the concern was about the number of interception controllers rather than surveillance operators. Accordingly, one of the independent variables was the number of controllers working together: three, two, or one. Load was varied using the index of interception difficulty established in the individual operator studies, with three levels derived from differences in target load, time available, and time overlap. In addition, there were three distinct teams of noncontroller personnel. The controllers had to defend both the AEW aircraft and the fleet, so there were two protected lines. The maximum interception load at any time was six. Eight alternate forms of the inputs for each load level were created by rotating the expected direction of attack and varying two temporal patterns. Other load parameters resembled those in the last two individual operator studies, as did performance measures and requirements. The order of the nine conditions resulting from the three load levels and three controller totals was randomized among the noncontroller teams. Five enlisted men were systematically rotated through two referee positions. There were twenty-seven sessions, one for each system condition. The experimental design was factorial. Analyses of variance revealed that interception-control performance was degraded by high difficulty levels, although co-ordinating operations by the noncontroller team were not. Thus, this experiment showed that the performance of controllers working together in a team resembled their performance in operating singly.

The final individual operator study (Connolly and Capuano 1954), which evaluated two interceptor vectoring aids, produced an unexpected but revealing result. One of the aids was a chart, the other a disc, a "circular computer," and both could be manipulated by controllers to make some of the necessary geometrical calculations to establish what the interceptor's course should be. A rear hemisphere tactic was required rather than a cutoff. The interesting experimental outcome was that in this situation, at least, determination of course "in-the-head" or "by-eye" proved superior to either aid.

The concluding experiment of the Project Cadillac program combined the surveillance subsystem with the interception subsystem (Connolly, Page, and Veniar 1955). It had a 3 X 3 X 2 factorial design with two replications. As the principal independent variable, the number of surveillance operators and number of interception controllers were combined in three different ways: two and one, two and two, and three and one. Further, there were three load levels and two crew compositions; each crew had the same subjects but in different positions. Results cannot be described here since the report has not been declassified. In simulation and measurement technology this experiment resembled those which preceded it.

## OVERVIEWS

Although the picture which has been presented of Project Cadillac suggests a systematic and coherent program, it should not be assumed that it followed some master plan adopted at the outset. In a pioneering effort of this nature, the researchers necessarily felt their way along, especially early in the game. Before the first multioperator experiment formally started, forty-four practice sessions helped the researchers understand the system they would investigate; some of these sessions also served as rehearsals for that first experiment. The practice sessions were essential for familiarizing the experimenters with their simulation and data-collection procedures, as well as experiment management. To those involved, the research seemed like the exploration of unknown land, without map or guide. Some idea of the conceptual organization which had to be accomplished can be gained from R. L. Chapman's *Experimental Methods of Evaluating a System: The Airborne C.I.C.* (1951)—the first published attempt to map out the domain of complex man-machine system experimentation. Among the topics it analyzed were "problems in research methodology," "the stimulus complex," "the system variables," "the problem of measurement," and "the meaning of human engineering."

From the preceding descriptions it should be clear that much innovative toil went into the design of the simulation inputs, the experimental designs, and the determination of suitable measurements. There was emphasis throughout on rigorous experimental control, on statistical analysis for assessing the significance of results, and on reporting the experiments in such a way that they could be replicated. These emphases reflected the influence of more traditional psychology experiments. It might be fair to infer that the researchers were so greatly concerned with experimental method that operational problems, though by no means disregarded, aroused relatively less interest. If there was such a bias, it was probably inevitable in a program that was pioneering in methodology, especially in its early stages.

It is not known to what extent experimental findings were translated into operational procedures or equipments. As noted earlier, after the program had been concluded a summary report was issued and could have supported such a transfer of learning, although the report's distribution was limited by security classification. Many of the experimental results dealt with surveillance and interception-control capacities of operators under various loads. These results could have been exploited to determine how many consoles to place in the airborne CIC and how they should be distributed between its two major functions. Such extrapolation might have removed any risk of over-equipping the aircraft, but they would have necessitated additional inquiry into the degree to which the capabilities of Navy officers performing surveillance and interception-control functions in the operational milieu matched those of the subjects in these experiments. If the experimental data had reached the generally accessible, open literature, they might also have contributed general knowledge about human capabilities in complex tasks.

One of the human engineering by-products of Project Cadillac was an increasing concern about the displays available to the airborne CIC's operators, and about displays in general in complex systems. This was not a unique by-product of such experimentation. Unfortunately, as has been the case in other, similar programs, with one exception (Parsons, Sinaiko, and McDonald 1952) it did not yield any separate reporting, so the treatment of display problems remained relatively unknown. Further, it is not known whether a number of display recommendations and innovations originating within the program were ever adopted for AEW operations by the Navy.

It is no trivial question whether this seven-year program yielded useful and exploited products. It was not an inexpensive undertaking. At times as many as nineteen persons staffed it, including engineering and junior personnel; and although, as mentioned earlier, there was a certain amount of internal turmoil as well as gaps in technical understanding between researchers and their management, much hard work was done.

Certainly one of the most striking aspects of the program was its attempt to experiment on a system while the system was still being developed, in order to influence its design. Although this may not be the only objective of man-machine system system experiments, it is an important one. Whether or not the attempt actually influenced the system design, to have made it at all was a break-through.

Possibly some of the findings might have been exploited by the Air Force which itself later operated similar AEW aircraft for air defense surveillance off the Atlantic and Pacific coasts; however, there is no evidence that Air Force AEW personnel were familiar with Project Cadillac. In fact, it would appear from the AZRAN Study (Chapter 11) that in the operation of Air Force AEW aircraft, insufficient human factors attention was given to communication procedures, displays, and arrangement of consoles and other operator facilities—some of the things which Project Cadillac investigated for the Navy.*

The Air Force did benefit from the program in a serendipitous fashion. A number of the Project Cadillac researchers undertook similar man-machine sys-

*However, for the pre-SAGE operations of Air Force aircraft the Project Cadillac reports were reviewed by the Aero-Medical Laboratory at Wright Air Development Center in 1954–55, according to G. J. Rath (personal communication).

tem experiments in Air Force-sponsored programs after they left the project. These included Chapman (Chapter 8); Parsons (Chapter 7); Sharkey (Chapter 6); and Connolly (Chapters 6 and 16). Others did the same for the Navy, Army, and other organizations; they included Sinaiko (Chapters 5, 12, and 21), Cusack (Chapters 9 and 27), and Veniar (Chapter 9). In many cases it is doubtful that these other programs could have been undertaken if some of their key personnel had not acquired, in Project Cadillac, the know-how to conduct this kind of complex, simulation-based experimentation. If this is indeed a fact, then perhaps the most significant accomplishment of Project Cadillac was to serve as a training ground in a new domain of applied science.

# 5

# Navy Laboratories, Facilities, and Contractors

Chapters 3 and 4 have shown the Navy's role in sponsoring man-machine system experiments concerned with shipboard combat information centers during World War II and with airborne CICs subsequently. The Navy supported more of this kind of research in the following years, and there was an extensive and pioneering early program at the Naval Research Laboratory (NRL). This work will be reviewed here, together with later experimentation and related research and simulation.

## NAVAL RESEARCH LABORATORY

Eight man-machine system experiments of varying scale were performed in a simulation laboratory at the Naval Research Laboratory in 1952–56. This laboratory was envisioned as a means for investigating CIC problems as early as 1950 by Captain (later Rear Admiral) C. Laning. Following a University of Michigan survey of World War II experiences of many individuals, an initial version of the laboratory was established at NRL and, beginning in 1951, was operated under contract by Tufts University. Most of the Tufts personnel moved to the university in Medford, Massachusetts, in 1953. A year later the remainder became the Systems Branch in an NRL Applications Research Division formed in place of Tufts to direct experimentation. Also in 1954 a CIC Facility Branch of NRL took shape at its Chesapeake Bay Annex, in Randle Cliffs, Maryland, where a number of tests were conducted, some in a simulation context.

In a shipboard CIC, information about aircraft (and other ships) was obtained from radar signals on plan position indicator (PPI) displays by trackers (and from radio reports from other ships) and telephoned by the trackers to plotters at a large vertical display. The plotters inscribed, with grease pencils, aircraft and ship paths on the display, thus aggregating the tracker information so that the officers in charge of the CIC could get a complete picture of a hostile attack and friendly elements. Much effort was devoted to improving such displays, so radar information about aircraft would be presented faster and more accurately. Various proposed solutions to the display problem figured in seven of the NRL experiments.

The first experiment has been described briefly by Sinaiko (1954). It examined a proposed expansion of the manually plotted display then in use by the Navy. Six-man teams, composed of Navy enlisted personnel as subjects, received target information over shipboard-type (sound-powered) telephones from other personnel reading data from scripts. The rates of information transmission by the latter could be systematically varied. The vertical display was photographed periodically, and analyses of the photographs indicated the accuracy of plotting; the study also measured the upper limits of a plotter's capacity.

The second experiment (Weiner and Sinaiko 1953) tackled a different aspect of the same general problem. A combat information center in a real-life Navy aircraft carrier would receive radar information about approaching hostile airplanes by radio from other ships in a task force, such as surrounding radar-equipped destroyers, as well as from its own radar. Some of the destroyers' position reports might pertain to a target which the carrier's own radar had not picked up, some might pertain to the same target but with differing data, and some might pertain to the same target with the same positional data. It was important to prevent false targets from being displayed on the vertical plot as a consequence of differing positional data. A horizontal filter plot was proposed to help eliminate such false targets. It would receive both radioed data and the carrier's own radar data about a target. It would correlate the data, passing along to the vertical plot the radioed data which represented only the hostile aircraft that the carrier's own radar had not detected; the other data would be filtered out.

Following six practice runs during a week of training, twelve experimental runs of 32 to 51 minutes each were conducted over a two-week period with a filter plot officer (FPO) and twenty-two Navy enlisted men acting as trackers, filter plotters, tellers (to the vertical plot), and plotters at the vertical display. Subjects rotated through task positions. The FPO position was filled on the basis of psychometric assessment to predict leadership and team compatibility characteristics. The experiment included two independent variables: input load, which was either light or heavy, and filtering procedures, which took several forms. The light load built up from zero to sixteen targets in 15 minutes, the heavy load from zero to twenty-four in 10 minutes. The filter plot officer was or was not required to make all filtering decisions; and the decisions were or were not relayed back to the trackers so that the latter could alter their reports. The trackers read data from scripts. Full loads were carried for only 15 minutes, but sessions lasted longer to preclude the effects on subjects that occur in expectation that a session is about to end and to disguise the part of the input from which experimental data were gathered.

The amount of filtering was measured by determining the number of "scrub" (eliminate) orders given by the filter plot officer or relayed by the tellers to the vertical plotters, and also by the number of tracks filtered at the vertical plot. Teller performance was also measured. The activities of the FPO were recorded systematically and his performance seemed to be the best single criterion of over-all system effectiveness. The tellers constituted a critical linkage. No statistically significant differences resulted from the various filtering procedures. The heavy load appeared to saturate the filter plot display and thereby reduced the proportion of targets filtered, delayed the relay of information to the vertical plot, caused failures to detect new targets, and degraded the maintenance of up-to-date tracks.

In later years, Sinaiko (1962) referred to the way in which data were gathered in this experiment as "a classic of inefficiency." He added:

> We took pictures. And we discovered two horrible things. First, nearly all of the vertical pictures we took, i.e., looking down over our subjects' heads and shoulders at the plotting table, were obscured by the subjects at work. Also most of the photos of a large, upright plotting board were of such a poor quality as to be almost beyond recognition. Second, our cameras were faithful but stupid; they uncritically recorded everything they saw. Needless to say, those of us who had to reduce the data had many hours of headache and eyestrain.

*Experiments on Mechanizing the Input to the Display*

Three studies evaluated the usefulness of introducing several proposed forms of automation to mechanize the transmission of data between plan position indicators and the vertical display. In the first of these (Scott et al. 1953), the mechanical linkage between PPIs and the vertical display was a camera and projection device, the Land Polaroid apparatus. A photograph of a PPI was taken and developed every 15 seconds and photographs were projected successively through twelve projectors to create tracks on a vertical surface, where plotters pasted sticky markers beside each newly photographed and projected radar signal (blip) to identify the track which it led. Thus, the trackers were eliminated and the vertical display plotters assumed some of their functions. This apparatus and the conventional manual system were operated by a dozen enlisted men as subjects at the same time and with the same simulation inputs in two areas separated by a visitors' observation area. This arrangement helped compare the two systems. There were eighteen runs, twelve capacity-test runs and eight operational test runs; the last included friend-or-foe identification and interceptor-request functions as well as surveillance. With heavy input loads the mechanical linkage produced faster plotting and more data, but also bigger course and speed errors and considerable clutter on the vertical display; its superiority was judged marginal.

The second study (Sinaiko et al. 1954) compared three systems. These were the conventional manual system, another camera projection device, the Kenyon Repromatic TPPI Camera-Projector (XW–5), and the Miller Optical Projection System, better known as "Mink," which had been developed at the Control Systems Laboratory of the University of Illinois (see Chapter 12). The area coverage and number of PPI scopes and plotters took two forms for the manual system, so this could actually be regarded as two systems in the experiment. The Kenyon apparatus projected its photographs to a large horizontal plotting board where plotters, located around it, marked each blip with a special, luminous grease pencil, as they would mark blips on a PPI, and added other information; then an image of this plotting board with its grease pencil tracks was projected to a vertical surface. The Mink was an electro-optical device presenting a horizontal, partially mirrored image from a single, large, upside down PPI. Plotters placed color-coded chips on radar blips instead of marking them with grease pencils. Mirror reflections of the chips were projected to a vertical surface.

Each set of equipment was emplaced in a separate partitioned part of the NRL laboratory, and two arrangements were run at the same time with the same simulation inputs. Thirty enlisted personnel subjects were assigned among the four arrangements for the course of the experiment; three CIC officers were rotated among them. Two input problems had a high target density—up to fifty targets at one time, and high target speeds; two others were easier, with approximately twenty target tracks and slower speeds. Each equipment arrangement encountered each level of problem load. In addition to number of raids designated and tracks held for 5 minutes, measures included time delays, intercommunication frequencies, and several ratings not only by the CIC officer subjects but also by visitors. Superiority of the Kenyon over the manual system was marginal at best, whereas the Mink outperformed all the others, especially in its capacity to delete indicators of tracks that had faded (disappeared); it was easy to remove the chips. One interesting finding was that the ranking of the three systems according to objective performance data was just the reverse of their ranking according to subjects' ratings.

However, the Navy had some misgivings as to what would happen to the chips on the Mink's horizontal surface when a ship rolled. So the Control Systems Laboratory came up with the "Sea Mink," in which a coating of silicone oil held the chips on the surface when it was tipped. Sinaiko et al. (1956) evaluated the Sea Mink experimentally, not for its reaction to a rolling ship but to determine whether, with a smaller surface and other design differences, it was as effective as another Mink version developed for the Air Force. In addition to comparing the two pieces of equipment, they varied the number of trackers between one and three with the Sea Mink and one and four with the Air Force Mink. Two naval officers, two petty officers, and six enlisted personnel were assigned among the four experimental conditions, each of which had twelve runs of 35 minutes each, with seventy-five targets in a run and a maximum of fifty at one time. There had been twenty-four training sessions. The blip/scan ratio was unity, that is, the radar "saw" every target with every antenna rotation, and no radar noise was introduced, so the simulated radar presentations lacked the degradation that the systems would encounter in the real world. Measurements were obtained for time delays, accuracy, and amount of information displayed. The results of this experiment have not been declassified. In a second phase of this experiment, the Sea Mink and a regular PPI were compared as instruments for conducting interceptions. Six subjects manned each set of equipment in a total of twelve runs involving 264 raids.

In all three studies, radar signals were simulated at PPIs by a device developed at Lincoln Laboratory (see Chapter 6), where a similar investigation of such mechanization had been undertaken for air defense surveillance in Air Force operations. This device was a flying spot scanner which converted light spots on 35-mm. film to electrical signals which were transmitted to the PPIs to represent radar signals; each frame of the film as it was fed through the scanner contained light spots positioned to represent the ranges and azimuths of aircraft detected during one rotation of a radar's antenna. Thus, the film automated the simulation, replacing the reading of a script and manipulation of switches and knobs; however, the film's contents reflected the same kind of design of simulation input.

Investigators at the Chesapeake Bay Annex (Plowman et al. 1956) also examined the Sea Mink and the shore-going version, which in this study had been renamed "Sky Screen." Three versions of the manual method and four of a summary display technique in the experimental Electronic Data System (EDS) were investigated at the same time, also with simulation input based on twelve scripts of four to twenty-five targets. Subjects were six officers and thirty enlisted personnel.

*Investigation of Transplot*

An alternative way to improve the PPI vertical display linkage had been evolved at the Lincoln Laboratory. It required no new equipment such as cameras and projectors. Instead, it was based on an ingenious conjecture. If the principal problem in the conventional manual system lay in the need for the tracker at a PPI to encode blip positions into co-ordinates and to telephone these to a plotter who then had to decode them back into positions on his display, why not simply eliminate the encoding-telephoning-decoding requirements? This could be accomplished by placing the PPI and tracker next to the plotter behind the vertical display. The plotter then could look at a tracker's markings on a horizontal PPI and pencil these at the same relative positions on his own larger vertical display. This approach, called "transplot," merely required placing equivalent rectangular grids for positional guidance over each surface and the reversal of east and west on the PPI to match the orientation of the back-plotted vertical display.

Sinaiko et al. (1955) experimentally compared three versions of transplot and two versions of the conventional manual method. The latter consisted of (1) a single tracker-plotter team handling all 360 degrees of radar coverage and (2) two teams, each with a PPI and each covering 180 degrees. Two of the transplot versions were similar, and in a third each of two operators acted as both tracker and plotter for 180 degrees, with each having his own PPI.

Every arrangement had a CIC officer responsible for over-all functioning, a radar control officer supervising the crews, and a status board keeper who computed and posted target courses and speeds on a display adjacent to the vertical plot. Subjects were four commissioned officers, three petty officers, and eighteen seamen. The commissioned officers rotated among arrangements, while the others were permanently assigned to a particular one. There were two independent variables in the simulation inputs, transduced by the flying spot scanner device. A heavy input, lasting 50 minutes, carried fifty tracks with a maximum of twenty-seven at one time; they were preponderantly high-speed tracks with some "split" tracks. A light input, lasting 40 minutes, carried forty tracks with a maximum of twenty-one at one time with fewer high-speed tracks and no splits. The blip-scan ratio was either unity or a lower, varying ratio—that is, a target's echo did not appear each time the rotating radar antenna scanned it. A third variable was the presence or absence of synthetic "sea return," the noise generated in radars by reflections from the ocean, simulated by means of another device. Factorial combinations of the three input variables each with two values and the five PPI-vertical plot arrangements were introduced in ninety-six runs, with two or three runs per condition.

Among the measures taken were the proportion of input targets actually plotted on the vertical display; latencies in plotting them there; latencies in making track designations there; latencies in detections of fades and course changes; plotting rates; accuracy of plotted positions and of course and speed estimations; and production of tracks and positional plots. Also, the CIC officers during every run rated the PPI-vertical plot arrangement for "system stability, tactical utility, and display appearance" on five-point scales. Most results favored the four-man transplot arrangement, generally with statistical significance; the two-man transplot arrangements did about as well as the four-man conventional method and out-performed the two-man conventional method.

In conjunction with the transplot experiment, a small study was conducted to compare the telephone transmission of data from trackers to vertical plotters with telephone transmission of data from the vertical plot to a summary display at a remote plotting station (Becker et al. 1956). The information being plotted on the vertical display in the transplot experiment was "told" by a teller to a plotter at a similar display in another room, much as such information would be relayed from a ship's CIC to another station. In twenty-four 40-minute sessions eighteeen subjects (Navy enlisted men) rotated through the positions in the simulated CIC and remote locations. In this experiment it was found that more information had been lost in the tracker-plotter link in the transplot CIC than in the teller-plotter link.

*Investigation of CIC Operations*

The most ambitious of the NRL experiments came in 1956, shortly before the program began to phase out (Chauvette, Sinaiko, and Buckley 1957). It investigated the entire air defense operation of a small-scale combat information center similar to that on a destroyer or destroyer escort, including the control of interceptor aircraft as well as detection and tracking of hostile aircraft at a PPI scope and plotting on the vertical display. It was small-scale in that there was only one tracker-plotter team, and only one air control officer to conduct interceptions. The simulated system was the conventional manual one in shipboard use, but one of the purposes of the study was to obtain baseline data about this system to help evaluate future semi-automatic systems. Although direct comparison of system performance was not the prime intention, such a comparison was made in a later study of an automatic system (see Chapter 12). The main objective was to find out how well measures of subsystem performance in a CIC were related to end-measures or total system performance. (No significant relationships were found.) Another purpose was to get information about the required activities of a CIC commander (such as his decision-making) and the differential effects of the parameters of an attack on these.

Two naval officers, with the ranks of commander and lieutenant and widely differing experience, served alternately as subjects in the role of CIC commander. Two ensigns alternated as air control officer (ACO), who was the intercept director. Two petty officers and six radarmen were the other subjects, changing positions to constitute eight combinations or crews. There were sixty-four runs of 37–58 minutes each, based upon an eight-by-eight Latin square design in which the rows were the eight crews, the columns were eight blocks of eight consecutive runs each, and the Latin letters were eight factorial combinations of the two commanders, two ACOs, and two input scripts which differed according to the direction from which one-half of the hostile tracks originated. Tests of statistical significance were applied to the results.

Each run had a 30-minute portion from which data were obtained and in which load was equivalent. Warm-up periods and terminal periods varied in length to conceal the basic input pattern from the subjects. In addition to the difference in directional origin of one-half of the tracks, the entire input was rotated through five orientations as a further method of disguising the equivalence of inputs between runs. The hostile aircraft were simulated by means of the flying spot scanner device, and the controllable interceptor aircraft were simulated with 15-J-1c target generators which were manned by Navy enlisted personnel trained to function as though they were pilots, operating their devices in response to commands from the ACO over a simulated radio. The 15-J-1c closely resembled the simulation equipment which had been used in the Cadillac Project (Chapter 4).

All interceptions were monitored by experimenters in a control room observing the synthetic radar signals of both hostile aircraft and interceptors on a PPI scope which displayed the same data as the PPIs in the experimental area used by the tracker and ACO. The experimenters refereed the outcome of interception attempts, applying a criterion of success based on the one used operationally in fleet exercises, but modified slightly to increase the number of critical tracks and the total of "tally-ho's" (radio code that the target was shot down). There was no objective method against which the referee judgments could be compared.

This experiment continued for approximately four weeks with an average of five laboratory runs each day. A week's training, including practice in the simulated CIC, preceded the experimental sessions. Although the results are still classified, it can be noted that a great many kinds of measures were applied. These included, as surveillance subsystem measures, time delays in detecting and displaying new targets, plotting accuracy, plotting rate, and accuracy of course and speed estimations. A number of over-all or "single-payoff" criterion measurements were obtained to compare with the subsystem results: interception attempts, interception attempts on critical targets, interceptions of all targets, interceptions of critical targets, penetration distance of all targets tally-ho'd, penetration distance of critical targets, and estimates of closest point of approach.

Various aspects of the experiment diverged from realism, some making the tasks less difficult, some making them more so. The former included a unity blip/scan ratio and absence of noise, the absence of non-air defense CIC functions such as antisubmarine warfare and navigation, assumptions of a stationary ship operating alone, omission of altitude and weather considerations, unlimited fuel and ammunition in the interceptors, and omission of the launch phase of interceptors. These missing aspects suggest the degree of complexity required for a complete simulation of an air defense situation. The features which made the subjects' tasks more difficult included the limited number of operating positions and a very heavy input load even for a fully equipped CIC.

According to Sinaiko (1962), in this experiment five years of experience resulted in more efficient data collection, especially compared with what had been done in the filter plot experiment described earlier. In this "single-payoff" study of a single measure of performance, Sinaiko wrote:

> We were interested in the relationship between the quality of a visual display and the goodness of decisions made by a CIC officer charged with allocating interceptors to hostile air targets. So, we observed only the display in question, i.e., we used humans trained to copy preselected tracks, and we recorded on forms only the behaviors of the decision-maker which were related to our purposes. In spite of this carefully preplanned effort to collect only selected, relevant facts, it took well over a year to reduce and analyze the data preparatory to publication.

The NRL researchers felt that despite its various limitations, due to the modest laboratory facility and equipment, this experiment showed the way to perform this type of experimentation on future systems, such as the Electronic Data System or the Naval Tactical Data System. But their projections were somewhat optimistic. Theirs was the first and last Navy laboratory study of combined shipboard CIC functions which could be strictly defined as an experiment.

### *Chesapeake Bay Annex*

The Electronic Data System was a set of radar target tracking and display equipment under development by the Naval Research Laboratory. Automaticity was introduced into the tracking function and PPI vertical display link by means of a conducting glass overlay on the PPI scope and a pencil probe which could position electronic markers, whereby $x$ and $y$ voltages could be generated to indicate a co-ordinate position and transmitted to a semiautomatic vertical plot or to gun-director stations.

Several studies of component functions of the EDS were conducted at the substantial Chesapeake Bay Annex facility (Nichols and Karroll 1955; Irish et al. 1955; Nichols and Plowman 1956). In their reports, descriptions of experimental designs and controls are conspicuous by their absence, and percentages and qualitative opinions tend to outnumber objective data in the results. It seems fair to characterize the Chesapeake Bay Annex facility as devoted primarily to demonstration.

The first study investigated the detection and tracking of actual, individual aircraft, seventy-seven propeller and sixty-one jet types, flying over Chesapeake Bay; both the "breadboard" EDS and the manual system were evaluated. However, this facility also had simulation equipment, both the flying spot scanner and the 15-J-1c devices. In fact, as a facility it was more elaborately equipped than the main NRL laboratory and was supplied with naval officers to direct the studies and officer and enlisted personnel for subjects (but only one individual to provide technical advice concerning experimentation). The second study was based on both types of simulation as well as actual aircraft. It investigated EDS surveillance operations with two manning configurations, two input loads, and two blip/scan ratios—unity and reduced—in twenty runs of 50–70 minutes each. The third looked at EDS tracker performance and the use of the electronic markers compared with grease pencil tracking and with long-persistence phosphors in PPI scopes. Another study tested broad band blue lighting in the CIC mock-up.

## NAVY ELECTRONICS LABORATORY AND NTDS

The Naval Tactical Data System (NTDS) is a computer-centered system developed by the Navy Electronics Laboratory (NEL) and currently installed on a number of Navy carriers and other ships for augmenting the capability of CIC personnel in the surveillance and intercept-direction functions of air defense ("anti-air warfare" in Navy parlance). Its experimental version underwent testing in 1959–60 at NEL's Applied System Development and Evaluation Center (ASDEC), and subsequently the system was service-tested. ASDEC was equipped with prototype NTDS consoles and other equipment to simulate an NTDS CIC and radar-sensed aircraft. As in the case of Cornfield (see Chapter 12) and SAGE (see Chapter 11), computer programs for the NTDS computer could generate simulation inputs within the computer itself, and the computer could record its own system performance and reduce the data.

ASDEC has been a relatively elaborate simulation facility. Although much of the NTDS evaluation there was undertaken with naval personnel operating NTDS equipment, this evaluation would more appropriately be called testing, in terms of goals, methods, and measures, rather than experimentation. ASDEC was also used for shore-based training of Navy personnel for NTDS operations on board ship. More recently, a new facility has been created for this purpose less than one-half mile from ASDEC in San Diego, at the Fleet Anti-Air Warfare Training Center. This is the Tactical Combat Direction and Advanced Electronic Warfare Trainer Complex (TACDEW), a large, multioperator, computer-based simulation facility for training Navy personnel primarily in NTDS operations. A similar facility has been established at the Fleet Anti-Air Warfare Training Center in Norfolk, Virginia. When the material for this chapter was being gathered, there was no indication that either will be the site of experimentation.

The ASDEC facility has also been used for simulation-based evaluation of a proposed Small Ship Command Data System for antisubmarine warfare and more recently for certification testing of the NTDS for antisubmarine warfare, but experimentation-oriented Navy scientists at NEL with particular interest in man-machine interfaces have not participated in this or the other ASDEC-centered evaluations. Although considerable component experimentation has been done in the human engineering group at NEL, the only man-machine system experiment resembling the studies described in this review was one concerning the manual CIC. This study dealt with a proposed innovation in the transplot arrangement that the NRL researchers had investigated earlier.

The NEL experiment (Coburn 1960) resembled that at NRL, except that a polar co-ordinate grid was placed over each display, instead of a rectangular grid, and a mirror arrangement helped the plotter see the track marks he had to copy.

The tracker simply marked pre-dotted tracks on a problem sheet; a color filter prevented the plotter from seeing these tracks. Twelve two-man teams, composed by pairing each of four subjects with every other, operated both the transplot and the manual ("telplot") method, in which a tracker at a PPI telephoned to the plotter. The plotting rate was much higher for transplot, and so was accuracy, as rated by four judges who compared photographs taken of the vertical plot with the marked problem sheets.

Although the Navy did not conduct man-machine system experiments on NTDS CIC functioning at NEL, it did participate further in man-machine system experimentation by helping to fund a contractor's experiment with a computer-based CIC (Chapter 12) and by providing experimental subjects to another contractor's self-funded, CIC-oriented experimental research (Chapter 18).

## TRAINING DEVICE CENTER AND OFFICE OF NAVAL RESEARCH

Various organizations have carried out a number of experimental studies in an anti-air warfare training research program which was sponsored by the Office of Naval Research (ONR) and the Training Device Center (TDC) and which was thought to be germane to shipboard CIC operations. For example, a set of ONR-sponsored studies at Princeton University in decision-making is described briefly in Chapter 21. In addition, external to this training research program, ONR has funded simulation-based experimental studies of individual operator performance with equipment for the control of submarines, in Project SUBIC at the Electric Boat Division of General Dynamics Corporation (e.g., Blair and Kaufman 1959), and at Honeywell, Inc. (e.g., McLane and Wolf 1965).

One of the TDC-sponsored studies in the anti-air warfare training research program has been conducted at the Electric Boat Division and has dealt with decision-making (see Chapter 21). Another study in that program, at Ohio State University, is described below. There have also been TDC-funded simulation-based experimental investigations of individual operator performance in other programs. One was an extensive study of simulation requirements for sonar operator training, at Human Factors Research, Inc. (Mackie and Harabedian 1964).

The Ohio State University research used some of the simulation equipment developed for earlier Air Force-sponsored experiments on air traffic control—a PPI-type display and a radar target generator with maneuverable aircraft (Chapter 10). Here one sees how expensive apparatus produced for one program in man-machine system experiments can be exploited in another, in which the simulated task differs. In the Navy-sponsored experiments the subjects took simplified roles of intercept directors rather than air traffic controllers. The task was to run simplified intercepts against hostile aircraft rather than guide aircraft safely to a feeder location for landing. Since the simulated radar signals were maneuverable, it was just as practical to control them to make near-collisions as to avoid them!

In the first three experiments (Briggs and Naylor 1964; Briggs and Naylor 1965; Naylor and Briggs 1965), interceptions were controlled with the PPI display as a "transfer" task, after training on a checkerboard display with checkers representing aircraft. Two intercept directors either independently or interactively conducted four interceptions each. With the relatively modest interaction requirements in these experiments, interaction conditions during training appeared to be less effective than independence conditions when interaction was required in the transfer task. In another experiment (Johnston 1966), it was found that an intercept director acquired co-ordination skills (for this task) just as well by training in co-ordinating his own two interceptions as in co-ordinating interceptions with another director. In a further experiment (Briggs and Johnston 1966*a*), similarity between the training and transfer tasks was varied in the type of display and the method of communication. Two more experiments examined the effects of changing the performance criterion conditions from simple to complex on the transfer task (Briggs and Johnston 1966*b*) and the effects of verbal communications on teamwork (Williges, Johnston, and Briggs 1966). This research program represents an attempt to obtain general findings from dyadic and triadic tasks considerably abstracted from a real-world operational setting. Thus it might be associated more appropriately with small group research (see Chapter 23) than with system experiments. Its interest lies in part in the very fact that it has drawn tasks from an operational situation and attempted to achieve findings which may be thought to be highly generalizable.

## SIMULATION FACILITIES

Although they have not been used for man-machine system experiments, the Navy possesses such large resources in shore-based simulation facilities (and so few afloat) that they should be mentioned just in case some might be exploited for experimental research in the future. The NEL ASDEC has already been noted, as have the two multimillion dollar TACDEWs. In addition, a Lockheed Electronics Company six-million-dollar (original cost only), Antisubmarine Warfare Coordinated Tactics Trainer has been installed at Norfolk and another at San Diego. Elaborate antisubmarine warfare trainers simulating airborne and ASROC operations have been developed by Curtiss-Wright and Honeywell. The Electric Boat Company has a large submarine trainer. An ASW simulation facility at the Naval Air Development Center has been testing the A-New system for antisubmarine patrol aircraft with the Real World Problem Generator (RWPG), built by Sylvania Electronic Systems. It can simulate eight preprogrammed targets, surface ships, or submarines. A ninth unit can represent a submarine employing reactive tactics, that is, responding dynamically to the actions of the hunter aircraft.

But perhaps the most elaborate of the Navy's simulation facilities, and one of the oldest, is the Navy Electronic Warfare Simulator (NEWS) at the Naval War College in Newport, Rhode Island. It was undertaken in 1947 and completed in 1957 at a cost of $7,250,000. Its name, adopted after earlier christenings as the Electronic Generator and Display System and the Electronic Maneuver Board System, is not intended to mean that the system simulates electronic warfare—possibly a defect is its limited capability in this regard—but that the equipment operates electronically. However, because it was designed more than twenty years ago it is an analog rather than a digital computer system.

NEWS occupies three floors of a wing of one of the War College buildings. The first floor houses control elements, umpire area, communications room, control room, and equipment room. Air conditioning equipment, stock room, and maintenance shops are on the second floor. The third floor is where the "players" operate in a NEWS exercise. The command center area is largely occupied by ten Green command centers and ten White command centers. Each of these rooms simulates, to some extent, a combat information center, in that it has a PPI-type display which can present simulated radar signals of aircraft and ships in the area surrounding the pretended ship or aircraft or submarine containing that center. A signal (and the center itself) can represent an aggregate of friendly or enemy units, e.g., a force, which can be identified by querying the signal or target with a probe device that produces a readout of target information on a panel elsewhere in the room.

By means of a control panel in each center, an operator can introduce and change the simulated course, speed, and altitude or depth of the unit represented by that center. A display of grid co-ordinates shows where the unit is geographically located at any instant. Operations in a NEWS exercise can take place in areas 40 by 40 nautical miles, 400 by 400, 1,000 by 1,000, and 4,000 by 4,000. When an enemy force is detected on the PPI-type display, the players can direct various simulated weapons against it, and indicators show whether there has been a hit and the extent of damage. Each group, Green or White, can distribute twenty-four units or forces among its centers.

The umpire area has a large vertical geographical plot which shows how the opposing forces of Green and White are distributed. Images of the units being maneuvered are projected on this display by forty-eight projectors, and plotters mark the tracks in fluorescent chalk or ink. On each side of the vertical plot, in a double-entry arrangement, is a status display showing Green forces on the left and White forces on the right, and embedded in each of these are lights representing opposing forces. When a particular force is detected, one of these lights goes on; the light flashes when that force is fired upon, and an indicator like that in the command centers shows the extent of damage, computed by a special analog computer. Various PPI-type displays, communications, and querying devices are situated on the floor of the umpire area for umpires to obtain additional information about the progress of a contest between Green and White. Umpires can play the role of a hostile force if a one-sided rather than a two-sided contest is conducted in the command centers. Exercises can be conducted in real time, or time can be speeded up by factors of two or four.

Several types of games are played with NEWS, according to McHugh (1961), whose account is the principal source for this admittedly sketchy description of this simulation facility. That account does not specify actual frequency of exercises or facility availability. One type of game consists of relatively simple situations prepared by a War College group to demonstrate NEWS to small groups of officers and thereby let them gain decision-making experience. In the second, more usual type, the students at the War College themselves prepare elaborate, opposing plans and become the players after extensive co-ordination. A third type consists of games played by fleet personnel to test or rehearse their plans. Regardless of type, the exercises conducted with NEWS fall into the category of war games described in Chapter 23.

# 6

# Research in Air Defense for the Air Force

In the early 1950's numerous schemes were proposed for improving the capability of the United States to defend itself against air attack by hostile bombers. Most of the proposals which were investigated aimed to introduce various degrees of automation into the current air defense system. This system consisted of a large number of radars, control sites associated with the radars, and interceptor bases throughout the nation, especially near its periphery. It was termed a "manual" system because human operators translated displayed radar signals into surveillance information, manually marking their tracks on display surfaces, and controlled the pilots of interceptor aircraft with voice commands over a radio link, making the required calculations for these commands in their heads. An Air Force center for conducting air defense and a Navy combat information center (CIC) for anti-air warfare (Chapters 3, 4, and 5) had many functional similarities.

Surveillance operators at plan position indicator (PPI) consoles detected radar signals on their cathode ray tube (CRT) scopes and tracked the progress of the aircraft from which the signals came by making marks on the scope face with a china-marking (grease) pencil. They communicated the positions, directions, and speeds of these tracks to plotters at a large vertical display of the geographical area surrounding the radar site. These plotters produced "the big picture," derived from a number of surveillance operators. Other plotters maintained status displays showing what interceptor aircraft were available and where. By examining the geographical display and the status boards, the officer in charge of the air defense team could make assignments of interceptors against those tracks which could not be identified as friendly. He would scramble an interceptor from its base and tell one of his intercept controllers, sitting at PPIs near him, to guide the interceptor pilot to a position where he could destroy the intruder.

A few projects were directed toward improving these manual procedures without introducing automatic aids or substitutions. For example, Bell Telephone Laboratories conducted analytical and observational studies for the Air Force, and the Air Force Personnel and Training Research Center made one venture toward experimentation (McKelvey and Cohen 1954). But relatively little was done in a systematic manner to increase the capacity of this manual system to conduct air defense, through training (or selection) of personnel, until the System Training Program emanated from the RAND Systems Research Laboratory (Chapters 8 and 11). Instead, proposals were based on the supposition that new equipment would do the job faster and with less error. Since data had not been assembled to indicate how well human operators accomplished or could accomplish various air defense tasks manually, this rationale had considerable appeal even to those whose livelihoods did not depend on developing new products. It was assumed at the time that very large numbers of Soviet bombers might attack the United States and that very large numbers of defending aircraft would become available for defense. The information load would increase and become more complex. The speeds of both the bombers and interceptors were expected to grow as a result of jet propulsion, so there would be less time available for tracking and interception. Perhaps even more important, the interceptor aircraft would traverse the coverage of more than one radar during a mission, so there might be a problem of "netting" the control sites, that is, interchanging information between them.

For the surveillance function it was variously proposed to track targets automatically or give the operator some machine assistance, and even to make automatic detections. For the intercept control function there were proposals to calculate vectors (paths to fly) for the interceptor automatically, send them directly from the machine to the aircraft as electronic signals, and even control the aircraft or missile with these signals—without pilot intervention. And there were also proposals to replace the manual plotting on the geographical and status displays with various kinds of "hardware."

As occurred also in the Navy (Chapter 5), some of these proposed innovations became the objects of man-machine system experiments for purposes of evaluation or design. One of these programs is described at some length in Chapter 7. Some experimentation of this nature was eventually applied to the innovation which came to dominate continental air defense, the SAGE system, described in Chapter 11 (see also Green 1963 and Sackman 1967). It was also brought to bear on new air defense equipment which was proposed for the air defense of tactical air locations in overseas environments. Some of the research not covered in other chapters is reviewed here rather briefly.

## LINCOLN LABORATORY

The Lincoln Laboratory of the Massachusetts Institute of Technology had a major role in research and development in air defense. Although its principal contribution was the SAGE system, in its early days (1951–53) a group of experimental psychologists investigated methods of improving the manual system. For example, they examined some new methods of generating the geographical display. These consisted of photographing a PPI display on which all tracks were grease-penciled, rapidly developing the photograph, and projecting it on a vertical surface displacing the old manually plotted map. It was conjectured that this technique would be more accurate and faster because it would eliminate the steps of telephone relay and replotting. This approach was the same as that investigated at the Naval Research Laboratory—reviewed in Chapter 5. The Lincoln Laboratory experimenters evaluated the Land Polaroid and Kenyon devices which figured in the NRL studies. Apparently no reports of these experimental evaluations were issued for distribution outside the Lincoln Laboratory.

One way of obtaining a photograph of all manually established tracks for such projection devices was to locate all the surveillance operators at a single large display—an expanded PPI—which then was the source for the photography. But how should the tasks be distributed among the operators? Should some have the responsibility of detection, others that of tracking? Or should each operator be required both to detect and to track some of the targets? The first arrangement is called "series," the second "parallel"; these became variables in a multi-operator system experiment. In addition there were alternative methods of dividing up the territory and the targets. According to Green (1963), one of the experimenters, the results "suggested that a parallel system was better when the operators were sufficiently intelligent to take advantage of its flexibility. The study also suggested that face-to-face communication was important for the successful operation of a parallel system."

A motion picture was made of operator performance in this "Pi-Sigma" experiment, as it was called, but analysis of the data was never completed and no report was published. According to some of the participants, the performance of the surveillance operators in this new manual configuration was surprisingly effective, surprising at least to influential system designers at Lincoln Laboratory who were proposing a virtually automatic system (the earliest version of SAGE) which would remedy the supposed defects of the manual one. Apparently little or no pressure was applied to the Pi-Sigma experimenters to document their findings.

### *Early SAGE Studies*

Lincoln Laboratory established a preliminary version of parts of SAGE (semiautomatic ground environment) in what was called the "Cape Cod System," which was evaluated in 1953–55 (Jacobs 1965). Some of the radars were situated at South Truro on Cape Cod, linked to a remarkable early digital computer, called "Whirlwind," at the Digital Computer Laboratory of the Massachusetts Institute of Technology in Cambridge, Massachusetts. It was the possibility that Whirlwind might be exploited for air defense that led its creators to propose SAGE (Martin 1959). Although some human operators were involved in the Cape Cod System, its principal objectives were to prove out automatic digital processing of radar video signals, digital communication techniques, automatic tracking using digital data, vectoring equations for manned interceptors, and display techniques (Jacobs 1965).

No attempt was made, then or later, to make an experimental comparison between the manned operation of SAGE and that of the manual system which it was destined to supplant (Jacobs 1965). It is also interesting that SAGE was designed and installed not with a parallel arrangement for surveillance operators, but rather a series arrangement where detection and tracking were handled by different sets of people. It was not until SAGE had been in operation for many years that the design was changed to the parallel scheme that the Pi-Sigma investigators had found could be superior for the manual system.

According to Jacobs (1965), two important test techniques were innovated in the Cape Cod System. One was to use the digital computer itself to simulate the air situation as a supplement to inputs from actual aircraft. The other was to use the computer to record system performance during tests and later to reduce the performance data off-line. The limitations and advantages of the first technique are discussed in Chapter 11.

Subsequently, an "Experimental SAGE Sector" was evaluated in 1957–60. It had a "partial prototype" direction center at the Lincoln Laboratory in Lexington, Massachusetts, connected to radars at South Truro; Bath, Maine; and Montauk Point, New York (Jacobs 1965). The goal was to optimize various subsystem functions, including tracking in the face of electronic countermeasures. A number of manned-system tests were conducted in this Experimental SAGE Sector. In addition, a System Development Corporation staff, headed by B. R. Wolin, conducted a number of human engineering studies (System Development Corporation 1959).

## WILLOW RUN RESEARCH CENTER

SAGE was not the only proposed innovation for automating air defense on a grand scale. It had a rival, Air Defense Integrated System (ADIS), under development at the same time by the University of Michigan at its Willow Run Research Center, with funding from the Air Force's Rome Air Development Center. ADIS differed from SAGE principally in the degree of centralization of functions and in the use of analog techniques for tracking. It proposed to use the AN/GPA-23 analog tracking equipment developed at the Electronics Research Laboratories (Chapter 7) for surveillance at the radar sites, whereas in SAGE the analog radar signals would be converted to digital at each radar and then sent to a central digital computer to be processed for detection and tracking. In other words, in SAGE a number of different radar sites would feed data to a common computer at a direction center where interceptor control would also be conducted (by joint man-computer operations), and a number of direction centers would be netted to a combat center. In ADIS, on the other hand, there was to be a direction center at each radar for both surveillance and weapons control. The direction centers were to be netted to a combat center, where the collective track information from the direction centers would be received and viewed, and where the interceptors would be selected, assigned to targets, and allocated to a direction center for control during their missions; ADIS would also employ digital computers. In 1954 a decision was made to adopt SAGE and scrap ADIS. (It is perhaps ironic that a new system, BUIC, which is likely to take the air defense function over from SAGE during hostilities, follows the ADIS pattern; surveillance and weapons control are accomplished with computers right at the radar sites.)

The projected ADIS combat center included a weapon assignment section, which directed and co-ordinated the activities of weapon assignment sections at the subsidiary direction centers. The combat center weapon assignment section was to consist of "a 'senior controller station' which makes major decisions regarding the over-all air situation and delegates responsibility, several 'assistant controller stations' which assume the responsibility for particular raids, a 'weapon distribution station' which supervises the maintenance of a weapon balance and reserve for the entire sector, and an 'air surveillance station' which supervises the summarization and display of the over-all situation, and several 'liaison stations' " (Davage, DeVoe, and Pittsley 1954).

A laboratory model of the combat center weapon assignment section was built and operated for almost a year at the Willow Run Research Center (at Willow Run Airport, Ypsilanti, Michigan). Two experimental programs were conducted there. One consisted of a series of tests to evaluate techniques and determine operational capabilities of the equipment used. The other compared ADIS operations with those at a neighboring manual combat center. In neither case, however, was the equipment for ADIS operations considered as a prototype of the final ADIS equipment. The Weapon Assignment Laboratory included two weapon assignment consoles, a digital central storage and associated access equipment, detailed status displays and a summary display, a tape reader, a kill target keyset (for feedback of interception information), manual status input equipment, automatic teletype inputs, and a simplified order storage and combat reporting system; these occupied three separate areas.

*First ADIS Program*

The first program looked particularly at the assistant controller. This operator received instructions from the senior controller, determined the most effective way to deploy the available weapons, formulated weapon assignment orders against the targets for which he was responsible, transmitted these orders to the proper direction center to initiate interceptions, monitored his assignments for success or failure, and issued new or revised orders as necessary. Five civilian technicians from the center served as subjects in a study of this station. After eight weeks of training (including practice) of two hours per day, there was a single experimental 30-minute run for each subject, divided into 18-minute and 12-minute sections. Inputs were simulated and consisted of sixty-two targets, forty-five of them labeled "hostile." The controller had to order weapons against ten of these in the first part and three in the second part of the run. By the beginning of the second part "the weapon supply was nearly depleted and enemy bombers were nearing the target cities. Hence, decisions for selecting weapons and planning interceptions were more difficult during the second section" (Davage, DeVoe, and Pittsley 1954).

Target data on paper tapes read by a tape reader were transferred to central digital storage and thence to the displays viewed by the controller. The tapes were generated with the MIDAC (University of Michigan Digital Automatic Computer). Three MIDAC input tapes were used for generation: one for control; a second containing the co-ordinates for the beginning and end points of each track leg in the basic air situation; and the third for each target, with its time of entry, velocity, initial rough height, raid size, identification, reporting direction center, and track number, as well as the legs along which it could travel and the changes in some of the parameters and the times these should occur. Most of the "hostile" attackers were approaching the Detroit and Pittsburgh industrial areas from the north, with numerous "doglegs" (course changes).

The experimenters measured the time needed to receive instructions, obtain information on an enemy target, formulate and transmit a weapon order, monitor, and make reassignments. From these measurements they concluded that a single controller could process approximately forty-six hostiles per hour for weapon assignment, or thirty-three when monitoring was added. On this basis it was estimated how many of these controllers and their supporting equipment would be needed in the system for larger totals of raids. These were regarded as minimum figures which could be increased if certain equipment improvements were undertaken. Thus, the experiment provided a basis for calculating hardware and operator requirements during this early stage of system design. (Since the system was canceled, it was never possible to test these results for validity; in the case of the competing system which was built, there were no comparable experiments with whose results operational experience could be compared.)

Error information was also reported. Errors were attributed to confusion in operating the console keyset to obtain data for display; failure to register the console number for the correct target (as when target tracks merged on the display); failure to check for changes in raid size and height; and failure to use the correct order number. There were also a few cases of selecting weapons which lacked immediate combat potential and of vectoring weapons inaccurately.

A number of design recommendations came from these results, both from observations during the runs and from ensuing analyses. It was proposed that 87 (60%) of the 154 buttons in the laboratory model of the controller's console keyboard be eliminated and the controller's order keyset be integrated with the console keyset, so it would no longer be necessary to copy target data from the console keyset to the order keyset. The experimenters commented: "Of the eleven information categories provided on the keyset, only six are required on the integrated console keyboard. Three of the five categories being eliminated, viz., target track number, order number, and console number, duplicate quantities already provided. Target altitude and identity are available at the direction center; hence, they need not be transmitted" (Davage, DeVoe, and Pittsley 1954).

"The controller's failure to select weapons having an immediate combat potential," they added, "can be attributed primarily to poor legibility and the vast quantity of information on the detailed status display even after preliminary modifications had been made. Much of the information on this display is never used by the controllers, and therefore serves only to confuse them. Performance would be improved by removing all information not necessary in formulating orders from the detailed status display and providing this information only to those stations requiring it, by using larger indicators, and by allowing more spacing between number indicators." In addition, some changes were proposed for indicators and switches, and a buzzer was recommended to accompany the telephone call light.

*Additional Studies*

In the same laboratory a number of component tests followed the controller experiment, or preceded it to establish equipment parameters for that experiment. Their role in the research program was to complement or supplement the experiment, so these studies will be mentioned briefly. In one a light pencil was compared with a cursor-knob device ("co-ordinate generator") for interrogating targets on the console CRT. Spatial separation among targets and the number of targets surrounding the one to be interrogated constituted independent variables. The cursor-knob device took less time, except in very fine positioning movements where the ratio of cursor movement to knob movement made such positioning difficult. Another study compared computer-generated vector lines with target trails on long-persistence phosphors to indicate target velocities. A third evaluated two types of number indicators on the console. In a fourth the subjects had to count fading target tabs on a CRT to establish the interaction effects of ambient illumination and persistence. A fifth examined discriminability among the available symbols or tabs which could code targets on the display; the eight best were selected through a confusion matrix. (Some of these, incidentally, do not appear in the published literature on coding.)

Still another study was an investigation of several versions of the status display. This display contained 550 Teleregister indicators, which were electromechanically activated. This was one of the devices which was being proposed to introduce more automation into air defense, and it was supposed to displace the grease pencil inscription of data on a lucite board. Apparently operators could extract information (either single entry or double entry) faster and more accurately from a status display with less information on it than they could from one with the entire range of information; and performance was somewhat improved when the latter display was divided into two parts by means of painted stripes.

The purpose of the last study to be described here was stated by the researchers as follows:

> It is desirable that controllers know the approximate interception point for any target while they are formulating the weapon orders for that interception. Accurate estimations of interception points aid the controller in carrying out SOP's such as using straight line, minimum distance interception paths and avoiding sending manned interceptors through restricted areas. At one time it was suggested that an interception-point computer be added to the system to aid the controllers. However, such a computer would be superfluous if the controllers could estimate interception points with sufficient accuracy.

It should be understood that the controllers at the combat center did not have to guide (vector) an interceptor to a contact with the target; this was done at the direction centers. So their estimates could be very approximate. Five subjects made estimates of vectors from four weapon bases to intercept twenty-seven simulated bombers with given speeds and courses. The average error was 17.1 miles, which did not seem to warrant automation.

One of the characteristics of all of this research was to try to get as much information as possible without excessive cost and effort. The studies were relatively simple and did not incorporate significance statistics; nevertheless, quantitative data were obtained by means of simulation inputs to replace sheer opinion as the basis of judgment. Some data on system performance at operational air defense sites were also acquired by the human factors personnel during exercises with actual aircraft, as in a heightfinding study by Bailey (1951).

### *Second ADIS Program*

The degree to which manual performance might be exploited without new automation has always intrigued a number of system researchers, including R. P. DeVoe and others at Willow Run. This interest characterized the program mentioned earlier as the second major effort of this Weapon Assignment Laboratory, undertaken by DeVoe and military personnel at the 30th Air Defense Division. The manual combat center was situated only a quarter mile or so from the laboratory. It was possible to enter the same simulation inputs into the ADIS experimental center and the manual center at the same time. Military personnel operated the manual center, well-practiced civilians its automated counterpart. From 200 to 300 tracks were displayed during a run in each system, and comparable measures of performance were obtained. In view of the somewhat informal manner in which this study was conducted and the differences between subjects, it may have been difficult to justify any rigorous conclusions. Nevertheless, the very small margin of superiority shown by the ADIS operations supported some of the recommendations of the experimenters, such as relying on a manual operation for standby instead of a second computer. In this connection, the question of back-up capabilities has been a persistent one in air defense; it might have warranted more frequent assessment of relative capabilities, where a decision to automate the primary system had already been made.

## OPERATIONAL APPLICATIONS LABORATORY, AFCRC

The first version of a semiautomatic system, designed not only for air defense but also for interdiction and return-to-base control in a tactical environment, was delivered to Air Force Cambridge Research Center in 1955–56. Thereafter, it underwent an extensive series of experimentally oriented tests conducted by AFCRC's Operational Applications Laboratory. Eventually this version, the TSQ-13(XD-1), was replaced by a second-generation system built by a different contractor; it was turned into a research tool for system experiments in 1960–61 by the same Operational Applications Laboratory (by then part of the Electronic Systems Division). This research is discussed in Chapter 16. What follows will simply outline events and omit any results, since the reports of the earlier research with this system have not yet been declassified (Connolly 1958, 1959; W. R. Fox 1960; Sharkey et al. 1958; Sulzer 1959; Sulzer and Cameron 1959). Unclassified descriptions of portions of the TSQ-13 are contained in Chapter 16.

The system was installed at Shaw Air Force Base and Myrtle Beach, South Carolina, in 1956, and two test program phases were completed, the first dealing with the tracking and reporting capabilities of the system, the second with GCI or ground control of interceptors and RTB or return-to-base (control of the interceptor aircraft back to their airfields after a mission). Then, early in 1957, the system was brought back to Hanscom Field, in Bedford, Massachusetts, and installed at the Katahdin Hill site for two more phases, one re-examining return-to-base and the interdiction-control function (control of tactical aircraft in missions against ground targets), the other the cross-telling (transfer) of track information from one system site to another. In the first three phases the inputs to the system came from real aircraft which flew many hundreds of sorties in accord with scripted plans. In the last phase the aircraft were both real and simulated (by means of 15-J-lc target generators).

*A Comparison Field Experiment*

Possibly the most interesting (for this book) and experimentally oriented aspect of the entire program was a concurrent examination, in the first phase, of the operation of the TSQ-13 and the manual tactical air control system which it was designed to supplant. The subjects in each case were military personnel of the Tactical Air Command, those in the manual system simply operating their regular equipment according to normal procedures, those in the TSQ-13 operating the experimental system after being trained by laboratory personnel. The same radar inputs from the same actual jet aircraft were provided to the two systems, along with the same auxiliary information, always at the same time. Photographs were taken at a PPI of the common radar inputs, to determine the paths the aircraft had actually taken; in tests with aircraft it is most unlikely that they will fly precisely as planned. The output of each system, that is, the processed tracks, then could be compared with these definitely established inputs. There were nineteen missions—loads increasing roughly with missions. Quantitative comparisons between the two systems and between load levels were tested for statistical significance. Data processing was examined for both speed and accuracy. In subsequent phases of the program, load levels were again varied but there was no comparison between systems, since only the TSQ-13 was operated, and there was no attempt to derive statistical significance.

Subsequently, two more relatively small experiments at Katahdin Hill examined the interdiction capabilities of the TSQ-13 with simulation inputs; and a third experiment, again with simulation, provided comparative data as to how well interdiction could be accomplished manually. As the program at Katahdin Hill was ending, the experimenters turned to component studies of system equipment—for example, an investigation of whether it would be advantageous to indicate that a displayed target was a new one by causing its signal on a CRT to blink. At this point the program was already beginning to phase into the system-independent research described in Chapter 16.

The same research group, headed by V. J. Sharkey, carried out investigations of other proposed semiautomatic systems at Hanscom Field, including a military air traffic control system called VOLSCAN and a tactical system called BADGE (base area defense ground environment). However, no reports of this work seem to have been distributed outside the experimenters' organization.

# 7

# Electronics Research Laboratories

The earliest computer-type venture in automation for air defense was a set of equipment called the AN/GPA-23, developed for the Air Force by the Electronics Research Laboratories of Columbia University. In a experimental investigation of this equipment in ERL's Engineering Psychology Laboratory in 1952-54, Air Force controllers conducted more than two thousand interceptions and in addition tracked more than seven hundred targets in three programs which laid the basis for the manufacture and installation of AN/GPA-23s throughout the United States.

The first program examined a developmental model in the laboratory with simulation, the second subjected this model to a field test with actual aircraft, the third centered around a production prototype back in the laboratory. The first program not only compared the new system with the manual one which it eventually supplanted but also yielded design guidance for the subsequent model; the second checked the results of the first in a more realistic environment; and the third verified the new design and originated methods for training Air Force personnel in operating the system.

The AN/GPA-23, which was installed and operated in air defense centers for a few years before SAGE replaced the "improved manual system" it helped consitute, in itself was not a comprehensive system like SAGE. It was aimed entirely at improving the function of interception. As noted in Chapter 6, in the manual system an air defense controller, following the detection and tracking of an unknown or hostile aircraft, would guide an interceptor aircraft to identify or repel the intruder. For this purpose he observed both the intruder's and interceptor's radar signals on his plan position indicator display, marking with a grease pencil their successive appearances on the scope face, and calculated in his head the guidance (vectoring) instructions which he gave by radio to the interceptor pilot. The AN/GPA-23, an analog computing system, was supposed to help him in these tasks by replacing the grease pencil marks with visible electronic tags and computing the vectoring instructions electronically on the basis of tag track data for both intruder and interceptor.

Guiding an interceptor to repel a bomber could be a very complex process. In addition to tracking each aircraft and calculating the compass headings the interceptor should fly, a controller had to make sure those headings would place the interceptor in such a position relative to its target that the pilot could use the interceptor's own radar for the last stage of the interception; the pilot could then himself see the target's radar signals and use them to operate his aircraft and its fire-control system so the interceptor would fly along the proper path until it fired a rocket. In other words, the ground controller did not try to complete the intercept himself. The controller's task would end when the interceptor still had about 10 miles to go on a closing heading which would bring it virtually to a collision with the intruder at an angle of about 90 degrees; this was the handover point for a beam intercept. Of even greater moment, a controller might have to guide a number of interceptors against a number of intruders at the same time.

Since it was expected that large numbers of defending interceptors would have to be ground controlled against large numbers of attacking bombers in any air attack on the United States, the Air Force felt it was essential to provide sufficient interception control capacity. The AN/GPA-23 was intended to enlarge that capacity per controller, that is, to increase the number of interceptions a controller could handle during the same time period. To be sure, there was no firm information about individual controller capacity in current manual operations, but it was assumed to be inadequate. The designers of the AN/GPA-23 believed that vectoring calculations were difficult for human beings to make and that tracking was the kind of task where the machine should aid the man. On the other hand, the task of guiding the interceptor to turn and acquire its closing heading and handover position was too complicated for a relatively simple computer, so it was left largely to the human operator, helped by a manipulable tag offset from the tag indicating the target. These basic design decisions had been made before the engineering psychologists arrived on the scene.

## FIRST PROGRAM: DEVELOPMENTAL MODEL IN THE LABORATORY

The first program consisted of two major experiments and six exploratory ones. The AN/GPA-23 equipment and a standard manual system console occupied neighboring but separate areas. The systems were generally operated at the same time. A bank of 15-J-1c electromechanical target generators (see Chapters 4 and 5) could produce synthetic radar signals of moving aircraft on the operator scopes at both locations. Availability of spare units and constant engineering attention kept this equipment functional. The capability of introducing the same targets concurrently into the two systems made it easier to compare them. A specially built intercommunication system enabled each controller to talk with the simulator operator ("pseudo-pilot") maneuvering his simulated interceptors. Other laboratory facilities included altimeter simulators, data link simulators, a timing system with incremental light clocks, and various voice, photographic, and operations recorders.

Three Air Force captains and two first lieutenants served as subjects. All had had operational experience as controllers. Since the criteria for their assignment to the laboratory by the Air Defense Command were undetermined, their former commanding officers were asked to rank them within the current population for technical proficiency in controlling interceptions. Four were ranked in the 70th and 80th percentiles and one in the 10th.

*Tracking Experiment*

The first experiment investigated the capabilities of the two systems for tracking. The systems received common inputs. Six simulated targets were tracked in a trial, a second set of three appearing after the first three had disappeared (faded). The performance criterion was how accurately each system predicted where the target would reappear after five minutes of fade. Such projections not only indicated how good a track was created before the target faded but also could be regarded as dead-reckoning made necessary by noise, including electronic countermeasures (methods by which attacking aircraft sought to conceal their radar echoes). In the manual system the controller simply advanced his grease pencil track for five minutes, while the AN/GPA-23 advanced its tracking tag (a small circle) automatically. This tag was initially placed over the target signal manually by the controller; a number of such manual positionings, if done precisely, gave the tag the same course and speed as the target's.

The experiment incorporated five independent variables for each of two tasks (establishing a track's course and establishing its speed): the two systems, five subjects, two run-throughs for each subject, the two sets of three targets each, and five durations for generating a track before the signal faded (1, 2, 3, 4, and 5 minutes). As shown in Figure 2, tracks could have three origins, twelve angular directions (courses), and six speeds. This diversity was intended to give subjects comparable inputs without effects from memorizing them. The experimental design was factorial, with 400 cells and three data in each cell. Trials were sequenced in an *ABBA* pattern for the two-valued variables. The subjects performed in each system, some starting with one for a particular duration of track generation, some with the other. Although subjects alternated between systems, their concurrent operation made equivalence of input more probable and shortened the total running time.

Briefly, the results (tested by analyses of variance) showed the AN/GPA-23 outperforming the manual system, the latter having twice as much error and three times as much variance. The latter measure is an interesting point of comparison; it indicates the relative incidence of large errors. What had more general implications was the finding that the best subject in the manual system performed no better on the semiautomatic system. Observation and subject protocols showed that this superior performer in the manual system adopted a procedure for judging speed and estimating future track position along the track course that the other subjects did not employ. Presumably he was the only one who used optimal methods for the current system. This finding raises an important question. If current operations are to be experimentally compared with a possible replacement system to determine whether the latter should take the field, should not the former first be optimized? It was also found that the difference between the best and poorest performers on the manual system was

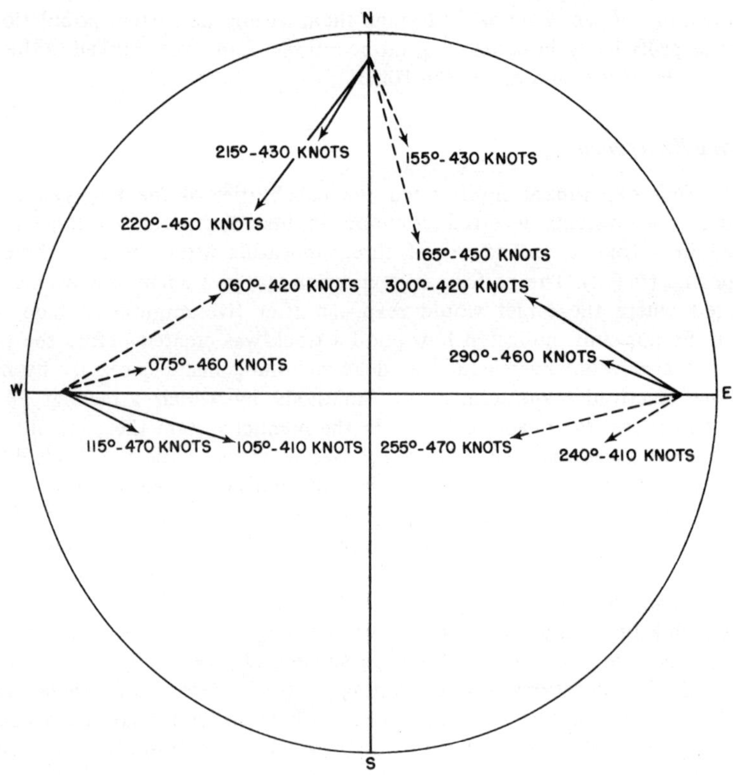


Fig. 2. The Stimulus Input for the Tracking Study (Parsons 1954*a*). (The shorter arrows show the courses and speeds of pips in Pattern I; the longer arrows show those in Pattern II. The solid arrows show the tracks in the first block of three within a trial, while the dashed arrows show the tracks in the second block.)

greater than the difference between systems, whether the system measures were the subject means or any one individual's scores. The experimenter noted in his report (Parsons 1954*a*) that the relationship between operator selection and system design is frequently disregarded in system studies. He observed further that the range among individual performances in the manual system showed there should be as representative a sample of subjects as possible in a study of this kind; it was obvious that the results from a single subject could have been completely misleading.

*Interception Experiment*

The second major experiment, also conducted by Parsons (1954*a*), concentrated on interceptions. The developmental model of the AN/GPA–23 had a maximum capacity of three interceptions overlapping in time. All trials were conducted at this capacity for both systems. Further, to increase load the operators in each had an additional task: to turn off intercom call lights which appeared every 20 seconds. The two systems were operated concurrently with common target inputs, as in the prior experiment, but now each system also had its own set of 15-J-lc simulation units as simulated interceptors. The Air Force officers functioned both as subjects and simulator operators, that is, as controllers and pseudopilots, alternating between roles.

There were seven independent variables: the two systems; two run-throughs of 240 interceptions each; five subjects; two distances (100 and 200 miles) between attacker and interceptor at the start of an intercept mission; two aircraft speeds; two time intervals between the starts of successive interceptions; and two patterns (crossing and reciprocal) relating the target's course to the bearing of the target initially from the interceptor base. After the second run-through, 112 more interceptions were conducted in two supplemental studies to examine performance with a single interception and with a particular method for guiding the interceptor in its turn to closing. As in the tracking study, the experimental design was factorial; however, there was some confounding of time intervals with runs and distances, and the two speeds could be compared in conjunction with only one of the distances. In the input there were four possible bomber origins for each of two interceptor bases. Input patterns for one of these bases are shown in Figure 3. To control for practice, *ABBA* sequencing characterized systems and distances. Subjects shifted between systems after every set of eight trials. A subject was almost always paired with the same interceptor simulator unit and the same simulator operator, with the result that the idiosyncrasies of any unit or operator were distributed equally among all variables except subjects.

There are two reasons for outlining the experimental design in this detail. First, it should be evident that a rather complex set of variables can be introduced as a matter of course into such an experiment without surrendering much orthogonality. Thereby a single investigation can generate a great deal of information. Second, four of the variables can be regarded as contributing to a general concept of input load, namely, time density, a term introduced in Chapter 4. The rate of new events with which a subject would have to cope was influenced by the distances between bomber and interceptor at the start, by their speeds, by the intervals between successive starts (scrambles), and by the patterns of a bomber's course relative to its direction from the interceptor base. In addition, the supplemental study of a single interception instead of three at a time afforded another time density comparison.

These variations in the time density of inputs requiring controller performance represented, by and large, the variations in actual operations. It was felt that the AN/GPA-23 and its effectiveness in relation to the manual system should be examined over such a range, rather than over a very limited set of conditions. Such a point of view is not always adopted in evaluations of new equipment. Instead, engineers are apt to ask, "Will it work?"—whatever that may mean. They may also ask, "Will it work better than the other system?" But these can be meaningless questions if the whole purpose of the new system is to cope more effectively with increasing loads on the military organization using it. One must then vary the load.

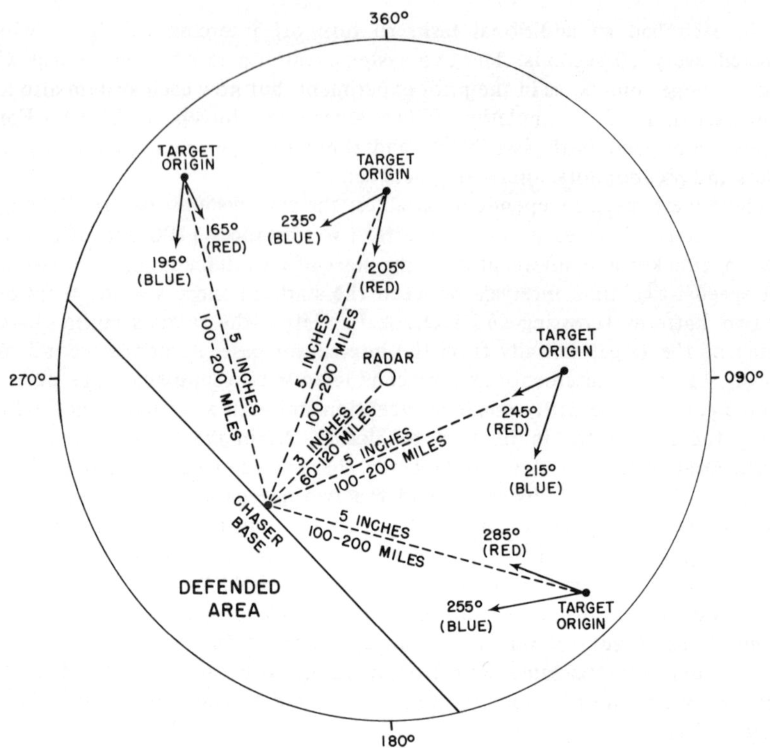


Fig. 3. Target Input Patterns with the Interceptor Base in the Southwest (Parsons 1954*a*).

By no means were all the pertinent input conditions covered in this experiment and its two supplements. Accordingly, a number of exploratory studies followed. One investigated what happened when the bomber changed its course, another the effects of requiring the interceptor to climb a considerable amount in a short distance, a third the probability of interception after loss of communication with the ground, and a fourth the effectiveness of communication by data link rather than voice radio. Data link was a new electronic communication technique for sending coded messages to aircraft.

The principal measures of effectiveness were the miss distance between the interceptor and its target as their paths crossed, and the divergence from 90 degrees in the angle between their paths (closing angle). These gave an index of approximate positioning accuracy at the point of presumed handover to the pilot and airborne fire-control system 10 miles away from the intersection of paths. The controllers were instructed to cease giving vectors at this point; they cheated, especially in the manual system, but not excessively. Other measures included penetration time, duration of closing phase, time after last vector, number of command headings, number of range and bearing messages, number of target course messages, errors and omissions in reponding to the intercom lights (the added task), and frequencies and duration of use of some of the AN/GPA-23 controls. About 5% of the data had to be discarded because of equipment malfunctions. Analyses of variance were conducted on the remainder.

The AN/GPA-23 performed well. It "worked," and under a wide range of conditions. So did the manual system. With two exceptions, differences in performance between the two systems were very small and lacked statistical significance. With the AN/GPA-23 fewer guidance vectors were transmitted to the interceptor "pilots." Of more importance, the error in the closing angle was considerably smaller in the manual system. Some of the other variables produced major, significant differences. Subjects varied greatly within each system. Increases in time density through increases in aircraft speed and reciprocal instead of crossing patterns degraded performance in each system, more so in the manual but not by much. The supplementary run with single interceptions showed better performance than that associated with triple interceptions.

*Consequences of the First Program*

On the basis of preliminary observations that the AN/GPA-23 was an effective instrument for conducting interceptions, a decision was made to proceed with its development before the data collected in this experiment were fully analyzed and reported. The good showing made by the manual system seemed to exert little influence on the decision-makers, either during the experiment or later. There were several reasons. One was an assumption that the AN/GPA-23 would principally help controllers whose levels of experience and skill in the manual system were much lower than those of the subjects in the experiment. Another was the belief that the production version of the AN/GPA-23 would include human engineering improvements, making its operation much more efficient; without doubt the arrangement of controls and displays in the developmental model had tended to stack the deck against it.

One of the purposes of these experiments was to generate human engineering improvements and, more particularly, to establish the most important design parameter for the final AN/GPA-23 console—the upper limit on its interception capacity. It was recommended that the production prototype console be built with a six-interception capability. The experiments had shown that with the AN/GPA-23 a controller could track six targets at the same time and could make three interceptions overlapping in time, even while handling successfully an additional task. An experiment was performed with the manual system by itself to ascertain whether the controllers could handle six interceptions overlapping in time in that system. Some could, but this load required almost continuous voice communication between controller and pilot. It was concluded that more than six might exceed voice communication limits also for the AN/GPA-23. The recommendation of a six-interception capacity was adopted on the presumption that this total would sometimes be needed, although generally a controller's capacity would be less.

In the production prototype console a single control stick and a few pushbuttons replaced a large number of tracking controls. Displays were redesigned and rearranged for better visual access and discrimination. If the six-interception capacity had been implemented with the panel arrangements in the development model, in which each interception had its own set of controls and displays, there would have been an even greater array of knobs, dials, and switches. The recommended, final design incorporated a single set for all interceptions, successively allocated among those being conducted.

Another recommendation from the large interception experiment was that the process of turning the interceptor to its closing heading for handover should be automated in some fashion. As noted earlier, this had been left largely up to human judgment, aided by an "off-set tag," because it would have required more complex computation than the relatively simple and inexpensive AN/GPA-23 could provide. Ironically, the superiority of the manual system with respect to the interceptor's closing angle resulted from its ability to produce better turns to the closing heading. It appeared that when the controllers had been making human vectoring judgments throughout the mission, they could make the complex judgment about the turn with greater accuracy than when vectoring up to that point had been accomplished by the machine. This was not a trivial matter, since the success of an interception depended on the extent of error in the closing angle. Although it was never possible to automate fully this aspect of interception in the AN/GPA-23, some additional aid was provided, as well as special training and rule-of-thumb procedures.

The results of the tracking experiment supported the original design assumption behind the AN/GPA-23, that the machine could help operators track targets. But the AN/GPA-23 designers had made a second assumption—that a human operator could not estimate vectors very accurately to guide interceptors, especially under heavy loads; hence automation was needed. What had not been ascertained was whether, with training, unaided operators could achieve accuracy in vectoring that was sufficient. It must be remembered that a controller had to guide the interceptor pilot to a turning point and then to a closing heading; shortly thereafter the airborne system supplied the precision guidance in the last phase of the interception. It appeared from the interception experiment that for the degree of precision required from the controller, the unaided but practiced human did well enough. Results of a brief further study suggested he could do as well as the machine, even if the number of vectors during the interception were limited to three in each system.

## SECOND PROGRAM: FIELD TEST

This laboratory program, as noted at the beginning of the chapter, was followed by a field test. This was not the sequence which had been originally planned. Immediately after the development version of the AN/GPA-23 was built in 1952, it was moved to a test site at Verona, New York, and for about two months strenuous attempts were made to conduct interceptions with actual aircraft. These attempts failed "because of insufficiency of aircraft, incapacities of ground radars, inefficiency of arrangements and inclemency of weather" (Parsons 1954*a*). Since a field test was still regarded as desirable, the equipment was installed in the summer of 1953 at an Air Defense Direction Center of the Air Defense Command near Saratoga Springs, New York, in the belief that an operational site would be more likely to yield data, and better data.

During a five-week period 100 attempts were made to carry out beam interceptions and 67 interception missions were actually conducted and measured; a substantial majority of these were evaluated as successful (Parsons 1954*b*). Some were arranged as single interceptions, some in doubles, and some in triples, that is, three overlapping in time. As might be expected, the usual vicissitudes of field testing occurred. Initially it was found almost impossible to detect the assigned aircraft on any radar scopes, and it was unclear whether the trouble lay in operational equipment at the site or in equipment in the aircraft. Co-ordination was difficult because the documentation authorizing the test did not arrive until after it was completed. But the ADDC personnel as well as three fighter-interceptor squadrons which supplied aircraft co-operated enthusiastically, and the weather co-operated about 60% of the time.

The AN/GPA-23 equipment was housed in a van outside the operations room of the ADDC and cabled to the site's surveillance radar. Two of the Air Force officers who had been subjects in the earlier laboratory experiments served as controllers, alternating in functions.

One manned a standard manual system scope in the ADDC's operations room and marshaled the airborne aircraft at designated locations before an interception trial began, giving voice-radio directions to the pilots. All of the aircraft were jet interceptors (of several types–F-86A, F-86D, and F-86F), but in any trial some acted as bombers and some as interceptors. The controller at the standard scope would start the bombers on scripted courses and at scripted speeds and altitudes. The other controller manned the AN/GPA-23 equipment in the van. He would start the interceptors and guide them to repel the bombers with the AN/GPA-23 equipment. The laboratory ground rules were in effect, and the van controller would cease vectoring when the interceptors still had ten miles to go in the closing phase of the interception. When the trial was completed, the controller in the operations room guided all the aircraft back to their base or, if enough fuel remained, to orbit locations for another trial.

Data were obtained from all possible sources: estimates of miss distance by the pilots of both interceptors and bombers; visual estimates on the AN/GPA-23 and standard scopes by the research personnel and the controllers themselves; and photographs of a PPI scope in the ADDC taken by ADDC personnel. The camera on this scope was able to record an entire interception on a single frame. Voice recordings were made of all ground-air communications. The research personnel who managed the test made certain that the AN/GPA-23 controller did not listen in on the directions which the operations room controller gave to the bomber pilots and did not have advance knowledge of what courses they would fly. It was not possible to follow an experimental design similar to that in the laboratory, but a number of different load variables were introduced so that a considerable range of conditions was sampled. Although it was not feasible to brief most of the aircraft pilots except by radio, they conformed most of the time to the scripted requirements, which included a variety of starting points for both bombers and interceptors and a variety of courses so that the AN/GPA-23 controller would not repeatedly encounter the same pattern. The techniques for conducting this program had to be developed almost in their entirety, because there had been no prior program of such a nature to use as a model.

Results were remarkably close to those achieved in the laboratory. The median miss distance was 2.0 miles and the median closing angle error was 19 degrees, compared with 1.5 miles and 15 degrees in the laboratory, so "it was concluded that field conditions had failed to produce any substantial decrement of system performance" (Parsons 1954*b*). New factors which had been absent in the laboratory included high-speed winds, variability in pilots' rates of turn, and low frequency of radar returns. The data were analyzed to describe quantitatively the load parameters of time density, appearance of radar signals on the scope, and pilot response, as well as such component-task results as tracking accuracy, frequency of computer-output transmissions, and turn-to-closing performance; and attempts were made to pinpoint the principal causes of interception failure.

## THIRD PROGRAM: PRODUCTION PROTOTYPE IN THE LABORATORY

Back in the laboratory, the third program in this research was undertaken with the production prototype of the AN/GPA-23 early in 1954 (Parsons and Sciorra 1954). Three Air Force captains from the Air Defense Command were assigned as new subjects. This program, based on simulation inputs, placed as much emphasis on training as it did on evaluation. It began with a series of training steps to explore a method of progressively increasing the operating requirements on the controllers; however, there was no control series to determine whether this was the optimal method. First, each controller learned component tasks: tracking, positioning the offset tag, and timing the turn to the closing heading. Then the subjects learned to conduct entire interceptions. They went through fourteen stages. In the first six stages they progressed from one to six interceptions at a time, in relatively simple situations. Next, they encountered more complex situations: multiple interceptors directed against one attacker, climb problems, wind effects, target course changes, and target fades. Each time a new condition was initiated, the total number of interceptions being conducted at the same time was at first reduced. All told, each controller conducted more than one hundred interceptions. The last phase of the program was an investigation of the possible use of the AN/GPA-23 for guiding missiles rather than manned interceptors against attacking aircraft. Since the report of the third program is still classified, results of this phase as well as the rest of the experiment cannot be discussed here. However, it can be stated that these results were evaluated as justifying the multi-interception capacity which had been designed into the production prototype as a consequence of the first laboratory program. In addition, the human engineering features of the console appeared to have been successful innovations. The subjects were able to shift rapidly from one interception to another and back again, since all of the displays pertinent to any interception were immediately restored as soon as the controller shifted to that interception with his interception selection switches.

There was only one feature which gave trouble. A stick actuated a multiposition switch to operate motors which moved tracking tags over the face of the scope. By moving the stick the controller had to place a tag precisely over a radar signal (blip) while the blip was stationary between antenna rotations. As noted earlier, after the controller did this several times, the tag and the blip would move along together, with the same course and speed. The tag's course and speed was transmitted to the vector computer. Not only was it critical to place the tag precisely over the blip, but the controller could track a number of blips and thereby conduct multiple interceptions only if he could position the tag in a very few seconds. Unfortunately, the production prototype was built with stick-associated motors which moved the tags so fast that tag positioning was imprecise and very time-consuming. This feature tended to subvert the principal purpose of the system—to increase a controller's capacity.

What was needed at this point was a component human engineering experiment to show the extent to which the system was degraded by this single deficiency in implementation. Instead, it was incorrectly presumed that human engineering analysis and recommendations of a slower tag movement would remove this problem in the manufacturer's production equipment. As it happened, however, the deficiency was accentuated. Thus, because a critical element was neglected in the system's human engineering, the AN/GPA-23 failed to live up to the promise it showed in the laboratory and field evaluations. In retrospect it seems it might well have been better to have eliminated this one defect than to have conducted the considerable experimental research which has been described. "For want of a nail, a shoe was lost . . . . "

Although this particular problem was not satisfactorily resolved, the Air Force did make a systematic effort to build some bridges between development, on the one hand, and production and use, on the other. Through a contract with the Operational Applications Laboratory of Air Force Cambridge Research Center, the Electronics Research Laboratories continued to provide services after the production protoype was delivered to Rome Air Development Center, which had funded the development of the AN/GPA-23, including the three research programs. Among the tasks completed under this new contract were extensive briefings at Air Defense Command headquarters concerning the system; filming of a motion picture; articles in an ADC periodical; development of rule-of-thumb procedures for making the turn-to-closing; assistance to Air Proving Ground Command in preparing a test program; recommendations to the Air Controller School at Tyndall Air Force Base for a training program; participation in training courses conducted by the RAND Corporation; analyses of interfaces between the AN/GPA-23 and data link equipment; analyses of data link communication requirements and limitations; human engineering recommendations; an operator's manual; and consultations with the manufacturer's engineers.

This later work covered a period of about two years (Parsons 1957). Although as a part-time activity it was not costly, it gave the operational user and to some extent the manufacturer some understanding of the design objectives and decisions of the original developer. It also emphasized the need for systematic training on new equipment despite the introduction of mechanization. This emphasis can be and was a partial counterpoise, at least, to the miraculous capability manufacturers may attribute to automation. It was truly an operational application supplement. Although such supplements to research and development may be infrequent, experience in this project indicated they might well be undertaken more widely.

# 8

# RAND's Systems Research Laboratory

Outstanding in the history of man-machine system research was a series of four experiments at the RAND Corporation's Systems Research Laboratory (SRL) in 1952–54. The locale was a former pool hall in Santa Monica, California, outfitted to simulate Air Force air defense sites. The program was characterized by scientific adventure, incessant effort, and rampant serendipity, not the least instance of which was the principal outcome: a multimillion dollar corporation and a vast training program spread around the world.

The four experiments, named "Casey," "Cowboy," "Cobra," and "Cogwheel," consumed 595 hours of session time, occupied 140 subjects, and cost one million dollars (according to an unofficial estimate). The purpose of the first experiment was simply to explore organizational behavior in an environment which simulated one that was pertinent to RAND's interest. It was not intended to achieve any particular applicational significance. But a potential application emerged as a by-product and began to dominate the research. The results of the first experiment prompted the second, in which military personnel replaced civilians as subjects. The second experiment induced the Air Force to undertake an on-site training program in air defense operations. The aims of the third experiment were to verify the results of the second and to educate new RAND personnel who would help create the training program. The fourth experiment sought mainly to orient the military personnel who would participate in the test of that training program. The series of experiments was not planned as such; one led to another.

Although these studies constitute one of the best known sets of man-machine system experiments, and justifiably so, their story has not been told comprehensively in a single, detailed account. The best source is an overview by the principal figures, R. L. Chapman, J. L. Kennedy, A. Newell, and W. C. Biel, published in 1959 in *Management Science*. This paper was a revision of part of a 1955 symposium described in RAND Papers 657, 658, 659, and 661. A shorter review appeared as a RAND report and likewise as another symposium presentation by Chapman and Kennedy (1956). Experimental results together with descriptive material were reported by the Staff Systems Research Laboratory (1953), Chapman (1956, 1960*c*), and Sweetland and Haythorn (1961). There have also been several limited descriptions of the experimental setting and operations (e.g., Chapman, Biel, Kennedy, and Newell 1952; Chapman and Weiner 1957; Kennedy 1962*a*; and Porter 1964).

## ORIGINS

How did it all start? In 1950 a number of psychologists attended a summer conference which RAND had called because its engineers and scientists were uncertain how to assess the contribution of human operators to the effectiveness—and degradation—of the future systems which they were studying for the Air Force. One of the attendees, Kennedy, had been heading a program at Tufts University for collecting human engineering data and had participated in the Applied Psychology Panel's development of applied research in World War II (see Chapter 3). After accepting an invitation to join RAND in 1951, he brought two other psychologists, Chapman and Biel, to the RAND Corporation. Chapman had been directing the technical program at Project Cadillac (see Chapter 4) and thereby had acquired know-how for creating a simulation laboratory and conducting complex experiments. Biel, whose experience during World War II (see Chapter 3) was likewise pertinent, came from human engineering research in the Aero-Medical Laboratory at Wright Field. The team was increased to four by the addition of Newell, a RAND physicist and mathematician who had been working on Air Force logistics problems.

The somewhat diverse composition of this team was important to its success, as each member contributed special talents, yet they were united in the spirit of scientific discovery. Kennedy communicated with RAND management as the spokesman, but the four members interacted with each other as peers, and plans and courses of action were usually based on consensus, often following considerable discussion.

Initially it was uncertain precisely what form their research would take. Even before the team was fully assembled one direction was explored through a five-man war game (see Chapter 23) called DORIS. But it seemed unlikely that one could acquire very wide knowledge from such an approach, beyond what had already been discovered in a rather abstract manner about information netting in small group research (see Chapter 23). The researchers had a common interest, the behavior of relatively complex organizations and their components. They also had an inclination to do experiments. This was not a traditional interest at RAND, where analysis was the preferred technique. On the other hand, RAND was interested in man-machine systems, particularly those in air defense. An air defense system could be regarded as the locus of organizational behavior which might account for system effectiveness—and degradation. Chapman had been doing air defense experiments: Project Cadillac could be construed as such. Chapman wrote a memorandum, Biel visited several operating air defense direction centers, and the consensus emerged to conduct experiments using a simulated air defense direction center (ADDC) as the organizational environment. It seemed amenable to experimentation because it offered objective measures of performance and controlled situations, and it involved complex human behavior. The role and functions of an ADDC in air defense will be described shortly; they have already been covered to some extent in Chapters 6 and 7.

The team could function in relative autonomy. They were members of the RAND Social Science Department, which had its headquarters on the East Coast. Before their project came under any extensive review, or even attracted much attention, it was well under way. A fast start resulted from hard work and several bits of good fortune. Kennedy found some available internal funding, inexpensive laboratory space (the former pool hall) was obtained, and a RAND engineer, M. O. Kappler, helped design and procure one of the more essential categories of experimental hardware, the communications equipment. According to their primary responsibilities Chapman was to put the laboratory together, Biel to gather support personnel and subjects, and Newell to organize the design and production of simulation materials, although "in actual practice these roles were played at various times by various combinations of people" (Porter 1964). A great deal of effort was expended. Within six months after the completion of the team the laboratory was in operation with practice sessions for the subjects of the first experiment.

The initial purpose of the laboratory was to determine what an organization, in this instance an air defense direction center crew, would do when it had to process a great deal of information—in this case extensive amounts of air traffic and hostile aircraft. In all the experiments the basic independent variable consisted of variations in load, with a number of subvariables in this category. The design of the first experiment organized this variable uncorrelated with time rather than in a stepwise fashion, as was done in the subsequent experiments. However, a step-wise manipulation occurred when the preplanned portion of this first experiment had been completed. Other independent variables were not systematically incorporated because the experimenters wished to derive concepts or hypotheses from what happened in the laboratory, rather than test preconceived hypotheses. Their only assumption was that a crew would learn how to function. They wanted to see how the organization under scrutiny, as a self-organizing one, would organize itself procedurally—not structurally. As an indication of the generality which the researchers wished to ascribe to their model, they initially referred to it as the information processing center (IPC).

## GENERAL ASPECTS

Aspects of the research common to the experiments will be described before summaries are given of the individual studies.

### *The Air Defense Organization*

The organization under scrutiny initially consisted of several sections and a commander, illustrated in Figure 4. Each section fulfilled an air defense function. One was surveillance: to detect, track, and report airborne objects, plotting tracks on a large map display which everyone could see. A second was identification: to identify such reported objects as friendly, unknown, or hostile. The

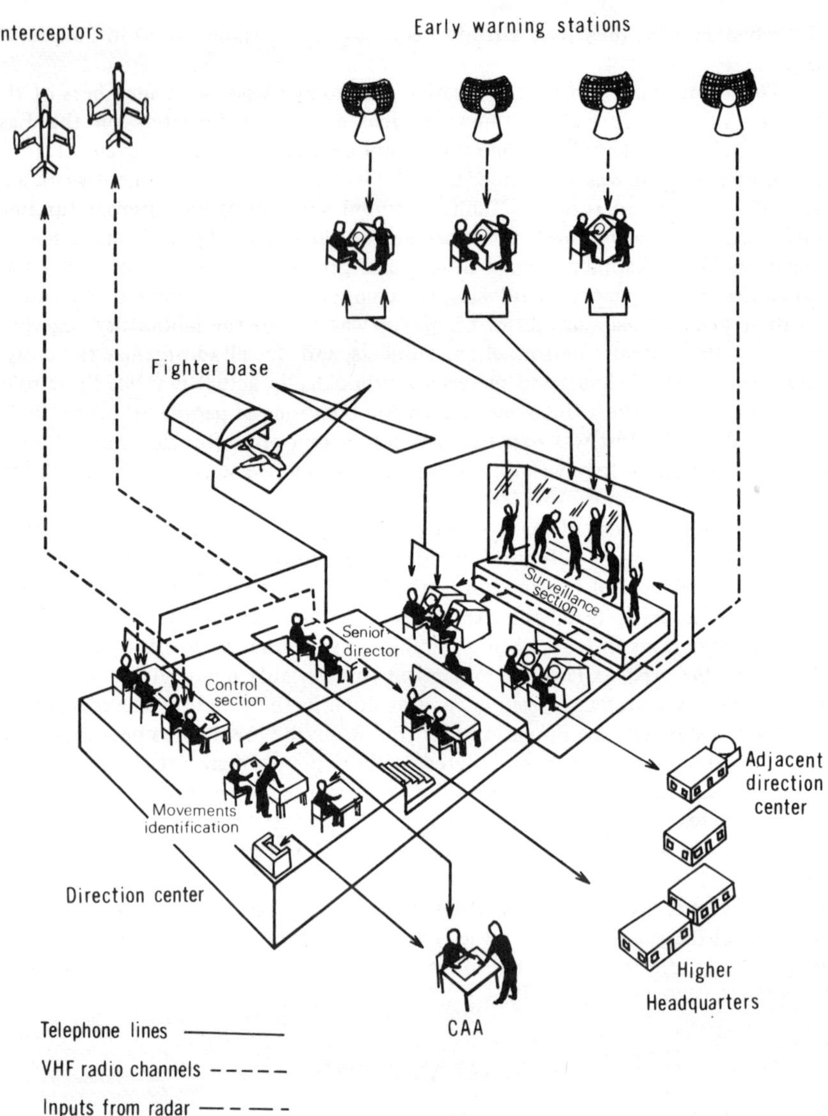


Fig. 4. Simplified Model of an Air Defense Direction Center (Chapman et al. 1959).

third was interception: to scramble and guide interceptor aircraft to intercept, visually identify, and, if necessary, destroy the unknowns and hostiles. Although such an organization resembled the combat information center in a Navy ship or AEW aircraft, described in Chapters 3, 4, and 5, there was more emphasis on the identification function because Air Force air defense direction centers, distributed around the continental United States, had to keep track of extensive civil and military air traffic. These centers, the operating sites of the air defense system before SAGE was built and installed, were supplemented by early warning (EW) stations which, like the ADDC, exercised the surveillance function with surveillance and heightfinding radar but which did not engage in identification or interception control. During the first experiment three EW sites initially represented by the experimental staff were added to the air defense organization under examination, so it came to consist of four sites and their operating personnel. This extension of system or organization boundaries resulted from the intensity and complexity of ADDC–EW interaction demonstrated earlier in the experiment.

Each of the ADDC sections had one or more simulated radar scopes (plan position indicators) on which they could see simulated radar echoes of airborne objects—friendly traffic, unknowns, hostiles. The EW stations also had simulation input devices. Simulations of internal and external communications included the intercom within the ADDC; telephone lines between it and the EW stations; telephone lines between the ADDC and an adjacent ADDC, a headquarters center (both added in the second study), the civil air traffic agency, and the interceptor bases; and radio links to the interceptor aircraft. The communication terminals outside the simulated ADDC and three EW stations were manned by members of the research support staff, who acted like the people who manned them in the real world. In this fashion the organization under scrutiny was "embedded" in a large environment, so it would have commerce not only within itself but also with the outside. The ADDC and EW sites, their radar coverage, and their relationships to each other and the rest of the world were actually modeled from operating air defense locations near Seattle, Washington.

The communication terminals and switchboards in the external world were, of course, outside the subject area. Also outside were the experimenters acting as managers and observers and the twenty-odd recorders that taped all the internal and external voice communications. The experimenters observed and recorded the actions of the subjects from a dais at the rear of the subject area. They were separated by a glass partition so the subjects would hear nothing in the experimenters' area. The laboratory also had a room where the subjects would gather after an experimental session for a debriefing. The ADDC surveillance map was a vertical display which the experimenters as well as the subjects could see and which could be photographed periodically as a way of recording data. Visitors, of whom there were many, could observe the subject area from the dais and listen to communications without disturbing the experiment. This turned out to be an important feature when high-ranking Air Force personnel wished to see and hear for themselves what was happening.

### *The Simulation*

The simulated air environment usually presented to the simulated air defense sites represented a peace-to-war situation. Commercial aircraft which had to be identified were detected and tracked, and there was considerable background traffic. "War" took the form of sneak raids or mass raids by enemy bombers.

Since the prime purpose of the air defense system was to repel any enemy attack, the commercial and background tracks could be regarded as readily filterable noise in the system, in which the hostiles were the targets. Unlike some air defense simulations (e.g., see Chapters 4 and 7), the air was not filled exclusively with hostile bombers and friendly interceptors. From the point of view of the experimenters their mixed picture was more realistic and could generate more complex organizational activity, although it omitted the procedure for grounding civil aircraft at the onset of hostilities.

During each experiment a large number of flight paths and associated radar echoes of aircraft had to be depicted, traveling at various courses and speeds over a very large area–approximately 100,000 square miles. To simulate radar echoes by means of manually controlled electromechanical target generators like the 15-J-lc (see Chapters 4, 5, 6, and 7) would have required a large array of these devices and a corps of operators, rendering it more difficult to manipulate load variables. Furthermore, the devices were noticeably imprecise and prone to malfunction, thus degrading experimental control over input variables. One of their major assets was their ability, in representing interceptions, to react to commanded courses and speeds contingent on the attack. Since in the RAND program interceptor aircraft were not simulated on the radar scopes, the synthetic inputs could be prepared in advance. The program concentrated on surveillance and identification.

Radar signals were ingeniously presented as digits on multifold paper which moved through a specially built device (illustrated in Figure 5) in which backlighting exposed the digits against a grid of the area. A new portion of the paper

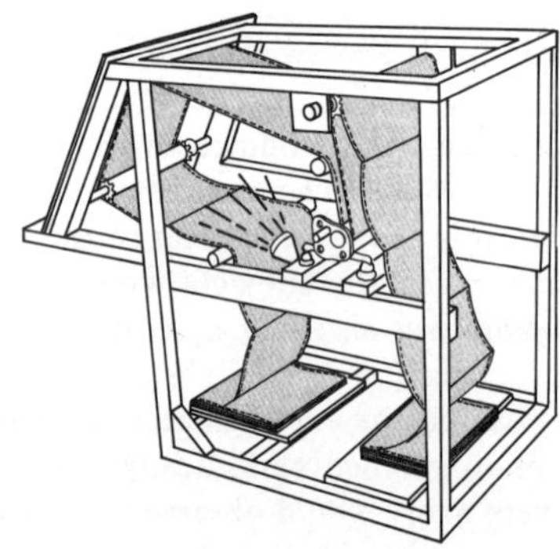

Fig. 5. The Simulated-Input Presentation Device (Chapman et al. 1959).

was displayed every 30 seconds under a plastic surface representing the external surface of a radar scope (PPI). The digit 1 represented signals (blips). These differed considerably from real blips in visual characteristics and frequencies of appearance, and there were no false signals due to weather or electronic countermeasures. On the other hand, the blips were reliable by being precisely positioned where they were supposed to appear.

The input devices at the EW stations were plotting boards, with the digit 8 as well as 1 representing radar signals. As just noted, radar tracks of interceptors did not show up on the scopes as did other tracks. Instead, a controller directed an interceptor from one checkpoint to another marked on his scope; the simulated pilot, using a similar checkboard, reported the estimated time of arrival at the next checkpoint to plotters at an interceptor movement board. Interceptor positions on this board were told to plotters at the common geographical display. Additional simulation inputs consisted of telephone and teletype messages. The simulation staff at the various points in the "embedding" environment worked from scripts.

The method of presenting surveillance tracks by showing a new digit every 30 seconds, while ingenious, occasioned much debate concerning its realism both before and after it was designed, because the signals, while numerous, precise, and realistically reflecting the coverage of the radars, did not look like radar blips, and their presentation rate deviated from the 12-second rate of the actual radar. Further, since the tracks were prepared in advance, one did not disappear if the hostile aircraft which it represented was shot down, nor could the hostile take evasive action if an interceptor approached. On the other hand, the experimenters felt that the important thing was to get the track information into the system so they could examine what then happened to it; the component discriminative behavior of surveillance operators in detecting and tracking was not an object of study. Content seemed more important than format. Furthermore, in the trade-off between simulation of heavy loads and simulation of target characteristics, it seemed more advantageous to concentrate on the effects of the former, especially since considerable research had already been done elsewhere on perceptual responses to the target characteristics.

These were persuasive arguments, especially in view of the actual goals with which the experiments were initiated. These did not include the evaluation of air defense effectiveness. For such an evaluation, it might have been useful to find out how much of the system's filtering of track data in fact occurred in plotting at the surveillance scopes as load increased, as well as later in the processing. Furthermore, the relative success of such filtering would surely have affected the subsequent processing of the data. To examine such filtering, greater realism might well have been required in the simulation.

During an experiment the printed digits representing radar signals of aircraft appeared at a rate of about three hundred per minute for as many as 180 hours. The sheets on which they were printed could total more than twenty thousand. Needless to say, their production benefited greatly from Newell's participation. It was a triumph of invention and a major operation which tied up RAND's computing facilities for days, even weeks, on end, all seven days per week, with three shifts per day. (This *did* bring the experiments to the attention of the rest of RAND.)

First the inputs had to be designed. (The input contents of each experiment are described further on.) The technology consisted of creating a "library" of about eight hundred flights, and also designing specific massed raids. The start and end point of each flight or track was hand-punched on IBM cards; then an IBM 604 computing card punch machine created decks of cards containing the blip positions within the tracks according to the radar coverage models established in the machine. For example, signals were omitted if the aircraft would be hidden from the radar by a mountain. The tracks were ordered according to time, and an IBM 407 printer transferred the card data to the multifold paper as digits, each in its proper geographical position and ready to appear in the proper half-minute interval. In the last three experiments the machine also prepared the "building blocks" by which the simulation inputs were organized. One reason why production took so long is that a stored-program electronic computer was unavailable for computation for these experiments. Such a computer was first employed later in 1954 for producing simulation inputs for the system training program that emanated from the research; production time dropped enormously.

*The Data Collected*

As impressive as the large quantities of data created for the simulation inputs were the vast amounts of data about crew performance collected during the experiments. Sixty file drawers held twelve thousand hours of recordings and other material when the experiments were concluded. The principal source of the data which were eventually reduced and described in reports of experimental results were the voice communications over the telephone lines connecting the subjects to each other and to the embedding environment. At first the voice recordings were transcribed by clerks, then coded and tabulated. However, with the second study the immense task of transcription was bypassed and the telephone messages were coded and tabulated directly from the recordings; coding quality was checked by a sampling technique. Another development was to obtain communication data while the communication was occurring. Starting with the second experiment, an IBM card was punched every 15 seconds showing which among certain lines were in use at the time. In the third and fourth experiments the communications behavior of the crew was coded on cards as it occurred, by means of special keyboards. According to the experimenters (Chapman et al. 1959), coding of such data for analysis was four times as fast in the fourth experiment as in the first, with only three steps instead of twelve. Voice records were obtained for that experiment but remained unexamined. "We collected as much data about the crews and their behavior as we could because we were searching for a framework rather than testing a hypothesis," the experimenters commented. "Only part of the data has been successfully coded or explored at any length although literally hundreds of very pretty hypotheses have been lost in it. Although much of this data has been used only to explain specific incidents, it should prove of more general value once we know the appropriate questions to ask of it."

Some of the other kinds of data collected during the first experiment included:

An activity analysis of key crew members: a coding of certain gross behaviors as observed from the dais every 30 seconds.

Photographs of the common geographical display every 2 minutes, during some of the sessions.

Written records prepared by the crew during a session: the crew chief's record of which positions were filled by which individuals and at which times as they rotated through positions; and logs maintained by identification, controller, and recorder personnel.

Microphone recordings of certain aspects of face-to-face conversations among crew members.

Pilot and umpire logs, and correlator logs.

Records of the crew's discussions before and after a session, together with diagrams showing how they grouped themeselves around the conference table.

The written reports including over-all performance summaries and critical track histories, by which the experimenters provided knowledge of results to the crew.

Administrative logs showing which experimental staff personnel performed which function.

Records of the experimenters' own regular one-hour postsession discussions.

Interviews with a number of crew members several months after the experiment.

In the second experiment some of these sources were expanded and new ones added: logs of embedding organizations; semistructured observations of crew activities, including some by experimental personnel with backgrounds in group dynamics and sociological experimentation, replacing the observations taken every 30 seconds; postexperimental interviews with all crew members; and examination of the five officers by one of Bales' standard situations for interaction analysis. The observation staff was larger, and in the third experiment larger still. In the latter the staff watched members of the crew for indications of involvement, reactions to the experimental conditions, and instances of problem solving. Data obtained by these incident recorders were later included among reported results. The camera now took pictures of the displays every minute. Expressions of attitudes by members of the crew were noted by coding "attitude" cards. Although the experimenters had considered a method of distinguishing between positive and negative attitudes and of indicating by whom, to whom, and about what the attitudes were expressed, only the fact that an attitude had been expressed was coded during the experiment. After the experiment the crew received a sociometric questionnaire, an attitude questionnaire, and one concerned with procedures; and several psychodrama sessions were conducted with some of the subjects.

In addition, the recorded information included the planning, preparation, and input materials for the experiment: the track presentations on multifold paper, flight plan scripts, summaries and descriptions of task environment, crew handbook, operating instructions, war plan, and lectures and special instructions. As already noted, most of the data exploited for analysis came from the verbal reports over the telephone lines, such as reports from scope operators to tellers at the geographical display, and reports between the EW stations and the ADDC. The photographs of the geographical display constituted back-up information which was little used because relatively few questions arose to require its use.

The observations from the dais (or "top deck," as it was usually called) were helpful to the experimenters for debriefing themselves and for arriving at hypotheses as to what was changing crew performance during an experiment, and they were the source of a list of various changes in crew procedures. These were important, but their reporting as experimental results presented certain problems in interpretation. Because they concerned detailed air defense operations, considerable familiarity with these operations was needed to understand the procedural changes; and to generalize from the procedural changes was a challenge. Further, the changes in procedures had to be regarded as dependent variables with respect to what preceded them and as independent variables which might have affected what followed. In neither case could cause-effect relations be specified with certainty.

Subjects' debriefing data also proved difficult to handle; they could serve only for searches as to whether what was discussed in the debriefing contained solutions which were subsequently implemented. The data from other sources, such as face-to-face conversations among the crews, were not used. A considerable portion of the principal data was analyzed, and a substantial amount of further analysis might have been undertaken. However, eventually the RAND Corporation apparently concluded there would be no analyses beyond what had been accomplished and reported, and the collection of data was destroyed. The discrepancy between the amount of data collected and amount analyzed for results, in this as in some other man-machine system experiments, has led to critical comment (e.g., Sinaiko 1962). In retrospect, it might be asked whether analysis should not differ according to the research strategy adopted. It may be more difficult to specify in advance what data should be gathered and reduced when the strategy is one of exploration, as in the case of the SRL experiments, than when particular hypotheses are being verified. On the other hand, it may be argued that is is still advisable to guard against too great expenditures of time and effort in data collection, regardless of the strategy, and against too large a ratio of data gathered to data analyzed.

Another set of data concerned the subjects. A battery of psychological tests was given to ninety candidates before the first experiment, covering aspects of intellect and temperament; sixteen test scores were obtained. Before the second experiment the subjects, this time military personnel, were similarly tested, and biographical information was acquired concerning military experience, both in air defense and otherwise, and qualification records.

### *The Subjects and Their Management*

The subjects of the first experiment were twenty-eight college students averaging twenty-three years old. The military subjects in the other three were thirty-nine, forty, and thirty-three Air Force officers and airmen supplied by the Air Defense Command. The students, most of whom had not previously known each other, heard twelve hours of lectures; later during practice sessions they were rotated through all positions and competed for the officer and noncommissioned officer positions. The experimenters based permanent assignments for the experimental sessions on test scores and performance. The subjects of the second experiment were supposed to have had air defense experience. Training was relatively short, and again the experimenters assigned positions according to the same criteria except that they took into account military rank. The crews for the last two experiments included five officers (as before); positions were assigned by the senior Air Force officer in the group.

The researchers gave considerable attention to strategies of managing the subjects so that their motivations and interest would tend to make them perform in a manner comparable to air defense personnel at an actual operational site during hostilities or prehostilities, and also to insure that their behavior would not be influenced by the fact that they were subjects in an experiment. It was observed in the Staff, Systems Research Laboratory report (1954): "The crew can easily adopt the attitude of the Hawthorne effect, that this is a 'special' job and that they are a 'special' crew. However, in both experiments, evidence was obtained that the crews did not perceive themselves in this way."

The researchers were particularly anxious about the "Hawthorne effect" because in the experiments there were no control (contrast) groups for whom conditions differed from those for the experimental groups. Since the major finding consisted of unexpectedly good crew performance under heavy loads, and since this performance was attributed to certain conditions, the possibility of a Hawthorne effect is still sometimes raised. One view is that a Hawthorne effect was inevitable.

Chapter 2 has discussed the Hawthorne effect and what its origins may be: changes as such appear to generate changed performance. To some the phrase most commonly means improved performance among experimental subjects attributable not to particular conditions but to the fact that they are experimental subjects receiving special attention from the researchers. The phrase "special attention," however, also needs more analysis. Does it mean that subjects know they are being intensively observed and analyzed? Does it mean that extrinsic reinforcement is provided to the subjects?

The SRL researchers were particularly concerned by the latter question but also worried about the former. In addition to noting that much the same change occurred from experiment to experiment—i.e., results were repeatable—the researchers discounted the likelihood of the Hawthorne effect by pointing out the various management strategies adopted to preclude or minimize it.

One strategy was to reduce interpersonal contact between experimenters and crew to an absolute minimum and otherwise cause the crew to disregard the experimenters. This was attempted by (1) making all arrangements through a crew leader; (2) preventing any unusual experimenter actions that would draw attention; and (3) obscuring any evidence of connections between experimenter observations of the crew and other experimenter behaviors. Another strategy was to get the crews exclusively concerned with the air defense goals they were supposed to achieve as system operators and to reinforce only their air defense behavior, so there would be less chance that they would be motivated to try to please the experimenters or regard themselves as something special. The strategy was implemented by emphasizing in various ways (sometimes dramatically) the importance of the air defense mission; by regularly sending a report to the crew after each session, describing what had been accomplished; and by abstaining from telling the crew at the start precisely what procedures to use—instead, it could adopt its own. The crew always had a debriefing session conducted by the senior officer after the experimental run; here the crew members could discuss what happened during the run. The experimenters carefully refrained from taking part, and they transmitted a feedback report about the run through the crew leader. This report was rigorously factual. As another aspect of this strategy, a crew's living conditions outside the laboratory were arranged to be similar to those prevailing before they became subjects—in the case of military subjects, for example, standard discipline, pay, and leave policy.

An additional strategy was to forestall the development of unwanted behavior arising from disbelief about the simulated task environment, whether this was the simulated ADDC and EW equipment, the simulated air picture, or the simulated personnel and organizations in the embedding environment. One way of doing this was to explain why, for example, there were constraints on complete realism and some inevitable artificiality. Another was to "maintain the creditability of the environments" sometimes by prevaricating, as in crediting to "a high-speed missile" an extremely fast-moving track which really resulted from a computational error. "By encouraging crews to learn and invent new procedures we further complicated the problem of maintaining realism," the researchers commented (Chapman et al. 1959). "The simulated environments had to 'snap back' realistically in response to any of their actions. To be able to do this in a way that maintained experimental control required a modicum of staff mechanism, judgment and coordination." Needless to say, because of this requirement and the need to forestall any motivational deviations, the crew members had to be kept under close observation every instant they were in the laboratory.

One of the interesting aspects of these strategies is that they were conceived and introduced largely in the endeavor to prevent competing motivations from contaminating the experiment. But it would appear that the experimenters also wanted to facilitate crew learning. By minimizing extrinsic reinforcement coming from the experimenters and by providing over-all and objectively derived knowledge of results, they felt that crew motivation would be centered around intrinsic reinforcement for everyone. Their arrangements also permitted a degree of subject participation in debriefings which presumably differed from such debriefings in actual military operations. Subsequently, it was conjectured that some of these strategies constituted significant methods of improving performance, and years later such conjectures became articles of faith, with numerous elaborations. For example, disassociation of the experimenters from the crew debriefings was elaborated into the concept that the proper way for a crew to evolve prodecures was to let members hold unconstrained discussions. This general topic will be resumed later.

## THE EXPERIMENTS

Management techniques as well as strategies were crucial to these large-scale experiments. Their preparation and conduct required tremendous drive, careful attention to detail, and long hours on the part of the experimenters. An actual experimental session was a major managerial undertaking, as Chapman (1955) has commented:

> Getting one of these experiments started, for instance, looks as involved as the takeoff of a superbomber. Formalization, cross checking, efficiency, and integration are all illustrated in this process. The extensive pre-run check lists, ringing of bells and flashing of lights, and down-to-the-second timing are all a part of getting 9 input mechanisms, 7 timing devices, 16 clocks, 22 recorders, and 65 people off and running at the same time. No mean trick.

At this point the various schedules, contents, and experimental designs of the four experiments will be reviewed, experiment by experiment, except that the second and third will be described together because of their similarities. Results will be treated at the same time. General conclusions and outcomes emanating from the research will be discussed subsequently.

### *Casey*

The first experiment, initially identified by the general project title "Project Simulator" and later named "Casey," began with practice sessions on February 4, 1952, and concluded June 8. The bulk of the sessions lasted from three to four hours each during Tuesday and Thursday afternoons and Saturday mornings. They followed the lectures which have already been mentioned. The first thirty-five sessions gave air defense training and orientation to the college student subjects. The crew members rotated through the crew positions and acquired the necessary component skills, familiarity with the simulated ADDC, simulated geography, and aircraft characteristics; they then learned to function as a team. Input load was progressively incremented as the experimenters also went through a learning process concerning operator capabilities in such a situation as well as the input production methods. It was the experience during these practice sessions that led to the incorporation of three EW stations netted with the ADDC as part of the subject organization. Initially these were manned by experimental support staff, but "It became painfully clear . . . that the experimenters' role as early-warning stations was untenable. Crew members on the other end of the telephone line queried the relevance of some of the information, wanted it given different priority, and wished to negotiate better procedures . . . . To rule out cooperation between the early-warning stations and the center was arbitrary and artificial. But, by helping to determine what information should be forwarded to the center, the experimenters would become, in effect, part of the organization being studied" (Chapman et al. 1959). The change made it necessary to build new presentation devices, revise the communication net, and revamp the design and preparation of input information. This development illustrated the value of preliminary sessions in such experiments, for bounding the system and for defining variables, and the consequences of such rehearsal.

Next came sixteen experimental sessions. Load level varied in a balanced manner; there was no progressive increase. Levels approximated those which the experimenters had been told represented a difficult task for an average operational ADDC. There were no massed raids; the threat consisted of sneak attacks and unknown aircraft. Tracks fell into two categories, penetrating and background. Between 14 and 28 of the former were in the air at any one time, while between 4 and 10 aircraft "penetrated" from "dangerous" directions during any 60-minute period. Altogether during the sessions 200 aircraft penetrated the defended area, of which 70 were unknown in that no flight plans had been filed for them, and of these 21 were "hostile."

During the sessions all of the 21 sneak attacks were successfully intercepted, as were all but 3 of the other unknowns. According to the experimenters (Staff, Systems Research Laboratory 1954), "the crew experienced less and less difficulty as it became more and more familiar with the task. The crew members grew more sensitive to load conditions; that is, they were able to relax quite completely under low loads." In consequence, it was decided to add two sessions of seven hours each to the experiment; these were run on successive evenings. In the first, 99 tracks penetrated from dangerous directions, 12 of them unknown and 3 of these sneak raids. The latter were all intercepted, as were 6 of the others and 5 more which had been called unknown because of misidentification. With background traffic considered, there were 20 to 80 aircraft within the defended area at any one time. In the second additional session, 159 aircraft penetrated from dangerous directions, 18 being unknown, 4 of these hostile; total aircraft in the area at one time varied between 20 and 90. Everything went well for the first five and one-half hours. Both of the hostiles that appeared were "shot down" and 6 out of 8 other unknowns were intercepted, as well as 6 misidentified as unknown. But in the last 90 minutes there occurred "progressive deterioration of the organization" to the point of breakdown; for example, 1 hostile bombed Seattle, and the other was identified as "friendly," while 3 nonhostile unknowns were misidentified and 3 more were never tracked.

The data from the two extra Casey sessions had far-reaching repercussions. That a crew of civilians could achieve as much success with such heavy input loads greatly interested a number of Air Force officers who heard of the results. The experimenters were invited to Air Defense Command headquarters to describe the study. The net consequence was a decision that RAND would fund and conduct another experiment, this time with Air Force personnel loaned as subjects.

### *Cowboy and Cobra*

In this next experiment, code-named "Cowboy," task load was designed to increase in a number of steps. "We realized," the researchers observed (Chapman et al. 1959), "that task difficulty was not the number of aircraft in the area but was instead the difference between the number of aircraft and the crew's load carrying capacity of the moment: the traffic load that was difficult to handle today might prove quite easy a week from now." The problem for the experimenters, of course, was "to estimate how fast the crew would learn in order to increase task difficulty fast enough to continue to challenge it but not so fast that the task would be too difficult." The description of the design and inputs for Casey also applies to the third experiment, "Cobra."

There were sixteen sessions of about eight hours each. Sessions, which made up four sets, were held on consecutive nights in Cowboy, but during the day in Cobra, except that the subjects had a night (or day) off between sets. Preceding these were six practice sessions of the same duration plus several shorter ones. Within each session were two three-and-a-half-hour problems, and within each problem were two periods of 75 minutes each. The periods were the basic experimental units. Within any problem the period was preceded by a warm-up phase, in which the air situation was built up to the experimental conditions desired. The period was followed by a "meshing" phase which accomplished a transition from one experimental condition to another.

As stated earlier, the independent variable was task load, but this varied in a number of ways and should be regarded as a group of variables. It was also possible to consider as independent variables the periods within problems and successive groups of periods with equivalent loads. The experimental design aspects of the input content have to be described to convey the complexity they can assume in this kind of experimentation.

A task load variable designated "intensity" consisted of various groups of tracks of traffic increasing in number as the experiment progressed. Any group of thirty-six was composed of three types: fourteen background friendly flights; twelve penetrating tracks, mostly from a dangerous direction and including some sneak attacks and other unknowns; and ten outgoing tracks toward the same dangerous direction. A second variable, "distribution," was the relative proportion of penetrating tracks among the three EW stations; this had two values, "even," in which the three stations had to handle only moderately differing proportions of these tracks, and "uneven," in which one station had twice as much to handle as either of the other two. A third load variable, also with two levels, was the absence or presence of discrepancies between actual positions of penetrating flights and the filed flight plan positions; discrepancies could generate "uncertainty."

Number and types of definitely hostile aircraft constituted another intensity variable. The intensity in this fourth variable also increased during the experiment. The types were massed raids of five to twenty bombers each; massed attacks with diversionary aircraft, ten to twenty-five in number; low altitude raids averaging two and one-half aircraft each; and submarine-launched missiles. The two types of massed raids totalled thirty in the first set of sessions, forty-eight in the second, seventy in the third, and ninety in the last. A fifth variable was the simultaneous appearance, or nonappearance, of a friendly flight without a flight plan in the vicinity of an attack; this was called "distraction." A sixth, "variety," covered the variation in enemy attack: sneaks, split sneaks, massed attacks, massed attacks with diversions, low altitude attacks, submarine-launched attacks, and "friendlies" without flight plan. Finally, a seventh, "redundancy of information," related to the number of radars with which certain attacks could be detected. These last four variables were called "rare event stresses"; the first three were "continuous pattern stresses."

The inputs were created by combining elements of several "decks" of tracks, a deck being a set of IBM cards which contained the information about various tracks. There were four equivalent 115-minute decks for background traffic; fourteen equivalent 115-minute decks of friendly tracks, including penetrating tracks as well as some from the background decks; and twenty-eight different 200-minute decks containing eight attacks each, for the definitely hostile aircraft. "A background deck, from one to five friendly decks, and a hostile deck were superimposed to make up the task environment. The background decks were equivalent with each other, as were the friendly decks; the hostile decks were each unique and constructed to reflect an intensity of stress as designated by the set in which they occurred" (Staff, Systems Research Laboratory 1954).

During each successive set of four experimental sessions there was a heavy track load and a light track load for each type of track intensity—friendlies (including penetrating aircraft) and definitely hostile. These alternated randomly among periods, with as many heavy loads as light within the set. Roughly the heavy level of one set became the light level of the next for the friendly tracks, while the heavy level on one set approximated the average level of the next set for all tracks and for the subclass of penetrating tracks within the friendlies. In any case, it should be understood that from set to set a kind of overlap occurred in the rising load levels. For all tracks the heavy load in the fourth (last) set was somewhat more than three times the light load of the first set. The distribution and uncertainty variables were distributed among the light and heavy loads of each set fairly evenly, and the distraction, variety, and redundancy variables were similarly distributed within sets or between the first pair of sets and the second.

Some lack of unanimity persists concerning the statistical analysis permitted by the experimental design. This was organized in a rather complex fashion, sufficiently complex, in fact, that a complete description is hard to come by in published accounts. Seven eight-by-eight orthogonal Latin squares were the vehicles for varying the stresses independently of one another—as labor-saving devices for obtaining a specified balance in the presentation of the values of the variables. Each of the cells in the squares could represent a characteristic of one of the experimental periods, and sixty-four different combinations of task elements were derived, one for each of the sixty-four periods. From one point of view the value of the design was only to "insure a counterbalanced presentation of variables" (Chapman et al. 1959). From another (Sweetland and Haythorn 1961), it made possible an analysis of variance to test the statistical significance of some of the results, as a consequence of the orthogonality among the following variables: three levels of load, three classes of track, two types of distribution, two periods (first and second part of each problem), and two amounts of "experience" (two successive groups of periods with the same load). It is apparent that such an analysis did not attempt to cover all the variation introduced into the experiments, such as the "rare event stress."

*Results*

Results of the defense against definitely hostile aircraft and other instances of "rare event stress" have not been declassified for Cobra, but for Cowboy it has been reported that of a total of 238 simulated bombers in massed attacks, including those with diversions, 206 were "shot down" (Staff, Systems Research Laboratory 1954). Effectiveness was 67% for sneak attacks, 75% for sneak attacks with splits, 59% for friendlies without flight plans, and 65% for the same type in the distraction category. Effectiveness against massed raids was recorded as 93% in the first set of sessions, 86% in the second, 97% in the third, and 94% in the fourth. As noted earlier, the totals which the subjects faced rose from 30 in the first set to 90 in the last. Effectiveness against combined totals of sneak attacks and friendlies without flight plans was proportionately higher for the lighter loads, but a result of considerable interest is that when the heavy load in one set became the light load in the next, effectiveness rose. "Group learning is most clearly indicated by the marked improvement with which the crew deals with the same penetrating load in successive sets: 19 percentage points improvement for load 3 between the second and third sets, and 36 points difference for load 4 between the third and fourth sets" (Staff, Systems Research Laboratory 1954).

Quantitative findings are not the predominant type reportable from this research, partly because of security restrictions, partly because even the data which were processed were not all summarized. Other published reports are reviewed here briefly. In Chapman et al. (1959) the number of tracks carried in Cowboy were plotted in comparison with tracks in the input for "all tracks," for "important tracks," and for "unimportant tracks," over the sixty-four periods of the experiment (but without actual values on the ordinate). Figure 6 shows these plots. The number of important tracks carried was shown as increasing, contributing some rise also to "all tracks," as the number of all tracks presented rose. Differential treatment of important and unimportant tracks, the principal point which the authors made, was readily apparent, but some caution might be advised in interpreting the rise in slope of the important track curve. In what sense do the data reflect crew learning? It should be recalled that load levels overlapped among sets. Sweetland and Haythorn (1961) plotted, as indicated in Figure 7, the number of penetrating tracks carried from period to period independent of sets, for five load levels. (They equated the heavy load of one set with the light load of the next to establish these five levels.) Although progressively more tracks were carried as the load increased, very little rise in slope is discernible within any load level. Further, they compared the first eight periods of crew performance handling each load with the last eight for the same load (in the following set), and wrote: "These comparisons showed tiny (but statistically significant) reductions of the number of tracks carried . . . . In general we were more impressed by the absence of evidence of learning, than the presence." (The fact that the number actually dropped may have been due to the fact that the input load for the second group of eight periods was, in fact, somewhat below that for the first. Another observation is that there was a tendency in the first

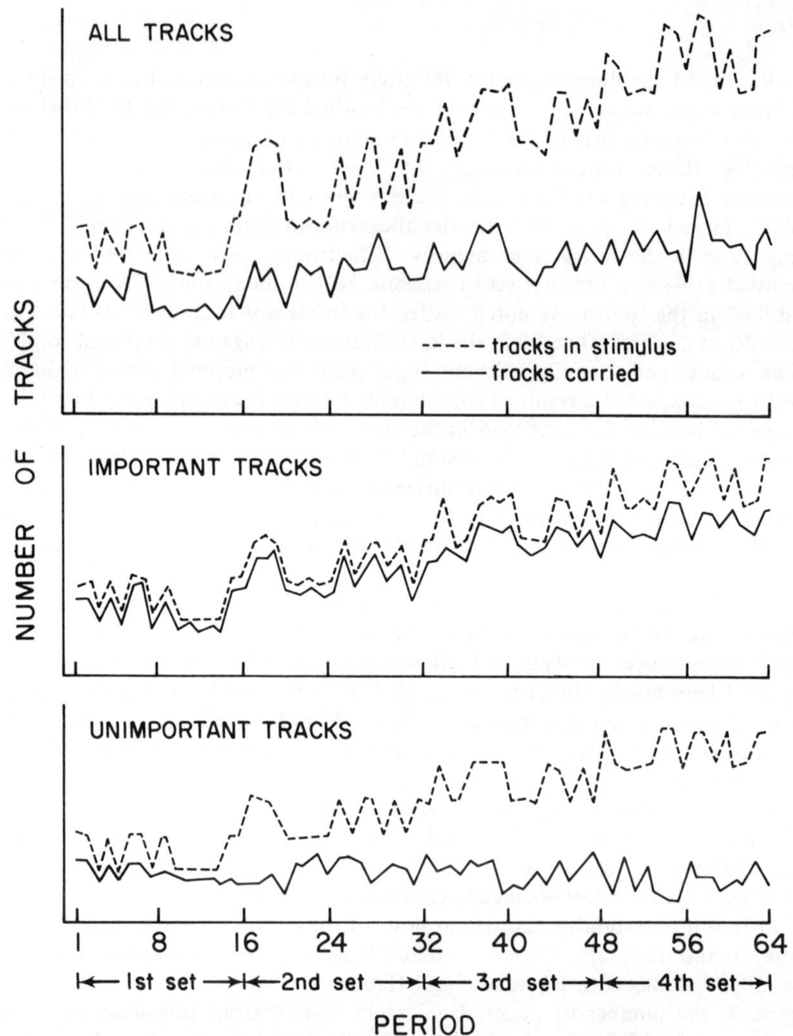


Fig. 6. Number of Tracks Carried (Chapman et al. 1959).

few periods for the number of tracks carried to increase, as if there were some initial learning.) The conclusion of Sweetland and Haythorn concerning the absence of "learning" in these experiments came to public notice only some time after the belief had spread widely that such "learning" had occurred.

Alexander (1955), one of the investigators added to the RAND staff for the Cobra experiment, made the following cogent comment about the question of "learning" and how this term should be characterized:

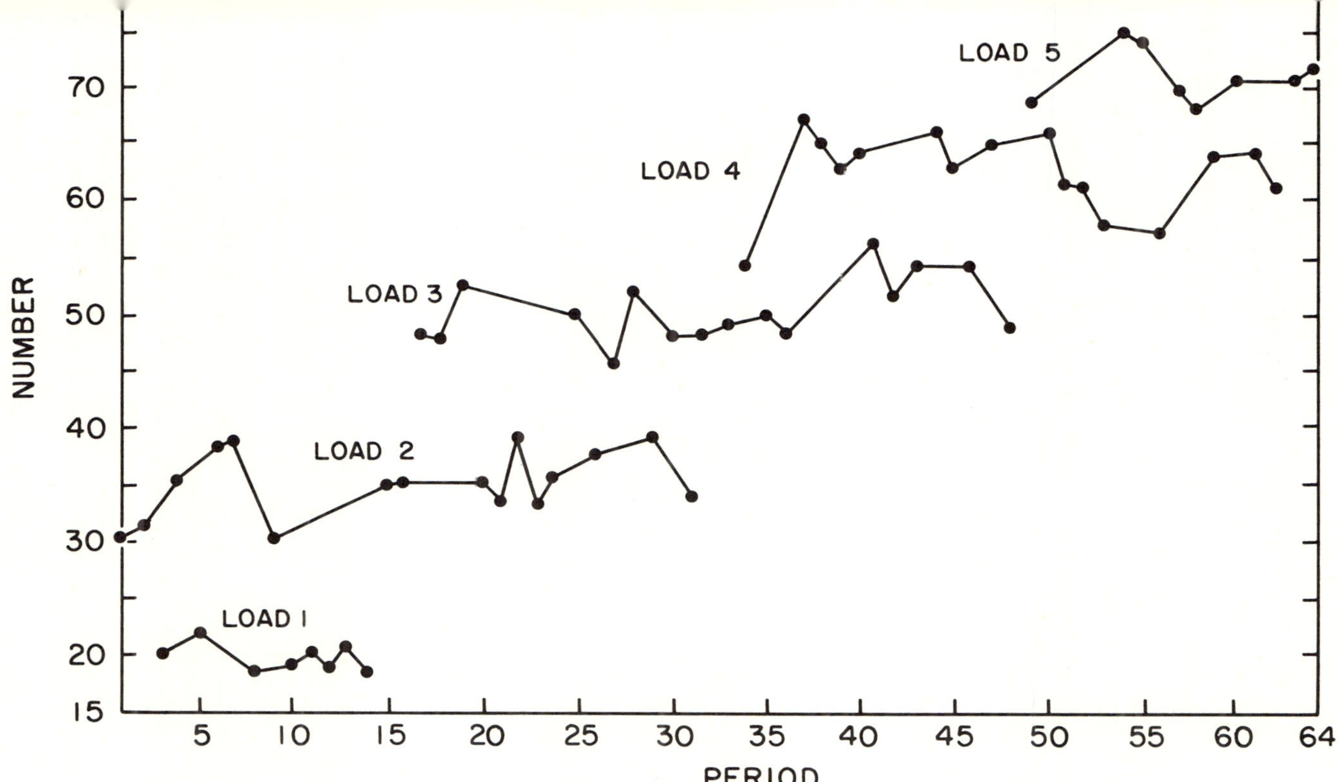


Fig. 7. Number of Penetrating Tracks Carried per Load, Independent of Sets (Sweetland and Haythorn 1961. Reprinted from *Behavioral Science*, vol. 6, no. 2 [1961], by permission of James G. Miller, M.D., Ph.D., editor).

In situations of this kind learning is represented in a somewhat different way than in classical learning experiments. In the usual learning experiment task difficulty is maintained at a constant level, and the experimenter makes inferences about the learning process by examining the changes in performance which take place. Because of our Casey experience, which impressed us rather strikingly with the fact that the capability of organizations increases markedly with experience, we used a design in the later experiments in which task difficulty was increased step by step throughout the experiment. When this is done, performance scores may remain nearly constant, and inferences about the learning process must take the increasing task difficulty into account.

The Sweetland and Haythorn analyses, the only ones in this research concerned with analyses of statistical significance, confirmed earlier findings that the crews did manage to carry heavier track loads as the loads which they had to carry grew heavier, and that "In general, crews maintained 'important' tracks and eliminated unimportant tracks. This process was called 'filtering'." They added: "Load increases finally caused a pruning of almost all behavior not critical to defending the area . . . the crews also (as load went up) tended to carry tracks for shorter and shorter times, and also with fewer and fewer reports." Their dependent variables was exclusively track-handling behavior, for which they used four measures: number of stimulus tracks carried, percentage of stimulus tracks carried, items of task-oriented behavior (each item of track handling information), and average number of responses (item of task information) per track. In their analyses of variance, Sweetland and Haythorn found that there was more activity during the first period of a problem than during the second, but this diminution could not be attributed to fatigue effects. They also found that the Cobra crew handled more nonessential tracks than the Cowboy crew, but the two crews performed similarly in carrying critical tracks.

Chapman's 1960*c* report, relatively unknown even to those familiar with the RAND SRL program, includes fifty-three pages of tabulated data and fifty-six pages of graphs, described as "an attempt to make available a portion of one of the most extensive collections of experimental evidence ever assembled about organizational behavior." The data came from all of the last three experiments, Cowboy, Cobra, and Cogwheel, and were coded in terminology which was derived from a conceptual model or organizational behavior rather than in air defense terms. Most of the tabulated data pertain to: (1) the number of task events presented to the subjects; (2) the number dealt with by at least one response; (3) the ratio of these two numbers; (4) the amount of information processing (position, speed, altitude, fade, etc. reports) for all events of a class of events or sum of classes; (5) the ratio between this last measure and the number of task events dealt with by at least one response; and (6) the product of the two ratios just described. These measures are shown for each class of tracks (local, penetrating, outgoing, and hostile), for each of a number of defended areas such as subdivisions of the ADDC area and the EW stations and their adjacent areas, for each of the sixty-four periods in all three experiments. There are also summary tables for all tracks. Additional measures are listed for a single category of tracks for one of the EW stations. The graphs depict: (1) actual processing rate, namely, all items of task-oriented behavior (also shown in the summary tables) for each period, for the ADDC and each EW station; (2) the first ratio described above for summed and selected classes; (3) the second ratio described above for summed and selected classes; and (4) some of the measures tabulated for a single category of tracks. The data are not aggregated in tables or graphs beyond the summaries per period, to indicate results in more general terms.

## CONSEQUENCES

The consequences of the experiments did not wait for the processing of the data. A senior general who observed an entire Cowboy session until 2 A. M. was impressed by what he saw. Cowboy was conducted in January of 1953. After its conclusion a study team met for two months at RAND to determine whether the techniques of the experiment's management and simulation might be converted into a training program. The feasibility finding was positive and even included estimates (later found to be over-modest) concerning command-wide installation: cost, about 1.5 million dollars; time required, about 18 months; personnel required, twelve professionals and eighteen technicians. It was decided that RAND would hire eight more professionals to design such a training program; these included L. T. Alexander (see Chapters 11 and 17), H. H. Harman (Chapter 17), and H. Sackman (Chapter 11). Cobra, which ran in February of 1954, was the experiment conducted to indoctrinate them.

### *Cogwheel*

The last experiment, "Cogwheel," was conducted the following June. It was clearly a consequence of the earlier ones. Like Cobra, it was funded by the Air Force, with its principal aim to indoctrinate the Air Force officers and enlisted personnel who would be manning four real, operating ADDCs and a control center where the training program would be tested and initiated. As with Cobra, the experimenters were happy to have another chance to run an experiment for checking the results obtained in Cowboy. From Cogwheel also came an educational motion picture which dramatized the simulation of air defense and the operations which led to improved crew performance (Chapman and Weiner 1957).

This fourth experiment closely resembled Cowboy and Cobra but was speeded-up: it was much shorter, requiring only fourteen four-hour sessions for both practice and experimental runs. The latter consisted of only sixteen periods, in contrast to the sixty-four of Cowboy and Cobra. The load variables and inputs for Cogwheel came from these other two. Eight periods were drawn from the first Cowboy-Cobra set, four from the second, and two each from the third and fourth. The results were never analyzed to the point where quantitative comparisons could be made between Cogwheel performance and the preceding experiments, except for some data in Chapman's 1960*c* report, but the experimenters judged that changes occurred in much the same way.

### *System Training Program*

The training program which resulted from this research was called STP, or system training program. When the Air Force tested its earliest version with the former Cogwheel crew at the operational sites, the test did not yield impressive evidence in pre-test, post-test comparisons to show that crew performance was significantly affected. Nevertheless, this training method using current operational equipment was welcomed as a presumably less expensive alternative to a proposed outlay of considerable magnitude for new equipment. This equipment involved automation of status displays and an optical projection system for transferring track data from a new type of surveillance PPI-type display to a common geographical display (see Chapters 5 and 6). RAND received a contract to design STP and help install it. A new division was created at RAND that shortly became larger than all the rest of that organization and then split off to form the System Development Corporation (SDC). Its president was M. O. Kappler, the engineer who had been peripherally associated with the experiments and was involved in the development of a new device for simulating the air picture for the training program. Its vice-president was one of the SRL experimenters, W. C. Biel. Both SDC and STP expanded with time in a way that no one thought would ensue from those two additional sessions of Casey. STP was eventually installed in the more than one hundred ADDC's in the manual air defense system, and when this system gave way to SAGE, a similar program was introduced into that computer-based system. The older STP was extended to Canada, Alaska, West Germany, Spain, Turkey, Norway, Hawaii, the Philippines, Okinawa, and other locations, and to tactical air locations. When air defense was semiautomated in West Germany with the 412L system, STP was converted to train its personnel, and when a stand-by system was developed for SAGE, called BUIC (Back-Up Interceptor Control), a similar training program was developed and installed; the BUIC program was named SETE—system exercising for training and evaluation.

Early during this expansion, RAND was asked to undertake the computer programming for SAGE as a consequence of its experience in air defense (from STP) and its pioneering in computer technology, including the computerized production of simulation input materials for the training program. SDC continued the SAGE programming and subsequently engaged in the design and production of computer programs for other command and control systems and then for other types of information systems.

To support STP, a rationale evolved embodying a number of "principles" which supposedly grew out of the Systems Research Laboratory experiments. Chapman and Kennedy (1956) had suggested that the experiments "enabled us to understand enough about how the organizations developed in the laboratory to formulate a useful principle: Train a team as a whole in an adequately simulated environment and give it knowledge of results." This last point was defined as "a factual critique which helps the organization identify its difficulties." Chapman et al. (1959) elaborated on this "off-the-top" finding as follows: "Our research indicates that these are the conditions necessary to promote organizational learning: clarify the goal, give the organization as a whole experience with tasks of increasing difficulty, and provide immediate knowledge of results." In subsequent years many versions of "STP principles" were published; as would be expected, new techniques emerged and old techniques received interpretations and emphasis which varied from those characterizing the experiments. One of the aims of the present account of these RAND SRL experiments has been to present their actual history in place of myths which may have gained some currency and may have created a mystique about the principal product of this research. One of the researchers, Kennedy (1965), has made an intriguing comment which may also promote more searching attitudes toward STP "principles." "It is possible," Kennedy wrote, "that the best summary statement about system training might be:

Something old (Law of Effect)
Something new (high fidelity environment simulation with computers)
Something borrowed (debriefings)
For the boys in blue? (specific to air defense environment)."

One consequence of the SRL experiments was, ironically, the dissolution of the laboratory. The report by the Staff, Systems Research Laboratory (1954) said that research on the science of organizations would continue in SRL as the system training program was being installed. Several ventures were projected, including a war game and the experimental investigation of a weather system. No more experiments were funded or undertaken, however, and the principal investigators dispersed. Yet there was some carry-over within RAND, in the establishment of a Logistics Systems Laboratory (Chapter 13). The experimental and simulation technology in that laboratory owed much to the SRL experiments.

*System Concepts and Principles*

What general concepts came out of the RAND SRL experiments? The experimenters believed that they had proved their basic, single hypothesis or assumption, that a motivated organization with a goal can and will adapt when it faces new situations and problems; it will solve its own problems. But why and how does it do so? According to Chapman et al. (1959):

> The members of each crew became an integral unit in which many interdependencies and coordinating skills developed. And each crew learned to perform more effectively. This learning showed itself in procedural short cuts, reassignment of functions, and increased motor skill to do the job faster and more accurately.
>
> We believe that "debriefings" following each session, where the operating results were reviewed, were crucial to the learning that led to improved performance. But we have been unable to relate the content of these discussions directly to crew development. Procedures were frequently changed without any sign that an operating problem had been recognized or a solution proposed. As a matter of fact, procedural changes sometimes moved in one direction while discussions went in another.

One procedural change was well demonstrated, as already noted: the filtering out or pruning of unimportant tracks as load increased. This could be otherwise stated as the filtering out of easily recognizable noise, so that the crew could concentrate on their real problem–repelling invaders. Chapman (1956) listed other procedural changes. But it would be difficult to categorize these for presentation to the reader, even if they were not classified, although Chapman et al. (1959) attempted to do so in such terms as using "redundancy in information input to rebalance the processing load" and to make cross checks, and "sensitivity to information patterns and awareness of action alternatives." Further it is impossible from the experimental evidence to determine the relative contributions to better performance from procedural changes, reassignment of personnel (there was a notable improvement in one of the experiments when one of the key supervisors was shifted to a different function), and increased skill, both individual and interactional. Yet this is an important distinction to make, as has been pointed out elsewhere (Parsons 1964), because the kinds of optimal feedback in debriefings could differ for developing new or different procedures ("procedurization") and for raising levels of skill in carrying out procedures.

Other concepts which later were elaborated into STP "principles," such as immediacy of knowledge of results, training the team as a whole, and presenting tasks of increasing difficulty, may have accounted for the experimental results. However, the experiments themselves could not provide evidence as to how much any one of these "principles" contributed, if at all, or in what proportion; and evidence of their individual value is just as difficult to find elsewhere. It is interesting that in later years none of the SRL researchers expressed as much enthusiasm about the "principles" as did many individuals who became converted to them. It must be recognized, however, that these concepts were instrumental in explaining the STP to potential or current military users and in widening its adoption, even though they were not principles in the scientific sense of the term and they were publicized in different versions. Chapter 11 describes SDC's attempts to verify STP experimentally.

Other concepts than these have been generated by the research. Two of them, failure stress and discomfort stress, were explained by Chapman et al. (1959) as follows: "As the task load increased, the crews were caught between two stresses—failure stress and discomfort stress. The first of these arises from the disparity between aspiration and performance; the second from the difference between the effort demanded by the task and that which can be comfortably afforded. The discomfort stress forces discriminations and short cuts in response; the failure stress guides the gradual acquisition of short cuts that do not degrade effectiveness."

This analysis can be interpreted as a pioneering attempt to structure "stress" in motivational terms rather than as something which degrades performance or physiological functioning. Rephrased in the language of operant conditioning, these concepts mean that operators were conditioned to avoid or escape failure and discomfort by performing in ways which successfully prevented or terminated these stresses. An analysis of the literature on stress (Parsons 1966) would suggest that this usage by Chapman et al. (1959) may have been the first of this nature to be published, as well as the first application to organizational behavior. It is also outstanding in that there have been few examinations of motivational variables in connection with man-machine system experiments. According to Chapman and Kennedy (1956), formulations using these stress concepts "should help to predict how fast and how far a system can adapt, to identify what is difficult in the task, and to define the conditions that help an organization use its resources most effectively."

Still other concepts suggested by the researchers relate to human engineering and personnel requirements (Chapman and Kennedy 1956) and could be considered as hypotheses for experimental inquiry:

> Once the importance of group learning is recognized it follows that equipment should be arranged not just to facilitate operation but also to help the men who operate the system learn to use its full potential most rapidly. Or, more practically, since specifying what these men are to learn is difficult unless the system can be operated under the emergency conditions it was designed for, doing anything that might hinder group learning should be avoided.
>
> Communication between members should be made as free and easy as possible. Facilities should be arranged so that each member of the group is given as complete a picture as possible of the task and how the organization is dealing with it—in central displays of some sort if these are feasible.

And: "Considering a system as an integral unit rather than as a collection of individuals says something about personnel selection. It suggests that, in manning a system, teams rather than individuals should be selected, that matching the individual to the job may be part of the organizational development process."

These various concepts are related to each other by another, namely, that it is profitable to think of the behavior of an organization as resembling that of an organism. Chapman et al. (1959) stated their belief that their simulated air defense establishment "profited from its experience to grow and adapt like a living organism." Among a number of possible derivative concepts, one is that since an organism must have the flexibility to adapt to circumstances, then so must an organization. But operational flexibility must then become an objective of the planners, designers, and managers of the organization. Although this approach seems to give the organization a "unitary" character, this is not necessary or perhaps even wise. In another report Chapman et al. (1952) suggested that much attention should be paid to this organism's "internal behavior" and "element interactions."

Chapman (1960*c*) has also drawn from the SRL research a conceptual framework of organizational behavior, in which he distinguishes between steady-state systems and changing-state systems. In the latter, the information processing capacity becomes "processing coercion," which together with an "effectiveness coercion" determines the way the state will change. The effectiveness coercion consists of a compelling pressure for task accomplishment, is necessary for performance to be optimized, and derives in part from the operators' motivation. Systems have "inertia," meaning tendencies to continue to operate with current attention and response practices and to maintain the same normative processing rate. The "attention practice" assigns relative importance to various task events. The response practice assigns the amount of information to be processed about those task events discriminated by the attention practice. Taken together, these practices determine what information processing will be given priority. The effectiveness coercion can overcome system inertia by bringing about changes in attention and response practices, and then the system state will change. This framework represents an attempt to describe in general terms how changes toward optimization come about in man-machine information processing systems. In comparing such systems and dealing with them in the abstract, some framework like Chapman's can have value.

Like the experiments from which it arose, such a framework should be regarded as heuristic. One way of looking at the RAND Systems Research Laboratory's research is in heuristic terms. Its Marco Polos explored new domains and originated significant concepts as guidelines for the solution of problems. Such contributions to discovery should be acclaimed. Rigorous and constrained analysis to determine functional relationships among variables is not the only face of science. The RAND Systems Research Laboratory program was scientific exploration on a massive scale.

# 9

# Studies of Army Operations

Complex man-machine experimentation in Army operations has been going on for many years, although much of it is not widely known. The research has been distributed among a number of organizations, some of them Army components, some not. They have included Psychological Research Associates, University of Michigan Willow Run Laboratories, New York University, Stanford Research Institute (see Chapter 14), Combat Development Experimentation Center (also see Chapter 14), Research Analysis Corporation, Combat Operations Research Group (CORG) of Technical Operations, Inc., Army Personnel Research Office (now the U.S. Army Behavior and Systems Research Laboratory), and Human Resources Research Office (HumRRO), formerly connected with George Washington Univeristy. By the nature of Army operations, the experimentation has been particularly concerned with evaluations, training, tactics, procedures, and manning. Among the objects of investigation have been rifle squads, battlefield surveillance, photo-interpretation, combat tactics of various types of military units, and tank operations. Some of the research has been done within four walls, but frequently the laboratory has been actual terrain designated and instrumented for experimental purposes.

## PSYCHOLOGICAL RESEARCH ASSOCIATES INFANTRY STUDIES

In the years 1950–57, M. D. Havron and his associates conducted a number of infantry studies of substantial scope, initially for the Institute for Research in Human Relations, then for Psychological Research Associates. These included: (1) development and standardization of field problems for testing scout squads of the reconnaissance platoon of an armored cavalry regiment (light); (2) development and experimental checkouts of tests of the effectiveness of infantry rifle squads and evaluation of effectiveness predictors; (3) development and experimental evaluation of four infantry rifle squad training methods and evaluation of a composite method developed from these; (4) experimental determination of the best size and composition of an infantry rifle squad; and (5) experimental investigations of infantry small arms fire (including its "psychological effectiveness") and communications.

*First Program*

The first two of these programs were undertaken for the Personnel Research Section, Personnel Research and Procedures Branch, The Adjutant General's Office. The field test problems in the first program (Havron, Fay, and Goodacre 1951) were evaluated for validity, reliability, and practicality. The test contents attained these objectives and were incorporated into a new field manual for armored units. Although Army field tests are primarily intended to determine whether training of Army units has resulted in their operational readiness, these tests turned out subsequently to be useful also as instruments for training. No definitive conclusions were drawn concerning a number of effectiveness predictors which were proposed, since these were tested with only twelve units, but some of their concepts entered into the second program. It was hoped to ascertain predictors of individual effectiveness through correlations with performance in field tests, in order to do a better job of selecting and classifying Army personnel.

*Second Program*

In the first of two parts of the second program, as reported by Havron, Fay, and McGrath (1952), a field test problem for an infantry rifle squad was constructed and pretested with six squads at Fort Benning, Georgia. It had an attack phase, a defense phase, a reconnaissance patrol phase, and a point of advanced guard phase. Then the problem was fully tested with thirty-seven squads at Fort Benning in April and May of 1952 and with sixty-three squads at Camp Atterbury in June and July. The four phases could be laid out on a circular course so four squads could be tested at the same time, each starting with a different phase and proceeding through the other three. This innovation made the test problem easier and more economical to administer.

Umpires rated forty-eight randomly selected squads, half of whom performed on one terrain course, half on another. Neither the difference between courses nor the serial position of phases within the test problem appreciably or significantly (according to analyses of variance) affected the ratings. Correlational analyses showed both high agreement between umpires and consistency between phases among the squads. However, squads drawn from one Army unit were significantly better than those drawn from two others. The researchers concluded that the field test problem could be standarized despite differences of terrain and that squads could begin it in different phases and go through it in different orders; and also that it was a reliable testing instrument. The researchers' success led the Army to ask them to evaluate a newly developed battalion field test, which they gave to three battalions of the 82nd Airborne Division at Fort Bragg, North Carolina, later in 1952 (Havron, Fay, and McGrath 1952). Subsequent analysis was based in part on observations of these exercises and on discussions with umpires and battalion officers.

The scene of the second part of the program was Fort Lewis, Washington, in 1954 (Havron, Lybrand, and Cohen 1954). First, three field test problems were constructed and pretested with rifle squads and umpires. One was for daylight firing of blanks, the second for daylight live firing, the third for night blank firing. The last two types of testing, which had not been covered in the first part of the program, posed new requirements. To simulate enemy fire, for example, long-fused firecrackers were placed in front of silhouette targets hidden by brush. Each problem yielded a squad leader score, a squad member score, and a total squad score; these were weighted to produce total scores. Then each of the daylight problems had four rehearsals and the night problem had eight. Each of the problems was tested with each of 112 rifle squads drawn from 3 regiments of the 44th Division, at the rate of 20 squads per week. There were twenty-four umpires and safety officers, most of the umpires being noncommissioned officers. Different terrain was allocated to each problem. As in the 1952 evaluation, umpire agreement was high; also, correlations were high between the scores of squads rated by different umpires on different days or different problems. Differences between terrains and order of problem phases did not significantly affect total problem scores. Because of these and other results, the field problem scores were deemed suitable effectiveness measures with which to compare predictor measures to see if these were valid. The assessment of predictors of effectiveness was actually the principal aim of the study. Sixty such measures of squad members had been obtained before a squad began any problem. A number of these were found to have high correlations with performance scores in the tests.

*Third Program*

In the third program (Havron, Gorham, Nordlie, and Bradford 1954), conducted for the Human Resources Research Office, thirty-two rifle squads were trained by four methods, eight squads per method, at Fort Jackson, South Carolina. One of the four methods was that currently in effect, with certain additions in technique resulting from the prior program. The others were called "group participation," "combat fundamentals," and "team training"; they were developed especially for the program. The control (current) method was administered to eight more squads after the others had been taught, to see whether the instructors improved. After training, the squads were given the rifle squad field test developed by the researchers in 1952 and another test called the "Leaderless Group Test." These provided scores to assess the training methods. The researchers interviewed the trainees and compiled the instructors' and their own opinions to evaluate major components of each method as well as specific techniques of instruction. Then they devised a "final training method" by integrating much of the combat fundamentals method and some of the team training method into the current method, together with some other new techniques which were innovated during the program. Forty more squads were trained with this final method, some by new instructors, and took the field test and Leaderless Group Test. Their scores were higher than those of squads trained earlier with the current method. In fact, average scores of squads trained by the final method were well above the 90th percentile of the trained squads tested in the second program; all squads made higher scores on the Leaderless Group Test than these earlier ones. This absence of overlap was a rather unusual finding.

### *Fourth Program*

The next program, investigating the optimal size and composition of a rifle squad, was initiated in 1954 for the Operations Research Office of The Johns Hopkins University (Havron, Burdick, Hutchins, and Buckley 1954; Havron et al. 1955). It was completed for the Combat Operations Research Group of Technical Operations, Inc., contracted with the Continental Army Command (Whittenburg et al. 1956). Experimental data were collected at Fort Benning, Georgia, in 1955 from eighty-eight rifle squads whose size and weapon assignment varied in eleven compositions. The principal question was how many riflemen or automatic weapons personnel one leader, unaided, could control; no assistant leader was designated. Squad size, not including the leader, varied between three, four, five, six, seven, and ten men (Havron and McGrath 1962). Each squad went through three days of training and a six-hour daylight field test which included both attack and defense aspects and which measured such squad capabilities as hits in firing, communications, vulnerable exposure of the squad leader, and speed of deployment. The variables of terrain, training, climate, and enemy action were controlled; others, such as weather, visibility, marksmanship, and physical condition of the personnel, varied randomly. Previous experimentation had explored verbal communication among squad members in action, with different methods of transmission, presence or absence of firing noise, and variation in foliage and wind direction.

While detailed results of this program are still classified, it can be stated that no one organization or weapon combination was greatly superior to another of approximately the same size and composition. Individual differences in leader capability, terrain density, and other factors had a marked effect, no matter which "table of organization" was being tested. A leader-to-man ratio of one leader to five men worked as well as any other, and significantly better than some ratios. A leader-to-man ratio of one leader to seven men taxed the leader, increasing his vulnerability and making control difficult. To test limits, some squads consisted of one leader with ten men. No leader was able to control ten men. (M. D. Havron, personal communication; Havron and McGrath 1962).

### *Fifth Program*

The last program, which concentrated on small arms fire and communications, was also conducted for the Combat Operations Research Group of Technical Operations, Inc., under the title "Platoon Organization Studies Research Program." It was preceded by field studies of the psychological effects of weapons. Data on individual performance of infantrymen were collected in experiments at the Combat Development Experimentation Center, Fort Ord, and Camp Roberts, both in California. Aggregated effects were calculated from the data from individuals (Havron et al. 1957). Representative of the kinds of experiments were those reported by Vaughan and Kassebaum (1957). In one study of how the amount of concealment degraded the ability to hit targets, twenty-four simulated bushes constructed of wood excelsior wrapped in chicken wire were placed in front of targets 200, 400, and 800 yards away on flat or sloping terrain. The width of the bushes varied. It was found that results could be fairly accurately predicted from the ratio of target size to bush size, together with knowledge of average mil error. In another experiment riflemen and machine gunners fired various numbers of bursts with differing numbers of rounds from different distances at subjects in covered pits. Subjects indicated which combination of firing volume and distance they regarded as most dangerous.

## UNIVERSITY OF MICHIGAN WILLOW RUN LABORATORIES

When the Air Force decided to place its air defense chips on SAGE (Massachusetts Institute of Technology) and to discontinue further development of ADIS (University of Michigan), as recounted in Chapter 6, the Willow Run Laboratories in Ypsilanti, Michigan, almost immediately shifted to research and development in tactical surveillance. Although initiated in 1954 for all three military services, this work became entirely oriented toward surveillance of the battlefield as an Army jurisdiction. Of particular interest were various sensors such as radars and photography and their integrated use, and especially the processing and interpreting of the data which they could provide. The Willow Run Laboratories, also identified at times as Willow Run Research Center and the Institute of Science and Technology, labeled this work "Project Michigan."

Some of the personnel in the man-machine system experimentation for ADIS found themselves in Project Michigan and were able to recreate man-machine system research, to a limited degree and with relatively modest equipment, on a few occasions during subsequent years. Some of their efforts will be reviewed shortly, to the extent information was obtainable; there have been few reports which were or have become unclassified or which became available outside the Willow Run Laboratories.

### *An Effort That Failed*

In addition to these efforts, another experimental unit existed during 1957–58 under the direction of S. Veniar. Much more elaborate objectives and equipment were projected for this activity. A facility was proposed which would require expenditures of $1,827,380 in 1958; $2,870,100 in 1959; $3,101,300 in 1960; and $3,239,300 in 1961. The proposed 1959 budget, for example, included $712,650 for digital computing equipment and space; $868,282 for analog computing and simulation equipment, displays, and maintenance; $855,000 for scientific personnel; and $475,000 for technical personnel. The proposal, which was unaccompanied by a detailed experimental plan, was never adopted.

One interesting piece of equipment, developed during this time period, did see some experimental use. This was ASITS, an automatic method for introducing teletypewriter inputs for simulation in laboratory experiments on combat surveillance or in experimental command post exercises (Kaufman, Payne, and Bailey 1959). The equipment accepted punched paper tape containing simulation inputs and sampled these, according to experimental plans, for distribution to receiving stations. In addition to teletypewriter transmitting and receiving units, it consisted of a clock-controlled sampler unit and a tape-controller sampler. ASITS figured in the only experiment in the 1957–58 program. W. T. Pollock and G. C. Bailey compared four arrangements of task allocation and equipment within two-man "receiver" and "collator" teams. The task was to receive dissimilar information from different sources, reduce it to a common form, and collate it according to the geographical locations with which the various inputs were associated so that multiple pieces of information could be combined for a particular point of reference. Eight military communications specialists were assigned as subjects to four two-man teams. Each team handled one hundred messages in each team-equipment arrangement (Pollock and Bailey unpublished report).

### *The Continuing Program*

In the continuing Project Michigan program mentioned previously, an Army control center was simulated at the Willow Run Laboratories in 1955 to represent manual information processing at divisional and regimental headquarters during wartime operations. In addition to considerable "exploratory" investigation in this facility, a series of studies, which admittedly left much to be desired in experimental control, provided information about delays and errors in routine logging, duplicating, plotting, and disseminating messages from a battlefront. Operation Husky (Mosimann, LaRoche, and DeVoe 1955*a*) recapitulated the first three days of the Allied invasion of Sicily in World War II. Twelve military personnel played the roles of the intelligence and operations sections of the 1st Infantry Division and 18th and 26th Infantry Regiments. Sixteen other "control" individuals functioned as message sources. There were three switchboard operators. The inputs, prepared by a group of Army officers to reflect events in the Sicily campaign, consisted of 165 telephone messages, 11 map overlays, and 116 documents and hand-carried messages. The exercise ran on three successive Thursdays for a total of twenty-seven and one-half hours. Although the extensive data collected gave gross indications of the speed and accuracy of processing information, the study suffered from variations in operator proficiency, changes in procedures, and fluctuations in input rate during its course. The researchers commented that "there were so many uncontrolled and varying factors that the greatest value of the exercise is in emphasizing the need for more detailed planning and control of operational parameters."

In consequence, two more multioperator studies followed, as well as a test of individual performance. Operation Slowdown (Mosimann, LaRoche, and DeVoe 1955*b*) and Operation Slowdown II (Mosimann, LaRoche, and DeVoe 1955*c*) incorporated the same subjects, control personnel, organizational context, and simulated situation as Husky. In the first, fifty messages from Husky's first day were divided into five groups of ten messages each; these groups went into the system at rates of one message per 20, 10, 5, 2½, and 1¼ minutes. Delays in recording and disseminating the messages progressively increased as entry intervals shorter than 10 minutes decreased. The second study copied the first by systematically varying the input rate (except for the 20-minute interval) with ten of the same messages for each rate, but the procedure was changed to connect telephones in conference loops or nets, instead of point-to-point, one loop for the division and one for each regiment; there was also a new procedure for dissemination. No appreciable change in complete processing time resulted. The test of individual operators determined how long they took and how accurate they were in four separate tasks: recording messages received over a telephone; plotting information on a situation display from written and spoken messages; logging messages in a journal; and extracting information from messages for worksheets. The rationale of the study was to provide "more useful measures" than those obtained in the preceding exercises; these, according to an unsigned laboratory memorandum, "were not very useful in analyzing the delays because various message lengths, message complexities, and operator procedures were present."

The researchers tried to obtain similar kinds of data about information processing in a large field exercise–Sagebrush–but the outcome failed to meet their hopes. They felt that there had been insufficient training of the subjects, that communications were inadequate, and that there seemed to be much concern as to how the experiment's outcome would affect an individual's standing. The experimental design they proposed encountered opposition, which was successful. However, some useful data were obtained concerning communication time delays.

Of considerable interest was the development in 1955–56 of a Surveillance Game by R. P. DeVoe and his associates (Brady, DeVoe, and Pittsley 1959) and its subsequent expansion at the Army's request into a Surveillance Station. With this, following a pilot experiment in December, 1958, D. H. Wilson and associates in 1959 conducted an extensive program of experimentation with six-man teams (Brown et al. 1960; DeCicco et al. 1962).

The Surveillance Game involved a single player who could view a vertical display of the status of sensor subsystems, a horizontal status display for weapons, and a vertical situation display. The sensor subsystem display listed the capabilities and limitations of four methods of reconnaissance: visual air, photography, infrared, and airborne side-looking radar. The situation display consisted of sensor data on acetate overlays superimposed on a map. The overlays were changed at fixed intervals to show new sets of data. The player could take one of three roles, as "postulator" only or, in addition, as assigner of missions to sensors or as assigner of weapons against targets. Postulation meant collating the surveillance data from the sensors on the overlays to summarize the data or to form a target on an evaluation sheet. An operator changed the overlays and scored the player's performance. The locale for the exercises was an area around the Italian–Yugoslav border; the simulated military situation could extend to forty hours of operations of an aggressor mechanized army, in much detail. In general terms, the player's task was to infer the location of hostile units from the sensor information available, and perhaps also to assign and fire weapons to destroy it. This game proved very educational. Although it was not used for experimentation as such, the approximately thirty players seemed to divide into two discernible types: those who wanted to collect complete information before acting, and those who gathered surveillance data to check on hypotheses of unit movement stemming largely from general knowledge about the kind of problem, the terrain, and time-space factors. The latter were more successful (R. P. DeVoe, personal communication).

In the Surveillance Station, some subjects plotted sensor data manually on a vertical target and detection display after receiving the data at consoles from overlays showing a related display. An overlay bore the content of messages from a computer-generated paper tape controlled by the ASITS device mentioned earlier. The subject would do some processing of the data at his console and transfer the result to the same co-ordinate position on the vertical display, where other subjects would perform postulation and sensor-mission assignment functions. The sensor data represented simulated tactical maneuvers of two aggressor divisions in a 20 × 20 mile square in the Hunter–Liggett Military Reservation in California. Experimental sessions lasted about three hours (two hours in the pilot experiment). The mission of the surveillance station was to track a task force of two tank companies and one rifle company. Subjects were scored on their ability to locate the positions of certain task force units at various times during the experimental session. The subjects were technicians and statistical clerks employed by the Willow Run Laboratories. They had had many hours of experience in the simulation operation before they served.

In the pilot experiment (G. C. Bailey, personal communication) there were seven runs in which eighteen subjects performed in various combinations. Measures included seven time or error scores which were related to seven factors, such as individuals, combinations of individuals, targets, and combinations of targets; the statistical significance of these relationships was determined by analyses of variance. In the experimental program proper there were 125 runs. More than one thousand observations of tracking error were related through analyses of variance to nine factors at five levels each, most of them associated with simulation input and crews. Thus, variations in the input, such as sensor capability, constituted a number of independent variables. The program was organized according to five blocks of five runs each, with a different set of six subjects for each block, which was a 5 × 5 Latin square. The results were reported for two of the blocks but are classified.

During this program the laboratories were asked to expand the simulated station to a much larger operation using an IBM 709 computer. Wilson pressed this development but it ran into a combination of high costs for equipment and programming, and divergent viewpoints concerning the degree of automaticity to be designed into Army surveillance systems. Eventually the undertaking was discontinued. The attention of the researchers turned to evaluations of radar equipment and problems in image interpretation.

## NEW YORK UNIVERSITY RADAR SURVEILLANCE STUDIES

For a number of years, beginning in 1954, the Research Division of New York University conducted experimental studies of human operations in radar surveillance systems which were precursors of the Army's Missile Master, a system for co-ordination and control of Nike ground-to-air missile batteries. Since reports of these have not been obtained for analysis here, only a brief review of this program is possible. Some of the studies were conducted for the Signal Corps, others in conjunction with Airborne Instruments Laboratory. A principal investigator was B. L. Cusack.

One project which had much in common with the Lincoln Laboratory's Pi-Sigma study (Chapter 6) investigated how to distribute surveillance tasks among operators. The detection and tracking tasks could be assigned to different operators; or the same operator could be responsible for all aspects of a track. Tracks could be assigned according to some rule, such as a track's geographical location; or they could be allocated sequentially from a pool of tracks. Another study examined task distribution among surveillance and heightfinding operators in countering electronic countermeasures (ECM). These studies used simulated aircraft radar signals and simulated ECM; to some extent they also compared simulation with actual radar presentations. Studies of individual performance were directed at types of cathode ray tubes (CRTs), priority schemes in heightfinding, comparison of individual console presentation with large-screen projections, surveillance and tracking radars, and effects on system performance of degraded tracking. In addition, six data processing and display configurations varying in extent of automaticity were experimentally compared for their influence on a task requiring threat evaluation.

## COMBAT OPERATIONS RESEARCH GROUP (CORG), TECHNICAL OPERATIONS, INC.

CORG provided technical assistance to the U.S. Army Combat Developments Command (and, before it was established, to the Continental Army Command) in many ways, including the preparation, conduct, and analysis of field experiments and troop tests. Distinctions between field experiments and troop tests have been set forth in CORG documentation which is not publicly available, and also by M. I. Kurke of CORG (1965), who in the same analysis sought to specify the distinguishing characteristics of "field exercise" in contrast to experiments and tests. Subsequently Kurke (1966) has recounted progress in Army troop test methodology.

Kurke's distinctions are:

Field Exercise—an exercise conducted in the field, under simulated war conditions, in which troops and armament of one side are actually present while those of the other side may be imaginary or in outline.

Field Experiment—an investigation to experiment with or evaluate new or revised doctrine and organizations, and new, modified or current material in order to develop combat capabilities.

Combat Development Troop Tests—a field investigation designed to test the ability of a prototype organizational structure to follow a specific doctrine, using specific equipment to complete a specific mission and/or test the concept of operations as limited by the structure and functions of a prototype organization.

In the field exercises, Kurke noted, the "method" is "free maneuver constrained by test events or problems incorporated within the scenario"; procedures are evaluated by military umpires, and the conduct of the test may be varied within broad guidelines by the decisions of the commander of the unit being tested. Accordingly, variables cannot be controlled. In a troop test the commander has less freedom of action within the scope of a more detailed scenario and within the constraint of a test plan to "collect pertinent data without undue contamination of test variables"; opinions and judgments of evaluators are systematically collected, may be scaled, and are supplemented by objective data. In field experiments, on the other hand, "controlled experimental procedures" are supposed to govern the collection of objective data which are supplemented by judgments of evaluators and participants.

Although field exercises and troop tests are outside the purview of this book, it is worth summarizing Kurke's description (1966) of a troop test called Water Bucket II at Camp A. P. Hill, Virginia, in 1965. Its objective was to evaluate employment and delivery of a "non-lethal riot-control munition in certain counter-insurgency tactical situations." The test directors were troop unit commanders. Squads representing protected and unprotected hostile and friendly troops were drawn from two platoons of forty-three men each and a test team of eleven men. The number of trials was limited by the learning factor and a requirement that any unprotected player be exposed only once. Tactical situations were varied. No two trials had the same set of conditions but many were closely related. A relatively uncomplicated scenario specified, for each trial, the troops' initial positions, area of the objective and assault line, and expected locations of munition impact, as well as positions of data collectors and instrumentation. The test produced various time measurements and information about munition malfunctions, troop actions, area of coverage, number of troops affected, and communication and control problems. Some data were collected in post-test interviews of participants and in questionnaires given to the chief evaluator of each trial. In the absence of pre-existing standards, conclusions were judgmental, based on the timeliness of munition delivery and its area coverage and also, secondary in importance, on the effects on unprotected troops and the ability of troops to "maintain command and control, move, acquire targets, and fire."

Kurke indicated that troop test methods, as exemplified in this instance, have been greatly improved in recent years. In an earlier report (Kurke 1963) he made a methodological survey of thirty-two troop tests conducted since 1955. He found that in about 40% of the tests no base data or comparative data were collected and that results "consisted solely of subjective impressions translated into narrative evaluations." Apparently virtually none of them included instrumentation to collect data. Most were subject to uncontrolled events, and generally there was a single run-through.

## ARMY PERSONNEL RESEARCH OFFICE (APRO) PROGRAM

The Army Personnel Research Office (later the Behavioral Science Research Laboratory and still later the U.S. Army Behavior and Systems Research Laboratory) was establishing a computer-based laboratory for complex man-machine experiments while the material for this book was being gathered; no studies had yet been conducted in it. Brief attention will be given to the programs which led up to the creation of this laboratory; it should be understood that these have been only a fraction of APRO's research, which has been directed by J. Uhlaner.

The approach was to engage in a substantial amount of component research before attempting large-scale experiments. One purpose was to define the important parameters of the elements of which the complex aggregate is composed, before trying to bring that aggregate under experimental control. In addition, researchers have acquired sophistication in the domains in which they are working. Measurement techniques were developed, criterion selection improved, and methods evolved for evaluating subsystem or system outputs through combining and interrelating measures of component responses in, for example, pay-off matrices. This latter aspect has been particularly important in view of APRO's mission of evaluation and assessment.

Initially APRO turned to an operational system, Missile Master, and conducted two experiments in individual tracking performance. But for investigating command and control systems the researchers considered it more fruitful to create more abstracted laboratory situations. Accordingly S. Ringel and his associates carried out a program on information assimilation and display coding, with emphasis on variation in information load in displays and on recognition of changes in items (called "updating") when such changes were or were not given conspicuity by a coding technique such as different lettering. Much of the material for the displays was drawn from the Army's prototype ARTOC (Army Tactical Operations Center).

Even before the coding studies began, J. Zeidner and his associates extensively examined the performance of photointerpreters. In one of their experiments, for example, they braved the biases of operational personnel by subjecting to experimental scrutiny the use of stereoptical techniques; they failed to find evidence of their presumed advantages. The photointerpretation work was broadened by Zeidner into a wide investigation of image interpretation in its many aspects. In this R. Sadacca of APRO was supported by a group from the System Development Corporation headed by R. S. Laymon and E. A. Waller. As the program progressed it began to encompass more complex situations, such as two-man and three-man teams (Doten, Cockrell, and Sadacca 1966), with a view toward eventual simulation of a TIIF (tactical image interpretation facility).

The new computer-based laboratory has been developed as a facility for increasingly large and complex experiments in both image interpretation and command and control operations, as well as for more experimentation on component performance. At some future time it may be possible to say whether the planning, installation, and initial operation of this laboratory have encountered difficulties met by similar enterprises in the past.

## HUMAN RESOURCES RESEARCH OFFICE EXPERIMENTS

The business of the Human Resources Research Office (HumRRO) has largely been research and development of training techniques for the Army's

Continental Army Command. HumRRO has done this work under contract with the Army Research Office. Although much of it has been the experimental evaluation of new training methods, generally these have related to individual military performance or courses in schoolroom curricula. However, there have been a few experiments which should be described here because of the simulation in which they were embedded, or the complexity of operations, or both.

Outstanding among these was the experimental evaluation of new methods for training tank platoons and platoon leaders in a simulation setting with miniature tanks (Baker et al. 1964). Adequate tactical training with real tanks or with opposing forces has been difficult to achieve, especially among Reserve and National Guard armor units; furthermore, it is expensive. One simulation solution to the problem, the "minature armor battlefield," was a triumph of ingenuity. Tanks were manufactured to order, on a scale of 1 to 25, to resemble the M48A2 model in both appearance and performance, such as grade-climbing capability. They were battery powered and controlled by radio. Turrets could traverse 360 degrees and the gun tube could project a narrow beam of light 20 to 25 feet. Sensitivity-adjustable photoelectric cells were mounted on the right and left sides just below the support rollers. When the beam from one tank hit a cell, the receiving tank was disabled and a red light appeared on its rear. In addition to the tanks, combat could be simulated by air rifles firing explosive pellets to represent artillery, by charges of magnesium powder to represent smoke rounds, and by firecrackers to represent mines.

A terrain model 28 X 76 feet was constructed in a barracks at Fort Knox, Kentucky with hills, vegetation, rivers, buildings, roads, and bridges also on a 1:25 scale and rearrangeable. At one end sat a platoon of five "aggressor" tank crews of three men each. In five compartments on a movable platform that could traverse the entire length of the model sat five "friendly" tank crews of three men each. Both groups had radio equipment with different sets of channels for maneuvering their tanks. Curtains operated by an instructor could limit visibility. An instructor controlled the operations of the aggressor personnel, while the experimental subjects were the friendlies.

In the crew training program fourteen platoons which had completed basic unit training were subjects. Seven platoons constituted control groups, the other seven receiving a week's training (forty hours) on the equipment, including familiarization exercises, briefings, critiques, and ten different problems in thirty-six runs. In another training program for platoon leaders, twenty-five second lieutenants encountered each of the three problems three times, rotating through the platoon leader role; twenty-five were control subjects, and another group of ten experienced officers served as an additional control group. In both programs the subjects underwent a HumRRO-devised field performance test after the training, as well as paper and pencil tests. Experimental crews and experimental leaders did considerably better than the control subjects on the performance tests; the differences were statistically significant except in the case of the experienced-officer control group.

Another form of simulation for the training of platoon leaders was somewhat less elaborate. In an Armor Combat Decisions Game small 2½-inch metal models of the M48A1 tank were moved by gunner-drivers with yard-long push paddles over a terrain board 8 X 16 feet, the features of which were commercially available plastic models on a 1:115 scale. Five friendly tank commanders directed the gunner-drivers by means of an interphone system and communicated among themselves and with the instructor by means of an actual radio net for tank platoons and companies, utilizing standard tank equipment. An instructor gave directions over a separate net for the movement of aggressor tanks. Tank movement was controlled by a metronome and a grid of 2-inch squares on the terrain board. For example, if a tank was moved one grid square every 2 seconds (as signaled by the metronome), its speed equaled 7.5 miles per hour of movement by a real tank. Tank gunfire was simulated with narrow-beam flashlights, artillery fire by balls of cotton, nuclear weapons by firecrackers, and smoke by rolls of steel wool. Twenty officers who were trained for forty hours in this setting performed significantly better on post-training tests, including one for field performance, than an equivalent number of control subjects.

*Other Studies*

HumRRO has paid much attention to infantry training and has updated the training methods which were developed under the HumRRO-contracted work of Psychological Research Associates, described earlier in this chapter. The culmination of its efforts in creating and testing a new training program in the tactical and patrolling operations of rifle squads has been detailed by Ward and Fooks (1965). Among this program's innovations were the use of opposing forces—pairs, teams, or squads—in training exercises; and the use of immediate-feedback stake courses for teaching combat formations, movement, selection of firing positions, and choice of cover and concealment. Trainees could learn from information posted on stakes whether they had made the correct choice of action.

Following some trial runs in 1962–63 at Fort Benning, Georgia, two companies totaling 324 graduates of advanced individual training received thirty-seven hours of tactical training and fifteen hours of patrolling training with the new methods while on a semitactical bivouac in the field. This training was conducted at Fort Ord, California, in 1964. It was followed by a field test which HumRRO had developed earlier (Nichols et al. 1962), and a questionnaire survey.

Twenty-six noncommissioned officers who had been instructors and observers filled out questionnaires rating this new program in comparison with the one it was designed to replace. The questionnaires contained fifty-six items concerning skills and knowledges, training time, realism, conduct of instruction, and motivation. The new program was judged more effective or much more so on every item. The trainees also filled out questionnaires; 54% stated, for example, that in their prior Army training either "quite a lot" or a "tremendous amount" of time was wasted during training, but only 4% felt this way about the new program, and 81% said they were more motivated by the new program than by "the usual Army training."

An instance of HumRRO research more experimental in emphasis and less directly oriented toward training has been reported by Berkun et al. (1962) and Berkun (1964). Since his accounts are readily accessible, the research will be reviewed here only briefly. It was concerned with the effects of "psychological stress." In one experiment, soldiers in a DC-3 aircraft were led to believe that a ditching was about to occur; effects were deduced from the way they filled out "standard" forms which burdened comprehension and memory. In a second study soldiers were left in isolated positions during alleged artillery firing. They had radios which failed; a subject could not summon assistance to guide him out until he repaired the radio. Artillery was simulated by nearby explosions of TNT. Speed of repairing the radio indicated the aversive effects of the situation. A third experiment also used TNT explosions; but in this instance the subjects were led to believe they had set them off by miswiring a switchbox and that thereby they had injured other soldiers. As possibly the most significant methodological aspect of these experiments, the subjects were not informed that they were involved in an experiment until it was all over.

# 10

# Ohio State University Air Traffic Control Experiments

Under the leadership of Paul M. Fitts, the Laboratory of Aviation Psychology of Ohio State University in the mid-1950's launched a series of nineteen simulation-based experiments on the human engineering aspects of air traffic control. The program ended in 1961, but was followed by the Ohio State University decision-making studies described in Chapter 21. The air traffic control experiments have been described in seventeen reports issued by the Behavioral Sciences Laboratory of the Aerospace Medical Laboratory (now Aerospace Medical Research Laboratories), which sponsored the research. Fitts et al. (1958) wrote an over-all description of much of the program, and the first fourteen experiments were summarized by Kidd (1959*c*). In addition, several of these experiments have been reported in journals (Kidd 1961*b*, 1961*c*; Kidd and Christy 1961; Kidd and Kinkade 1962; Kinkade and Kidd 1962).

The research was oriented toward military rather than civil air traffic control—the guidance and separation of aircraft—and toward particular aspects of such control. One of these was reliance on assistance from ground-based radar. The other was the phase of flight in which aircraft approach a landing area. The research was not particularly intended to provide technical aid to civil air traffic control (see Chapter 15) which, at least at that time, depended less than the military version on ground controllers using ground radars. Civil air traffic control instead vested more responsibility in the pilots of commercial aircraft. The military problems of air traffic control had become obvious during the Berlin airlift in 1949 and the Korean War. In 1950 the Committee on Aviation Psychology of the National Research Council "sponsored a planning and field study of human engineering problems in air traffic control with funds from the Air Navigation Development Board"; its report "provided the basis for a planned program of laboratory experimentation" (Kidd 1959*c*). This program started formally at Ohio State University in 1952.

Virtually the first order of business was to design and build a simulation capability. Attempts to make use of the 15-J-lc device (see Chapters 4, 5, 6, 7) revealed so much variability and unreliability that new equipment was deemed essential. The electronic target simulator which resulted (Hixson et al. 1954) was analog-type equipment able to display signals of thirty aircraft of various types on a plan position indicator (PPI) type of display, of which there came to be four. The electronic target simulator permitted human operators, representing pilots, to insert aircraft positions, headings, speeds, altitudes, and turn rates. It also incorporated altitude effects on speed, wind effects on speed and heading, and various aircraft identification methods and coding arrangements. In the first fourteen experiments the simulator generated 17,269 flights (Kidd 1959*c*). The simulator was used some years later in another program (Chapter 5).

It did not present on the displays the clutter, "noise," and "fading" that are characteristic of real-life, radar-based air traffic control systems. (Noise may consist of echoes from hills, clouds, and even birds. Fading is the disappearance of radar signals for various reasons.) This intentional simplification was rationalized as follows (Fitts et al. 1958):

One of the major tenets that we have followed is that human capabilities should first be determined under optimal system conditions (e.g., with "idealized" displays and reliable information) and then determined for nonoptimal or degraded systems. Only data obtained under idealized conditions permit an estimate to be made of the upper limits of system performance that could result from future improvements in the machine aspect of the man-machine system. One of the gratifying results of our policy of first studying human performance under idealized conditions is that on several occasions it has been unnecessary to go on to the study of degraded systems. In each case, by the time a series of human factors research studies has been completed, engineering progress has made it possible to eliminate many of the deficiencies of existing systems, and hence had rendered unnecessary the study of the effects of such deficiencies on human performance.

This viewpoint, which recurred in many of the program's reports, is a persuasive one. It would be more so, however, if the reports had described the various examples where it became unnecessary to go on to the study of degraded systems, or if there had actually been any follow-on experiments at all in which the radar presentation was degraded. It would also be interesting to know whether any of the results were mistakenly accepted or used by system designers as reflecting either representative or *minimum* human performance in a real system. The results of many of the experiments might well be evaluated to determine whether the researchers' conclusions would be valid if the operators received nonidealized inputs.

Prior to the nineteen experiments in the program, J. C. McGuire and C. L. Kraft did some nonexperimental studies at the RAPCON Center (Radar Approach Control) at Wright-Patterson Air Force Base. These included activity analyses, communications flow analyses, position and console descriptions, and questionnaire surveys of controllers. In other words, a real-world investigation preceded the experimental research in order to disclose operating methods and problems.

In addition, in an experimental study in January 1955 (Kraft, C. L., Chenoweth, E., and McGuire, J. C., unpublished report), two real aircraft were piloted in close proximity to each other in twelve approaches to the Wright-Patterson base. Ground control alternated between two controllers at the RAPCON Center. The researchers made a "microanalysis of the flow of information in the ground-air loop" by visually observing and recording pilot and controller activity, obtaining voice recordings of radio communications, and making video recordings by means of manual Skiatron plots and frame-per-sweep photographs of a PPI display. In addition to content analyses of the communications, ratings by the controllers were examined for consistency of judgments of the separate runs; the investigators also analyzed controller and pilot variability on successive runs.

The same twelve approaches were then reproduced with the electronic target simulator, in a second part of the study (Alluisi 1956). It had been decided that information analyses of actual operations were too time-consuming and costly for the research program; there also were hazards. In this laboratory phase of this study another approach was investigated: the ranking of visual records of the aircraft tracks.

## SUBJECTS: PATTERN-FEEDER CONTROLLERS

The air traffic control function central to the research consisted of the tasks of the pattern-feeder controller. These tasks included the acceptance of incoming flights from a pickup controller acting as intermediary between the en route and the terminal systems, and handover of the flights to a GCA (ground control approach) controller guiding aircraft through the GCA gate, a sort of funnel. Pattern-feeder controllers would have to direct all incoming aircraft within the area, illustrated in Figure 8. It extended fifty miles between the handover from the pickup controller and handover to the GCA controller. The aircraft would enter this area from different directions, at different points, and at different altitudes. A pattern-feeder controller would try to keep the aircraft separated as prescribed by safety rules while at the same time he attempted to hold both flight times and fuel consumption to a minimum.

In eleven of the experiments there was a single pattern-feeder controller; in eight there were two acting as a team. One of these eight experiments compared three-man as well as two-man teams with single controller operations. The pattern-feeder controllers were regarded as the experimental subjects. But there were other personnel in the experiments representing the operating personnel with whom the pattern-feeder controllers interacted. Beginning with the sixth experiment, these quasi subjects included a pickup controller and a GCA controller, whose roles were generally played by well-trained university students; the thirteenth experiment had, in addition, a departures controller. The various stations are shown in Figure 9. Other quasi subjects were the pilots flying the simulated aircraft which the pattern-feeder controllers were directing. There could be as many as fifteen of these pseudopilots. They also were trained university students. The published reports do not indicate the numbers required for particular experiments.

Except in one instance, all the pattern-feeder controller subjects in the first eight experiments were professional controllers. In all but two of these eight experiments there were four subjects. In one experiment, two subjects constituted a single two-man team. In another, two two-man teams were made up from a single professional controller and two novices. In the remaining eleven experiments the controllers were nonprofessionals, such as college students, who

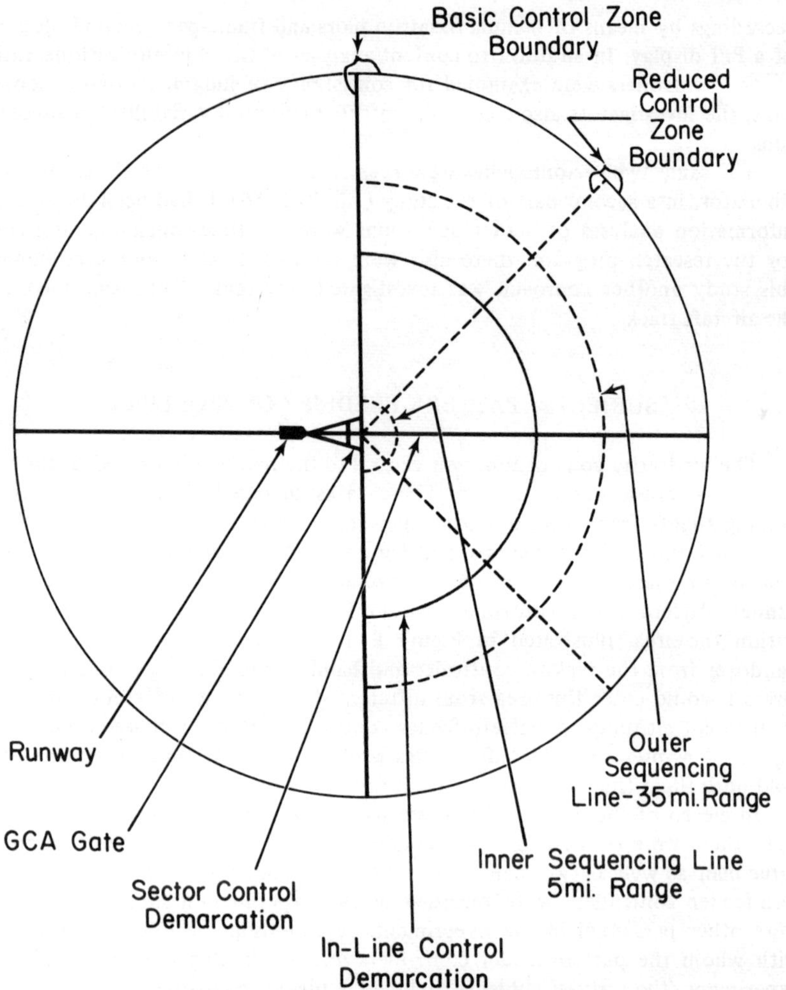


Fig. 8. Area Coverage of the Pattern-Feeder Controller (Kidd 1961*a*).

received extensive training in the control tasks before they started an experiment. Their number in any study varied between six and twenty, averaging about eleven.

## INDIVIDUAL DIFFERENCES AMONG SUBJECTS

The experimental designs generally were such that not only could measurements have been reported for the performance of individual subjects but also analyses of variance could have demonstrated any statistically significant differ-

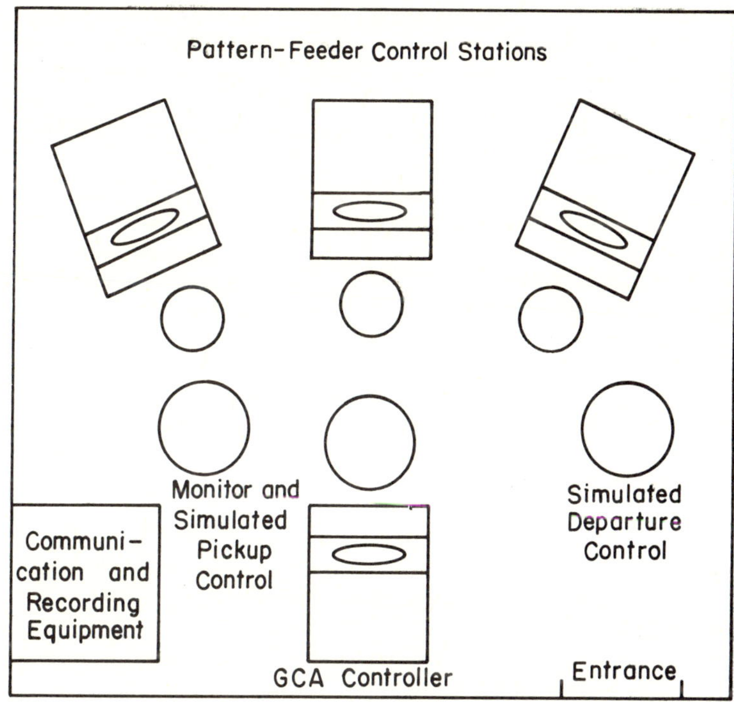


Fig. 9. Stations of Subjects and Quasi Subjects (Kidd 1959*b*).

ences between subjects. The following summary shows the extent to which individual differences and their significance were reported.

Experiment IV was the study in which the professional controller was paired with one of two novices. Although the differences between the novice controllers were generally not significant, there were consistently significant interactions between novices and procedures. In fact, two of these interactions yielded the only results from the experiment reaching the .01 level of statistical significance. The experimenters properly interpreted this outcome to mean that "one controller's performance is consistently better when he is using one procedure while the other controller is consistently better when he is using the other procedure" (Schipper et al. 1956*b*).

In the report on Experiment VII (Kidd et al. 1958), performance differences among six two-man teams composed by pairing four professional controllers were so large that they reached the .01 level of significance, the only other source of variance to do so being the sequence of trials; none of the independent variables that the experiment was explicitly investigating reached even the .05 level. Although this interesting finding about individual differences received a paragraph of comment within the report, it was regarded as "somewhat aside from the main purposes of the study" and was not mentioned in the report's summary or abstract.

In Experiment XIX (Kidd and Christy 1961), three supervisors each took three different roles–laissez faire, active monitoring, and direct participation–in supervising two-man teams. These roles, and loads, were the independent variables of interest. But individual differences were the major outcome. The experimenters reported: "The influence of supervisors as individuals yielded a mean square variance 3.7 times greater than that derived from the role factor. The interactions of role and supervisor and of role and load were statistically negligible." (Results from two different load conditions differed at the .05 level of significance.)

In the report on Experiment IX (Kidd and Kinkade 1958), the only published analysis of variance shows that "subjects and order" (for the nine nonprofessional subjects) was the only source of variance to reach the .01 level; the text contains no comment on this finding. The single analysis of variance in the report of Experiment XVII (Howell, Christy, and Kinkade 1959), indicated that the performances of the six nonprofessional subjects did not differ significantly. On the other hand, the report on Experiment XVIII (Kidd 1961*a*) showed that six two-man teams made up from twelve nonprofessionals differed significantly among themselves at the .01 level, a finding left undiscussed. In the report on Experiment VII (Schipper, Kidd, Shelly, and Smode 1957), individual differences received somewhat different treatment. In the four analyses of variance, the differences among the four professional controllers would appear to have been the only sources of variance reaching the .01 level; however, in each analysis the "F" (ratio of within and between mean squares) for controllers was simply recorded as "not evaluated."

In summary, differences between experimental subjects were noted and analyzed for statistical significance in the reports for only six of the nineteen experiments. They were found to be highly significant in five of the six cases; these five included both professional and nonprofessional controllers. In a seventh instance, highly significant differences can be inferred from the data, but the authors of the report omitted the analysis. It seems probable that such results received little emphasis from the researchers because their experiments were directed at human engineering problems rather than manning (staffing) or selection. Although the research staff discussed this matter of individual differences, disinterest predominated. Further, initially it seemed wise to forego comparisons among the professional controllers, on whose participation the project depended, lest they be alienated. Nevertheless, it would appear that emphasis on one field of human factors–human engineering–can reduce the interest in another–manning, including personnel requirements. Yet if individual differences seem to have such importance in system effectiveness, should they not receive more attention from system designers and developers?

## DISPLAY VARIABLES

What were the human engineering problems of overriding importance in this research? Fitts et al. (1958) divided them into display (or information) variables, load variables, and procedural variables. Display variables "involve the type of information made available to controllers, the degree of precision in the information, and the way in which the information is encoded and displayed." Ten of the experiments were concerned with display problems.

Experiments II and III (Schipper et al. 1956*a*) either identified incoming aircraft continuously with a "clock code" on the controller's PPI display or enabled the controllers to call up an alphanumeric identity with a light pencil, while Experiment VI (Schipper, Kraft, Smode, and Fitts 1957) compared the use of the clock code with no identification at all for the radar signals. (The clock code was composed of different positions of a clock's hands.) Experiment VII (Schipper, Kidd, Shelly, and Smode 1957) provided aircraft altitude information to the controller either through a visual display or by requiring him to interrogate the pseudopilots on the simulated radio link.

Experiments IX (Kidd and Kinkade 1958), XII (Kinkade and Kidd 1959*a*), XIV (Kinkade and Kidd 1959*b*), and XVII (Howell, Christy, and Kinkade 1959) all examined the use of a supplementary display, not for the controllers but for the pseudopilots. This was an airborne position indicator (API), a representation of a possible future display for aircraft cockpits to show pilots where they were in relation to the landing strip. Under an alternative condition pilots could initiate their own descent and speed adjustments rather than respond to a controller's instructions, and in still another condition they could also initiate changes in heading. One could call this display variable also a procedural variable since it altered the distribution of tasks between controller and pilot.

Experiment XII investigated the use of the API with or without ground reference points displayed on it, with or without fixed approach paths displayed on it, and with or without aircraft identification being furnished to the controller. Experiment XIV varied the proportion of aircraft equipped with the API (none, 33%, 67%, or all).

## LOAD VARIABLES

Fitts et al. (1958) defined the second category of variables, load variables, as those which "define the input to the air-traffic-control systems, such as the traffic with which the controller must cope." They were manipulated in a number of ways.

In Experiment I (Schipper and Versace 1956) the variable was the time available for controller action to avoid incipient collisions, with five levels ranging from 4 to 8 minutes. In Experiments II and III (Schipper et al. 1956*a*), IV (Schipper et al. 1956*b*), V (Versace 1956), VI (Schipper, Kraft, Smode, and Fitts 1957), VII (Schipper, Kidd, Shelly, and Smode 1957), XIII (Kidd 1959*b*, 1961*c*), XVIII (Kidd 1961*a*), and XIX (Kidd and Christy 1961), the load variable was the average time interval between aircraft that the pattern-feeder controller had to accept. In Experiment XVIII control zone area and arrival area were also varied, with two area sizes. Experiment V included simulated aircraft emergencies in some of the flights; these could be regarded as a load variable. In Experiment VII (Kidd et al. 1958) the load variable consisted of conditions of regularity-irregularity within the time interval between arrivals and within the spatial area of arrival. In Experiment IX (Kidd and Kinkade 1958) the types of incoming aircraft constituted the load variables, with one, two, or four degrees of heterogeneity. The principal form of load variation—the entry arrival interval—had four levels in two experiments, three in three, and two levels in four experiments, the levels varying from as low as 30 seconds to as high as 144 seconds.

In addition, in Experiment X (Kidd 1959*a*, 1961*b*), load was varied not in terms of entry rate but according to the number of aircraft required to be continuously under control. This experiment investigated methods of training pattern-feeder controllers and thus did not fit any of the three human engineering categories. One group of controllers was trained under three increasing levels of load, while for another the number of trials was the same, but in all the trials the load level was the highest of the three; this was also the level at which both groups were tested. The three levels were four, five, and six aircraft, over nine learning trials of 30 minutes each. Those with constant high-load practice did better on a tenth test trial than those with graduated-load practice.

In another training study (Experiment XV) reported by Kinkade and Kidd (1959*c*, 1962) two groups of subjects, selected by random sampling, both had ten 30-minute training trials on the simulator. But before these, one group also had six hours of practice—twelve trials of three games each—on an "operational game" which was developed "as a highly abstract embodiment of the basic features of the radar air traffic control situation." As in Chinese checkers, the subjects moved metal tokens across a board over specified routes. The subjects with this game practice did better in the air traffic control training with the electronic target simulator than the others.

## PROCEDURAL VARIABLES

In specifying the third category of variables as "procedural," Fitts et al. (1958) said they included "communications procedures, procedures by means of which two or more individuals make joint or complementary decisions, and procedures governing the types of instructions that controllers are permitted to issue to aircraft pilots. An important subclass of procedural variables is the way in which two or more men divide responsibility." From this description, and from the actual studies which could be placed in this category, it becomes clear that procedural variables are diverse, to say the least. If they tend to constitute a grab-bag, this may be so partly because there exists in the human engineering literature no taxonomy of procedures, nor even much of an attempt to create one.

Experiment IV (Schipper et al. 1956*b*) compared "in-line" with "sector" control; as already noted, two-man teams performed the pattern-feeder control function. With the in-line method an "outside" controller first picked up an incoming aircraft and guided it. Then he turned it over to the "inside" controller, who delivered it to GCA control. With the sector method one pattern-feeder controller controlled aircraft in the northern portion of the control area and the other did this in the southern area. In Experiment V (Versace 1956) again there was a two-man pattern-feeder control team. The controllers either worked side-by-side with direct communication between them or they were separated by a wall through which they passed data strips down a chute, and they had to communicate by intercom.

Experiment XI (Kidd and Hooper 1959) was still another two-controller study. It had two procedural variables, each with three conditions. One was the method of task division. Incoming aircraft were assigned in alternation to the controllers, or by sector as in Experiment IV, or according to which of two landing fields the aircraft had as its destination (with a different controller responsible for each field). The other procedural variable (at least so it might be called) was the option of exchanging control responsibility between the controllers. The option was unrestrained, absent, or could be exercised only after an aircraft had been guided thirty miles by one controller.

The number of controllers in the pattern-feeder control team varied between one, two, and three individuals in Experiment XIII (Kidd 1959*c*). When there was more than one, responsibilities were assigned according to the sector method. In Experiment XVI (Kidd and Kinkade 1959) the variable of primary interest was the replacement of one controller by another, following four degrees or methods of orientation prior to change-over. In Experiment XVII (Howell, Christy, and Kinkade 1959) two subjects began functioning together, one controlling and the other monitoring another PPI display. Then both displays were blacked out as though they had suffered a breakdown and a synthetic display was substituted. One or the other of the subjects had to take over control with this display. As in the case of the regular PPI display, it could provide two levels of control flexibility with regard to the number of fixed flight paths along which the controller could guide the incoming aircraft.

In-line and sector methods were again compared in Experiment XVIII (Kidd 1961*a*). In addition, as illustrated earlier in Figure 8, there were three alternative procedures for sequencing aircraft while they were under pattern-feeder control. Controllers accomplished the required separations between aircraft 35 miles from turnover to GCA, or 5 miles from it, or they simply made sure that the separations would be in effect at the turnover point. Method of supervision was a procedural variable in the last of the nineteen experiments (Kidd and Christy 1961), in which six two-man teams were formed by random pairing from a twelve-man sample; each of three supervisors supervised each of the teams once. As indicated earlier, there were three types of supervision: laissez faire, in which the supervisor was a passive monitor; active monitoring, in which the supervisor initiated instructions when he detected errors or difficulties; and direct participation, in which the supervisor himself took corrective action by communicating with the pseudopilots rather than acting through the controllers. Each team experienced each condition twice, and each supervisor took part in each condition four times.

From the foregoing descriptions it should be apparent that five of the procedural experiments also varied loads. As previously noted, the API display experiments might also be regarded as procedural. Load was a variable in four of the non-API display experiments. Thus, more than one category of human engineering variables appeared in a majority of the nineteen experiments.

## EXPERIMENTAL OPERATIONS

Aside from subjects or teams and order of presentation, the experiments contained two or three, and sometimes four, independent variables, with two to five values each (e.g., 2 X 4, 2 X 3, 2 X 2 X 2, 2 X 2 X 3, 2 X 4, 2 X 2, 2 X 2 X 5, 3 X 3, 3 X 3, 2 X 2 X 2 X 2, 2 X 2 X 2 X 3). Typically the variables were organized in a factorial design—usually by means of one or more Latin squares which controlled for order of presentation—and statistical significance was tested by analyses of variance. Customarily subjects served as their own controls. The two training experiments used matched groups. "In terms of statistical power," Kidd (1959*c*) observed, "the designs employed have consistently allowed the rejection of the null hypothesis when differences in performance in the range between 5% and 10% have been observed. This level of differentiation is quite compatible with the engineering realities associated with the system being studied."

An analysis of fourteen experiments shows that the total number of trials or problems in an experiment ranged from 32 to 160, half of them being 64 or more. The number of trials per subject or team was between 4 and 24, half of them 12 or more. The number of trials per session was between 4 and 6, generally with 10-minute intervals between trials. The number of sessions ranged from 8 to 32; one-half of the 14 experiments had 9 sessions or more. Trials or problems usually lasted 30 minutes, some being as short as 25 minutes and a few more than 50 minutes. In addition, the sixteenth experiment (omitted from the foregoing summary) investigated the effects of extended controller activity. Each session lasted 3.5 hours, and there were two sessions per subject.

A battery of performance measures reflecting both safety and efficiency criteria was used in this research rather than a single measure. Efficiency measures included over-all flight time, percent delay, fuel consumed in flight, and frequency of missed approaches. Safety measures included frequency of separation errors at various stages of an aircraft's approach and maintenance of specified intervals between landings and departures. In addition, the researchers analyzed the content and frequency of communications, measured delays in responding to emergencies or major system disturbances, and obtained job satisfaction ratings from the subjects. The measures and methods of collecting data were adopted on the basis of experience in the experimental study, mentioned near the beginning of the chapter, which preceded Experiment I.

### *Experimental Findings*

The following experimental results and conclusions have been adapted from the summaries by Kidd (1959*c*) and the reports of the last four experiments. First to be considered are the display variables. The data showed that the display of a target's identity increased a controller's capacity. The clock code and identification by call-up with a light pencil were equally effective; and clock code identification was better than none at all, especially under high-load conditions. System performance was relatively unaffected by the mode of obtaining altitude information—auxiliary display or querying the pilot—but the display reduced the time the controllers talked by 8%. Controllers tended to shift from using the display to making a radio inquiry when the load grew heavy. When controllers used either the API display or an identification method, the system became maximally adaptable. In addition, the redistribution of work load through using the API improved system performance, such improvement being proportional to the number of aircraft so equipped.

With regard to load variables, it was found that system efficiency decreased as entry rate increased, especially when the interval between aircraft became less than a minute. Although in earlier experiments differences between lower rates failed to achieve any statistical significance, in the last experiment a 90-second entry interval increased delay 12.1% over that with a 120-second interval. Irregularity of entry in time or space had no significant effects. Kidd (1959*c*) drew the conclusion that a skillful pattern-feeder controller can easily control eight aircraft at the same time with near-minimum flight delay and virtually complete safety when radar presentation and communications make complete information available. Loads exceeding ten aircraft definitely introduce delays and hazards.

In addition to the finding about distribution of tasks by means of the airborne position indicator display, various results were related to procedural variables. For example, it appeared that face-to-face communication could be distracting, and that assignment of aircraft within a two-man controller team according to aircraft destination (when there were two airfields) was superior to either in-line or sector assignment, those other two methods being equally effective. It was also found that co-ordination and integration between team members placed an additional load on the system. With a constant load, only a slight advantage came from increasing the size of the team. It was concluded further that "some functions such as emergency procedures and inter-controller communication procedures should be standardized while other functions such as routing during the approach should be kept flexible."

Kidd (1959*c*) suggested that a single underlying factor could be related to most of the findings, namely, the susceptibility of human short-term memory to interference:

> Thus, with regard to distribution of responsibility, it is the short-term memory capacity that is burdened when input sources are multiplied without a commensurate reduction in net input load. Input organization likewise stresses the loss of memory content that occurs when an operator switches his attention from one display to another. Insofar as procedural flexibility is concerned, the lack of extrapolative capacity seems to be but one facet of the memory problem in that extrapolation requires the simultaneous synthesis of a number of discrete items of information; any momentary loss leads to a wrong prediction.

According to Fitts et al. (1958), the Laboratory of Aviation Psychology planned to continue its studies with "(a) the study of simulated automatic control systems in which people will be asked to monitor the system and to handle emergencies, (b) the study of different kinds of procedures for attaining a high level of effectiveness from a group of men who are working together, (c) the development of displays, work stations, and communication nets suitable for use by groups of three or more controllers, and (d) the study of the optimum number of controllers for performing different functions and handling various loads.'' However, these plans were not realized.

## ASSOCIATED RESEARCH

In a brief continuation, three experiments were sponsored by the Operational Applications Office of the Electronic Systems Division of the Air Force, to investigate not the design or management of an air traffic control system but rather the effects of noise in degrading communications (Kidd 1961*d*, 1963). In one experiment the signal-to-noise ratio was varied in the voice communication links between a pattern-feeder controller and pilots. In the second, noise masking was kept constant and limitations were imposed on the frequency characteristics of the signal. In the third, transmissions were interrupted to determine the effects of intermittency. As possibly the most interesting aspect of these studies, the measures of communication degradation were not conventional intelligibility measures but the kinds of measures of system performance obtained in the preceding program.

Extensive supporting research accompanied the large experiments. Experiments in this "related technical research" were directed at visibility and lighting, specific display principles, information coding, and information-handling ability. One of the products was a method of "broad-band blue lighting" of rooms containing radar displays (Kraft 1956; Kraft and Fitts 1954); this was widely adopted. Kraft exploited this lighting innovation in redesigning flight-progress strips, communication indicators on consoles, and facility status displays for the RAPCON at Wright–Patterson Air Force Base. McGuire and Kraft also evaluated horizontal displays and twin-microphone, split-headset voice communications. Much of the component research yielded human engineering data of general import as well as experimental findings about human information processing. It had the further advantages that it provided additional channels for graduate work among many talented students and for publications in journals. Thus the larger air traffic control studies were part of an over-all research program. The scope of the program may be grasped from the fact that forty-one individuals were associated with the project, as supervisor, research associate, or research assistant, for one month or more on regular appointments between 1952 and mid-1956; many were part-time or short-term appointments (Alluisi 1956).

One stated objective of this over-all air traffic control program (Fitts et al. 1958) was to "provide human engineering principles that can be used by the engineers who will design future air traffic control systems, and by the operational personnel who will devise the procedures to be employed in operating these systems . . . . From a psychological viewpoint, another goal of the research is to provide quantitative estimates of human capacity for performing the different types of functions which may characterize future air traffic control and similar complex man-machine systems."

It is tantalizing to be unable—because it is so difficult to assemble the evidence—to establish how widely these objectives were realized in the use of data from the nineteen air traffic control experiments, whether by system engineers and operational personnel or in human engineering exploitation of the knowledge those studies yielded about human capacities and limitations. For they were indeed productive, and they advanced the state-of-the-art of man-machine system experimentation.

# 11

# System Development Corporation Field Experiments

The account in Chapter 8 of the experiments in the RAND Corporation's Systems Research Laboratory described how these led the Air Force to establish the system training program (STP) to train personnel for air defense systems in North America and later overseas. This training program in turn gave rise to a number of experiments by the System Development Corporation (SDC), and some while that organization was still part of RAND. Most of the experiments associated with STP were at field locations. The few conducted on SDC premises are described in Chapter 17, along with other SDC laboratory experiments. The present chapter divides the field experiments into four categories. Some were embedded in the pre-SAGE manual air defense system, some in the SAGE computer-based system. Within each of these groups were experiments intended to evaluate or improve system training, or some feature of it, and others which tried to evaluate or improve the system itself in some fashion, using the STP simulation and exercising capability for experimentation.

## MANUAL AIR DEFENSE

As Chapter 8 indicated, the STP was adopted to train crews at operational Air Defense Direction Centers (ADDCs) (Goodwin 1957). One of the first requirements was to create a simulation capability that could introduce simulated radar signals of moving aircraft into plan position indicator (PPI) displays manned by surveillance, identification, and intercept-control personnel. The make-believe consoles and digit-printed IBM paper of the Systems Research Laboratory were obviously unsuited to this purpose, if only because the operators were supposed to be trained at their regular equipment. The capability that resulted may be best understood as composed of two parts, the production of simulation data and the translation of the data into electronic signals for the simulated display of radar-detected tracks.

The production process was complex. Hostile and friendly tracks were designed and scripted or tracks were selected from a "library" to compose an exercise lasting one to three hours, generally two. Decks of cards were punched to contain the track data. These or pre-punched library decks were fed into a computer which followed programs made up of various models that would enable the computer to process the tracks to show the effects of aircraft and radar characteristics. The outcome of processing was a magnetic tape that contained an exercise's track inputs. By means of especially designed equipment these inputs were transferred, in a digital-to-analog conversion, to 70-mm. film on which spots represented aircraft positions; the co-ordinate positions of the spots in any frame specified the range and directional (azimuth) positions of the simulated aircraft as seen by the radar viewing them. Each frame of the film contained signals for all the aircraft seen by that ADDC radar during one radar antenna rotation. The film included indicators of aircraft altitudes. A different film was made from a different tape for every ADDC participating in an exercise. Since two or more ADDC radars might see the same aircraft, these inputs had to be co-ordinated originally in the computer so that every simulated aircraft seen by two or more geographically separated radars at different ADDCs would appear at the same geographical locations.

As the second part of the simulation capability, a device was installed at each ADDC radar to translate the spots on the film into electronic signals which would enter the radar as though they were radar echoes, so the radar's processing would result in their presentation on the ADDC's PPI displays in the same fashion as real radar signals (blips). This device, the AN/GPS-T2, was designed by RAND but produced by a contractor for the Air Force. It resembled apparatus developed at Lincoln Laboratory (see Chapters 5 and 6). It served its purpose—up to a point. It transduced the film spots into radar signals, which moved across the PPI displays as actual aircraft signals might. But there were several drawbacks. Although the device required considerable maintenance, as so often occurs with training devices provisions for such maintenance were often overlooked. The tracks were fixed; no changes could be made in them during the exercise. For example, a track could not be prevented from continuing to appear on PPI displays even if the enemy bomber it represented was shot down in the exercise, nor could such an aircraft take evasive action. The signals (blips) did resemble actual blips, but they frequently lacked the fidelity that would be preferred for training in visual discrimination. That level of quality was not deemed a requirement in the original plans for STP; its need did not become pressing until training in ECCM (electronic counter-countermeasures) began to emphasize visual discriminations of signal from noise. Simulation of electronic countermeasures was omitted, except in a rudimentary form, until a modification called the anti-countermeasures trainer (ACTER) was developed and added to the AN/GPS-T2. In spite of these drawbacks the program developers felt that the AN/GPS-T2 sufficed for training. It was a relatively simple, inexpensive device that could be produced and installed quickly—one that could get the signals into the system.

The experiments had to depend on the simulation capability developed for training, not too unreasonable a situation in experiments which investigated the training itself. In addition, to simulate interceptor aircraft the experiments had to rely on 15-J-1c devices (described in earlier chapters), which were already installed as training devices at air defense field sites. These were too few and too unreliable to simulate noninterceptor aircraft.

*System Training Experiments*

These experiments illustrate the problems that so often occur in experimental research in an operational setting. They may be problems beyond the control of the experimenters, and they usually do not originate from the subject matter of the experiments. They are accentuated if the operational system is young, developing, or being installed during the experimentation.

*The 85th Division Experiment.* The first major field experiment to investigate the system training program (Jaffe 1958) was a large-scale study in the 85th Air Division of Air Defense Command. It involved twenty air defense crews, four at each of five Air Defense Direction Centers. Ten crews, two at each location, received system training in two STP exercises per week for six weeks. The other ten simply experienced day-to-day operations. Thus, the essential comparison was between STP and no STP. All crews took the same pre-test before and post-test after the STP period. The two tests were the same and consisted of two problems, as the simulation input for an exercise has been called in the system training program. They contained high-load, difficult situations. The six STP training problems were medium load. In addition, all crews practiced with two light-load problems in shakedown exercises before the pre-test. All problems were especially designed for the experiment.

The experiment was conducted between November 26, 1956, and February 28, 1957, in the course of installing the system training program in the 85th Division. This training program was introduced into the Air Defense Command division by division with the help of the organization at RAND (later SDC) that developed the program and did the experimenting. Originally the experiment was even more ambitious, but a variety of circumstances made it impossible to include two additional air divisions as planned.

In all of the exercises in the 85th Division experiment, all five ADDCs took part in an exercise together and received co-ordinated simulated inputs. Thus each of the four crews at each ADDC was, so to speak, part of a division-wide crew. However, there was relatively little interaction between ADDCs in this experiment, because of a lack of telephone communications. Following the post-test, STP was suspended for the two crews at each site that had been receiving it and given to the other two crews in twelve exercises. Then all four crews took a third test. These operations constituted a supplementary experiment in March and April of 1957.

Since the four crews at each location had already been formed, it was impossible to compose them on the basis of a preliminary exercise. However, the pre-test showed they all had approximately the same ability at the outset. It was impossible to keep them from exchanging information about the experiment or to prevent turnover within crews. There was a minor amount of interchange of personnel between the two STP-trained crews and between the two non-STP crews, but very little between crews with different training. The researchers estimated that about 25% of the officers and airmen taking the pre-test did not participate in the post-test.

Loss of "crew integrity" was not the only difficulty. Although core elements of the crews had been trained in Santa Monica in STP principles and in such operations as debriefing, crews did not always follow prescribed debriefing methods. Exercise schedules had to be revised at times due to actual (live) operations. The inputs of simulated aircraft altitude, which were supposed to come from the AN/GPS-T2, usually failed to reach the heightfinding scope reliably. Only two of the five ADDCs were equipped with 15-J-lc devices for the simulation of intercept control, so a makeshift map-plotting substitute had to be developed for dead-reckoning simulated interceptor aircraft. This did not yield very good results.

In addition to extensive qualitative, "critical incident" data, six measures of performance were ultimately analyzed, all based on frequency data; latency data seemed to be both unreliable and insensitive. From pre-test to post-test the STP-trained crews improved 17% on initial plots, the other crews only 6%; 18% on track establishment, the others 4%; 17% on track classification, the others 3%; and 24% on track correlation, the others showed no improvement. The experimenters reported that the differences in improvement were statistically significant. Two measures concerned with tactical action (intercept control) did not yield statistically significant differences. In the third test, the crews for whom STP had been suspended performed as well as they had on the post-test; the research report omitted results for the other crews.

*The M-130 Experiment.* Another field experiment was conducted in 1957 at an ADDC code-named M-130 (Alexander, Kepner, and Tregoe 1962). Since the air defense site had not yet entered its operational phase, the experiment was not hampered by and did not interfere with operational requirements. Equipment was functional. Crews were on hand that had received individual training but had not worked together. The subjects were four thirteen-man crews equated by assigning personnel according to air defense experience, rank, skill classification, and scores on an operations information test.

All four crews had two shakedown runs on a practice problem and then one run on two pre-test problems, which were high-load situations containing forty-two and forty flights, as well as ten "critical" flights involving "difficult" situations, such as hostile mass raids and deviations from flight plans among friendly aircraft. Each crew then exercised twice with each of six training problems, before encountering two post-test problems which had the same inputs as those in the pre-test problems. Exercises were run twice daily, five days a week, for two months. Before the post-test problems all the crews were given one exercise, which included situations and requirements missing from the training problems, to determine how the crews would react to novel demands. The main objective of the experiment was to evaluate, in combination, the two STP practices of conducting a discussion-type debriefing after every exercise and presenting, at this debriefing, information about the crew's performance (knowledge of results). The experimental design did not attempt to differentiate the effects of one practice from the other. Two crews received knowledge of results in a debriefing session after each training exercise, whereas the other two had no debriefing sessions. Other experimental arrangements have been summarized by Alexander, Kepner, and Tregoe (1962) as follows:

> Crew performance information was collected by a military team which had been trained by and worked under the supervision of the experimenters. This information was obtained in three ways: from observation of the vertical board

on which were displayed all tracks processed by the crew; from logs maintained by experimenter personnel who simulated adjacent sites; from logs maintained by experimenter personnel who simulated interceptor pilots. The objective criteria for successful completion of each stage of information processing of tracks which were used were obtained from military regulations.

In order to minimize the possibility of transmission of information among crews, the following procedures were utilized: differential sequencing of exercises, crew rotation, coding of problem numbers, fostering of a spirit of competition by instituting a "Crew of the Month" award. Observation by the experimenters and interviews during the course of the project indicated that very little information of value was passed among the crews.

Ten types of system performance (and several combined types) were measured for the following four functions: detection of radar tracks and tracks reported from another ADDC ("cross-told") at PPI displays; recording of each of these two types of tracks on the multiviewer vertical display; maintenance of critical and noncritical tracks on the vertical display; and system output for adequate tactical action and reporting of data to other ADDCs. Except for the measure concerned with tactical action, the crews that had received knowledge of results and debriefings improved in their post-test performance over the pretest, whereas the other two crews did not. Analyses of variance indicated that seven of the thirteen differences reached the .05 level of statistical significance, and four others reached the .10 level.

When the crews encountered the problem presenting a novel situation (the adjacent ADDC was destroyed), the crews which had received the STP post-exercise treatments performed considerably better than the others.

The experimenters were especially interested in examining why in the post-test problems the crews which received knowledge of results in post-session debriefings improved more in some functions than in others relative to the other crews. They concluded that such feedback led to more improvement in those functions where less information about its performance level was available to the crew *during* operations. "The data indicate," they said, "that there is an inverse relationship between the visibility of a function and the amount of performance improvement demonstrated for that function." Although no measuring scale for "visibility" was developed, the researchers suggested that visibility ingredients included (1) the display of the results of his actions to an operator; (2) their display to his supervisor; (3) direct communication to an operator from persons affected by his actions; (4) observation by an operator of the activities of persons affected by his actions; (5) reception by an operator of information about other inputs reaching those personnel affected by his actions; and (6) an operator's awareness of "the kind and distribution of information needed by" those other personnel. The experimenters further advised that "consideration should be given to the possibility of designing or redesigning increased visibility into a system function so that more feedback is available within the operating situation."

A retest was conducted at M-130 four months later (Kepner and Tregoe 1959). In the interim the two crews which had not received the post-exercise STP training in the original experiment had been receiving it intensively for four months. They showed decided improvement. Conditions during the intervening period had been uncontrolled, and the earlier STP-trained crews had received less system training. Their retest performance failed to match their scores on the post-test of the original experiment.

*The M-96 Experiment.* An experimental study was undertaken at the M-96 ADDC in 1958 as a sequel to the M-130 experiment (Jensen, Tilton, and Anderson 1958). It was hoped to assess the different contributions to training of STP debriefing (discussions) and feedback (knowledge of results), the two features investigated jointly at M-130. Four crews served as subjects. One crew received both debriefing and feedback, a second only debriefing, a third only feedback, and a fourth neither. When the project was scheduled, the site was not expected to enter operational status until the end of the experiment. Unfortunately for the researchers, the ADDC became operational shortly after the experiment began, because a state of alert was declared. Experimental control was lost. Crews received variable amounts of operational experience during the experiment, and in all cases such experience exceeded STP experience. Crew membership changed due to turnover. One no-debriefing crew, it was later discovered, had had discussion sessions during slow periods of regular operations. The debriefing-only crew had had intensive training sessions on component tasks.

Other difficulties troubled the study. Its start was delayed by a lag in the installation of operational equipment. Special communication facilities had to be rearranged. The multiviewer vertical display was weeks late in arriving. When it did arrive, the painter assigned to inscribe the reference information was on leave. When he returned a week later, the job took still another week. The alert (due to a Mid-East crisis) interrupted all experimental activities for ten days. The AN/GPS-T2 frequently malfunctioned—a disaster in the evenings, when no maintenance man was on hand. Data were lost for one of the crews because the photography of the vertical display was poor. Six members of this crew were absent from its post-test and four new men were present, due to a military policy of crew rotation.

Each of the four crews had two pre-test exercises and (apparently) two post-test exercises, all involving a war-time problem—all the flights were "critical." There were ten system training program exercises for each crew between pre-test and post-test. However, in the scheduling of these and the pre-test and post-test exercises, various divergences occurred between crews.

Results were reported for three of the crews, but these results were not regarded as highly trustworthy. The experiment was, however, particularly productive in showing what can happen to experimental research in an operational setting. In addition to the misfortunes already noted, the SDC experimenters found that it had been unwise to rely on the relatively untrained operational personnel to collect data and perform simulation. Such tasks, they reported, should be handled by highly skilled personnel from the research organization itself. Data reported by monitors gave results contrary to those based on data derived from photographs.

*Other Studies.* A number of other smaller studies investigated "manual" STP, but their status as well-controlled experiments seems questionable. One was a team competition study in 1957. Crews received the results of other crews' performance as well as their own; the one with the best performance over a period of time received special passes (Cranston, Holmes, and Maatsch 1958). Still another examined feedback furnished by supervisors who monitored voice communication lines and informed crew members about their performance (Berkowitz, Best, and Rockett 1958). There was also a study, discontinued because of an alert, to investigate the value of specific feedback information about a particular operations function; communications between two sites were carefully monitored at each end and then the sites exchanged monitoring data (Bughman and Jaffe 1958).

The pre-SAGE system training program had been criticized by the Air Defense Command for its concentration on the surveillance and identification functions and relative neglect of intercept control. Accordingly, a study was done in 1957 in the 27th Air Division to determine whether STP could be expanded to train intercept directors (E. H. Holmes and T. R. Wilson, internal SDC publication). This field experiment was a fairly modest one. Two groups of intercept directors were trained at one site—one a group of four intensively trained in five sessions, the other a group of four partially trained. A group of three at another site received none of the special training. All received a pre-test in May and a post-test in June with an STP problem requiring intercept control; the extent of improvement matched the extent of training.

*System Improvement Experiment*

The principal experiment by SDC researchers with the goal of evaluation or improvement of the manual air defense system dealt with the effects of electronic countermeasures. Since it was associated with a laboratory-centered program, it is described in connection with that program in Chapter 17. Manual operations were involved in three others studies, the COIN, AZRAN, and Mode III projects, but since they came about in the SAGE era and concerned interfaces with SAGE, they are described in the next part of this chapter.

## SAGE (SEMIAUTOMATIC GROUND ENVIRONMENT)

As developed for the air defense of the United States, the SAGE System consisted of approximately a score of direction centers and a smaller number of combat centers which were higher headquarters for co-ordinating the actions of the direction centers. All the centers had digital computers. The larger and more powerful computers were in the direction centers, which were netted to each other and to the combat centers, radar sites, and interceptor aircraft bases by data link or telephone communications or both. Teletype, telephone, and eventually data link provided communications between the centers and the headquarters center of the North American Air Defense Command in Colorado Springs.

Analog-to-digital conversion computers at the radar sites ("long range radars") digitized radar signals which were transmitted by data link to the direction center computers for processing and display to surveillance, identification, and intercept-control sections. The intercept directors sent guidance and information messages to interceptor pilots by radio and data link. The interceptor aircraft and bases were not part of SAGE. Neither were the surveillance and heightfinding radars which provided the basic data about the geographical locations and altitudes of airborne objects. The systematic SAGE development, as such, did include the selection and the individual (as well as some crew) training of most of the operating personnel, but not of the highest-ranking decision-makers and their staffs. As SAGE was turned over unit by unit to the Air Defense Command, the organizational boundaries of a sector containing a direction center encompassed these other entities and personnel. They came to be widely regarded as parts of SAGE. Accordingly, although in its design and development SAGE was almost exclusively a computer-based information-processing system, the term as related to operations has included the sensor, effector, and decision-making functions (radars, weapons, and commanders) to which the information processing was related, as illustrated in Figure 10.

### *System Training Experiments*

The system training program for training crews in the manual air defense system was adapted by the System Development Corporation to SAGE and installed location by location. Most of the adaptation initially concentrated on methods of simulation. The STP in SAGE has been described by Rowell and Streich (1964) and Rosove (1967), and a detailed critique has appeared in a Navy-sponsored report (Parsons 1964); Sackman (1967) has given some of the particulars about simulation techniques. In STP as first practiced in SAGE, the long-range radars (the sensors) and their personnel were excluded from exercises, as were the effects of electronic warfare, that is, electronic countermeasures (ECM) brought to bear by the enemy against the radars and ground-air communications. Eventually system training was extended to the personnel at the radars, largely to incorporate simulation inputs of electronic countermeasures there (Parsons 1960*b*). But the primary emphasis has always rested on training the operations personnel within the direction centers.

Such emphasis has been closely associated with the methods of simulation for SAGE STP exercises. Essentially there have been two methods, other than the simulation inputs at the radars mentioned above. In one simulation technique, personnel acting as pilots but sitting at consoles in a special room in a direction center could, through button pressing and other other switch actions, insert aircraft signals into the computer and maneuver these signals to represent aircraft positions, courses, and speeds. The computer displayed these on the PPI-like "situation displays" in the surveillance, identification, and intercept-control sections. This simulation resembled—but greatly improved on—the maneuverable target simulation of the 15-J-1c device used for interceptor simulation in the manual air defense system, noted earlier in this chapter and described in previous chapters. Intercept directors communicated with the pseudopilots by radio-simulating telephones; or computer commands were executed by simulated interceptors through programmed simulation of data link within the computer itself. This method of simulation was employed almost exclusively in SAGE STP for the control of manned interceptor aircraft and unmanned Bomarc missiles.

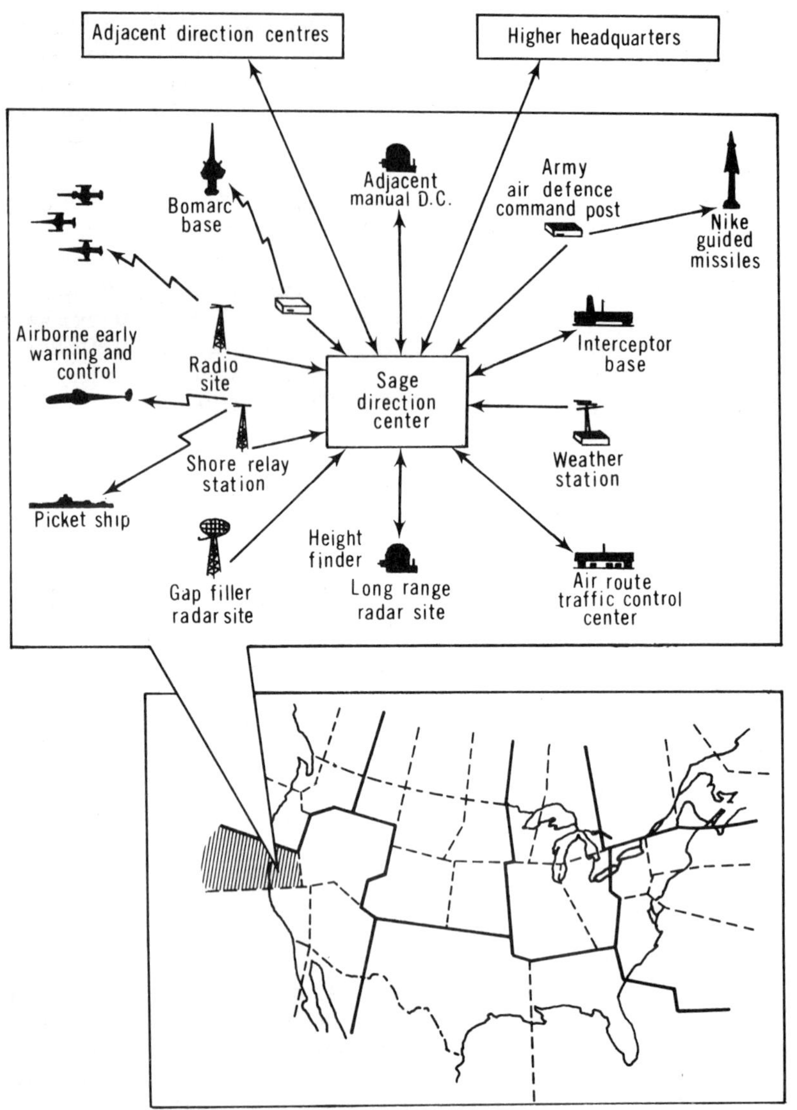


Fig. 10. Relationships between a SAGE Direction Center and Other Air Defense Elements (Rowell and Streich 1964).

The principal simulation technique in SAGE STP exercises represented hostile aircraft–i.e., bombers–and friendly aircraft, which could be commercial traffic or Strategic Air Command (SAC) flights. Problem inputs were prepared in much the same fashion as they had been for manual air defense STP (described earlier), up through the point of computer production of a magnetic tape, on which they were placed. However, the eventual outcome of the production process for this kind of simulation in SAGE was not storage of signals on film, but storage on a magnetic tape which could be processed by the computer in the SAGE direction center. The inputs were introduced into that computer as through they were digitized radar signals. Eventually, SDC developed methods whereby the direction center computer itself could produce the magnetic tape for a computer-based exercise.

Because the direction center in SAGE carried out the functions of a number of the ADDCs in the manual system, covering a much larger surveillance area and directing many more interceptor aircraft than a single ADDC, a direction center duty crew and staff personnel in an exercise might total fifty to eighty individuals rather than a score. The contents of an input tape would be correspondingly more substantial and complex. Additionally, an exercise required the simulation of various nonradar inputs, such as weather information, information from higher echelons, intelligence data, aircraft flight plans from the FAA and military units, and interceptor base data. SAGE also interfaced with, and could exert control over, Army centers which co-ordinated antiaircraft (Nike) missiles.

A SAGE STP exercise might involve a single direction center, two or more in a division operating together, two or more divisions, or the entire system together with all other North American air defense units and NORAD (North American Air Defense Command) headquarters. In the preparation and conduct of all exercises, SDC training representatives and computer programmers provided help to the Air Force personnel who were responsible for the training program.

Since preselected surveillance inputs were being introduced directly into the direction center and were being processed by that computer, it became possible also to program the same computer to collect data concerning the joint performance of console operators and the computer itself. It was also possible to program the reduction of these performance data and the printout of summary information shortly after an exercise. However, because some performance information even within the direction center could not be automatically recorded in this fashion, SAGE operators were observed by skilled Air Force monitors (in STP jargon called a TOR team–training operations report). The automatic recording method also was unable to register what might happen within the sensor (radar) and effector (interceptor) elements, since these were poorly represented, misrepresented, or unrepresented in a Direction Center exercise.

*Washington Air Defense Sector Study*. A single experimental attempt was made to evaluate the over-all effectiveness of the system training program in SAGE. This 1958 study, embodying forty exercises, was initiated and reported by J. T. Rowell, who headed the SDC team of training representatives in the Washington Air Defense Sector (WAADS) at the direction center at Fort Lee, Virginia. Rowell (1962) wrote:

Data were collected over a two-and-a-half month interval which extended from the completion date of positional training by the Air Training Command to operational date for the Sector. This period offered an excellent opportunity for such a study since there was no turnover of crew personnel, the crews were not engaged in activities other than system training and each crew, consisting of approximately fifty officers and airmen, was equally inexperienced in operating the SAGE system.

Each of four crews went through ten two-hour exercises, based on five input problems, two exercises per problem. No crews could be assigned as non-STP crews for purposes of experimental control, because all were required to receive training within the time available for the experiment. The problems were progressively more complex. They were those which had been designed for installing STP in this SAGE sector; one might describe the experiment as measuring the effects of installing the system training program.

The study assessed system performance through five measures: the proportion of "critical" input tracks for which computer-generated tracks were correctly initiated; the proportion correctly identified as hostile or unknown; the proportion of these against which interceptors were dispatched; the proportion of those which were intercepted; and the proportion of "critical" input tracks which were intercepted. Although the last problem was much more difficult than the first, the crews averaged 96% success on the fifth measure in the two exercises using this problem, compared with 32% success on this measure in the two exercises based on easier problems at the beginning of the experiment. The researcher reported the difference as statistically significant. The first two measures showed almost perfect performance from the start. Substantial improvement was evident in the other two measures.

*Weapons Director Study*. The SAGE direction center in the Kansas City Air Defense Sector (KCADS) in 1959 was a nonoperational site used primarily for testing SAGE computer programs and equipment and for some initial crew training by the Air Training Command. It also had advantages as a location for conducting experiments, one of which was a study of the SAGE weapons director function (Ford and Katter 1960*a*).

A weapons director (WD) in SAGE is the officer who makes selections (1) among interceptor aircraft and Bomarc missiles to intercept or interrogate hostile or unknown aircraft, and (2) among intercept directors to guide the designated interceptors using the computer's output of directional commands (vectors) and other information. The WD, who has a technician to assist him, must process much displayed information, operate many switches at his console, handle extensive communications within the direction center and with external elements, make many crucial decisions, and monitor the ongoing actions of the interceptors and intercept directors he has assigned.

The WD and his technician constitute a nodal position responsible for much of the interactional performance of the intercept-control section. It has been occasionally conjectured that training such nodal positions might be a relatively simple and inexpensive method of achieving many of the effects of the system training program. Although this heretical notion was never fully tested, it provided the rationale for this KCADS experiment.

The subjects were nine two-man teams of weapons director and technician, three of them with considerable operating experience, four with little experience, and two with none. (Level of performance was found to be correlated with level of experience.) Each team performed three times in an 80-minute exercise which had a portion characterized by low input rate followed by one with high input rate. However, due to scheduling and computer difficulties, complete data were not obtained for three of the less experienced teams. In addition to the input tape, as few as eleven individuals were required to "man all other positions needed to maintain the system operating context and to provide simulations, pilots, airbases, etc.," according to Ford and Katter (1960*a*). "One or two observers could be used depending upon how much detail of operator skills and procedures were required." These researchers came to some further conclusions beyond a finding that weapons director training was feasible:

> Reliable observations of operator performance can be obtained; independent sets of observations made on the same run averaged better than 95 per cent agreement. Such high observer agreement was obtained by (a) gearing the observational schedule to programmed inputs, in order to help cue observers, (b) making the observational categories very explicit, and (c) taking care to apportion the observers' jobs so none would be overloaded . . . .
>
> Operators exhibited a high degree of motivation, some continuing to work and take switch actions after the computer had stopped cycling . . . . All understood clearly that performance results depended completely on their individual capabilities, and . . . were unanimous in acclaiming the exercise as a challenge which mobilized their best efforts since all results were seen as being under their individual control . . . . There was considerable performance variability among the teams. The most inclusive measure of performance is the raw number of flights successfully handled during a run. The best single run score on this was seven times as great as the poorest! (The best was a very high percentage of the total possible.) Even among the first-run scores of the three experienced teams, the best was about 40 per cent better than the worst. This much variability would have serious implications for system performance and reliability, if it were compounded throughout the system . . .
>
> A good overall measure of learning effect is the average percentage improvement per hour of exercise, using the number of flights successfully handled as the score. Over the three experimental exercises, the average percentage improvement per hour was about 15 per cent . . . .
>
> Several behaviors that we have called Planning and Team Coordination Activities showed a markedly similar pattern of development; their frequencies rose from an intermediate value in the first run to a maximum in the second, and then diminished again in the third. The behaviors showing this pattern were: asking questions, answering questions, monitoring and prompting teammate, planning and conferring, giving directions to each other, and clearing excess information from displays. Planning and Coordination Activities are increased whenever a new situation forces extra attention and extra communication in order to share knowledge and reach agreements about changing or initiating procedures. After the novelty of a situation has been overcome by learning, the relative frequency of such activity decreases . . . . That the six behaviors listed actually represent learning activities is strengthened by another finding: experienced teams showed lower frequencies on all six behaviors than did inexperienced teams.

The experimenters deduced that although no performance criteria had been specified, the subjects apparently selected a quantity criterion, such as the number of critical flights neglected, rather than a quality criterion, such as use of the most appropriate weapon. The experimenters warned that such a short-sighted tendency might be accentuated by too much emphasis on high input rates in training.

They specified four operator techniques as helping the subjects reduce and smooth quantitative load. One was simply working faster, by deleting nonessential, though helpful, actions. Another was selection of alternative actions which required less time, such as assigning an interception to an airborne interceptor rather than scrambling one. Sequencing actions according to their required durations and associated delays was the third. The fourth was called "anticipatory planning"; the operator might take an action before it was necessary because he had some slack time.

*Electronic Countermeasures Study*. Another investigation can be regarded only as quasi-experimental. It as an attempt in the Kansas City Air Defense Sector in 1959 to get information on which to base electronic counter-countermeasures (ECCM) training in SAGE. Since electronic warfare is highly sensitive to security considerations, all the material here concerning this study has been taken from an unclassified report on the development and installation of SAGE ECCM training (Parsons 1960*b*).

The study addressed itself to a major uncertainty at the time. Who should be trained? Particularly, should ECCM training include personnel at the long range radars, even though these were usually some distance from the direction center to which they supplied sensor data? Opinions were many and divergent, whereas hard information was scarce since no task analyses had been performed to examine the effects of electronic countermeasures on SAGE and the steps necessary to counteract them. (Later, extensive testing with actual aircraft—for example, SAC bombers—which generated electronic countermeasures contributed valuable information about the kinds of ECCM activities required in SAGE. The reports of these tests by the MITRE Corporation and various Air Force units are classified.)

To provide the empirical evidence for task analyses, an ECM environment was created by means of simulation at three locations in the KCADS sector: the direction center in Grandview, Missouri, and two direction center-tied long range radars at Olathe and Hutchinson, Kansas. Simulated aircraft signals came from the AN/GPS–T2 and simulated ECM from the ACTER (anti-countermeasures trainer) at the radar sites; these devices provided inputs to these sites and thus to the direction center. Data were gathered during approximately 150 hours of test time in the course of four months. Two standard sets of inputs stored on film were created for the purpose. One was a "tactical" problem representing a mythical land-sea environment. The other consisted of 7-minute portions covering such ECM and aircraft parameters as number of jamming aircraft, distance between jammers and radar, power of jamming, combined use of chaff and jamming, etc. This second set of inputs was intended to show the relationships between various inputs at the radars and diverse tasks that personnel would have to learn to perform both at the radars and at the direction center. The study not only furnished information for such analyses but also enabled the tryout of training techniques, tested various aspects of equipment compatibility, and furnished an invaluable fund of direct experience to the SDC personnel who would subsequently install the ECCM training program throughout the United States. The same two sets of inputs were used for that installation. During the study the inclusion of radar personnel in exercises required debriefings over telephone nets connecting the direction center and the radar. A technique of loudspeaker commentary during an exercise evolved to explain at one location what was occurring at the other.

A major finding resolved the prevailing uncertainty about training. The radar personnel clearly had to be included in ECCM training but preferably in surveillance-only exercises. The Air Defense Command was persuaded of this necessity by the data collected.

The Air Defense Command took advantage of the study to request SDC to make an experimental test of some proposed operational equipment (Parsons 1960*a*). The requested data were provided, but probably the test should never have been attempted. It appeared that electronic equipment designed and produced for the purpose of simulation in a training program is likely to lack the precision, fidelity, and reliability needed for engineering-oriented testing.

*The WEST Test.* When the system training program was devised for the manual air defense system, much emphasis was placed on including all elements of the ADDC in an exercise. The adage, "train the system as a whole," was reasserted for SAGE, along with the rest of STP. But eventually doubt arose whether some sections of a direction center would be optimally or even sufficiently trained if this concept monopolized the program. The training which concentrated on the ECCM functions, just reviewed, was the first major deviation. Shortly afterward two new programs were developed, one for training the direction center's surveillance section separately, the other for separate training of its weapons (intercept-control) section.

The rationale for both has been related by Okanes (1962) to the need for tighter control over inputs. Although the surveillance training program was never adopted, it included some interesting aspects, such as specifying inputs which the automatic features of the computer would be unable to process, thus making human intervention essential. Then the computer could be exploited to collect data about such intervention and to relate these data to the inputs to provide measures of operator performance. The weapons section had a different problem with inputs. The computer-processed information about hostile aircraft which reached that section in an exercise (or in actual operations) would depend on the operational actions taken by the surveillance personnel in pushing the numerous buttons on their consoles. Since these actions were unpredictable, there was no way to make sure that a particular input for training would reach the weapons section or to replicate the input reliably. A computer-processed listing of surveillance outputs had to be furnished after an exercise to show what the input had actually been. Maps and scripts prepared in advance might display to monitors of the intercept-control function tracks which in fact had never been detected by the surveillance personnel.

In consequence, SDC developed a training program called weapons evaluation and subsystem training (WEST) which circumvented this difficulty by putting simulated surveillance button presses ("switch actions") on the magnetic tape that stored the processed radar signals. Every time a particular tape was run, the same surveillance outputs would be given to the weapons section. It was possible to simulate error-free surveillance operations or introduce and reliably repeat surveillance mistakes or omissions. As another innovation of WEST the tapes could be produced by the direction center computer. New tape inputs to the weapons section, based on the outcome of a prior exercise, became readily available.

This solution to the training difficulty posed by serial information processing in computer-based systems (see Chapter 16) was evaluated in 1961 in a series of exercises in the 30th Air Division (Cockrell and Murphy 1961).

At the Sault St. Marie Air Defense Sector a single weapons crew went through eight training exercises of increasing difficulty, first with sixteen, then twenty-four, then thirty-two, and finally with forty flights. For each flight total the problem was either simple—just straight and level tracks—or complex—many changes in heading, speed, and altitude. At the Chicago Air Defense Sector three crews performed in three demonstration exercises and four training exercises.

The Air Defense Command made an over-all evaluation of the WEST training technique by getting opinions from the crews and other 30th Division personnel. Queries probed WEST's usefulness as a training and measuring tool; the amount of computer time needed for training and on-site problem production; the amount of personnel time needed for these purposes; and the ease with which WEST could be integrated into the on-going training program. Favorable responses led to WEST's adoption.

In the WEST test an observer judged the outcome of an interception and recorded various aspects on a form. However, programs to automate the collection and reduction of interception data by means of the operational computer as an umpire had been in development as part of the WEST effort, and, as already indicated, a similar program had been created for the surveillance subsystem training program. Modifications of these programs were eventually adopted for use in SAGE training and evaluation.

*Project NORM*. For many years, interest in methods of system evaluation failed to match interest in system training, but the rise of subsystem training and the development of associated measurement techniques using computer programs began to bring evaluation to the fore. Some studies of SAGE system performance criteria were done at SDC as early as 1961, but the trend really began in 1963 with the development of BUIC (back-up interceptor control) as a system for war-time air defense. For BUIC the term STP was replaced by SETE, meaning system exercising for training and evaluation. The new term represented an effort not only to give evaluation equal weight with training, but also to take note that an exercise by itself was neither training nor evaluation, simply the vehicle for either or both.

The SETE development made it necessary to establish agreed-upon criteria, and measures derived therefrom, for air defense system and subsystem performance. One result was heightened interest in SAGE performance criteria and measurement, leading to Project NORM and eventually the widespread use of its products within the Air Defense Command.

Project NORM (normative operations reporting method) was initially aimed at determining the relationships among system and subsystem performance measures, their relevance and consistency, and effects of situational variables on performance. A pilot study took advantage of a simulation-based training mission at the Phoenix Air Defense Sector (Cunningham, Sheldon, and Zagorski 1965). Measurements were obtained for seventeen situational variables, nineteen subsystem performance variables, and five system performance variables. These were subjected to linear factor analysis and linear multivariate regression analysis.

Subsequently, the objectives of Project NORM were stated as the derivation of improved SAGE performance measures and the development of normative scales to assess changes in crew performance and differences between crews (Sheldon and Zagorski 1965). Analysis and consultation led to the specification and description of eighteen performance variables, thirty-two "mission difficulty" variables, and fifteen "crew-influenced mission difficulty" variables. Two simulation-based missions were conducted at each SAGE sector to collect data bearing on these 65 variables; the data came from 719 flights. A computer program was developed to extract almost 47,000 different items of information from the recordings of the missions and the input tapes. The application of descriptive statistics led, for various reasons, to the elimination of some variables. Factor analysis studies then sought to determine "what measures best reflect the quality of crew performance," that is, crew effectiveness. For example, three measures were designated as defining a general crew performance factor derived from the analysis; a tracking performance factor was associated with three other measures.

As in the pilot study, multiple regression analysis produced predictors of crew performance. Such predictors, together with an adequate data base, made it possible to develop equations whereby expected performance could be determined from the circumstances of a simulation mission and the characteristics of the SAGE sector. After each actual performance measure was compared with an expected value, the deviation was converted into a measure of relative performance. The final product, a scoring procedure based on relative scaling, was thereby independent of the difficulty of a particular mission and of inalterable sector characteristics. Crews could be compared even though they received different mission inputs and did not face equivalent environmental circumstances.

### *System Improvement Experiments*

The distinction between system training and system improvement can be nebulous if one of the objectives of system training exercises is to create or alter system procedures. One of the notions in the rationale for the system training program has been that relatively unconstrained discussion in a debriefing following an exercise would produce proposals for procedural innovation or change which could be tried out in a subsequent exercise. This process of "procedurization" (Parsons 1964) has, in fact, been called "system learning." Although SDC field representatives frequently asserted its occurrence, no documentation exists to bear out their informal testimony, possibly because of security restrictions but more probably due to lack of interest in proving STP's effectiveness.

Whether or not any system improvements resulted from training exercises, some have come about through procedurization of one kind or another as a consequence of field experiments. As indicated earlier, some of these experiments involved manual operations associated with SAGE and occurred in the SAGE era. Their descriptions will precede those of two research efforts which dealt explicitly with SAGE.

*The AZRAN Study*. As already mentioned, most of the radar data entering the SAGE computers came over data links from long range radars where special SAGE equipment converted the data from analog to digital form. However, a certain number of radars did not have this conversion equipment. Their data arrived at the direction center by teletype in alphanumeric form and operators punched the data into IBM cards to enter the computer. The geographical position of the radar signal of a radar-detected aircraft was described according to a rectangular grid co-ordinate arrangement called "GEOREF." The SAGE computers processed surveillance data stated in GEOREF terms.

Among the units providing radar target positions thus relayed in GEOREF co-ordinates were those in Air Force airborne early warning and control (AEW&C) aircraft. These patrolled continuously off the Atlantic and Pacific Coasts to extend radar coverage out to sea and thereby give advance warning of an enemy bomber attack over the water. They resembled the Navy AEW&C aircraft described in Chapter 4. In these Air Force radar-equipped aircraft two surveillance operators sat at PPI scopes, detecting and tracking airborne objects seen by the radar. By intercom they reported each target signal's position to a plotter, describing the position in miles of range (distance from the aircraft) and degrees of azimuth (direction from the aircraft). The plotter recorded the position on a vertical geographical display carrying range and azimuth reference marks. The other side of this transparent display was marked with a GEOREF grid. On that side a scanner recorded the position in GEOREF terms on a special form, which he handed to a Dualex operator for transmission to the direction center. Under heavy load conditions the Dualex operator had to sit in the aisle so two plotters could work at the display.

The layout of the aircraft indicated scant human engineering attention. Whether or not the human engineering and experimental results of the earlier research on AEW&C aircraft for the Navy would have been applicable, Pacific Coast AEW&C Headquarters at Mather Air Force Base appeared to be unaware of that research. But this was not the only or primary problem. Partly because positional data had to be converted from azimuth and range figures to GEOREF co-ordinates in the aircraft, the information arriving at the Direction Center was inaccurate and late. Errors were made in the time-consuming conversion. It seemed possible that if the data were reported in azimuth and range to the Direction Center and converted to GEOREF by the computer, surveillance information from AEW&C aircraft might be more reliable and faster, especially if some of the airborne processing were also better human-engineered. An experiment was performed in the early months of 1961 at Mather Air Force Base to test this possibility (Freed 1961*a*, 1961*b*; Wiechers 1963).

The experiment was carried out in the trailer in which simulation-based system training exercises were conducted on the ground for AEW&C crews.

Inputs came from a medium-heavy STP training problem. Fourteen crews of nine experienced Air Force military personnel operated in the conventional manner, with two plotters at the vertical display and the Dualex operator in the aisle. Fourteen equivalent crews operated in what was labeled the AZRAN mode. The scope operators reported target positions in azimuth and range not to a plotter but to the scanner. He recorded the data on the special form in terms of azimuth and range and placed the form in a "buffer basket" next to the Dualex operator. The latter sent the data as it appeared on the form to the direction center—in azimuth and range co-ordinates, not in GEOREF. The single plotter with his central display was no longer a relay point in the information flow. The plotter obtained azimuth and range data by listening to the intercom transmission from the surveillance operators to the scanner; if he fell behind, he could also acquire information from the forms in the buffer basket. The display was needed to supply information only to individuals within the aircraft. Since only one plotter was required, one man was freed to monitor a third scope and to make raid assessments and estimates of track speeds; these had often given trouble.

In the experiment itself it was unnecessary to send the position reports to the direction center; reports could be taken from the Dualex tape as if they had reached there. The results showed that the AZRAN mode greatly surpassed the conventional method in accuracy, in speed of processing from signal detection to Dualex transmission, and in number of transmissions per unit time. Assessments of raid size and speed improved. The SAGE computers were thereafter programmed to make the conversions from range and azimuth to GEOREF co-ordinates; in other words, the computer assumed the task previously done less effectively by human operators in the aircraft. The AZRAN mode was adopted as the standard method.

Freed (1961*b*) offered the following comments, which have some generality:

> Why does AZRAN provide this improved performance? First the number of information channeling functions in the aircraft are reduced. Reducing the number of functions decreases the possibilities of error as well as decreasing the time involved in processing. Next, the largest area for error is eliminated by requiring the computer to make the GEOREF conversion. Provision is made for observing information flow and controlling it. Queuing is smoothed, bottlenecks reduced, loss of information minimized, and accuracy of information processing improved.
>
> . . . receivers of information seem to be more effective when transmissions are one-way only. Plotters, for example, seem able to double their effectiveness when they do not have to respond and when they listen to only one person transmitting at a time.
>
> . . . it seems less efficient when the same persons both record and relay information, than when each has one of the two functions.
>
> . . . The idea of the use of buffering techniques in a manual system, derived from computer systems, seems to have merit in smoothing and speeding the flow of information, preventing its loss, and reducing errors.

*The COIN Study*. The Royal Canadian Air Force suspected that too much time was required to forward air defense reports from subordinate headquarters to higher commands and that the relayed data lacked sufficient accuracy. With SDC support, a number of Canada-wide system training program problems were run by the RCAF's Committee on Information Needs (COIN) in 1960, with observers at all echelons monitoring the dispatch and receipt times of all messages. Latencies were indeed found to be too long, and errors too frequent. Then the information-forwarding system was revised to include only the information required for efficient system operation. For example, several hundred reports from subordinate headquarters were reduced to a half-dozen. Another series of STP exercises followed. The processing rate from the lowest echelon to RCAF Headquarters was greatly reduced, and the smaller number of reports proved to be sufficient (Wiechers 1963).

One of the ways in which it was possible to reduce the quantity of reports to higher headquarters was to eliminate the successive positions, courses, and speeds of the component aircraft in reporting hostile raids. It appeared adequate just to report the number and approximate location of such aircraft to higher echelons, since the latter could in any case do nothing about intercepting the raids; the intercept-control function was in the hands of the lower echelons which were doing the reporting. The COIN study, which concentrated on a manual system environment, never became well known south of the Canadian border. Its recommendations were not entirely pertinent to the SAGE system; because of interfaces between Canadian air defense and SAGE there were also limitations on the extent to which they could be put into practice south of the border.

*Mode III Study*. Prior to the development and installation of the BUIC System, if a SAGE direction center were put out of action in hostilities there were to be two recourses. These were important, because the SAGE system was not designed to withstand nuclear attack, and many direction centers were built near prime targets of hostile ICBM's, namely, SAC bomber bases. One recourse was Mode II, in which adjoining direction centers took over the responsibilities of the one which had been destroyed. The next recourse was Mode III, in which the personnel at the long range radars would do their best to conduct air defense. One of the radar sites within a group would become a master direction center (MDC) handling identification and weapon assignment functions for all, the others becoming subordinate sites with limited functions. There were numerous procedural and structural questions as to how to implement this MDC concept—questions concerning manning, communications, and training.

An SDC experiment, consisting of 192 exercises in the 20th Air Division between October 13 and December 19, 1958, sought to provide some answers (Bumpus 1959). Each of four crews of inexperienced personnel encountered each of eight conditions in six exercises. The crews were located at three sites—one Master Direction Center and two subordinate locations. The exercises employed the simulation methods of the manual system training program. Eight two-hour high-load problems (input sets) provided three wartime situations and five peacetime situations.

The eight experimental conditions resulted from three variables, each varied two ways in a factorial arrangement (2 × 2 × 2). As one independent variable, an operator plotted or did not plot surveillance data on a vertical display at the subordinate sites. When the board was not used, operators at PPI displays communicated their data directly to the MDC; when it was used, the information went to a plotter and was relayed to the MDC from the display by a teller. The second variable consisted of the two subordinate sites. In the two states of the third variable, the subordinate sites either forwarded information on each track every two minutes until ordered to cease, or they sent only an initial position and one additional plot, unless the MDC requested further information.

Inexperienced personnel were selected as subjects because it was presumed that in the SAGE era surveillance and weapon assignment would be handled primarily at the SAGE direction centers. The personnel at the radars would lack operational practice in these functions. The manning varied among experimental conditions, one of the effects of different conditions being to require different numbers of personnel; the total crew for the three sites ranged from twenty-five to thirty-one. Results are still classified, but they seemed to be conclusive. Data were collected by military personnel carefully trained by the seven-man SDC research team. The principal source of objective data was the multiviewer vertical display at the MDC; in addition, this display was rated by expert observers.

*Senior Weapons Director Study*. In SAGE, the position of senior weapons director (SWD) was a critical one. This officer co-ordinated the activities of a number of weapons teams of intercept directors, each headed by a weapons director (WD). (An experiment in training the WD position was described earlier in this chapter.) The SWD and WD were decision-making positions. To the extent that personnel selection, procedure specification, and training for these positions were incorporated into SAGE development, those who created the system explicitly developed decision-making along with other functions. But the extent of this development was limited. For example, procedural questions remained after SAGE became operational. A two-part experiment in 1959 at the SAGE Kansas City Air Defense Sector investigated some of these (Hall, R. W. and Levine, R. A., internal SDC publication).

First, each of two sets of two weapons direction teams operated together without a senior weapons director, an abnormal arrangement contrived to reveal trouble situations for further investigation. One set of two weapons directors was structured so that they agreed to divide assignment of target tracks by area, whereas in the other set the WDs agreed to take tracks assigned randomly. Each set engaged in four two-hour exercises under one of the conditions, the seventy-seven tracks in the inputs coming from a problem tape prepared for another project. Next, four weapons direction teams operated with a senior weapons director in charge of sets of two. One SWD had charge in four exercises in which assignment was by area, another SWD in four exercises in which assignment was at the SWD's discretion. All positions in the direction center not in the weapons teams were simulated by operators with scripts.

Trouble situations which emerged in the first part of the experiment were communicated to the SWDs before the start of the second part. These were the transfer of hostile tracks between teams and efficient use of interceptor aircraft, as in transfer of control between teams. In the experiment's second part, in which SWD's took part, over-all success was approximately the same regardless of method of assignment, but area assignment required the use of fewer interceptor aircraft and led to more of their transfers between teams.

*Computer Processing-Time Studies*. By far the most extensive and innovative field research on SAGE as a system took place in 1961–63. It has been well described by the principal investigator, H. Sackman (1963; 1964*a*, *b*; 1967). Although this experimental investigation was directed principally at the amount of "frame time" (duration of a program cycle) required by the SAGE computers to process radar inputs and operator actions, it resulted not only in considerable additional information about SAGE operations but also in the concept of "regenerative recording" for the analysis of computer-based systems, and in the implementation of that concept.

The first of three data-collecting studies in the field was a pilot study in 1961 at the Sault St. Marie Air Defense Sector (SMADS). In the second study, also in 1961, the participants were nine SAGE direction centers and two combat centers in the 26th and 30th Air Divisions; these covered the northeast part of the United States. The third, in 1962, was at the Phoenix Air Defense Sector (PHADS). In addition to data analyses performed immediately after each study, four follow-on studies were based on subsequently replaying the record of the second and largest of the field studies. These replay or play-back studies exploited the "chronicles" computer program which had been created to implement regenerative recording.

The pilot study at SMADS had two objectives. One was to determine what would happen to SAGE computers if they received very heavy system inputs to process. The SAGE computer went through recurrent processing cycles called frames and subframes. There was some concern lest heavy inputs would extend the duration of a frame to a point where the system might not be able to meet its air defense requirements. Radar data might be irretrievably lost and responses to the tactics of hostile aircraft might become excessively slow. The second objective was to try out the three programs constituting a program set called "chronos," composed of "chronicles," "chronometer," and "chronograph." Chronometer permitted measurement of subprogram operating times and overall frame time; it also recorded track composition, radar returns, operator switch actions, track merit, and other timing and load variables (totaling eighty-nine) automatically collected for every frame. Chronograph processed the chronometer data into a printout of one page per frame and produced punched cards and magnetic tape for further data reduction as desired. The pilot study used a high-load simulation tape which had been produced for the system training program. The subjects were the men in one of the direction center's operational crews, carrying out their various functions.

The eleven-site study also made use of the STP context. A planning conference for Air Defense Command and SDC participants preceded it to prepare the considerable co-ordination required for so many locations to take part in an integrated exercise embodying some new data-collection requirements. A single three-hour exercise followed a four-hour emergency warning build-up of defensive weapons in the area involved, with all battle staffs at the combat centers and direction centers taking part. As had been planned, two of the direction centers were put out of action in the last hour of the exercise and their functions were handled in the Mode II fashion mentioned earlier. Details of the exercise, including those which showed the extent to which the simulation really covered wartime eventualities, have not appeared in an unclassified publication.

In addition to the automatic recording of exercise operations, all military participants filled out a questionnaire. The questionnaires yielded data about personnel experience levels (which turned out to be approximately equivalent at all sites for parallel positions), manning (also fairly uniform), and simulation team support (extremely variable, this variability being associated with numerous simulation difficulties). Military personnel rated the simulation inputs for degree of realism on a four-point scale. The average rating was the point labeled "fairly realistic," the next to highest point. None of the raters, to be sure, had had an opportunity to experience an actual, full-scale air defense battle.

Every ten minutes during the exercise the military participants also rated individual operator loads from "light" to "breakdown" and rated computer system performance from "very good" to "breakdown," in each case on a five-point scale. According to twenty-four analyses of variance, "individuals gave consistently different ratings, and ratings varied systematically with different time periods" (Sackman and Munson 1964). There were very low correlations between the two scales but substantial (and statistically significant) correlations between ratings of work load and number of operator switch actions. (By "switch actions" are meant messages operators sent to the computer by pressing buttons and moving other switches.)

The questionnaire also contained the question: "What was the most serious operational problem occurring at your particular station at your heavy load period during this test?" Most replies were concerned with equipment, simulation, console actions and displays, and human communication. It appeared that as frame time increased, not only did computer responses to switch actions take longer but also operator load rose, so "the computer system responds more slowly at a time when the operator has more work to do and needs a faster response."

The analysis of objective data collected by chronometer and chronograph led to the derivation of "multiple regression equations estimating overall frame time and component subprogram time as a function of a small set of input load predictors." These predictors were derived in part from factors isolated through a factor analysis; they accounted for most of the frame time variance. A regression analysis supplemented the factor analysis. Six predictor variables were determined. Further, since the regression coefficients were similar for different direction centers and different load levels, a single set of regression equations appeared satisfactory.

As noted at the outset, four studies were done later by means of the chronicles playback capability on the basis of the eleven-site exercise. The chronicles program recorded initial computer core content to establish the starting condition on magnetic tape and then recorded all computer program inputs up to 30,000 words per frame. "Then," Sackman and Munson (1964) observed:

> by resetting initial core memory conditions and playing all inputs back into the operational program in the rerun mode, the computer will follow the same steps in the same sequence as occurred in the original test. This amounts to total recording of the entire computer operation. The investigator may analyze any part of the test run as often and at any time that he wishes . . . . In regenerative recording, the inputs are tagged with real-time identifiers. This permits the investigator to delete, modify or add any program or input changes in a rerun in nonreal-time and yet maintain the timing integrity of the original real-time test operation.

One of the four studies was an investigation of a new set of SAGE computer programs to determine how the new programs handled the frame time problem in processing the eleven-site exercise operations. Another study "involved dynamic instruction counting to develop statistical norms of how frequently single instructions and certain strings of instructions empirically occur in the SAGE program" (Sackman 1964*a*). The other two, described below, were more closely related to man-machine considerations.

One study analyzed the 58,131 recorded switch actions made by 592 console operators in the eleven-site exercise (Sackman 1964*b*). Various frequency totals and distributions were ascertained. Switch actions averaged two thousand per combat hour in a single direction center. There were very large individual differences in rates among types of operators; some typically had heavy loads, some light. An exploratory analysis of operator error was limited to a sample of 250 frames from one sector. This sample was found to contain 290 "illegal" switch actions (12.5% of all switch actions) which the computer recognized as incorrect. A correlation analysis indicated that the number of "illegal" actions rose as the load increased on an individual operator. In a further analysis of a sample of 4,979 switch actions, it was found that approximately 30% of all switch actions were useless repetitions of previous actions.

*The Feedback Problem*. The fourth study examined how frame time was affected by the current method and by two proposed methods of switch-action processing for providing displays in response to operator switch requests. With the current method there could be a delay of many seconds, even up to a minute, between the time that an operator requested a display and the time the computer provided it. There was no way for the operator to know that the computer had received his request and was acting on it, until the display appeared. Accordingly, operators frequently initiated a new request before the computer's response to the prior one was completed, with the result that the computer response which showed up was incorrectly assumed to be the one appropriate to the new request. The delay in response overloaded the operator's short-term memory, especially in high-load situations. Many of the illegal and repetitive actions found in the analysis previously described apparently originated from these conditions. Of the two proposed methods which were compared with the current method, one was expected approximately to double computer processing speed, the other to triple it. Both included another new feature, the appearance of "OK" on the operator's display to indicate that a switch action was "legal." However, despite the divergences among the three methods, the regenerative-recording analysis revealed no major differences in effects on frame time.

The issue was again attacked in the third field study, which directly compared the current and two proposed methods for effectiveness (Sackman 1963). A crew of thirty-six experienced Air Force operators manned most of the operational and simulation positions in the PHADS Direction Center in three one-hour exercises, one exercise for each method. Unavailability of computer time made it impossible to counterbalance with more sessions to control for possible practice effects. The simulation inputs came from the first hour of an STP problem tape.

Observers with stop watches ascertained computer response times. Average response times for the three methods were 11.39, 7.51, and 5.75 seconds. Further, total system response time was analyzed into three components: computer processing time, display equipment cycle time, and human recognition time. This study led to reprogramming the switch action/display makeup portion of the SAGE computer's operational program.

In actuality, this was not the first experiment which investigated SAGE computer response time. Murphy, Katter, Wattenbarger, and Pool (1962) had conducted an experiment at the PHADS Direction Center (which had replaced KCADS as a test location) that incorporated a READI light indicator to notify the operator that the computer had accepted his switch input. This indicator greatly reduced errors attributable to slow computer responses to successive operator actions. These researchers found that approximately 40% of all recorded switch-action errors had this origin. They also found that under certain conditions when switch-action rate was high, such as one every ten seconds, as many as one-quarter of such actions might be in error.

It may be of some interest that these investigations of feedback to the console operator in SAGE, and the necessary programming changes they instigated, occurred when the system had already been operational for four years. For a system of such size and import, relatively little human engineering analysis was involved in its design, development, and early operation.

# 12

# Coordinated Science Laboratory

The locale of two notable man-machine system experiments in 1957–59 was the Coordinated Science Laboratory (CSL), formerly the Control Systems Laboratory, of the University of Illinois, Urbana, Illinois; three all-computer studies followed and supplemented them. This program, carried out by H. W. Sinaiko, is documented in three reports (Sinaiko 1958; Sinaiko and Cartwright 1959; and Sinaiko and Shpiner 1960). The laboratory program was funded by the Army Signal Corps and Army Ordnance Corps, the Office of Naval Research, and the Office of Scientific Research, Air Research and Development Command of the Air Force.

The primary objective of the research holds particular interest. The laboratory had developed a computer-based combat information center (CIC) called the Cornfield System to demonstrate that naval air defense (antiair warfare) could be carried out automatically. (Combat information centers have been described in Chapter 3, 4, and 5.) The experimental program sought to determine the relative effectiveness of various degrees of system automation, that is, various allocations of functions to man and machine. At one extreme was virtually complete automaticity, i.e., almost no human intervention. At the other extreme, a completely manual system could be represented by data from an earlier experiment done at the Naval Research Laboratory (see Chapter 5.) In between were three man-machine combinations.

The first CSL experiment, called Artful, probed the three man-machine combinations and the almost wholly automatic condition. The second, Careful, investigated whether similar results would develop if the input loads—hostile attacks—were considerably heavier. The supplementary all-computer studies, in which the system operated in the fully automatic mode, examined a number of aspects, including such experimental tactics as the number of runs and length of run.

The Cornfield System, schematized in Figure 11, consisted of a special-purpose tracking computer processing data from one or two radars, a general-purpose computer, the ILLIAC, and Charactron displays of processed track data. These displays, together with a keyset, could link human operators to the computer. The system, envisioned for defense of a ship against attacking aircraft, could perform automatic tracking, threat evaluation, weapons assignment, and

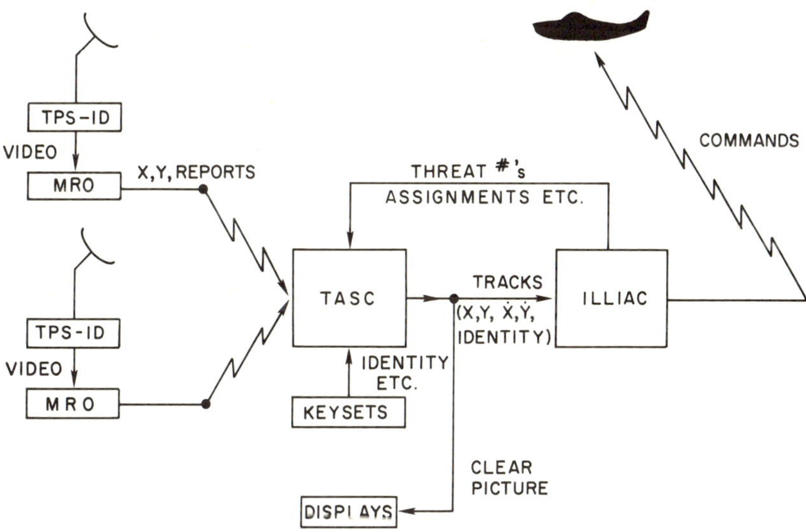


Fig. 11. Simplified Cornfield Schematic (Sinaiko 1958).

interceptor control, through the ICON II control program. The threat of each hostile aircraft (any aircraft not identified as friendly by a human operator) was computed on the basis of the distance from the ship the aircraft had reached and its relative (closing) velocity in approaching the ship. Interceptor aircraft were selected for repelling an attacker possessing at least a certain degree of threat, as long as the computer judged such interception possible. The computer found the best unoccupied interceptor and the best busy one according to the times it calculated were required for interception. It chose between interceptors and it could make reassignments. It could guide the assigned interceptor with vector messages along a collision course toward the hostile aircraft, to a computer break-off five miles away. The only operation which the Cornfield System could not be programmed to perform automatically was to reidentify lost tracks; these were instances where the signal for a hostile aircraft became disassociated in the tracking computer from the position stored in the computer. The very few that occurred were identified by operators, whose actions accounted for the slight tinge of manual intervention in the automatic mode. No noise or other confusion factors were introduced to disconcert the tracking computer.

Tracking was automatic in all the CSL studies, except for the human intervention just noted. In the fully automated mode, threat evaluation, weapons assignment, and interceptor control were also automatic. In one of the three man-machine combinations, the computer executed these three functions, but operators could intervene and handle them manually. When they did intervene, they would evaluate in their heads the threat of a hostile aircraft, assign interceptor aircraft against it before the computer did so or reassign after it did so, or calculate and transmit voice commands to the simulated pilot to guide him toward the target. In another combination, the operators handled the threat evaluation and weapons selection functions, leaving interceptor control to the computer. In the third, the humans took over this last function also. (The Careful experiment omitted this condition.)

To complete the picture, it should be pointed out that the comparison experiment at the Naval Research Laboratory (NRL) required tracking by humans instead of a computer. Thus, in that experiment all system functions were manual, as they were in real operations in the days before computers.

## THE ARTFUL AND CAREFUL EXPERIMENTS

In the Artful experiment the three subjects in the mixed conditions, acting as two CIC officers and a keyset operator, were two Navy officers and an enlisted man, respectively, all ROTC instructors; in Careful there was one CIC officer (one of the Artful subjects) and the operator. The same subjects performed in all conditions. The NRL study had employed an entirely different set of experimental subjects. Prior to Artful the subjects were trained for six weeks, some of the time in practice runs with low-load inputs. During that experiment the two officers alternated their positions. To detect and control for any practice effects during the experiments, each unique combination of variables occurred twice in Artful and four times in Careful.

A secondary objective of the studies was to determine the effects of input load, especially in the second experiment in which, as noted above, it was sought to discover whether heavier loads would alter the relative effectiveness of the various man-machine conditions. As shown in Figure 12, in Artful a heavy load consisted of twenty-two attacking aircraft, ten of them "critical" in that they were programmed to fly close enough to the ship to score a hit; and an average of ten attackers were being tracked at one time. There was also a light load in which these figures were approximately halved. Careful also had heavy and light loads, the former with sixty attacking aircraft, all critical, the latter with thirty-seven of which seventeen were critical. All runs lasted about 30 minutes. The Careful script included omnidirectional, realistic, and radial attack patterns. In both experiments different but equivalent versions of the basic script were constructed through such techniques as rotation and position interchange so that subjects would not re-encounter the same apparent total attack.

Artful, incorporating four degrees of human intervention, two levels of load, two crews (officer alternation), and two replications, ran thirty-two times. Careful had twenty-four runs, reflecting three intervention modes, two load levels, and four replications. Order of runs was randomized, except that all the automatic mode runs in Careful occurred at the end.

Attacking aircraft were simulated in both experiments by punched paper tape input to the tracking computer, the tapes containing the scripted tracks. An innovation in the second experiment was the disappearance (fading) of raids which were judged "killed" (shot down). The generation of interceptor aircraft signals also showed some technological evolution. In Careful the general purpose computer, ILLIAC, generated them, and the tracking computer moved them in

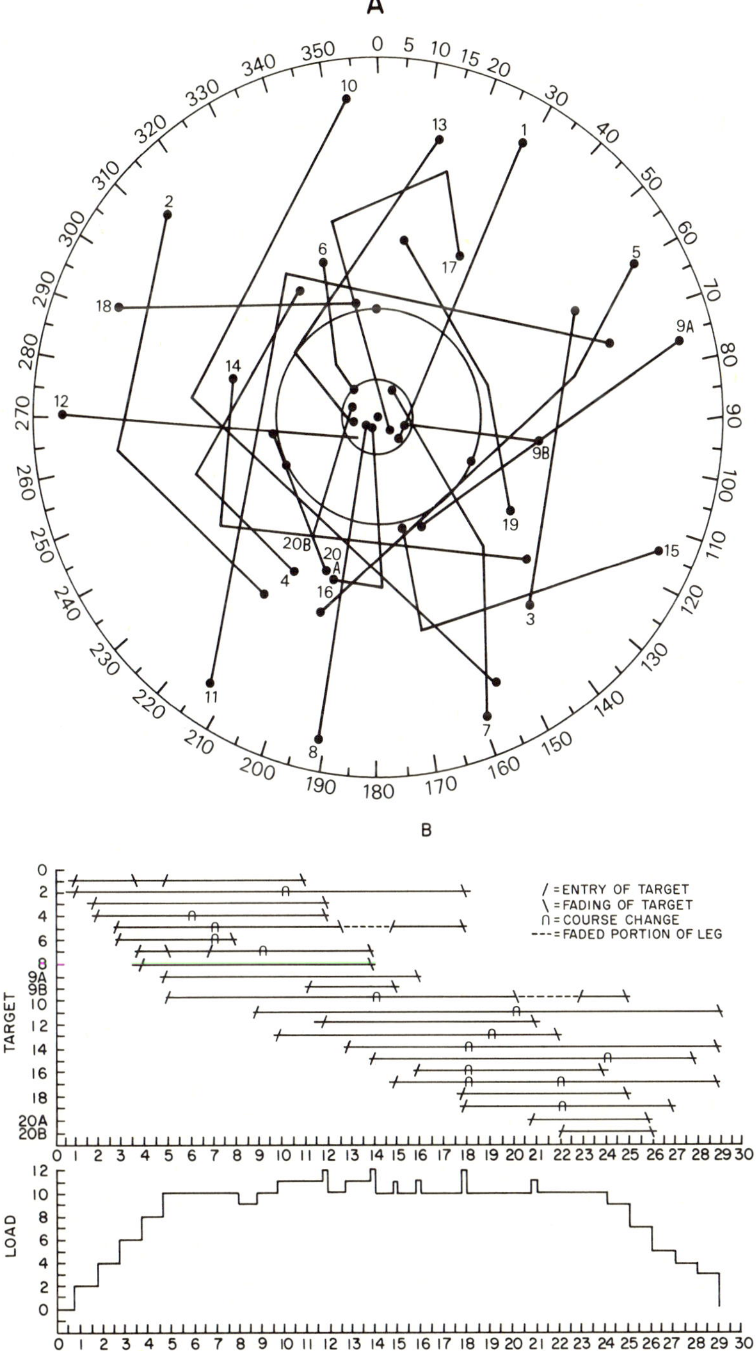


Fig. 12. Heavy Load Input in Artful. *A*. Hostile Air Target Configuration. *B*. Time Profile (Sinaiko 1958).

response to ILLIAC's control commands; thus there was no manual interceptor control or manual simulation. In Artful the simulated interceptors had been produced and moved by standard 15–J–lc (described in earlier chapters) target generators manipulated by technicians who responded to control orders from the computer or, in the manual control conditions, from one of the CIC officers.

In each experiment one class of performance measures related to kills of attacking aircraft, while a second concerned assignments and deassignments of interceptor aircraft. The first included judgments of intercept success and such measures as time and range of kill, with averages and standard deviations. These were derived through processing by the ILLIAC itself, in conjunction with visual observations and reports by the CIC officers. Data in the second category were gathered and processed in the ILLIAC by a special program called DOPE (Data Obtaining Program Evaluation).

## ARTFUL RESULTS

The results for the kill class of measures in Artful were sufficiently similar in the four differing conditions of automaticity (by inspection—no significance statistics were reported) that it would not be justifiable to ascribe superiority to any of the conditions. The automatic mode did tend to kill its attacking aircraft at somewhat greater ranges from the ship in the heavy-load condition; on the other hand, it killed fewer critical targets. Differences among hits on the ship and the durations over which it survived were negligible. Sinaiko (1958) speculated that results might have differed if each experimental run had ended when the ship was hit. "The men, because of their underlying need to survive and fight again, might have sacrificed range for assurance that they would not be hit; the computer, failing to perceive a difference between being bombed early or late, would have played the same, unvarying game." Needless to say, survival time and range at kills both decreased as the load increased.

The two more automatic Artful modes (fully automatic and human intervention option) showed considerably more interceptor assignments than the modes in which the operator was obliged to make the assignment himself. This was especially the case with increase in load. Of even greater interest were the choices by humans to do the assigning themselves in the intervention option condition. When the input load increased, the human operators intervened relatively more often, rather than less as had been expected, and more kills were made after human than after computer assignment! Sinaiko said it seemed that "When loads were light, the men appeared willing to let the computer carry most of the assignment responsibility; when loads were heavy, the men much more often stepped in, over-rode the computer, deassigned, and re-assigned interceptors," as though they were competing with the computer.

The foregoing results must be considered, of course, in connection with certain conditions of the experiment, for example, near-perfect tracking, and a criterion of kill which disregarded the bearing and heading of the interceptor relative to the attacker. Due to these aspects it would not be entirely valid to compare the Artful data with those from the NRL all-manual experiment, which involved more stringent kill criteria and more difficult tracking. In that study, which employed the same input loads, more attackers penetrated, the kill range was shorter, and the ship suffered more hits. In the ratio of assignments to kills the all-manual experiment produced approximately the same results as the two Artful conditions in which assignment also was manual.

## CAREFUL RESULTS

The Careful experiment resulted from allegations in some quarters that the more automatic conditions would really show superiority if the input load were very heavy. However, in Careful the outcome was just the opposite. A smaller proportion of attackers was killed in the wholly automatic mode than in the other modes under the heavy load; differences were trivial under the moderate load; so were differences in average kill range under both loads! As in Artful, in the option mode the human operator took more assignment responsibility away from the computer as the load increased—"as stress on the system builds up human decision-makers tend to take over more and more of the total task from automatic system elements" (Sinaiko and Cartwright 1959).

Several other observations came out of Careful. One was that the CIC officer resorted to pattern recognition and tended to deal with attacking aircraft as groups in making assignments; he would send a single interceptor against three raids in the same vicinity, whereas the computer would deal with each separately. There also seemed to be a tendency for the human operator to pursue marginal targets or those which could not be intercepted in time and which the computer decided to forego. When the human decision-maker was faced with what to him (but not to the computer) was uncertainty, he apparently evolved an additional response—"maybe." The experimenters added: "Perhaps, too, man's sensitivity to his own survival needs was operating here. That is, in the marginal area of choice the man's decision to continue an interception would be affected by his knowledge of what might happen if the raid penetrated the vital zone. Our computer has not had this type of sensitivity built into it." Thus we encounter in this study, in addition to the human's exploitation of his capacity for pattern recognition, an apparent demonstration that motivational variables may intrude into decision-making.

A third observation has some implications for human engineering in the design of display systems. In the Cornfield System it was possible to classify and display attacking aircraft according to certain attributes of identity and the status of the engagement. The CIC officer never once used this feature. Schemes for categorizing data are built into many computer programs, not merely those for air defense. It seems possible that sometimes the data are classified in certain ways because it is relatively easy to do so rather than because the classification is useful. Classification schemes in displays should be subjected to human engineering analysis.

## ALL-COMPUTER STUDIES

The genesis of the three supplementary all-computer experiments was set forth by Sinaiko and Shpiner (1960) thus: "Since the performance of Cornfield was generally good in the Artful experiment irrespective of whether the system operated in a fully automatic mode or with some degree of human intervention, questions about the way the simulated air defense battle was fought were raised. How would the system perform if a different defense strategy, i.e., disposition of weapons, were employed? How much of the behavior of the system was due to the particular script . . . used in Artful? Would Cornfield's ability to 'fight the battle' have changed if more defensive weapons had been available? And, what sorts of interactions, if any, would there have been between these variables?" Reading between the lines, one might infer that somewhere there existed a suspicion that in some way the Artful and Careful experiments had been unfair to the computer.

In the first of these all-computer studies it was shown that changes in attacker load, in number of interceptors, and in disposition of interceptors did affect system performance. However, since only the fully automatic mode was employed, it was not possible to show whether similar results would have occurred in the less automatic modes. The second experiment inquired whether four runs per condition (the number in Careful) would provide reliable data. Results of ten runs were compared with samples of four from among the ten; the differences were not statistically significant. However, as Sinaiko and Shpiner (1960) point out, one cannot extrapolate to samples involving humans, where "factors such as learning, boredom, fatigue and motivation would operate to increase the variability of performance, and therefore necessitate larger samples." The third study explored run duration. In Careful three different patterns of attack had been introduced sequentially during a 30-minute run. Now these were also introduced separately as 10-minute runs and their effects compared with those from a 30-minute run in which they occurred in sequence. Only one significant difference was found among twelve comparisons.

## NOTEWORTHY ASPECTS OF THE CSL PROGRAM

These supplemental all-computer studies constitute only one of a number of interesting aspects of experimental strategy and methodology in the CSL program. Another is the attempt to relate one of the experiments in this program to an experiment in a previous program at another location, even to the extent of using the same inputs. Still others include inferences of motivational variables in operator decision-making performance; concern with the interactions between input load and system design, i.e., the mix of automatic and human functioning; reliance on only a few trained subjects; multiplicity of measures; and conduct of a follow-on experiment to explore the generality of the results of the first.

Innovations meriting particular emphasis were automatic insertion of input by computer; data recording and reduction by the same computer that is part of the system under observation; and computer refereeing of action (interception) outcomes. Such exploitation of the system's own computer for the assessment of system performance could and did greatly simplify and expedite the assessment task. The Naval Tactical Data System and SAGE would incorporate such a capability when they became part of the operational scene (Chapters 5 and 11).

On the other hand, when the system's computer is the exclusive gatherer and reducer of performance data, the only data which are gathered and reduced are those which describe what the computer receives, processes, and emits. Actions in the system external to the computer are excluded, no matter how significant they may be. Certain limitations can characterize system experiments that deal with a computerized context, if an experiment oversimplifies the interfaces between the central data processor (and its human equivalent) and the input and action domains with which it deals. For example, in the CSL program the perfect tracking and undemanding requirements for interceptor positioning surely were not representative of actual air defense. Might not greater realism here have differentially influenced the effects of the principal variables under investigation, namely, the kind of mix of man and machine?

It is tempting to ask whether anything learned from the CSL programs was later used in the development of the computer-based system which was actually introduced into Navy antiair warfare operations—the Naval Tactical Data System described in Chapter 5. It does seem likely that one carryover was the operational computer's collection and reduction of its own performance data. But it has not been feasible in preparing this account to trace contributions to NTDS design from Cornfield design or the results of the CSL experiments; Sinaiko himself (personal communication) has wondered about the impact of his research.

# 13

# RAND's Logistics Systems Laboratory

The RAND Corporation's Logistics Systems Laboratory was established in October, 1956, and has produced four very large experiments having to do with Air Force logistics policies and procedures. A number of RAND reports have credited the RAND Systems Research Laboratory experiments (Chapter 7) as one of the reasons why this new program was undertaken, but the logistics studies differed in some important characteristics.

The Logistics Systems Laboratory was an integral part of a large RAND entity, the Logistics Department, which was investigating many logistics problems. The objectives of an experiment were developed within this larger milieu, and the results and experience from an experiment to some extent spread through that milieu.

The Laboratory had formal Air Force authorization and funding from the start and Air Force participants and consultants for every experiment.

Each experiment ran in the laboratory for many weeks but represented an even longer real-world time span. There was extensive time compression: a laboratory hour might deal with the events of several hours or a day or more in the system being simulated.

The principal simulation media were pieces of paper delivered to the subjects. These represented reports, such as reports of equipment malfunctions and inventory items. The subjects did not interface directly with simulated system hardware (except for a few status displays) or deal with any symbols generated by that hardware other than alphanumeric symbols—letters and numerals.

The simulation inputs were produced by a digital computer in a partially contingent or reactive fashion. The inputs for any session depended in part on the actions in the preceding session. Further, the computer's models simulated a number of interacting subsystems or environments, so there was considerable compression of system operations.

Activities of a large number of people in the real system had their parallels in those of relatively few subjects in the laboratory. Instead of one-to-one correspondence, a set of personnel was aggregated into a single individual in what might be called "organizational compression."

Not all the content or schedule of an experiment was preplanned. Much was based on what was found out during the experiment. The independent variables in only one of the studies were organized according to an experimental design. (However, this one was very complex.)

The studies have been referred to as "experiments" (and in one instance a "pseudo-experiment") in some of the RAND reports, but their collective title more commonly has been "simulation-game." This nomenclature probably reflects the operations research-economist-applied mathematician orientations and techniques of the milieu in which the laboratory was embedded.

Each of the four experiments was called an LP, for "laboratory problem." Thus, they have been name-coded simply as LP–I, LP–II, LP–III, and LP–IV. Each has been described in one or more RAND reports. However, since the primary interest has been the substantive knowledge acquired from the LP, and this has been directed toward an audience interested more in such knowledge than in how it was acquired, none of the study reports fully describes an experiment in a way which would permit even approximate replication; in other words, no report has comprehensively described the experimental scheduling, procedures, laboratory layout, subjects, and simulation. Nevertheless, with the co-operation of existing and former RAND staff and by examining numerous documents, it has been possible to put together a reasonably complete picture.

Reports giving a general account of the research program and its rationale, with illustrative material from one or more studies, were published during that research by the earlier laboratory managers (M. A. Geisler and W. W. Haythorn) and other Logistics Department personnel; these are listed in the References. The most complete overview of the research, covering the first three experiments, has come from Geisler, Haythorn, and W. A. Steger (1962); these have also been summarized by Walker (1962). The fourth experiment has been described by the laboratory manager responsible for it, I. K. Cohen (1963) and by Cohen and Van Horn (1964).

## THE LABORATORY'S ORIGIN

The Logistics Systems Laboratory initially occupied the former pool hall which had been transformed into the Systems Research Laboratory; some of its staff had been associated with that laboratory as professional or support personnel, or they had worked in data analysis following the SRL experiments. Air Force Regulation 20–8 stated the following laboratory objectives (Rauner 1958; Haythorn 1959*b*):

To study the organizational and functional interactions of the logistics system.

To test and evaluate alternative data-flow systems, logistics policies, and organizational and management structures, and to explore required equipment characteristics in order to facilitate selection of the most efficient, complete, and integrated systems.

To compare and evaluate partial and entire logistics system changes in a laboratory environment representing realistic peace and war situations prior to service testing.

To provide an opportunity for Air Force personnel to broaden their logistics perspectives by participating in logistics research work.

To explore the man-machine relationships in data processing.

To develop the steps necessary to accomplish a transition to an advanced system utilizing modern automatic data-processing equipment with minimum disruption to operations.

During the time period of the laboratory's inception, the Air Force logistics system was buying, storing, distributing, and computing the requirements for more than one million different items to support Air Force organizations (Haythorn 1957). In a single year the fifteen Air Force supply depots of the Air Material Command (which had assets five times those of the General Motors Corporation) stocked more than 800,000 different items of supply and processed more than 41 million items varying from a 2¢ washer to a $200,000 engine (Rauner 1958).

As its scope grew even greater, the Air Force sought ways of improving its logistics network, or system, and the RAND Corporation provided analyses and policy recommendations. But "the translation of the broad findings of RAND research studies into detailed procedures required by an operating system raises many important questions for which the previous research provided virtually no guidance." So the laboratory was founded "to conduct experiments to discover how the proposed changes in the Air Force logistics system would work (Geisler, Haythorn, and Steger 1962). Since Air Force personnel would continue to function in any logistics system in decision-making and other roles, it was felt that RAND's recommendations would become more meaningful if RAND research included such human roles; but it should probe further than simply how much human beings degraded system functioning. "The positive or beneficial effects on performance that human factors may produce are frequently ignored," two RAND investigators noted, "mainly because of the difficulty involved in estimating such positive effects. However, it is fairly well known, if not precisely measured, that humans can improve system output as they learn better how to plan and control various resource combinations in pursuit of a given goal" (Rauner and Steger 1962).

## SUMMARY OF THE RESEARCH

The first experiment, LP–I, took place in 1957. Two preparatory experiments, Prolog I and Prolog II, preceded it. It simulated the operations of two supply systems at a supply center and ten aircraft bases. The two systems were the then current system and an imaginary one created out of three proposed new policies for supplying equipment items for the aircraft at the simulated bases. The experimental sessions ran for about three months and represented a three-and-one-half-year period.

The second experiment, LP–II, simulated maintenance (and operations) of a future ICBM squadron to see what would happen due to various assumptions about degree of centralization, missile system reliability, variations in manning and equipment resources, different operational requirements, and the type of management system employed by the missile squadron. The design of LP-II began in 1957 and the runs were conducted in 1958, taking about one hundred working hours to simulate about three hundred hours of system operation. A related all-computer simulation accompanied LP–II, and two others followed it.

The design of LP-III began while LP-II was running in the laboratory. LP-III occupied the laboratory from mid-1959 into 1961 with seventeen runs of differing durations, some lasting three weeks and representing seven to nine months of real time. The simulated environment consisted of two weapon (missile) systems, each with nine bases, and two inventory classes. The experiment compared two methods of managing the provision of spare parts, inventory management and weapons system management; it also varied the stresses on the management system and examined the effects of different levels of system response capability, or responsiveness.

Instead of commencing during LP–III, the design of LP–IV was delayed while the Air Force studied whether it should organize its own laboratory to develop the LP–II technique for the Minuteman ICBM system; that undertaking, if it occurred, was expected to involve a substantial part of the Logistics Systems Laboratory staff (Geisler, Haythorn, and Steger 1962). "When the Air Force decided in late 1960 not to proceed with the Minuteman study because they felt it came too late to be of significant value for the program of that weapon system, work was then begun on LP–IV." This experiment, actually a set of studies, dealt with the maintenance function at an Air Force aircraft base. During 1962–64, laboratory examination of current maintenance operations was followed by similar examination of an improved manual system, a field test of that system, two supporting small experiments, and an all-computer simulation; a laboratory study of an automated system was projected in addition.

## FIRST EXPERIMENT

The first experiment, LP–I, investigated the provisioning and distribution of airframe spare parts, in response to need, for two types of fighter-interceptor aircraft. The experiment has been described in detail by Enke (1957), Geisler (1957, 1958), Haythorn (1958), and Rauner (1958). The RAND analysts had concluded that this supply activity could be improved if (1) the provisioning of expensive parts could be deferred as long as requirements permitted; (2) cheaper parts were procured and distributed under a more comprehensive formula which, for example, took into account the costs of reordering, of holding inventories, and of parts shortages; and (3) resupply was automated and record-keeping was centralized through an electronic data processing center. These proposed innovations constituted Logistics System 2. The supply practices current in 1956 were Logistics System 1. Each system is diagrammed in Figure 13. The main burden of LP–I was to test the feasibility of No. 2 and compare it with No. 1 as a "benchmark."

### *Simulation*

The two systems were simulated in the laboratory and operated at the same time, in neighboring but separated spaces. Each was staffed by a number of

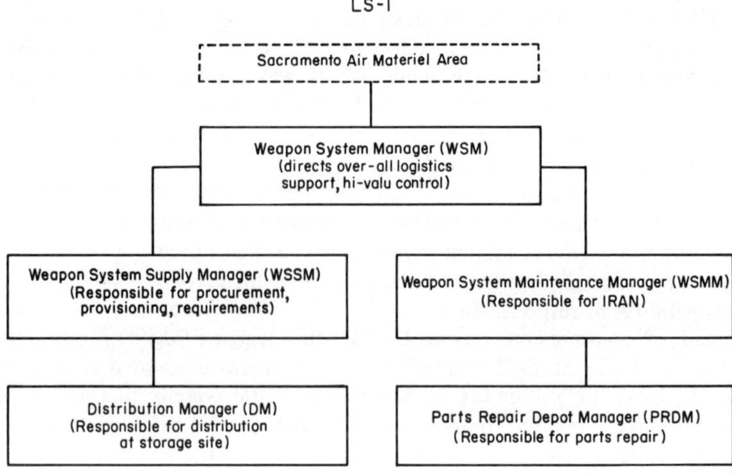


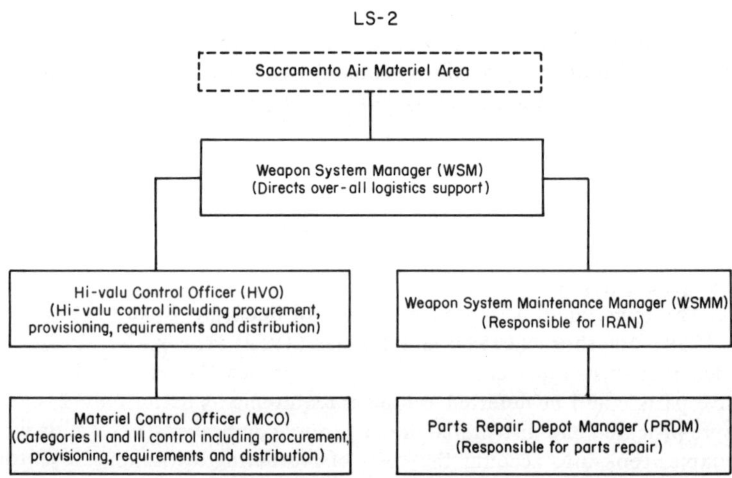


Fig. 13. The Two Logistics System Organizations (Rauner 1958).

managers and a few clerks. Five of the managers represented a centralized grouping. At the head was a weapon system manager at an air materiel area (Sacramento). Supporting him were managers at a parts repair depot and an IRAN (inspect and repair as necessary) depot, and two supply and distribution managers whose roles differed between the two systems. The other managers represented base maintenance and supply officers at simulated airbases, which varied in number during the experiment between four and ten, with the same total for each system. The roles were taken by Air Force civilians with similar real-life occupational experience, assisted by enlisted personnel; all were loaned for the experiment by Air Force organizations. The subjects did not change during the experiment, that is, each system was staffed throughout by the same personnel.

As indicated in Figure 14, additional organizations, "embedding" and related, were simulated by individuals from the RAND experimental and clerical staff. These functioned outside the subject area, particularly on a balcony or dais from which the subject area could be viewed. They included USAF headquarters, two major Air Force commands, a factory, and a transportation office. Parts might have to be ordered from the factory, and transportation factors were important because the airbases were at various distances from the air materiel area.

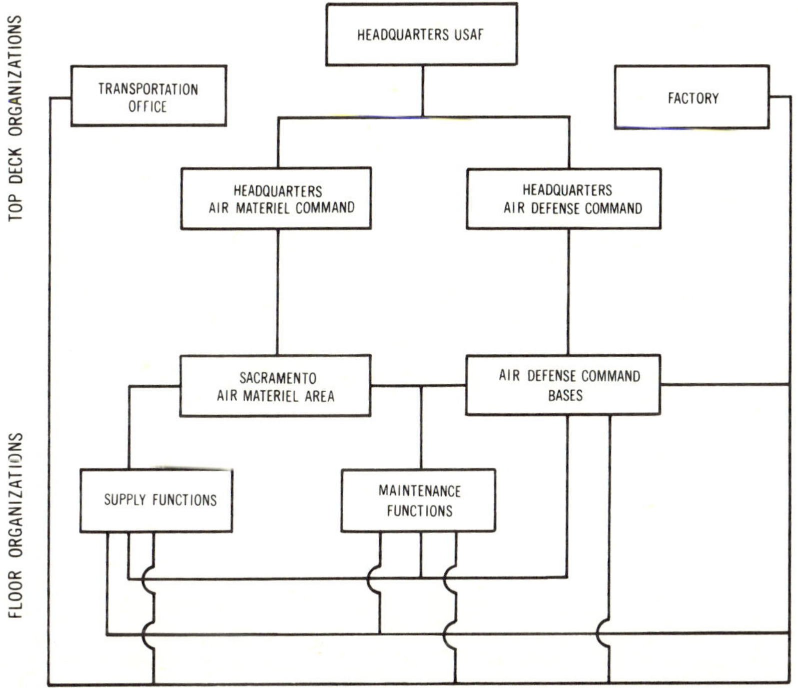


Fig. 14. Organizations Simulated in the First Experiment (Rauner 1958).

An IBM 704 digital computer also played a major role. The subjects often communicated with each other face-to-face or by telephone, depending on real-life propinquity. They also were in contact with the embedding organizations by telephone or mail. But a great deal of their interaction occurred with and through the computer. They received messages from it in the form of printouts (delivered by clerks), and transmitted messages to it on forms and as reports (which were converted to punched cards by RAND clerks). The computer and its processing represented all of the other events, people, and their transactions at the air materiel area and airbases pertinent to the supply operation; for Logistics System 2 it also represented the computer in that system. Each airbase manager received daily status reports which identified the aircraft awaiting parts and the parts themselves and which indicated the criticality of failures, time required to repair failures, aircraft availability and status data, and similar information. These managers prepared and sent status and requirements reports and forms daily to the weapon system manager at the air materiel area, as well as monthly, quarterly, and annual reports, all via the computer. They also engaged in extensive problem-solving and planning activities. The managers in the centralized group received, from the computer, printouts which included both the airbase reports and various status reports and listings about the operations for which the managers were responsible. A number of programs regulated the computer's processing, notably a failure model. It generated, as a function of flying hours, a variety of failures in an 800-item sample of airframe parts in the aircraft at the airbases, thus triggering stochastic demands on the supply systems. The systems coped not only with daily requirements but also with those arising from monthly, quarterly, and annual cycles. The computer programs included more than 25,000 instructions.

*Experimental Operations*

At one point it was expected that all five years of a fighter aircraft's life would be represented in the experiment. Plans settled on four years, but the experiment terminated with six months to go. Two brief wars were simulated toward the end of the runs. Altogether the laboratory gathered data during about two hundred hours in half-hour segments, each representing a day of real time. However, in order to exercise monthly, quarterly, and annual cycles more frequently, real time was also compressed, in the sense that each month of real time was presumed to transpire in ten real-time days. It was this arrangement which made it possible to simulate the passage of three and one-half years during three months of laboratory occupancy.

Between the half-hour segments representing real days, the computer processed the various forms and reports (about one hundred different ones during the experiment) which had been made out by the subjects and converted into punched cards; and it produced the printouts for the subjects to deal with in the next time segment. What it produced for one time segment depended in large part on what it received from the preceding time segment, since the requirement of one day depended in part on actions taken the previous day at the airbases and at the air materiel area. As already noted, the card inputs were punched and card decks were made up by RAND clerks at the laboratory. The computer was at another location about ten blocks away; residents of Santa Monica became accustomed to messengers rushing back and forth, carrying punched cards in one direction and printouts in the other. Turn-around time for the computer was supposed to be about a half-hour but averaged close to an hour, and additional time was needed at the end of quarterly and annual reporting cycles. As a result the experiment required approximately five hundred hours of computer time. Needless to say, computer failures occasioned more than a few delays and reruns.

To manage the laboratory demanded effort and skill. One of its features was a RAND-manned "control point" at which the forms and reports were punched into cards, the printouts were received and distributed, and both of these voluminous sets of paper were checked for error. It included the transportation office, which delayed the transmission of reports and punched cards to simulate the span of mail delivery, repair time lags, and transport durations. The control point personnel also attempted to discover and correct all violations of ground rules about the simulation. Other RAND personnel from Statistical Services were on hand to answer queries about the origin of numbers on printouts and to trouble-shoot the data system when it malfunctioned. A third RAND group was an observation staff who registered critical incidents, managed an array of voice recorders connected to telephone lines and microphones to pick up face-to-face conversations, administered questionnaires, and conducted interviews. These tried to assess the subjects' comprehension of system procedures, their degree of satisfaction with the experiment, their problems with the simulation, and their views about its realism.

During the peak activity of a run as many as one hundred persons were involved, including more than thirty subjects, the nine Numerical Analysis Department programmers who developed the computer programs and staffed the Statistical Services group, as many as twenty clerks, fourteen Air Force long-term consultants, and the professional staff of fifteen psychologists, economists, mathematicians, and systems analysts. It was a big job to co-ordinate and train all these people. In preparation for the experiment two 300-page manuals were compiled, one for each logistics system. The subjects went through two to four weeks of training. A rehearsal or mock-up phase which preceded this and subsequent experiments was far too modest, with the result that the first sessions progressed somewhat slowly and in a sense constituted mock-up sessions. The mock-up had been underemphasized, partly due to a certain haste in getting the experiment under way—within nine months after the establishment of the laboratory, and partly because two exploratory experiments had already been conducted.

The aims of Prologs I and II were not only exploration but also education for the experimental staff (Logistics Systems Laboratory 1957*a*, 1957*b*). They were shorter than LP-I and smaller in scope. The first, covering five weeks in October–November 1956, incorporated only three airbases and one type of aircraft. Various schemes for compressing time were tried out, together with the failure model in the computer for generating simulated malfunctions in aircraft parts. In Prolog II the Laboratory was occupied for about six weeks, representing two real years. Logistics System 2 was operated as well as System 1, and each dealt with a different aircraft type. Other time-compression ratios were explored. The scenario included one war. These prologs produced a number of "methodological lessons for future studies" (Logistics Systems Laboratory 1957*b*): there should be a permanent, informed, and trained staff of laboratory supervisors; a detailed and specific preliminary design would be preferable to generating methodology in the course of the run; when personnel representing embedding organizations develop rules and strategies, these should be standardized and recorded; enough personnel should be assigned to make sure that subjects abide by the rules and to develop adequate and reliable techniques for data collection. Reading between the lines, one can see what some of the problems must have been in putting this kind of laboratory into operation.

In the experiment proper, as well as in the prologs, several kinds of data could be gathered. It is illuminating to examine what was done with them. One type was "critical incident" observations by the experimental staff; 1,462 incidents were recorded in Prolog II. These proved difficult to categorize and such data are not described in the reports of LP–I. Another type was frequency of communications of various kinds, among subjects, between subjects and embedding organizations, and between subjects and the computer. Messages were coded for analysis according to content. Although some of their distributions were reported (Beverly 1958), these data apparently were not considered to have enough significance to be singled out in the reports of over-all results of the experiment.

A third type of data consisted of (1) dollar expenditures for spares and maintenance; (2) transportation costs; and (3) rates of aircraft out of commission or not fully equipped. On the basis of these measures it was concluded that the proposed Logistics System 2 policies for cheaper parts proved superior, and also, when the data were adjusted to allow for a misestimate in requirements stemming from malfunctions, the proposed new procurement policy for expensive parts also was judged much better. Due to the experimental design, or lack of it, any benefits of centralized data processing and record keeping in System 2 could not be isolated; and it was impossible to apportion some of the effects of System 2 between the two policy variables.

The experiment also had some by-products, noted by Enke (1957):

> Indirectly, the Lab has already proved extremely useful in several ways, some of them rather unexpected. One, it is an integrating force, pulling together much of the rest of the research in the Department. Two, it compels specification, forcing the authors of ideas for improved logistics to state them in sufficient detail for simulation. And three, it imposes deadlines, threatening to waste the time of numerous people if needed inputs and models are not available on schedule.

Enke also commented that by embracing parts of a logistic system the laboratory helped ascertain "whether an improvement in one part of the system has been obtained at the cost of impaired performance in some other part" and "provides a reasonable guarantee against the introduction of new policies that rob Peter to pay Paul." He also noted that the new policies tested in the experiment "promise to provide savings in the procurement of airframe spares approaching 50 percent, and this percentage is potentially applicable to Air Force expenditures of about $400 million a year."

### *Problems of Experimentation*

A number of problems in conducting this kind of research were also brought to light by the experiment.

1. Although Logistics System 1 was supposed to serve as a benchmark for comparison purposes, it was modeled from the Air Force's 1956 operations and did not fully characterize the operations in effect in 1958 when the final reports were published.

2. Geisler (1959) observed that "the experiment did not provide the opportunity to study the interaction among weapons systems in their demand for common spare parts, which is not a trivial problem to the real world logistics system."

3. It was difficult to demonstrate just what the human actions were that justified this kind of laboratory examination instead of an analytical study. Presumably the various managers made a number of decisions (and in fact it was felt they should have to make important decisions so they would not be bored during the experiment, as they had been in Prolog II). But policy rules continued to dictate how many of these decisions should be made. Further, their significance and number were not clearly specified, although Geisler (1959) reported that decisions about procurement and distribution of expensive parts "amounted to several thousand elements."

4. Other kinds of human actions that might have influenced system performance, such as clerical errors in reporting transactions, were ruled out when transactions were simulated entirely within the computer. Thus, much potential human-generated noise was missing. On the other hand, some noise entered that might not have characterized the real world. Many of the subjects never did succeed in learning the proposed Logistics System 2 procedures adequately.

5. One of the advantages of running the two systems at the same time was to conserve time and funds. But, as might occur in any reactive simulation, the systems could go out of phase with each other if a manager made a procedural error in one of them. This did happen, although a technique was developed for partially counteracting such imbalances by arbitrarily constraining the number of input-producing events, in this case aircraft flights in a base's flight operations.

6. A major problem occurred in the model of the manually simulated factory. The factory was not nearly as responsive to requirements placed on it for expensive parts as such a factory would be in the real world, it was subsequently concluded. Another misestimate occurred in the failure model's rates of malfunction in airframe parts. "About half way through the run, it was discovered that the demand data supplied to RAND by the Air Force for the purpose of constructing the failure model did not reflect *actual* demand rates that had occurred in the real world, but were merely *estimates* for procurement. These estimates of demand rates proved as much as ten times higher than actual demands" (Rauner 1958).

7. Among other troubles, the transaction load occasionally exceeded the programmed computer capacity. Several essential reports and computations were not available early enough. Subjects used wrong procedures that resulted in program stoppages. Some, but not all, of these problems could be solved during the experiment, and of course on-the-job repairs caused delays and difficulties in scheduling.

8. Problems also arose in data analysis, which was sometimes "hampered by the lack of proper data" (Rauner and Steger 1960). "Often this situation arose

because the question of all the factors that might have to be analyzed was not raised early enough in the design period, or because there was uncertainty about what might have to be analyzed. And once the experimental design was set, it was often difficult to change it to meet analysis needs that developed later."

9. According to Geisler (1959), among the most important procedural difficulties were "the programming errors which had not been fully eliminated before the run began. Although a little checking of the program was done before the start of the run, not enough time was allowed in our schedule for this activity." Also, the subjects made numerous input errors in submitting data for the computer, and many of these were not detected in time.

In summarizing methodological lessons learned from this experiment, the same author took note of the need for allowing enough time for programming and also for the rehearsal of laboratory subjects, through realistic scheduling. He also advised dividing an experiment into several runs instead of one long one, both to sustain subjects' interest and to allow for errors that could ruin an experiment if it consisted of only a single, long run. Further, he suggested making ample provision for adequate analysis of the data, with participation from the simulated organizations. With respect to the laboratory organization, he counseled that it be very flexible since it would have to change continually as the experimental study went through various phases from modeling to analysis; and there should be close communication between the laboratory and the research organization which specifies the policies to be studied there. One must be grateful to this laboratory manager for his sharing of problems arising in this kind of research.

In addition, some further information about methodological problems encountered in this experiment is contained in documentation circulated only to RAND staff. Although this information will not be discussed here, it has contributed to the general treatment of methodology in Chapter 2.

## SECOND EXPERIMENT

Much of the methodological experience gained in the first experiment could be exploited in the second, LP–II. Apparently LP–I induced considerable caution, expressed in flexibility of scheduling and a strategy of organizing the experiment while it was in process. The laboratory was moved from its LP–I site to the main RAND Corporation building. About eight months were devoted to preparation, nine to laboratory operations, and twelve to analysis of the data. The laboratory operations actually covered only seventeen weeks of the formal experiment. Air Force personnel were the subjects in two time-separated phases; then RAND personnel functioned as subjects in a supplementary phase. There were also several all-computer simulations. Two further phases which had been planned failed to materialize.

As indicated earlier, this experiment was concerned with future maintenance and operations activities of an ICBM squadron. Maintenance was of most interest, since by comparison the operation of such an organization is less demanding. In contrast to LP–I, this experiment did not examine a current system and its possible improvement, because at the time no operational ICBMs existed. Rather, it sought to create a superior organizational arrangement for a new weapon, to be established in a future time period, 1963–65. Descriptions of this experiment have come from Haythorn (1959*a*), Rauner and Steger (1960), and Sweetland (1961).

*Objects of Inquiry*

The Air Force at the time was wondering how many launch complexes to net together with a squadron co-ordination and support center, and how many missiles to place in a complex. Some of the then current Air Force thinking favored a decentralized approach, with only three complexes of three missiles each, plus a spare. This was the context of the first phase of the experiment. But there were differences of opinion within the Air Force. On the basis of two mathematical models, RAND analysts reasoned that a more centralized arrangement could be less costly, or more effective, or both. After the first LP–II laboratory phase, an all-computer simulation tested an arrangement of nine launch complexes of four missiles each, with four spares. The second laboratory phase was undertaken to test the favorable results of this study for their validity. The more centralized and the decentralized versions were operated in parallel. The supplementary phase with RAND subjects investigated the extent of manning at the launch complexes. The phases which were dropped would have enlarged the scope by looking at multisquadron and forcewide contexts. A second all-computer simulation examined the effects of a ratio of one missile per launch complex. A third simulated the decentralized arrangement of the first experimental phase, but with a maintenance policy which required (1) no preventive maintenance and (2) missile replacement every six months—a "low activity" policy which had received some Air Force attention.

In addition to the difference in organizational arrangement, a number of other variables figured in the laboratory simulation. They included an information system for squadron co-ordination; three unit-simulated combat missions for "commander-confidence testing"; variation in maintenance requirements imposed by higher authority; changes in estimates of missile reliability; helicopter transport; a snowstorm; and at the end of the second phase a reduction in personnel and equipment resources. A major variable was a set of procedures which the Air Force subjects themselves evolved. At first they were constrained by well-defined procedures, with multiple goals which included certain tasks in associating missiles with targets. Then they were given a single goal, to maintain as many missiles on alert as possible, still conforming to general maintenance-cycle requirements, technical order compliances, and operational countdowns, but they were allowed to generate their own procedures within the launch complex and within the squadron co-ordination and support center.

The new procedures, reported by Sweetland (1961), were some of the most interesting developments in the experiment. One technique was "opportunistic scheduling." Instead of holding to a rigid schedule of preventive maintenance through checkouts, for example, the responsible personnel arranged for the necessary resources and then waited at least for a while (two to four days) for a critical malfunction. When it occurred, the scheduled maintenance was performed at the same time that the malfunction was corrected. The subjects called a second technique "packaging." Instead of conducting scheduled maintenance on a one-at-a-time basis, several maintenance operations were performed together. The third technique combined these other two: packaging several accumulated maintenance operations and waiting to accomplish them until there was a critical malfunction. Since the missile was given an over-all reliability of only about 0.6, enough malfunctions occurred to do most of the scheduled maintenance in this fashion, and much of the daily and weekly scheduled maintenance could be eliminated because that which occurred after a malfunction was more comprehensive. Sweetland also noted:

In LP–II, the observation that most impressed both participants and experimenters was the need for a hand-in-glove relationship between operations and support (maintenance and supply). A marked pay-off occurred when all the participants united their efforts in the single objective of maintaining the maximum number of missiles in T–15 status. This meant that the participants ignored some of the traditional divisions of responsibilities throughout the course of the experiment.

*Experimental Operations*

The experiment was based on a ratio of 1:3 for time compression. Five minutes in the laboratory represented a 15-minute increment at the real-world launch complex and squadron center. Thus, eight hours in the laboratory could represent a twenty-four-hour day, and the total of one hundred laboratory hours represented three hundred in real time. Geisler (1959) said there was a one-week training period.

Three subjects represented the activities of a commander and a number of aides at the squadron co-ordination and support center, and a subject played the role of each of the officers in charge of a launch complex. A majority of the subjects were assigned to the experiment from the Strategic Air Command; some came also from the Air Defense Command, the Air Training Command, and the Air Research and Development Command. RAND clerical personnel served as assistants and log-keepers, and other RAND personnel represented embedding organizations—several command headquarters, a depot, and a factory, as shown in Figure 15. Most of the simulation was instrumented through IBM cards. These represented resources. There was one card for every man in the unit manning document (at the center and launch complexes), for each spare part, for each kit of test and support equipment, and for each item of ground handling equipment. The simulated center handled those cards representing its resources, and each launch complex also had its appropriate set. Instead of moving actual men, spares, or equipment, the officers at these simulated locations dispatched IBM cards together with instructions. A RAND-manned control point (similar to that in LP–I) received these and eventually returned them to signify that the task specified in the instructions had been completed. The time of return was computed according to the time required by the task. This in turn was based partly on an aptitude index for the particular type of personnel who would have to perform the task in the real world.

The simulated tasks which had to be performed at the launch complexes originated from two sources: a daily operations order from the squadron center

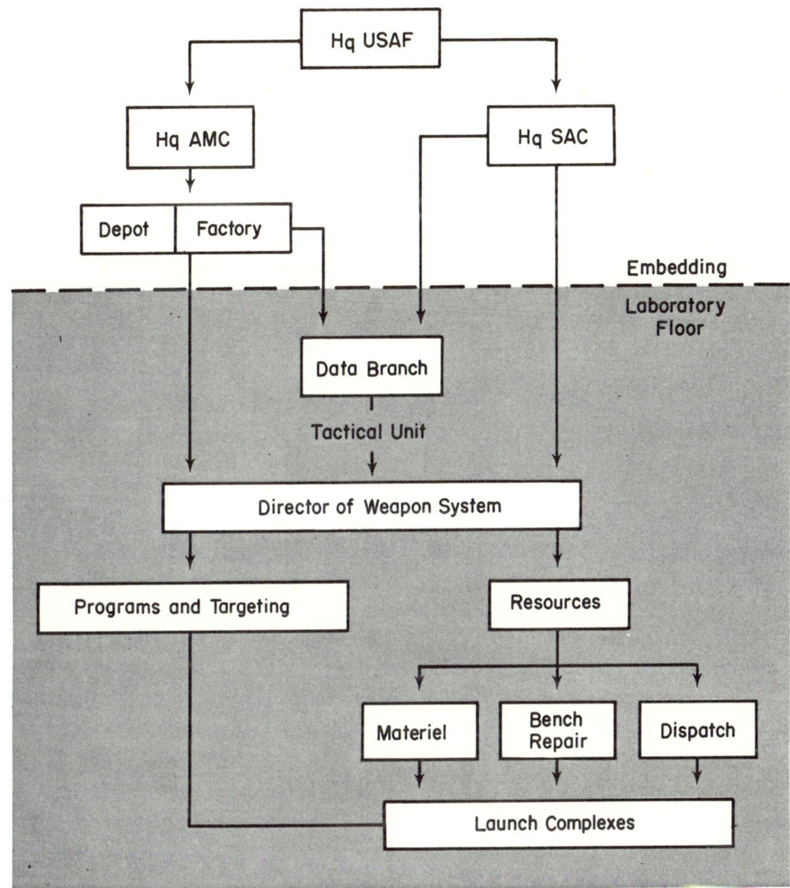


Fig. 15. Activities Represented in the Second Experiment (Rauner and Steger 1960).

which required the launch complex to initiate a particular "situation"; and malfunctions occurring in the missiles, their ground support equipment, and related hardware. When the control point determined that a malfunction had developed, it signaled the launch complex officer by activating a flashing red light at a console where he was stationed, and then it sent him a card specifying what had malfunctioned and what resources were required for troubleshooting. The control point personnel had a computer-generated listing of potential malfunctions which had been randomly drawn from distributions of failure rates and stress factors. The stress factors arose from the kinds of situations into which the launch complex might be placed. The stresses included "turn on," "operating," "standing," "countdown," "handling," and "in-flight." There were forty possible situations, lasting between one and fifty-nine 15-minute time segments. A "stress book" indicated which stress factor applied to each of 1,541 equipment units in each time segment of each situation. The control point was informed of every situation which the launch complex officer initiated and thus could determine which stress factor might be generating malfunctions. This was the guide to the selection of actual from among potential failures.

Thus, the control point had a precomputed set of all possible malfunctions, against which its clerks had to match the stress factor which actually developed for each 15-minute time segment (for each missile). A RAMAC 305 generated a new set every day on the basis of what had occurred the preceding day. This was the reactive aspect of the simulation. It was stochastic in that particular malfunctions were randomly selected, within the constraint of the reliability estimates which had been made for the 1,541 units of hardware. Malfunction rates of each unit were also derived for each stress. These estimates resulted from RAND engineering analyses and the assiduous collection of data about test and manufacturer experience; they gave an entire missile and its support equipment the over-all reliability of 0.6 out of a possible 1.0. The missile was an entirely mythical one called ATLAN (a combination of Atlas and Titan). In the latter part of the experiment the computer itself performed the stress-matching operation previously done by the control point clerks.

Although the nature of the experiment did not permit the use of significance statistics in the analyses of results, it was concluded that the centralized system was able to maintain the same number of missiles on probable alert and place the same number on target as the decentralized one. But the cost of the centralized system was found to be about one billion dollars less—striking evidence of superiority. Because the three new scheduling procedures evolved by the subjects were used in both systems virtually throughout the experiment, there were no data from a control condition in the experiment itself for comparison. However, in preliminary analysis and mock-up runs with the prior procedures, it was predicted that only 65 to 70% of the missiles would, on the average, be on T–15 alert. But the actual average with the new procedures was 80% and rising toward 90% in the last part of the experiment.

Possibly this was an instance where the run should have continued. It is not clear from any of the RAND reports how it was decided when to terminate a phase of the experiment, or any condition within it. It is believed that decisions when to shift or terminate conditions resulted from intuitive estimates as to when a condition had reached a steady state. How to optimize such estimates was one of the major methodological problems which LP–II posed.

## THIRD EXPERIMENT

The third Logistics Systems Laboratory experiment, LP–III, had many points of interest. One was the incorporation of a very complex experimental design in four of its runs. Another was a shift in experimental objectives after the results of this portion of the experiment became apparent. What happened can be partially pieced together from reports by Nelson and Peterson (1962), Haythorn (1963*b*), and Walker (1962). But some aspects of the operations of LP–III and its reporting are still obscure.

The setting consisted of two alternative organizations for providing logistic support to two different missile systems at nine different bases, as diagrammed in Figures 16 and 17. Support involved two different property classes of thirty-two different spare parts each, a total of sixty-four. It was possible to investigate the two organizations, two missile systems, and two property classes in their eight combinations in a 2 X 2 X 2 experimental design. Each of these combinations could be subjected to three kinds of stress, each kind having two levels. One stress was an increase in the time a missile had to be on alert, another was a reduction in repair capacity, and a third consisted of varying aggregates of time lag in information transmission (e.g., preparing routine repair reports). The three stresses of two levels each were arranged in a 2 X 2 Latin square design to constitute four stress conditions. When each of these was associated with the eight conditions already described, thirty-two different conditions resulted.

In addition, the nine missile bases varied according to such characteristics as size, precedence, and maintenance capacity. These characteristics among the nine bases were related to each other through a 3 X 3 Graeco-Latin square design. Finally, the sixty-four spare parts varied among themselves according to nine characteristics: failure rate, variance in the rate, initial error in estimating it, cost, shop flow-time, base reparability rate, wearout rate, procurement lead time, and depot repair man-hour requirements. An 8 X 8 Graeco-Latin square contained the sixty-four parts with different combinations of these nine parts characteristics. It was then possible in the experimental design to "nest" the two squares, and thereby vary the base characteristics and parts characteristics, within each of the thirty-two orthogonal conditions noted above to obtain a factorial

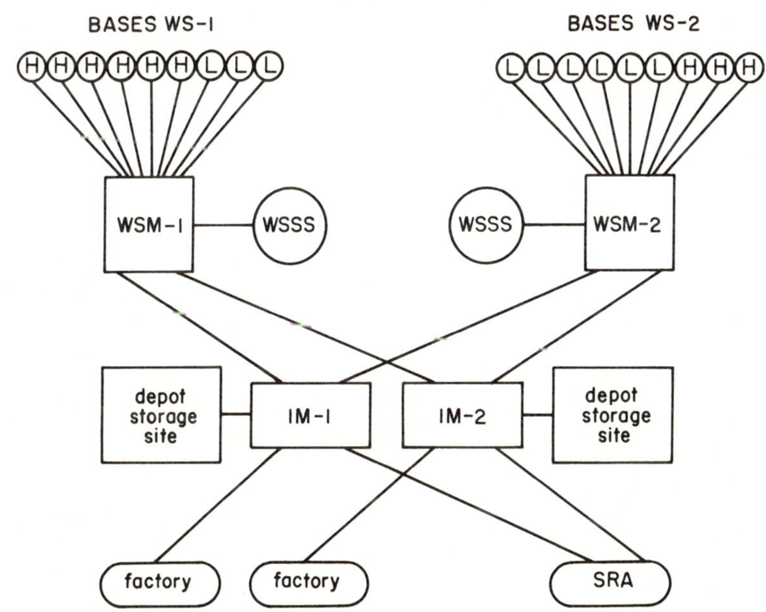


Fig. 16. Inventory Management (IM) Organization in the Third Experiment (Nelson and Peterson 1962). (Each IM is responsible for a group of parts common to two weapon systems.)

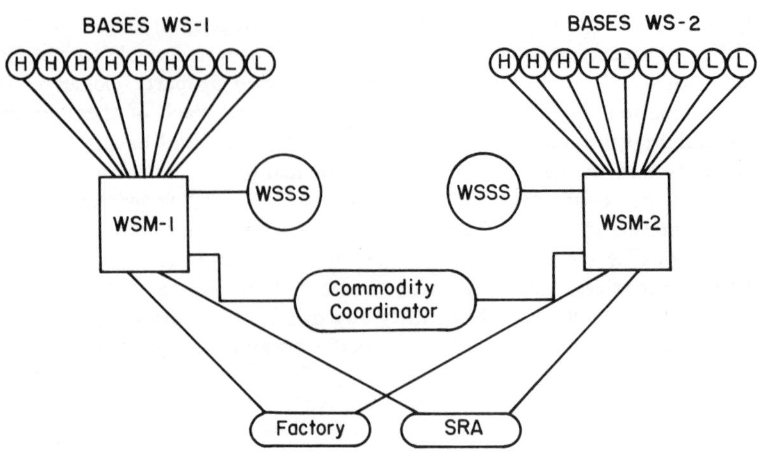


Fig. 17. Weapon System Manager (WSM) Organization in the Third Experiment (Nelson and Peterson 1962). (Each Weapon System Manager is fully responsible for the complete range of parts for his weapon system. The Commodity Co-ordinator is responsible for weapon-system balance in contract termination approval and directing interweapon-system asset transfers.)

design of eighteen independent variables; this design, to be sure, had the limitation that nonorthogonality existed among the interactions of the variables in the three squares. This was the very complex experimental design referred to earlier (Haythorn 1963*b*).

In the time scheme for the four runs in which this design was executed, a two-and-one-half-hour (half-day) period in the laboratory constituted one week of real system time. One period occurred in the morning; then the associated IBM 709 computer processed the data which had been produced, and it generated new inputs for the subjects; then another experimental period occurred in the afternoon, and the computer processed that period's data at night. (RAND clerks prepared IBM card inputs with card-punch equipment right in the laboratory, as in previous experiments.) Within each set of four periods (a month of real system time) all of the parts characteristics conditions and base characteristics conditions were introduced. There was a shift from one of the thirty-two other conditions to another between each set of four periods, and all thirty-two were covered in thirty-two sets (thirty-two months of real system time). However, these were divided into four runs of eight months each, this being the maximum duration of any real system time span.

By this arrangement, each of the four runs required by the experimental design occupied the laboratory for about a month. The two organizations constituting the organization variable were operated in parallel in the earlier runs. The Air Force Logistics Command supplied forty-two technical specialists who rotated in and out for one-month or two-month tours as experimental subjects. Any one subject was equally involved in each of the simulated organizations.

The total number of runs in the experiment has been stated as fifteen (Geisler, Haythorn, and Steger 1962) and seventeen (Nelson and Peterson 1962;

Walker 1962). The discrepancy arises from the fact that the earliest runs could be regarded as a shakedown exercise, and another early run had to be repeated because of a computer model error. Subsequent to the four runs of the experimental design (whose detailed results have not been included in any report distributed outside RAND), additional runs, some of them with RAND personnel as subjects, investigated the effects on the supply organization of repair response, resupply cycle response, and distribution precision. The runs in the experimental design had failed to show any statistically significant difference between the results which were achieved with the two types of organization (weapon system management and inventory management) providing logistic support to the two missile systems, although it had been expected on the basis of analytic studies that one of them would be superior; indeed, the experiment arose out of such studies. The subsequent runs explored, in part, the consequences of holding the stresses constant and varying the response capabilities of the organizations. Since such variation tended to submerge the differences between organizations, the variable of organizational structure was no longer regarded as focal. Rather, RAND inquiry was directed toward how the support system should be integrated and how responsive the system should be (Nelson and Peterson 1962).

In LP-III the computer not only performed the computations which the organization managers had to have to make their decisions but also handled numerous complex interdependencies between models, such as the factory model, distribution model, and parts failure model (which triggered demand). The output cards from one model became the input cards to another. As already noted, the computer processed inputs and provided printouts twice every twenty-four hours. There was a special computer run after every four laboratory periods, that is, every other night, to prepare monthly reports. An innovation in this experiment was the computer analysis of data after every period, and particularly in the special run every other night to show the results of a simulated month of operations.

## FOURTH EXPERIMENT

The methodology of the fourth experiment, LP-IV, more closely resembled that of the second. The researchers made no attempt to create and adhere to a rigorous experimental design. They emphasized (1) a "benchmark run" simulating then current methods and (2) subsequent comparisons between its results and those of proposed methods. They assigned different runs to the benchmark and proposed systems, rather than operating them in parallel or in alternating segments of the same run.

The interest of the laboratory turned to problems of weapon base maintenance and to the possibility of developing a fully automatic system for managing such maintenance. Although an experimental examination of such a system was projected, it had not been undertaken when the material was assembled for this account, which covers only the benchmark run and a run of a proposed improved manual maintenance system, together with a number of associated studies.

Experimental subjects staffed the job control organization of a Strategic Air Command heavy bomber wing of B–52s and KC–135s, namely, the one at Beale Air Force Base, rechristened "Byess Air Force Base" in the experiment. Figure 18 shows the control cycle. Personnel functions in job control were aggregated into three positions, a senior controller and two dispatchers, one for armament and electronics, the other for field maintenance. The field maintenance dispatcher scheduled jobs and communicated with the simulated maintenance shops. Responding to malfunction and repair requirements, the senior controller received "job discovery" information from a simulated flight line, transmitted messages to the flight line, monitored aircraft status, and coped with crises. In the benchmark run these three positions were manned by Air Force officers with appropriate experience. The benchmark system is diagrammed in Figure 19.

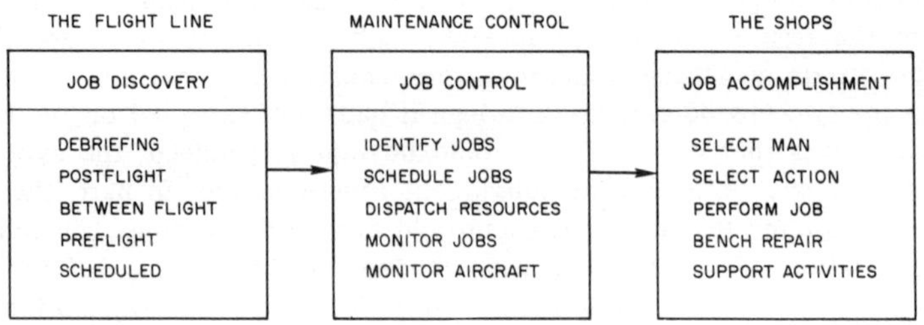


Fig. 18. The Control Cycle, Fourth Experiment (Cohen and Van Horn 1964).

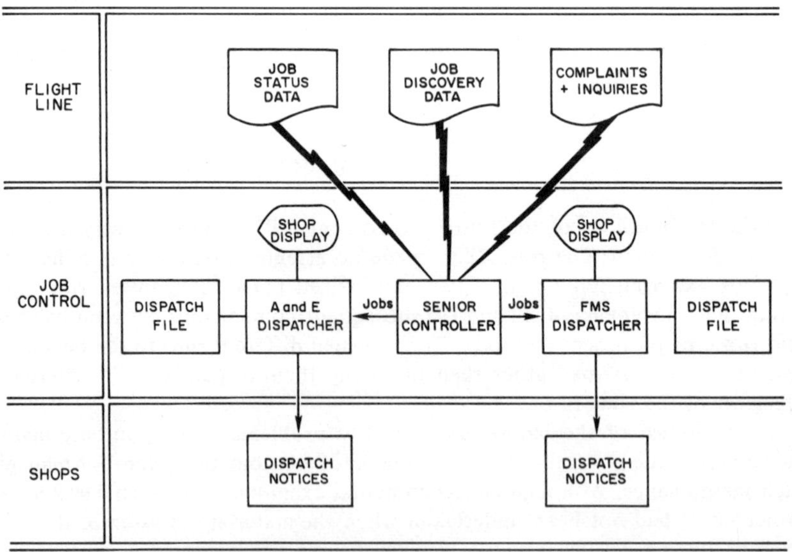


Fig. 19. The Benchmark (Current) System in the Fourth Experiment (Cohen and Van Horn 1964).

Programs in an IBM 1401 computer in the laboratory contained models representing the environment with which job control interacted. An aircraft model simulated (1) the major systems of the aircraft and their components; (2) more than twenty situations in which the aircraft might find itself, including scheduled maintenance; and (3) malfunctions which would lead to unscheduled maintenance. Bomber wing personnel, spare parts, and equipment were simulated in a resource model. A maintenance model identified the resources needed to perform a particular maintenance job and the job's elapsed time. Operations and maintenance plans were also stored in the computer and, in interaction with the maintenance requirements, generated daily, weekly, and monthly schedules made known to job control.

When a simulated B-52 was dispatched on a mission in accordance with an operations plan, it was matched by the computer against a pre-run output of the failure portion of the aircraft model for the particular situation in effect. Any malfunctions calculated by the computer were made known to the control point where RAND clerks, playing the roles of maintenance personnel, passed messages about these and other malfunctions and maintenance actions to the job control subjects. The latter in turn sent messages of action requirements to the control point for translation to computer inputs for processing by the maintenance and resource models. Of course, this highly simplified account hardly does justice to the complexity of actual laboratory operations, which in large measure reflected the complexity of actual maintenance control at a bomber wing. A great deal of effort preceding the experiment was devoted to collecting data from the real environment. The models also profited from a prior RAND study of maintenance at an Air Force base.

*LP-IV Content*

The benchmark run occupied the laboratory for five weeks of eight-hour days, representing five weeks of twenty-four-hour days in the real organization. In the LP-IV experiment, laboratory time was divided into 15-minute segments, each of which simulated 15 minutes or more of real system time, according to the work load and capabilities of the real organization. When the latter's work load was light, time was further compressed. Every 15 minutes the computer would receive a batch of inputs from the control point and process these and previously stored inputs from other sources. The computer produced printouts and a history tape. The control point clerks telephoned the printout messages to the job control personnel. The computer processing and telephoning frequently took longer than anticipated and cut into the next 15-minute segment of job control operations, reducing it by an average of 7.5 minutes.

During the first three weeks, Byess job control activities were designed to resemble those at Beale as closely as possible; after the run the experimenters attempted to compare the two performances to determine whether they had, in fact, simulated the real system. This comparison for validation proved more difficult than expected. Cohen and Van Horn (1964) later commented:

> Several system measures which the staff thought important for such comparisons were either poorly defined or nonexistent in the field. For example, aircraft turnaround (clock-hours from start to end of maintenance) is an important system measure, but is not recorded in the field.

Furthermore, most (field) data are collected and aggregated over intervals of time—monthly, for the most part. Monthly aggregations require an inordinately large sample of data to permit statistical analysis. In the management of aircraft maintenance, the sortie would appear to be a reasonable unit of measure to assist management and to permit statistical analysis, but the month is mainly used instead. This state of affairs not only complicates both laboratory and real-world validity studies, but indicates how important the unit-of-measure problem is for research appraisal in a laboratory.

During the last two weeks of this first run, noise was introduced in the form of delays in job starts ranging from a few minutes to 90 minutes. The subjects detected these delays only about half the time, and the noise degraded system effectiveness. This result bore out some observations at Beale. In the fifth week, when additional alert sorties were required, the subjects managed to reduce turn-around-times to meet the heavier requirement, although their expressed goal was to meet scheduled take-off times rather than to reduce turn-around times.

Consequently, in the next run this other goal became paramount, and procedures and displays were created to help achieve it. One innovation in this run was "event monitor and action reporting." A displayed file of time-sequenced event records enabled job control to monitor such job events as start, stop, and delay; the simulated mechanics had to furnish the event data for this file, which "effectively meshed maintenance data with the mission or sortie experience." A second innovation was job scheduling and weapon display. The required time for the longest job on an aircraft established a deadline by which all other jobs must be completed. Deadlines were established and revised as necessary for all jobs on the aircraft, essential and nonessential. They were posted on a display "which effectively focused Job Control energies on the aircraft's turnaround progress." This "long job rule" permitted flexible scheduling of other jobs.

This run, in which RAND personnel served as subjects, had three phases, all with the reduction of turn-around time as the single and common objective. Throughout the subjects coped with noise like that in the last part of the first run. In the first two weeks an event monitor helped them. The third week marked the advent of the new scheduling procedure and display. In the fourth week both of the innovations operated and additional sorties were required. Results showed a substantial reduction in turn-around times.

### *Associated Studies*

A field test was undertaken at an Air Force base to validate these experimental findings but its outcome was equivocal. Because the experimenters could not exercise control, the reduction of turn-around times could not be established as the primary goal, nor was the same workload placed on the job control personnel.

Back in the laboratory a small supplementary experiment with college students as subjects investigated the effects of four different goal instructions to the subjects: to do as well as they could, or meet the take-off time, or turn around in eight hours, or use the long job rule. This rule (which bears a resemblance to the PERT technique for scheduling system development) had favorable effects, as it had in the second laboratory run. The final comment of the experimenters was: "The orthodoxy of system design stresses the critical role of objectives but provides little other guidance. The LP-IV experience strongly suggests that finding a 'good' objective is a complicated process that deserves a great deal more study" (Cohen and Van Horn 1964).

The environment models developed for LP-IV and the data collected for them were exploited in other studies. Several were analytic studies of trends and correlations in the workload generation process. Of particular interest was an all-computer simulation called base operations maintenance simulator (BOMS) (Geisler and Ginsberg 1965), programmed with Simscript. BOMS was run both before and after the second LP-IV experimental run in the laboratory. Initially it was used to (1) select from a large number of possible dispatching rules one or two to be examined in the laboratory; (2) "perform sensitivity tests on the various resource levels so their realistic resource levels could be set for the lab runs"; and (3) aid in interpreting laboratory results by determining the sensitivity of the system to various operations schedules. Subsequently it became a method of testing the long-job rule and other laboratory innovations. Since the time compression in the all-computer simulation was about 20,000:1, many repetitions could provide the large sample size that permitted the statistical testing which had not been possible with the laboratory findings.

## OVERVIEWS OF THE RESEARCH

One of the outcomes of all this RAND research has been the construction of two models of the kind of project exemplified by the LPs. One of these (Geisler, Haythorn, and Steger 1962) describes the chronological evolution of the experiment. A design phase comes first. It defines the experiment's objectives, "identifies policies to be tested, specifies the organizations to be simulated, determines the experimental design, and develops the general goals and characteristics of the analysis." It may take from six to twelve months. The new policies or system characteristics to be examined are the products of analytical research elsewhere in the embedding department, and the experimental research team is composed of both laboratory personnel and some of these analysts.

Next comes a modeling phase requiring six to eight months. (Actually, phases overlap.) The environment is defined in mathematical terms for computer simulation and the data of which it is composed are collected. One can then prepare descriptions of the simulated organization and its interactions with other organizations; and decisions can be made as to which functions should be simulated on the computer and which by the subjects. Computer programs and a computer data base are created, manuals are written for conducting the experiment and instructing the subjects, and the simulation input reports and subject output reports are formatted.

Before one initiates the operation phase, that is, the actual experimental runs, it is essential to have a mock-up, or rehearsal. "At this time the laboratory staff makes an effort to run the computer models and programs, receives training in the operations it will perform during the experiment, develops the necessary experience to determine whether the Laboratory resources assigned to the experiment are sufficient for the tasks required, and obtains estimates of the computer time, elapsed time, etc., needed to accomplish the various steps in the experiment." The staff personnel role-play the subjects. About a month is needed for the mock-up and the readjustments which it reveals are necessary.

The operation phase may take four to six months. Each set of subjects must receive indoctrination in the simulation techniques, the object system, and their particular assignments, which may also require specialized training. A number of relatively short runs are preferred to fewer longer ones, partly to maintain subjects' interest and partly to be able to change experimental conditions and/or sets of subjects from run to run as indicated by "efficient experimental design" and achievement of experimental validity. (As a by-product, more military personnel are exposed to the proposed system or procedures, can critique them, and can take part eventually in their implementation.) "In the early experiments, the Laboratory made very little use of formalized experimental design and more or less adapted its experimental course to results as they unfolded." But experimental design can result in more sophisticated analysis, and it also helps in the assignment of staff personnel and the scheduling of computer time and subjects. Nevertheless, it may be necessary to resume the operations phase as a result of what is learned in analysis—to conduct more runs "to fill in gaps in knowledge, to test new hunches, or to recheck an unexpected prior result."

The analysis phase actually goes on throughout. It may involve side studies and field tests and may yield results before the experiment is completed. Recommendations result from the comprehensive reduction of vast amounts of data and the resulting conclusions. Briefings and clearly written reports inform the military sponsor of the results, the system features that were investigated, and the system procedures being recommended.

The second model (Haythorn 1963*b*) consists of a matrix prepared for LP–IV in which functions of the system being investigated (eighteen in this case) are matched against twenty-five research functions. The latter are: field research and written description (comprising "description"); modeling, information system representation, and computer programming (comprising "laboratory representation"); demand generation; justification, laboratory implementation, and real-world duplication (comprising "policy"); data requirements, data collection, and laboratory data production (comprising "input preparation"); experimental design, crew training, and manual preparation; performance and cost (comprising "evaluation"); and analysis, documentation, spin-offs, laboratory operations, implementation aid to the Air Force, Air Force briefing, interface specifications, and final reports. This matrix can be regarded as stemming from a task analysis of research operations. It provides a method of assigning responsibilities to members of the research staff.

In addition to these guides to this type of research, the RAND experimenters have felt the experiments made significant impacts on Air Force logistics policies and methods. Not the least of such impacts has been the education which Air Force officers have received through serving as experimental subjects ("participants," in LP language) and as consultants and visitors during the experiments. According to Geisler, Haythorn, and Steger (1962), "LP–I has had considerable influence on the direction of the Air Force supply system." Air Materiel Command personnel designing the inventory control system for the ICBMs "showed particular interest in the deferred procurement and responsive prediction techniques used, and in the use of automatic resupply for low-cost items," and Sacramento Air Materiel Area personnel who had taken part in the experiment put much of Logistics System 2 into practice for the F-100 aircraft series, while AMC used LP-I policies in their development of an inventory system for low-cost items on a wide scale. "Many of the policies adopted for the Air Force in missile system management and operations since LP-II reflect the findings of that experiment," these researchers added, citing Minuteman wing and squadron organization, inspection and checkout policies, and Minuteman adoption of the LP-II management system to a considerable degree.

In some contrast to effects on hardware (and software) design, it can be difficult to trace the effects of research findings and recommendations concerning policies and procedures—unless the organization which adopts them publicizes their origin, an unlikely event. Even if the precise extent remains indeterminate, the statements by the RAND researchers indicate their earlier work did have a valuable impact. Their reports did not discuss the effects of the third and fourth experiments.

# 14

# Combat Development Experimentation Center

Between 1957 and 1966 approximately twenty-five large-scale field experiments were conducted by the Army's Combat Development Experimentation Center (CDEC), rechristened toward the end of that period the Combat Developments Command Experimentation Command (CDCEC) after it became part of the newly created U.S. Army Combat Developments Command. The rate of experiments in 1965 was about four per year. In addition, there have been "side" experiments connected with the larger ones, as well as all-computer simulations and smaller field studies. CDCEC headquarters is at Fort Ord, near Monterey, California, but most of the experimentation has been carried out at the 268,000-acre Hunter–Liggett Military Reservation fifty-five miles to the south.

After CDEC's establishment in November, 1956, then as part of the Continental Army Command, it received technical support from Psychological Research Associates, The Johns Hopkins University's Operations Research Office, and Technical Operations Inc. (see Chapter 9). In 1958 Stanford Research Institute (SRI) became the on-site contractor to supply the professional support needed to convert what might more properly be called field tests into experiments. Its efforts to achieve better experimentation by objectivizing data and controlling variables started to pay off about 1963. In association with those efforts, human factors expertise increased in the SRI field staff. In 1966 Stanford Research Institute was replaced by another contractor. This chapter will review some of the Stanford Research Institute's attempts to help the Combat Development Experimentation Center match its field investigations to its name.

The ambitions in the Army's first, and relatively young, attempt in field experimentation have been expressed by Murdoch and Edmondson (1962), who wrote that the purpose and role of CDEC were

> to prepare, conduct and evaluate, with maximum objectivity and scientific control, experiments with concepts, organizations, tactics, doctrines, and procedures for future combat . . . . This field laboratory provides for the execution of experimental concepts by men and machines under realistically simulated combat conditions. These exercises produce both quantitative and qualitative data under conditions that reflect operational degradations, human behavior, functioning of machines, enemy measures and countermeasures, and other elements that influence the system under evaluation. Instrumentation and simulation means have been developed and the methodology of field experimentation is continually being improved.

To date, the program for field experimentation has been concerned with evaluating the tactical application of new concepts of equipments, organizations, weapons systems, and surveillance and target acquisition systems. CDEC is not an equipment-testing agency. However, we do measure the performance of equipment in a simulated war. By simulation, concepts for new equipments are evaluated in terms of their tactical application. Thus, the need for, and capabilities of, proposed equipments may be determined prior to expending time and money on the development of an item.

The advent of the tactical nuclear weapons to the battlefield has stimulated new concepts of organization based on increasing dispersion, surveillance, mobility, and more critical command and control. Experimentation has been conducted on the controlability, mobility, target-acquisition ability, vulnerability, sustainability, and the destructive force of new organizations. Experimentation has also been conducted on the vulnerability of low flying aircraft to ground fires including the REDEYE infrared homing missile. In addition, many data from logistical tests have been collected on the problems of supply and resupply, maintenance, and medical evacuation.

Among still other objects of investigation in recent years have been the preferable composition of rifle squads and platoons (a continuing problem due to changes in weaponry and tactics; see Chapter 9); operations and composition of infantry, vehicles, armor, weapons, and communications in a new type of battalion; relative advantages of light observation helicopters and fixed-wing aircraft for various operational tasks; organization of a rifle company for antitank action; battalion antitank operations; effectiveness of a missile-armed helicopter against tanks and antiaircraft, and vice versa; evaluation of various types of small arms; effects of toxic environments on battalion operations; methods of locating battlefield casualties; devices for improving night-time vision in combat operations; and the number of men, type of equipment, and tactics required for an advanced-concept, all-purpose unit capable of independent operations on the 1970–75 battlefield in a nuclear environment. Thus, the range of interest has been wide, as has been the time scale to which experimental results might be applied.

Continuously supporting more than one hundred officers and a score of scientists in CDEC itself has been an infantry battle group larger than three thousand officers and men, as well as military personnel from other Army units on occasion. The SRI contingent has included mathematicians, statisticians, physicists, psychologists, chemists, engineers, and representatives from other professions. At Fort Ord the supporting facilities have included a mock-up of a communications net and an accurately detailed terrain model of some of the Hunter–Liggett area on a 1:1000 scale. No longer in use, this model, approximately 20 × 60 feet, served for preplay in planning some of the experiments, but it could not duplicate the detail, complexity and subtle variability of actual terrain, where the outcome of operations depends greatly on line-of-sight visibility, concealment by vegetation and other small terrain features, and the effects of such features on maneuverability.

The Hunter–Liggett "laboratory" which provides the actual terrain is mostly mountainous with rugged ridgelines running north and south rising to more than 3,200 feet, but it also has extensive hilly and flat, open areas. Ranges of temperatures from freezing to 100 degrees and of climate from heavy rainfall to very dry (and dusty) conditions furnish a variety of environments, although there are no rice paddies or parallels to jungle or arctic conditions.

## PLANNING OF EXPERIMENTS

As matters stood in 1966, to initiate an experiment a directive was issued by the parent command (Combat Developments Command) or developed by CDCEC at the parent command's request or in response to a CDCEC proposal, all in consonance with a five-year experimentation schedule of the parent command, reflecting its analysis and planning. CDCEC and its technical support organization then took the following steps: (1) a project analysis identifying specific problem areas or obstacles to accomplishment, evaluation concepts and criteria, methods of collecting data and types of instrumentation available or possible, type of field combat activities and resources required and available, time frame, and methods of performing the experiment; (2) an outline plan; and (3) a detailed plan. These steps have been described in an *Experimentation Manual* (U.S. Army Combat Developments Command Experimentation Command 1966).

The outline plan stated the objectives and essential elements of analysis, concept and scope, location and schedule, and various other aspects such as budget and security considerations, experimentation design, instrumentation requirements, and materiel requirements. The experimentation design was supposed to narrate the events constituting a trial or run; specify independent variables, dependent variables, variables held constant, and uncontrolled variables; list the sequence of trials; state the hypotheses and the mathematical techniques for testing them; and describe any computer simulation or other supplementary research to be included.

The detailed plan contained the specific instructions for supporting and conducting the experiment and was subject to continuing updating, revision, and addition. It was supposed to include scenarios for players and controllers, task organization, detailed data-collection plans (including required levels of accuracy), detailed instrumentation methods, the timing and conditions of and within trials, replications and variations of individual features, detailed schedules, requirements for training the participants, a communications plan, detailed lists of personnel and materiel, and plans for safety and emergency procedures to deal with contingencies.

### *Example: "Operations at Night"*

These plans were exemplified in the Operations at Night experiment. The outline plan (Headquarters, Combat Developments Command Experimentation Center 1964) said the objective was to determine the capability of reinforced battalion-size units of the 1965–70 time period to operate at night; to indicate the improvements necessary in organization, doctrine, equipment, and training; and to develop techniques and equipment for monitoring, controlling, and collecting data about night combat operations. Although experimentation was to go through three phases, only the first phase received specific attention. It was intended to investigate the capability of individuals and squads to move cross-country at night; capability of troops to detect and acquire target and intelligence information at night; and capability of individual observers to call for and adjust indirect fire on targets. Experimentation was planned to cover three and one-half months, starting five and one-half months after the beginning of planning.

Under the heading "experimental design" were descriptions of purpose, objectives, field test arrangements, variables, data requirements, personnel requirements, and critical personnel requirements. As variables the outline plan specified illumination (varied between conditions of normal night-time, daylight, searchlight, and artillery illuminating shells); night-vision devices and equipment (a number of infrared, radio, and other devices); terrain (open, close or hilly, and a combination of these); and soldier proficiency (differences to be partly reduced through a training program and measured by a special investigation of human factors). In addition, the variables section stated: "The experimentation will be sequential in nature, i.e., what is learned on early trials will affect the conduct and number of subsequent trials. This phase will produce comparative measures of performance under various conditions of illumination, with or without night vision aids. Such a program calls for a balanced design within the constraints of individual variation and the increasing familiarity of the soldiers with the equipment, with night activities, and with special techniques for achieving good performance measures." The plan did not explain precisely what was meant by these last phrases.

Preceding the section on variables, a section on field test arrangements described the general characteristics of units, courses, and measurements in the three phases. Under data requirements the plan stated that for cross-country movement, the duration, deviation from a given path, and distance error at a terminal point would be measured; for target acquisition, data would be obtained on the number of targets acquired, length of time required to acquire them, and accuracy of their locations by the subject; for fire adjustment and control, the accuracy and speed of sensing and adjusting would be measured with respect to arrays of explosives emplaced around targets and set off to simulate indirect fire systems. Personnel requirements were estimated as 13 officers and 154 enlisted personnel for player and player support personnel; 6 officers and 161 enlisted personnel for operations and control; 2 officers and 84 enlisted personnel for aggressor personnel; 2 officers and 40 enlisted personnel for instrumentation; and one officer and 12 enlisted personnel for data collection and analysis. "Critical personnel" were technical representatives not available in CDCEC to assist with some of the special night-vision devices.

The instrumentation section specified that the subjects (players) would all wear blinking infrared beacons on their helmets, so that their positions could be continuously photographed from nearby hills and also observed by controllers through infrared devices. Another method of keeping track of their positions was to be implemented by battery-powered radar transmitters carried by each soldier. Each of two spaced radar receiver units was to obtain and print out the azimuths of receptions on a common time base, so the intersections of these directions could indicate the soldier's position. Vehicles were to be tracked either with this Rawin Set or the Direct Range Measuring System (DRMS). The latter consisted of *A* units which sent high-frequency radio signals received and returned by *B* units; the *A* units determined the ranges of the *B* units by the elapsed times; and the intersection of the arcs from two spaced *A* units established the position of a *B* unit.

*Detailed Plan*

In the detailed plan (Headquarters, Combat Developments Command Experimentation Center 1965), the section on "experimental design" and an annex specified how to relate to each other the three phases, the various terrain courses and "lanes," the elements (e.g., individuals, squads, tanks, forward observers), and the number of players per element. Figure 20 is a schematic diagram of one of the courses. Design tables allocated trial numbers, order of start, and three night-vision conditions: two different devices and no device. Daylight trials were

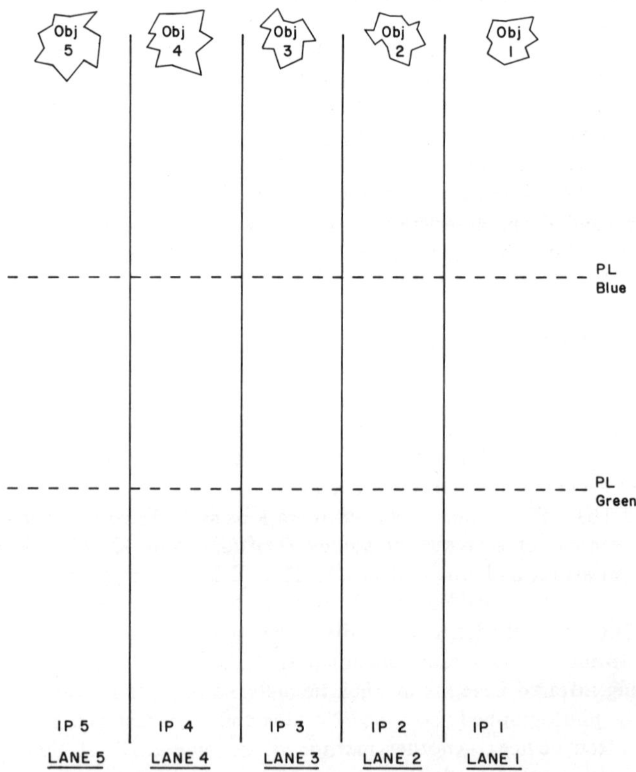


Fig. 20. Schematic Diagram of a Cross-Country Movement Course in the Operations at Night Experiment (Headquarters, Combat Developments Command Experimentation Center 1964).

to follow night trials, and conditions other than normal night illumination were deferred. For fifty-four individual cross-country trials, subjects were assigned to six groups of one to three soldiers. The combinations of number of soldiers per group, device, start order, and lane were systematically varied between blocks of six trials in an attempt to make the main variables orthogonal with each other and to control for learning and individual differences. Similar designs were indicated for some of the other parts of the experiment.

A training plan annex identified the units to be involved and the training content and schedules for player, aggressor, and control personnel, who were to get 30, 27, and 27 hours of night training respectively. Pre-experiment proficiency tests and performance tests with the night-vision equipment were projected for player personnel. A player operations plan annex contained the scenarios and pictorial descriptions of the courses, and the instructions which the players were to receive. A controller operations plan annex specified the duties of the controllers, a function of considerable importance in this kind of experimentation. For example, some of them were to brief and start the players, others to observe and record the movements of the players from concealed positions, still others to require that the players conform to the rules, and some to set off demolition charges. An aggressor operations plan annex made it clear that this experiment was one-sided. Aggressor forces would not undertake offensive operations but would only simulate defensive positions and actions which the players would try to detect and identify. Aggressor riflemen were required to run, crawl, lie prone, dig foxholes, fire rifles; aggressor tanks and antitank squads had to fire simulator rounds; several aggressor personnel might sit together and smoke cigarettes. In all, the plan specified a great variety of nonoffensive aggressor actions.

A communications plan annex set forth the various radio and wire nets for all the personnel involved, with their components, frequencies, and call signs. An emergency procedures plan annex provided, among other things, that a helicopter would be standing by for emergency evacuation, as well as an ambulance at the field command post. An instrumentation plan annex told how to establish a field operations center. Plotters at charts were to receive azimuth and range reports from tracking sites that obtained the data through infrared sighting instruments and the DRMS equipment, respectively; they would make intersects to show player positions on the charts. It was indicated that the Rawin radar equipment would also be used if this was available. It was further indicated that infrared photography might be foregone due to various problems; for example, the camera locations were ineffective and the correct film was not available.

Hopefully, this account has indicated some of the ramifications of this kind of experimentation. Yet this was a relatively simple and inexpensive experiment, with a cost projected in the neighborhood of $100,000 (including personnel per diem and travel expenses). The individual cost of some of the field experiments has been calculated at more than $2,000,000, even where there was only a single instance per experimental condition and where data were obtained without benefit of instrumentation. The instrumentation cost could be very high and so could the cost of personnel and munitions, especially in experiments where large military units (and a large aggressor force) were maneuvered.

## PROBLEMS IN EXPERIMENTATION

Even in relatively modest experiments which attempted to follow good design, problems could arise which should have been foreseen, or were unexpected. An example was an investigation of devices for locating battlefield casualties (Combat Development Experimentation Center 1961). These devices included a helmet-mounted radio (worn by the casualty) and head-mounted infrared binoculars (on the searcher). In daylight runs each of three rescue teams operated three locating methods in three different types of terrain; and night runs included several additional methods. To control for learning, the subjects received advance training in the methods, and the order of conditions was randomized. Unfortunately the results were inconclusive. The situations were insufficiently realistic, because the detection of casualties was unimpeded by hostile action or heavy underbrush (or jungle vegetation)—what might be called "noise" in other situations. The nights happened to be so clear and bright that unaided vision proved better than the head-mounted infrared binoculars, which, due to poor human engineering design, the soldiers had to keep in place with their hands as they moved about. There was still another problem. Many of the subjects had difficulty reading maps and orienting themselves.

### *Selection and Training of Subjects*

The problem of the selection and training of subjects received particular attention in connection with the Basic Mounted Unit experiment (Mills, Vollmer, and Anderson 1961). This 1959 experiment evaluated four compositions of a very mobile tactical unit as the primary combat component of a combined arms force capable of sustained, dispersed operations on future battlefields. A Red force and a Blue force opposed each other in attack, defense, reconnaissance, and counter-reconnaissance postures. The four unit compositions, four postures, and two forces were arranged in sixteen experimental conditions. A one-day simulated battle took place in each condition. However, both the Red and Blue forces changed personnel assignments among the unit compositions from day to day, so a unit composition did not carry the same personnel through the entire series of sixteen conditions.

To find out how the manning of the forces might have affected the outcome, one-third of the tactical personnel in the experiment were investigated through Army personnel records and two questionnaires, one administered before the experiment and one after. It was found that more Red force than Blue force vehicle commanders and assistant commanders had had combat duty, held combat decorations, had had civilian work experience, preferred experiment duty, and received best and next best choices for team leader. Further, it was found that nine personnel measures were related to unit performance scores. These personnel measures were two leadership scores, combat experience, civilian work experience, civilian supervisory experience, identification with experiment duty, army career identification, billeting arrangements, and educational level. All or virtually all were related negatively to their unit's "relative invulnerability" measures in the attack and defense postures, and positively to this measure in the reconnaissance posture and counter-reconnaissance posture, particularly the former. "Relative invulnerability" was the ratio of own to enemy's invulnerability (duration of tactical survival).

The investigators concluded that the reliability of the experimental findings was subject to question because unit compositions "constituted from the Red force enjoyed advantages of command talent, particularly in leadership ability" and that "command ability imbalance was associated with experiment findings" for the reconnaissance posture, where commanders had the most opportunity to be effective. The investigators reviewed possible methods of personnel assignment that an experiment could incorporate to prevent confounding: (1) random assignments of companies to each condition; (2) alternating companies between conditions; (3) random assignments of individuals; (4) precision matching of individuals between groups; (5) ex post facto matching through analysis of covariance; and (6) frequency distribution control, that is, matching groups in terms of over-all distribution of a given factor or factors within the group rather than individual by individual. They preferred alternating companies between conditions, since this method does not break up existing units, makes no assumptions about normality or homogeneity of variance, makes it likely that the groups are equated on all variables—known and unknown alike—and does not require equality of training.

The commanders were asked in the questionnaires how to improve the experiment. Whether comments came before or after the experiment, the largest proportion of answers called for improvement in training, especially detailed skill training in, for example, map reading, radio operation, and radio discipline. They also emphasized how important it was for the player personnel to understand clearly the objectives of the experimentation. "Some aspects of experiment orientation might be most efficiently handled by civilian scientific personnel who have taken a role in devising experimental procedures," the investigators added.

### *Observers' Data and Evaluations*

In addition to its concern about subject selection and training, Stanford Research Institute attempted to improve the methodology of CDEC's field experimentation through devising better techniques for collecting data. In some of the experiments, such as the one which compared light observation helicopters with fixed-wing aircraft and another which examined a new type of battalion, the data consisted entirely of judgments by observers and participants or their factual statements that something did or did not occur. The need to reduce or supplant individual human fallibility in gathering data has been well recognized by CDEC (Murdoch and Edmundson 1962); the problem has been how to do it.

In field experiments much of the performance data must come from human observers, called umpires or evaluators, for two reasons. Instrumentation to collect data may be expensive or unavailable, and instruments frequently cannot do the job. The performance of troops and vehicles and their aggregates in reaction to terrain and to an opposing force may become so complex and variable that much of it can be adequately discriminated and reported only through the capabilities of a human to recognize complex patterns and verbalize them. But it is possible that this human discrimination and reporting can be improved. Rittenhouse (1962*a*, *b*, 1966) developed a technique for such improvement. The following description was taken from his reports.

The setting for this development was an experiment in 1961 on squad and platoon organization. Three different squad and platoon organizations each went once through three different terrain courses. Subjects were rotated through the three organizations to control for variation in individual performance. Their actions were governed by scenarios which had been written to create a reasonable sample of tasks "with respect to the variety and difficulty of the activities the units would be expected to engage in under combat conditions." These tasks were then "broken down into clearly defined segments small enough to be observed closely by evaluators in the field. . . . In the case of squads, for example, such things as changing formation or direction, starting, stopping, or shifting fires and the assault appeared to be the types of activities to which observation and evaluation could be applied. Most of these occurred a number of times in each scenario. By focussing on these small segments of action which were essentially repeated many times during the course of the tactical sequences, it was hoped that a form of acceptable replication would be obtained."

In preparation for the experiment, a rating item was prepared for each of the smaller segments of action. It listed instances of possible difficulty in performing the action. Evaluators had to rate on four-point scales the degree of each difficulty during the action. Space was provided for the user to list and rate additional difficulties. In connection with each item the evaluator also had to mark a checklist of the possible causes of the specified problem or difficulty; there was space for him to note down any cause not on the list. The form containing the set of items further required him to judge for each item "whether or not leadership or other individual deficiencies might be governing factors in the instance cited. . . . The rating forms thus directed the observations of the evaluators to the significant aspects of performance. . . . They provided for the recording of observations in systematic fashion so that when summarized such organizational differences as might exist were revealed."

To create the instances of difficulty in the first place, military experts studied segments of scenario tasks and stated what would indicate to them, as evaluators, that the organization was performing either well or badly. The indicators selected as difficulty items were those thought to be observable by evaluators under field conditions. Possible causes of observed difficulties were derived from analysis, military experience, and judgments of the experts.

In a limited pretest of the technique three evaluators accompanied a squad of each platoon over the courses and made independent ratings, which were substantially in agreement. "In general, they perceived the same problems at various times in the sequence of actions and ascribed the same causes of these problems to the same features or organizational structure or to personnel failures," the investigator reported. "There were only a few cases of grossly different ratings."

Then came the experiment, in which the evaluators used the rating forms. Ratings identified by task segment and organization were tabulated in such categories as control and mobility to yield the frequencies of occurrence of observed difficulties according to the degree of each, together with lists of probable causes and their frequencies. Some senior military personnel had been assigned as additional evaluators. Although they had more extensive and varied military experience, they were less familiar with the scenarios, the terrain, and the rating technique. When their rating forms were compared with those of the regularly assigned evaluators there were greater divergences than in the pre-test "but there was still enough agreement to support the notion that the evaluation scheme could reasonably be employed by individuals of varying military background without destroying its usefulness."

In addition, fourteen SRI technical personnel acted as additional evaluators during a few runs of the experiment and commented on the technique. They encountered some difficulty in rating some problems phrased in such terms as "too much time was taken," and it was felt that more explicit scale definitions should be provided in the future, although the difficulty of doing so was acknowledged. It was suggested that by "walking through" each scenario the items could be more specifically tailored to the scenario situation, although this approach might limit generalization across scenarios.

This experiment acquired additional kinds of data. The evaluators provided more general judgments about particular features of organizational structure and made comparative ratings of these. Observations were recorded by the players themselves and by the aggressors and controllers. Duration measures were taken of the time required to accomplish the action segments on which the specific ratings were based, such as the time after the issuance of an order by a squad leader to shift fires until the last man shifted. The number of observations of enemy activity by each organization was recorded for each action segment. However, due to the unreliability of these measures, an attempt to correlate them with the specific ratings "was only moderately successful."

At an earlier date, Fend and Cloutier (1958) of Technical Operations, Inc. studied umpiring and evaluating in a 1957 experiment which pitted four infantry rifle companies against two aggressor companies. The experiment involved four weapons systems, four terrain courses, and four phases. The combinations of variables and their order were organized through a Graeco-Latin square which permitted analyses of variance of the objective data. (Tests of statistical significance do not otherwise seem to have been frequent in the CDEC program.) Many problems of umpiring and evaluation were uncovered. Collusion between participants and umpires might occur unless the assessment process was well isolated from the participants. "For example, should a rocket launcher crew know that, from an umpire table, the next mission will be successful if they fire at the side of a tank, but must be umpired a miss if they fire at the front, the crew would certainly be induced to wait until the most favorable target is presented, for umpire rather than tactical reasons." It was suggested also that "individuals trained in unique umpire operations should be retained throughout an experiment."

The investigators further commented: "The primary purpose of a military evaluator's report is that of explaining WHY a particular action succeeded or failed. . . . Experience indicates that military evaluators must, in the future, be trained to determine *why* each major decision was made and to explain the essential reasons for the success of each facet of the action. Forms provided them must reflect the WHY question in each phase of action to be expected from the scenario." As noted above, Rittenhouse came to a similar conclusion four years later.

*Instrumentation for Collecting Data*

Another approach to obtaining better information from field experiments has been through instrumentation. This received such emphasis at CDEC that instrumentation was often the pacing factor for an experiment; delays in such instrumentation delayed experiments. CDCEC had a five-year plan for developing and procuring "building blocks of compatible multi-purpose instrumentation." The DRMS system and other methods of telling where people were have already been described. Another set of equipment was the "Hit Count Skin/ Acoustic Miss Distance Indicator" to measure live-fire hits and misses. Omnidirectional microphones recorded misses by measuring supersonic shock waves of projectiles as they passed through the air. Miss distance was determined by measuring shock-wave amplitude. Miss data and hits, automatically recorded, were telemetered to a fixed receiving station. In addition, direct fire could be simulated with a direct fire simulator system which could be installed on tank guns, recoilless rifles, and other weapons. It provided immediate indications of a hit, and it processed, recorded, and transmitted data required for hit assessments. Equipment has also been devised for co-ordinated timing of data received through hardware and human sensors; such timing is a complicated enterprise.

An interesting kind of instrumentation was proposed by J. K. Arima (personal communication) for a study of Army aircraft survivability. Arima was interested in how ground gun crews would be affected by the uncertainty of the direction of an approaching aircraft, the uncertainty as to whether it was friend or foe, and the uncertainty as to whether a line-of-sight opportunity would ever exist. Sounds of approaching helicopters, for example, might not provide the cues to eliminate such uncertainties. Arima proposed an arrangement of various types of miniaturized fixed-wing aircraft and helicopters traveling on wires strung between towers surrounding a live-firing machine gun position. Loudspeakers were to broadcast recordings of various actual aircraft engine sounds. This arrangement could provide a great variety of situations, including areas masked from the gunners, to create uncertainties which could not be easily established in runs of actual aircraft. Gunners could fire real bullets at relatively inexpensive targets. (As a psychologist, Arima has called attention to other important human variables to which field experiments may not or cannot give enough emphasis, such as fatigue and threat in combat.)

Of all the instrumentation problems at CDEC, probably the most difficult has been the co-ordination of kills and casualties in two-sided combat situations. This problem was investigated by Jones et al. (1959) and Twery (1961). In such situations there can be high probabilities of false kills or casualties—those made by a soldier or vehicle that has already been put out of action or those made at impossible distances. In an analysis of an actual experiment, Twery found high percentages of false kills of the first type, even including many cases where the weapon which fired not only had been out of action but its operators had received notification to that effect. But most instances resulted from an average communications delay of 4 minutes in receiving notification. It was also probable that players would fire at a target which had already been "killed" and would kill it a second time, would kill the wrong target due to false target identification, or would fail to register a legitimate kill; in addition, kills might be registered erroneously due to mistakes in viewing or communication. Added to the over-all problem was the requirement that targets must be identified and casualties assessed for indirect as well as direct fire.

A report concerning an experiment about tank-antitank combat (Twery, Barson, and Johnson 1963) described a casualty assessment system involving a central station and radio net whereby a kill notification could be radioed to each tank. But even with this arrangement some tanks continued to fire after they had been killed. Some vehicles lost their radio communications. A number "committed suicide"—they terminated their action after they had been notified erroneously they were dead. In one instance a tank followed a lunch truck across a field and was killed; but it continued to where the lunch truck had parked.

The same report asserted that among the problems of field experimentation were the realism of firing weapons and killing of targets, adequate motivation of troops to provide realistic tactical behavior, inadequacies of umpiring (both human and instrumental), limited replications for making inferences, learning factors, inadequate tactical control, and cost in time and dollars. It suggested that "the major difficulties concern the prompt and appropriate killing of units" and that "just the necessity of making kill assessments (through a time-consuming human control system) during the running of a field situation necessarily contaminates field results. This is particularly true for rapidly-paced action such as in a tank attack."

The control problem in two-sided combat situations was stressed in a number of CDEC reports. It extended beyond the problem of assessment. As indicated in the description of the Operations at Night experiment, large numbers of controllers might be required to make sure that the players follow the scenario or react to enemy action according to some general rules. For example, if a subject was "hit," what might his companion do? Run? Crawl? Lie down? A controller might have to be on hand to make sure he did what he was supposed to do. A field experiment might need as many controllers as subjects.

## COMPLEMENTARY ROLE OF ALL-COMPUTER SIMULATION

One of the important aspects of the SRI–CDEC program was the interplay between field experiments and all-computer simulation. The former could both provide useful data for the latter's models and test the validity of computer solutions; the latter could furnish the many replications needed to acquire assurance about results and could explore additional variables. Murdoch and Edmondson (1962) observed that in the aircraft survivability experiment, field data were collected concerning actual flight paths, weapon locations, in-view patterns for each weapon, and gunner performance (e.g., time required for lock-on with the REDEYE missile and time of fire). "These data were then used as the basis for a computer simulation study of the system saturation problem of the effect of varying factors such as the single-shot kill probability and firing doctrine."

The Twery, Barson, and Johnson (1963) report indicated how a computer simulation followed and complemented the field experiment. It incorporated the weapon placements, tank movements, and invisibilities which had been occasioned by the terrain in the field experiment, but which would have been difficult or impossible to originate realistically in the computer program. Computer replications of the initial field situations resulted in event histories and casualty averages which were compared with those obtained in the field experiment. There followed many computer runs in which weapons system specifications and tactics could be varied.

Although the experiments cited illustrate only the role of all-computer simulation in following a field experiment, SRI staff have asserted that such simulation has also usefully come first. No doubt examples can be found in the reports of the CDEC and CDCEC programs. However, most of these are classified or otherwise unavailable for public consumption—one reason why more of the experiments at the Hunter–Liggett Military Reservation have not been described here. The CDEC–CDCEC experience suggests it might be wise to devote further study to the ways in which man-machine system experiments and all-computer simulations can strengthen each other and the research programs which exploit both approaches.

# 15

# CAA and FAA Air Traffic Control Research

The two types of systems which have figured most often in man-machine system experiments have been air defense and air traffic control. The two varieties of experiments have had much in common. They have simulated airborne objects in a three-dimensional, radar-viewed air environment. Prominent in their operations have been communication with and control of pilots, played by quasi subjects. Team members and frequently teams (subsystems) have interacted with each other in exercising different tasks. But the similarities between the two areas of research seem to have been less manifest to most of the experimenters in each area than might be expected, if cross references in the literature of either are any index. Inevitably, it must be asked whether one might have taught the other some useful lessons.

Experimentation in air traffic control (ATC) has taken place at Ohio State University (Chapter 10), System Development Corporation (Chapter 17), and MITRE Corporation (Chapter 19). The University of Illinois built a simulator to study the interaction between cockpit instrument displays and traffic approach control systems (Johnson, Williams, and Roscoe 1951). But most air traffic control experimentation has been done by the Technical Development Center of the Civil Aeronautics Administration (CAA) and its successor, the National Aviation Facilities Experimental Center (NAFEC) of the Federal Aviation Agency (FAA)—later the Federal Aviation Administration.

The System Development Corporation's single air traffic control experiment followed SDC's air defense research and benefited from some experimenter experience. The MITRE studies drew directly on programming experience for the SAGE system—more so than on experimental research in air defense. But the CAA Technical Development Center's work proceeded in virtual independence of the simulation of the air environment elsewhere or other man-machine system experimentation. The FAA National Aviation Facilities Experimental Center did somewhat better. It obtained some contractor and in-house help based on first-hand experience with such research. Due in part, at least, to this assistance, its studies began to achieve an experimental sophistication justifying the name of the center where they were conducted.

It is difficult to state precisely how many different man-machine system experiments have been conducted at either center, especially since they were still going on at NAFEC when the data for this account were assembled. The estimate depends in part on the definition of "experiment." The reports of the earlier studies at the CAA center omit many of the particulars about the conduct of the investigation on which the conclusions were based; sometimes the time and extent of an investigation are not made clear. An estimate based on available documentation places approximately thirty studies at the CAA center during the nine years of such research there, although almost all of these lacked some of the salient characteristics of an experiment. According to another estimate, during the first two years of its existence some eighteen studies were conducted at NAFEC, and they progressively acquired merit as actual experiments. During the next two years there were approximately eighteen more studies. The average per year would seem to be between eight and ten. A maximum NAFEC capacity of about sixteen per year can be calculated from an average laboratory occupancy time of six weeks per study (Slattery 1965) and the availability of two simulation complexes.

In recent years, the picture has been complicated as a result of projects involving air traffic control operations and simulations with digital computers. In any case, it seems safe to estimate that together the CAA and FAA centers conducted more than sixty multioperator, simulation-based studies of air traffic control between 1950 and mid-1965. This record certainly establishes this program of man-machine system experimentation as the most extensive which has been undertaken. It is also distinguished as being the most long-lived.

Additional kinds of studies have included "slow time" (graphical) simulation; more recently, fast time (digital computer) simulation; experiments on individuals viewing displays or operating data-entry devices (e.g., Paul and Buckley 1967); and experiments on individuals in aircraft cockpit simulators (e.g., McKelvey, Ontiveros et al. 1961*a*, *b*).

## TECHNICAL DEVELOPMENT CENTER

The Technical Development Center of the Civil Aeronautics Administration, initially the Technical Development and Evaluation Center, was in Indianapolis. The simulation research there has been described by T. K. Vickers (Vickers and Miller 1956; Vickers 1959), who had much to do with its execution and growth. It began with the acquisition of the first simulation apparatus in 1950 and ended when the center was disestablished in 1959; the work then moved to NAFEC under the newly created Federal Aviation Agency. The research in Indianapolis on air traffic control was supported technically by the Franklin Institute Laboratories; FIL developed the technique of graphical simulation (see Chapter 24) and contributed significantly to the dynamic simulation studies based on real-time simulation of air traffic, of equipment, and of personnel (Berkowitz and Fritz, 1955; Berkowitz, Fritz, and Miller 1957; Miller 1958; Brinton and Miller 1961).

### *Simulation and Methodology*

The initial apparatus for what was called "dynamic simulation" was the Navascreen, similar to an Australian development in 1948 (apparently the first air traffic control simulator). The Navascreen equipment consisted of a translucent screen on which a variety of maps could be projected, together with six target generators. These were motor-driven light projectors each capable of projecting a spot of light and moving it in a straight line across the screen to simulate radar echoes from an aircraft. Personnel (who frequently were women) controlled the light projectors, thus playing the roles of pilots in the simulated aircraft. Air traffic controllers sat on the other side of the screen, observed the movements of the light spots within the mapped area, spoke to the pilots over a telephone simulating a radio, and told them, as necessary, how and where to move their aircraft. Due to its limited capacity, only small-scale, terminal-area studies could be conducted with this equipment (Vickers 1959). The principal objectives in 1951 were the evaluation of air traffic control procedures for the Washington, D.C., terminal area and the use of terminal area radar in a multistack approach system. (A "stack" is a location where arriving aircraft are required to remain in orbit at individually assigned altitudes awaiting their turn to begin the approach to the airport.)

Air traffic control not only occurs at terminal areas for arriving and departing aircraft but also has expanded to cover aircraft which are en route. The transition between en route and terminal control is a significant interface. So is that within the terminal area between control of aircraft seen by radar prior to landing or subsequent to take-off and control at the airport itself over landing, departure, taxiing, and queuing. Terminal control was originally the main concern, but en route control has become increasingly important during the jet era.

In 1952 the simulation capability of the Indianapolis facility was extended when Navascreen was replaced by Teleran (Baker, Grant, and Vickers 1953). Projectors continued to place moving spots on a screen, but the spots could make turns. Of greater importance, television cameras photographed the screen, now opaque, and the video was processed through a device called a "flying spot scanner" to give it the appearance of intermittently-appearing spots in conjunction with a rotating sweep. This had the appearance of a radar plan position indicator (PPI) display. Air traffic controllers no longer viewed a single large display; they had individual TV monitors which resembled individual PPI consoles. By 1955 (Vickers and Miller 1956), there were eighteen projector units, to simulate eighteen aircraft at one time. The simulated aircraft could turn at 1.5 degrees per second as well as 3 degrees per second. Wind drift was simulated by moving the map projector's platform with a motor. In one of the rare efforts to introduce "ground clutter" into air traffic control simulation, a supplementary projector could superimpose radar clutter on the mapped area. (Controllers using radar to see the aircraft they are directing can be greatly troubled by the radar reflections from surface objects and terrain near airports. These reflections may hide or even appear to be the radar signals from the aircraft.) The improved equipment also simulated radar beacon signals. (Aircraft equipped with beacons send back a strong signal that can be read through clutter; and the beacon signal can be coded to indicate the aircraft's identity or altitude, and other information.)

By 1958 the Indianapolis facility had an ARTC control room for air route traffic control as well as one for terminal approach control. Four TV cameras could pick up separate portions of the screen as though each camera were a separate radar. Each could be moved about to represent different radar locations in the total area. In two rooms were forty-two positions for simulated pilots, each with its own target-generator projector; these were enough to represent all arriving and departing aircraft within fifty miles of Washington, D.C., at that time. Each simulated pilot could talk to controllers over a 20-channel "radio." The screen display presented to the pilots their navigational information (e.g., the geography they could see from the aircraft, or a map). The building had been expanded for briefing, training, and data processing. A year later, the simulation equipment moved to NAFEC.

Vickers (1959) summarized the questions which he and the Franklin Institute researchers investigated as "the convergence problems," "approach systems," and "airport design." He also expressed interest in "human factors," as involved in workload simplication, data acquisition (display), and decision-making. Reports of studies contained numerous conclusions and recommendations; but virtually none of those published by CAA personnel described the study itself and its methodology. Possibly these omissions occurred because familiarity with methodology was limited. By way of illustration, the CAA researchers (Baker, Grant, and Vickers 1953; Vickers 1959) mentioned their discovery in the earlier part of the program that controller personnel memorized the single, short "traffic sample" that constituted the input. They concluded that such samples should be "longer and that there should be more than one sample, if possible." Researchers more knowledgeable about experimental method might have understood this problem from the start. In addition, the reports provided subjective, qualitative data rather than objective, quantitative data. For example, the report of one study (Anderson, Armour, et al. 1957) said, "The conclusions in this report are based primarily on observations by some twelve air controllers who worked on the simulation project."

An apparent preference for exploration and subjective analysis was reflected in the following reminiscence by Vickers (1965):

> I actually believe we made our most significant contribution to air traffic control when we had only twelve, maybe eighteen targets. Later, when the whole organization grew so big and unwieldly, we found that we were spending much more time on administrative, housekeeping, and personnel problems, and there was much less time for anyone to be dreaming up new concepts. In addition, our schedule was getting too tight to explore any new ideas which did originate. As a result, quantity won out over quality. This is why I am very skeptical of any "research" program in which all the methodology and possible findings are spelled out in meticulous detail in advance.

*Specific Studies*

On the other hand, when Franklin Institute researchers published the report of a study, the approach was very different. Berkowitz and Doering (1954) described in considerable detail (not entirely matched by clarity) an experiment to examine the traffic-handling capabilities of three proposed procedural arrangements at the Washington, D.C., National Airport. More than five thousand simulated flights were included in fifty-four runs. Three random (and presumed equivalent) traffic samples were constructed from data obtained in a recent survey, "with the traffic arbitrarily increased by 15 per cent to account for increased aircraft operations with no jets included for a few years hence." (Predicting future traffic loads has always been a somewhat hazardous enterprise, especially for the Washington airports.) In this study the "samples" (the ATC term for "scenarios," "scripts," or "problems") were constructed from records of actual traffic rather than outright replicas. Some of the CAA and FAA studies have used replicas; in almost none have inputs been wholly synthetic.

The samples represented peak traffic periods and were initially three hours long. They were shortened to two and one-half hours to fit two runs into an eight-hour work day, and also because "It was too fatiguing for the simulator controllers and the aircraft-console operators." There were eighteen experimental conditions, each run three times. They were composed by combining factorially the three traffic samples, two traffic situations (arrivals only, and arrivals plus departures), and the three procedural configurations. A plan to double the size of the experiment by adding a 20-knot head wind condition was dropped because the wind feature in the simulation apparatus was not yet operable.

The controllers and pilots were rotated from run to run. Apparently, there was only one set of subjects. The report did not state how many subjects there were or how they were assembled. The procedural configurations differed "in layout of the inner feeding stacks and in their associated procedures," which depended on particular geographical and equipment features in the Washington area. Each configuration envisioned two-sector control.

The report contained tables and bar graphs of quantitative results for average delay per aircraft, number of aircraft delayed, and maximum delay. An analysis of variance was performed but not presented. The three procedural configurations were equally effective, or rather, ineffective, since the main finding was: "With prolonged peak traffic and present semi-arbitrary safety rules, use of the CAA uniform-radar-separation rule of three miles will result in an abnormally high percentage of probable wave-offs both on the glide-slope and on runway No. 36 at Washington National Airport unless some form of speed control is used."

One interesting feature of the study was a comparison between results obtained on the dynamic simulator and results obtained with graphical simulation. Differences were presumed to indicate the effects of workload on the controllers, "since in the graphical analysis perfect knowledge, perfect execution, and unlimited time to make the best decision in each instance dictate that the graphical delays are the lowest possible that are consistent with the particular rules, samples, and conditions being used."

The report recommended further simulation studies to investigate the effects of introducing jet aircraft, very slow aircraft, and helicopters; conditions of headwind and downwind approaches; airport surface control; airport shutdown; changes in acceptance rate with weather; changes in descent rules; and the use of intersecting runways for landing and takeoff. It also urged that the simulation be augmented and improved in various ways (many of which were subsequently accomplished).

During the Indianapolis studies the air traffic control experimental program described in Chapter 10 was going on at Ohio State University. The Ohio State program concerned Air Force aircraft, pilots, and procedures, the CAA program their civilian counterparts. In addition to some natural rivalry, a major difference in viewpoint about procedures and division of control impaired communication between the programs and to some extent made the research in one inapplicable to the other. As explained in Chapter 10, for the landing phase in terminal air control during low visibility conditions, the Air Force relied heavily on ground radar control of approach (GCA), in which the pilot was directed by a ground controller observing a radar scope. On the other hand, in civil aircraft the pilot retained more responsibility, guiding his aircraft by means of a cockpit display which was part of the Instrument Landing System (ILS). The divergence in method stemmed in part from requirements which the Air Force encountered in parts of the world where the ILS method seemed to be less applicable.

Accordingly, the CAA researchers were less interested than they might have been in an investigation based on the Air Force viewpoint. One important point of contact did develop. As indicated in Appendix II, one of the Ohio State experimenters (J. S. Kidd) was brought to the Indianapolis facility by a CAA contractor to give lectures in experimental design and control. These might have borne more fruit at that facility had it remained in existence; news of its impending demise arrived two-thirds of the way through Kidd's course.

During its nine years the Indianapolis facility used dynamic simulation to investigate air traffic control procedures and arrangements for Washington (four occasions), the New York metropolitan area, Chicago (two occasions), Fort Worth–Dallas, Norfolk, Baltimore, Indianapolis, a proposed Davidsonville Naval Air Station, Los Angeles, Detroit, Jacksonville, Seattle–Tacoma, Miami, and Denver–Colorado Springs. Undertaken at the request of the Office of Federal Airways, the New York Metropolitan Area Study compared present and rearranged navigational aids, with and without radar, for LaGuardia, Idlewild (subsequently Kennedy), Newark, and Teterboro airports (Anderson and Dowling 1954). One of the Chicago studies evaluated various runway configurations at O'Hare Airport and investigated the route structure required to handle the large volume of traffic predicted for the Chicago area (Armour et al. 1958). In this two-part study, interrupted by laboratory alterations and other investigations, seventy-seven tests contained 5,647 simulated flights, 2,793 of which were jets.

In addition, three studies examined proposed displays. One was the "sky-screen" display (see Chapter 5), another a three-dimensional display. The third study compared a proposed "panoramic" pictorial display with the standard flight progress board and with full radar control, as these were related to unsafe traffic separations (conflictions), delays, altitude changes, and communication loads (Vickers and Miller 1956). The study used five one-hour samples based on actual traffic in the Indianapolis air route and approach control sectors.

Other studies attacked general control problems independent of particular locations. For example, one was instigated by the Airways Modernization Board to explore "smoothing" at fixes fifty miles from the airport with aircraft being slowed down by either speed control or "path stretching," and control being either rough or fine. Another investigation studied how the placement of holding patterns, the feeding altitude, and an aircraft's rate of turning would affect the flow of jet and conventional aircraft arriving in a terminal area. A third dealt with a data processing central concept, with emphasis on the use of remote holding fixes. A fourth investigated procedures for coping with missed approaches, including the reintroduction of aircraft into the arrival sequence. A fifth studied methods of feeding two parallel runways, comparing three arrangements: one approach controller and one fix (position from which to make the approach); one approach controller and two fixes; and two and two. A sixth study (Astholz and Vickers 1958) developed procedures for controlling civil jet aircraft. Three studies sponsored by the Army explored a number of traffic control procedures, traffic patterns, and control displays for tactical airlift operations (Vickers 1954, Anderson and Vickers 1955) as well as airway structure, airport design and navigation and scheduling procedures in the control of a large number of logistic and support aircraft (both rotary-wing and fixed-wing) in conditions of virtual radio silence (Vickers 1957).

## NATIONAL AVIATION FACILITIES EXPERIMENTAL CENTER

The Federal Aviation Agency's National Aviation Facilities Experimental Center in Atlantic City carried on the Indianapolis program. The simulation apparatus transferred there was used at NAFEC until its retirement in 1962. Franklin Institute Laboratories had studied and stated the requirements for a replacement, the Universal Air Traffic Control Simulator (Grubmeyer 1956). In April 1960, a Model A simulator embodying many of these requirements and built by Aircraft Armaments Inc. was installed at NAFEC; fifteen months later a Model B simulator from the same firm began operation. These were generally employed in separate studies but could be linked to provide combined capacities.

The Model A and Model B simulators were electronic analog devices. This chapter will not attempt to cover real-time simulation at NAFEC by means of digital computers. Such simulation of the air environment has been closely related to projects for introducing digital computers into air traffic control operations. At the time the material for this chapter was assembled, the history of such developments was difficult to trace even though it was relatively recent.* In any case, although future simulation may rely heavily on digital computers, NAFEC studies were based essentially on the analog devices through 1966, when the Model A and Model B simulators were still in constant operation.

These simulators generated simulated radar echoes of aircraft in response to manual inputs and displayed them on cathode ray tube (CRT) scopes. The manual inputs came from the personnel acting as pilots, each sitting at a radar target generator with a panel by which the simulated aircraft could be made to move in specified directions and at specified speeds and altitudes. Each pseudopilot had a map display showing the land area around his simulated aircraft. The control panels and displays were easy to use, having benefited from human engineering in their design. Telephones representing voice radio connected the

*A new digital simulation facility has twelve displays, seven for controller consoles and five for pseudopilot consoles, each with data entry devices to communicate with an XDS Sigma 5 computer.

pilots and controllers. Model A simulated three radars covering 200 X 200 miles, Model B four radars covering 400 X 400 miles. Both had large arrays of CRT scopes in a simulated terminal approach control area and an ARTC control room, with associated flight data boards and communications equipment. Model A had 48 pilot positions, Model B, 60. Model B had beacon simulation. Neither simulated ground clutter and weather noise. In each a simulated aircraft could fly as fast as 2,500 knots at altitudes up to 80,000 feet and turn at rates up to 20 degrees per second. Four variations of wind could be inserted. Several outputs could be converted into digital form for processing by a digital computer and rapid printout of data: aircraft positions, communications activity, and settings of switches at pilot panels.

Responsible for most of the experiments with this apparatus has been either NAFEC's Experimentation Division or its Evaluation Division. Two experiments were done under contract for special objectives. One of these was part of a program by Aircraft Armaments Inc. (Kidd et al. 1963*a*) to familiarize the FAA personnel at NAFEC with experimental methodology for man-machine system experiments; one product was a guide excerpted extensively in Appendix II. The program had considerable impact. Partly due to it, and counsel which preceded it, the studies undertaken in 1961 and later began to look more and more like careful experiments, although it might be argued that further progress was needed. Support had also been provided by a human factors group at NAFEC available for technical help on demand. This group engaged in human engineering research and application, including experimental studies of displays, control panels, and airport features; it had no direct responsibility for the man-machine system experiments.

According to Slattery (1965), requests for experiments came from FAA's Air Traffic Service, FAA's System Research and Development Service, the Air Force, the Army, and foreign governments. A typical study required eight weeks of planning and preparation, six weeks of simulation, and twelve weeks of report preparation and review. The planning portion included trips, observations, acquisition of traffic samples, establishing equipment and communication requirements, development of experimental design (specification of variables, scheduling of runs and teams, determination of measures, selection of analysis methods), preparation of maps and handbooks, and set-up of communications and simulation apparatus. A planning outline included the following major subdivisions: problem definition and background, major objectives, assumptions and ground rules, personnel requirements, and operational plans. This last item simply meant the ground environment and procedures to be simulated, such as runway configurations, facility equipment configurations, communications, maps and charts, sectorization and areas of responsibility, special strip marking symbology, and sector fix posting arrangements. The traffic samples and operating procedures had to be specified as well.

During the simulation period, exploratory runs took four weeks and data runs two weeks. One-half of the reporting period was required for review and approval.

Slattery (1965) listed the following as areas of NAFEC experimentation: simultaneous dual approaches; combining of approach control facilities; control equipment arrangements; control positions; traffic-flow patterns; terminal radar service area; positive control area; final approach spacing; pictorial display usage; supersonic transport control procedures; airport site selection; Euro-control upper airspace jurisdiction; and movement of large numbers of Army helicopters under instrument conditions. Among the localities with particular air traffic control problems which were experimentally analyzed were San Francisco; Phoenix; Los Angeles Extended Area; Detroit–Chicago (en route); Palmdale (California) restricted area; Anchorage; Atlanta; Kansas City; San Diego; Chicago: Indianapolis–Chicago; Washington, D.C.; New York; Honolulu; Rome; Frankfurt; Berlin; Athens; and Western Europe.

One of the experimental programs has been called STARE, for single terminal and runway experimentation. It investigated a future semiautomated terminal control in which aircraft arrival times at a terminal area twenty-five miles in radius would be rigidly arranged. STARE simulated computer tracking, processing, and display. Another interesting study dealt with the arrangement of equipment in ARTC centers; in particular, the study compared the arrangement of controller consoles in in-line, "peninsula," and "island" configurations. Still another study, the "Hub Feeder" project, sought to develop procedures for expediting and simplifying short haul, intercity air traffic control capabilities to reduce the workloads of controllers, pilots, and support personnel.

### *Specific Location Studies*

A sample of five NAFEC studies will be described here to illustrate the program; it would be too space-consuming and repetitive to review them all. The five were directed at air traffic control problems in particular geographic locations.

*San Diego*. A 1959 investigation reported by Faison et al. (1960) studied instrument flight rule (IFR) operations in the San Diego area. Three possible arrangements for the Miramar Radar Air Traffic Control Center (RATCC) were compared with respect to sector boundaries, controller positions, and equipment distribution. Two airway systems were compared for their ability to handle traffic for a proposed Brown Civil Airport. Lindbergh Tower was compared with the Miramar RATCC as the location for Lindbergh–North Island radar approach control. A proposed route structure for the current San Diego complex was evaluated for capacity and control effectiveness, with separation standards and availability of additional airspace as constraining factors. Thus, the study incorporated numerous objectives, comparisons, and evaluations. The San Diego area was regarded as particularly complicated for air traffic control due to the terrain, proximity of the Mexican border, presence of many Navy jet training flights, and the almost daily incursion of low stratus clouds or fog bank from the sea.

Presimulation planning and study required about two thousand man-hours, during which more than fifty exploratory maps were drawn and five traffic samples were developed. These samples were based on flight progress strips for three busy days at Miramar RATCC and approach control strips at Lindbergh Tower, with percentage increases in traffic to allow for future growth. The simulation required seventy runs involving approximately eleven thousand flights over a five-week period. Data concerning acceptance rates, delays, and communication loads were obtained from pilot simulator logs and communication counters; they yielded quantitative results in most aspects of the study. In addition the participating controllers gave judgments in questionnaires, progress critiques, and opinions expressed at the end of the program.

In the Miramar RATCC portion of the study fourteen control positions were filled by controller subjects, ten within the RATCC and four representing ground and tower control. Some changes were made in the arrangements as the simulation progressed; in fact, one of the three control arrangements was developed as a consequence of simulating the others. An interesting measurement was an attempt to register the duration of direct verbal "coordinations" between controllers, as well as their frequency. Observers used stop watches but encountered "difficulty in discerning when the coordination actually began and ended" (Faison et al. 1960). Considerable reliance was placed on controller opinions as well as quantitative data. The study report stated that except for the Brown Civil Airport investigation, various portions of the study were embodied in experimental designs that permitted statistical analysis of the significance of differences in results; but no significance data were included and a statement concerning significance was attached to only one of the studies (Lindbergh Tower vs. Miramar RATCC). The simulation apparatus was the Indianapolis system.

*Washington, D.C.* Between October 9 and December 9, 1961, 140 simulation sessions of one hour and a quarter each were held at NAFEC with the Model B simulator in an experiment to "study, evaluate, and modify a proposed terminal area procedural plan" for the Washington, D.C. metropolitan area (Bottomley et al. 1962). The experiment was requested by FAA's Eastern Region through the Air Traffic Service (ATS). The area procedural plan encompassed Washington National, Dulles International, and Andrews Air Force Base airports, in an area 120 by 120 nautical miles. Projections of traffic density included the following for Dulles:

> Since Dulles Airport was under construction at the time of this evaluation, no actual traffic was available for study. Little information could be obtained regarding the types or volume of aircraft that would utilize the airport, other than that it was primarily a jet airport. At the exploratory meeting in Washington, between ATS and Aviation Research and Development Service (ARDS) personnel, it was agreed that a 60-per-hour aircraft movement, comprised of 75 per cent jet and 25 per cent conventional types, would approximate the anticipated activity.

For Washington National Airport, 36 arrivals and 30 departures per hour were projected; for Andrews AFB, 60 IFR aircraft movements per hour.

The experiment embraced six conditions deriving from two variables in a 2 X 3 design. An Andrews AFB climb corridor did or did not exist, and the experimenters simulated three alternative equipment configurations (and accompanying procedures). One was a common approach control or RAPCON-type IFR room at Washington National Airport, the others were "in-line" and "butterfly" arrangements of consoles and communications located at and integrated with the air route traffic control center (ARTCC) at Leesburg, Virginia. The first placed the burden of separating the arriving, en route, and departing aircraft on the ARTCC facility. The in-line arrangement required co-ordination between transition arrival controllers communicating by telephone rather than in physical proximity with each other. In the butterfly configuration the transition arrival controllers and their consoles were six feet from each other, in one case with another controller between them. None of the configurations was regarded as ideal.

As in many of the NAFEC experiments, the existence of an experimental design did not necessarily mean that the control which it implied was achieved. Bottomley et al. (1962) set forth some of the problems:

> It should be pointed out that, due to logistic considerations, one sample was used throughout the evaluation and the controllers became rather proficient by the end of the simulation. An attempt was made to compensate for this situation by changing the identities of the aircraft during the evaluation and a slight improvement was noted.
>
> For the purpose of statistical analysis, these six conditions were studied using a minimum of six runs for each condition. In this experiment, the six runs for each condition were derived by having three crews of controllers work twice under each condition. This resulted in a total of 36 one-hour test runs. The three crews or teams were not independent, nor were they different controllers each time. Instead, the same controllers were rotated through different positions of operation under each team set-up at each airport. This rotational arrangement was made in order to establish the three teams, as it was not possible to obtain any additional controllers to be used as independent subjects.
>
> Another compromise had to be made in the running of this experimental design. In order to eliminate, or at least minimize, the learning effect in the experiment, it is standard practice to run the experimental conditions in random order. However, in this particular case, the equipment configurations were too large and complicated to make quick changes between them. Since there was a limited time allotted for the dynamic simulation, the only way to proceed was to run each configuration in its entirety and then go on to the next configuration.
>
> This learning effect seems to be reflected in the data . . . . The runs under each configuration were performed in random order, and practice runs were made before the start of the data runs to acquaint the controller teams with the control procedures used with the different configurations.

It should be realized that the controller crews in this experiment were sizable. They varied between thirty-six and forty-four individuals with associated consoles and communications. In addition, eight to ten interfacing controllers were represented. Although the study report does not specify the number of pseudopilots required, it can be presumed on the basis of the traffic sample that the Model B apparatus was used to capacity, that is, the study employed the maximum of sixty pilot simulators. The simulator personnel completed a data sheet for each flight, recording departure time, arrival time over final approach fix, holding time at fixes, total flight time, number of altitude changes, and number of radar vectors. Communications activity data were obtained automatically. Controllers had opportunities to hold critiques; and controllers at the end of each run filled out a narrative questionnaire and a questionnaire with seven-point rating scales to indicate how difficult various operations seemed to them.

Data took quantitative form in fifteen measures in this experiment, and from some of these came eight submeasures or summary measures. The results were presented in an analysis which assessed the effects of airports and teams as well as the specified independent variables. The statistical significance of differences between equipment/controller configurations and between Andrews AFB climb corridor vs. no corridor was tested by the Wilcoxon Matched-Pairs Signed-Ranks Test and the Colin White Signed-Ranks Test, the .01 level being required as an indication of a significant difference. Virtually no differences were significant for the Andrews climb corridor variable and only a few for the configuration variable. In the latter case the amount of difference was small and the differences could be attributed to the order in which the configurations were tested. However, the controllers clearly rated the butterfly configuration first and the in-line one last.

The researchers also correlated controllers' questionnaire ratings against four objective measures of component performance and one of system performance. This last correlation was +.45. The others ranged from +.68 for average speed changes down to +.03 for average communications duration. The researchers attributed such a low correlation to the great "spread" of observations for this measure.

During three weeks of exploratory runs prior to the data runs, and again after these, small studies of three to sixteen runs each investigated such features as radar outages, traffic saturation at the start of a run, traffic handling in the area between a 40-mile and 70-mile radius of Washington National Airport, and the number of controllers and type of radar equipment required to handle visual flight rule (VFR) proficiency flights from Andrews AFB.

*New York City.* Another experiment involving the simulation of three airports in a complex air traffic situation investigated the control of helicopters in the New York City area (Sluka et al. 1962). As in the case of the Washington study just described and the Honolulu study to be reviewed next, the NAFEC FAA researchers were supported by the Franklin Institute Laboratories. This experiment was requested by the System Management Division of FAA's Systems Research and Development Service (SRDS), of which the Experimentation Division that did the study was also a part. (The Evaluation Division that conducted the Washington and Honolulu studies was similarly part of SRDS—earlier known as the Aviation Research and Development Service.)

This two-phase experiment looked at current airway structures and navigational aids and then future structures and aids as well as traffic to and from a proposed helipad at the New York World's Fair site. The phases ran in sequence early in 1962 and were not intended to be compared. Within each phase were three variables each having two states. Either the same controllers provided IFR separations to both helicopter aircraft and fixed-wing aircraft over common radio frequencies, or each type of aircraft had its own discrete radio frequencies and controllers. Helicopter aircraft were either phased into the same flight patterns and approach sequences as fixed-wing aircraft, or they had separate and nonconflicting patterns and approaches. Two sets of control standards differed in the parameters of airway width, obstruction clearance altitude, vertical separation, radar separation, and holding pattern buffer. In each phase the common frequency condition ran in its entirety prior to the discrete frequency condition. The order of conditions for this variable and for phases was not randomized—as were the other variables—due to "the complexity of the control procedures" (Sluka et al. 1962). Practice runs preceded the data runs. There were three traffic samples, each 65 minutes long but varying in load: 24, 34, and 50 helicopter operations per hour. These were factorially combined with other conditions to make forty-eight experimental conditions; since each occurred twice, there were ninety-six runs.

Only one crew of controllers was available. Its size differed slightly between the discrete and common frequency conditions; the maximum number was eighteen. Five represented the Newark approach control facility, six the Idlewild facility, four LaGuardia, and three more the control towers at Newark, Idlewild (Kennedy), and LaGuardia airports. Although the subjects all worked in the same experimental location for the study, the facilities they simulated were widely dispersed. The simulation (on the Model B apparatus) covered an area sixty miles square. Twelve target generators were modified to include the helicopter performance characteristics of vertical lift and slow forward motion. Approximately 154 fixed-wing aircraft were simulated in each sample.

In addition to communications measures, data included the number of radar vectors for both types of aircraft, the number of holding occasions for helicopters and the durations of helicopter holds, the number of airport arrivals and departures of each type, helicopter delays, and mean excess helicopter time in the system. Within each phase differences between the experimental conditions were subjected to analyses of variance to assess statistical significance at the .05 level. It was found that not only discrete radio frequencies and controllers but also segregation of helicopters from fixed-wing aircraft improved performance when traffic density increased. Controller opinions paralleled these findings and also showed a preference for segregation when the traffic load was light. The two sets of control standards had little differential effect.

*Honolulu.* Two studies, one concerned with Honolulu terminal area radar control, the other with alternative en route airway structures to the east and west of Honolulu, were conducted at NAFEC with the Indianapolis simulator in 1962. Intriguingly, at no point in the report of these studies (Cassell et al. 1962) were they called experiments, although the report's appendices described an "experimental design" for each, analyses of variance for one, "t" tests of significance for the other, and significance levels. Rather, they were called "tests." Perhaps the choice of nomenclature stemmed from authorship by personnel in the Evaluation Division of NAFEC rather than its Experimentation Division.

The terminal area study compared three arrangements of aircraft control along with three levels of radar assistance, 60%, 80%, and 100% of radar participation, in a 3 X 3 design. Two teams of nine controllers, each from the Honolulu area, took park in two runs of each of the nine experimental conditions. Thirty-six data sessions of 65 minutes each followed fifty-six practice and exploratory runs. The three arrangements arose from the weather peculiarities of Honolulu. Two arrangements were adapted to the prevailing northeast wind; the third was suitable for a southwest wind. The two prevailing wind arrangements differed in the method of achieving separation between the aircraft under control and unidentified radar targets; the final approach course passed over a Naval air station airport at which there was moderate to heavy activity. The en route study covered a distance of about 160 miles out to sea, eastward and westward, where there were no navigational aids. Four traffic samples were developed, each lasting 75 minutes. Two proposed route patterns were compared for the east area; in the west area the existing pattern was compared with a proposed one. Twenty-four data collection runs followed fifteen others.

*Studies of General Problems*

In addition to the studies of particular areas, NAFEC experiments have investigated general problems of air traffic control. For example, between June 13 and July 29, 1960, 5,500 simulated approaches were "flown" in 111 sessions with the Indianapolis simulator, in a study of required procedures for simultaneous approaches to parallel runways (Balachowski et al. 1960). Eight subjects made up four one-man teams and three two-man teams, each team participating in two data runs; twenty-four aircraft were simulated in each run.

*En Route Control.* In May of 1960 NAFEC's Test and Experimentation Division (1960—no individual authors listed) studied a plan of FAA's Bureau of Air Traffic Management "for the positive control of high-altitude air traffic on an area basis," meaning en route control in contrast to terminal (including approach) control. Some high altitude en route control had been initiated in 1958, together with a radar flight following and traffic advisory service for jets. The scene of the study was the area between Indianapolis and Chicago, covered by three long-range radars. The aims were to determine the capabilities of en route positive control, evaluate procedures for implementing it, and find out how it might affect the aircraft being controlled. Certain requirements and conditions in the FAA plan were designed into the simulation.

NAFEC used the Model A simulator for the first time in this study, following an extensive "slow time" (graphical) simulation to determine the adequacy of the traffic sample for the dynamic simulation and to estimate how much control was needed to provide safe separation between aircraft. A single 90-minute traffic sample with several variations was constructed to simulate an area 200 by 200 nautical miles, with Dayton, Ohio, at its center. It contained civil jet airways, positive control airways, active Strategic Air Command bases, a restricted area, active Air Guard bases, underlying approach control and RAPCON facilities, and radar and nonradar centers and sectors. Thus, it included problems of control and co-ordination within centers and between centers, within sectors and between sectors, with or without radar.

First there were eight simulation sessions to observe the current flight-following operations in action in the specified area, four with radar and four without. Then came six measurement runs (and apparently some practice runs) simulating the proposed en route control plan, with a 30% increase in traffic within the sample. Initially the en route control team consisted of three controllers, but during the simulation it was found necessary to add a fourth, "coordinator," position. A third phase, which included five measurement runs, added more SAC traffic to see what it would do to the plan's operations. Finally, two series of relatively short tests examined particular features of the positive en route control arrangement.

The quantitative data obtained indicated the number of holds of aircraft at approach fixes, reroutings, undesired altitudes, en route holds, and radar-derived directional commands (vectors) given to aircraft. The data came mostly from sheets filled out by the pilot simulators during each of their flights. At the end of each run, the controllers filled out questionnaires and discussed the run. Their comments helped determine the advantages and disadvantages of various control procedures.

*Civil Jet Aircraft.* Another study examined operational procedures for controlling civil jet aircraft in a transition terminal area; New York was chosen as a suitable locale (Eichenlaub, Conway, et al. 1961). In addition to investigating the current method of control, the study compared procedures which involved straight-in vs. base leg routing, 20,000 feet vs. 15,000 feet altitude, and single stack vs. dual stack. Two traffic samples differed by the percentage of turbojet aircraft among the fifty-seven aircraft in the sample (30% and 65%). Since each of the 18 experimental conditions had 6 runs, there were 108 sessions in a six-week period. Three largely independent and representative crews of nine controllers each participated twice in each experimental condition; a few of the controllers in noncritical positions were common to all the crews. The order of conditions was randomized. Practice runs preceded the data runs. Measures included arrival rate, number of "conflictions," number and duration of holding pattern delays, delays in vectoring, number of vectors, time in the system, interval between arrivals, and communication workload. Results were examined for statistical significance at the .01 and .05 confidence levels through analyses of variance.

*Aircraft Pictorial Displays.* A series of studies in 1960–61 investigated how well pilot simulators could use (simulated) pictorial displays in their (simulated) aircraft to keep their aircraft on prescribed routes and terminal paths, instead of relying exclusively on instructions from ground controllers (Faison and Sluka 1961; Sluka 1963). This technique was tried out in a high density terminal area (New York) in one study; in a medium density terminal area (Kansas City) in a second; in a low density terminal area (Salt Lake City) in a third; and in three en route areas (Miami, Jacksonville, and New York) in a fourth. The Model A simulation apparatus included, at the console of each pilot simulator, a map display that could serve as a simulated pictorial display (PD) in an aircraft. A spot of light moving across the console map indicated the aircraft's geographical position in relation to routes and paths marked on overlays, which also showed VOR-navigable routes. (VOR is a system of ground radio beacons.)

Considerable variety in route and path patterns, other aspects of routing, controller workload, and proportion of PD-equipped aircraft characterized these studies, which emphasized quantitative data. The experimenters drew conclusions favorable to the airborne display from the first two studies, as follows:

> The tests of this simulation study consistently demonstrated advantages to control derived from the use of PD techniques. The exploratory nature of the test

design precluded sufficient control of variables to prove or disprove strictly proportional relationships of specific percentages of PD to degrees of efficient traffic management. However, measurable improvements were established and favorable trends were shown.

This program had several interesting aspects. One was a human engineering finding about the labeling of routes and paths on the pictorial displays. Area arrival and departure paths close to each airport were letter-coded for the airport and also numeral-coded. Confusion was lessened if the numeral of a path that led into a route was the same numeral as the route's. It was also asserted that two-character and three-character labels of this sort helped radar departure controllers ascertain quickly the airport of departure, the route and heading on which the aircraft could be expected, and the anticipated point of handoff.

Although these Model A simulator studies were intended primarily to learn how the functioning of controllers would be affected by the aircraft pictorial displays, it occurred to the researchers to ask also how well pilots could navigate simply with such displays. A C-11 jet instrument trainer was equipped with an AVION RT-1 pictorial navigation display and also with an AN/ARN-21 course line computer. Four pilots made a total of eighty runs in a three-phase experiment to evaluate and compare these two devices. The researchers concluded that both devices could enable pilots to accomplish precision instrument flight. However, the pictorial display seemed not only more effective for quick computation and display of flight-path timing requirements but also more versatile in lending itself readily to both path-stretching and speed-control techniques. In contrast, the four pilots differed widely in their opinions about the relative advantages of the two devices.

In the Model A simulator studies of the pictorial display the pilot simulators carried much of the burden of navigation; and in all the NAFEC experiments these simulator personnel were required to represent, in their performance in operating switches at consoles and in communications, what actual pilots might be expected to do. But probably only a few had ever flown an airplane. Many were women. It was a role which citizens in the Atlantic City area were paid to assume on a part-time basis, and some were undoubtedly more gifted than others.

Yet in none of the many FAA reports scrutinized for this review was there any mention whatsoever of the characteristics of these quasi subjects. Perhaps the omission is not too surprising since the reports also omitted any characteristics of the controller subjects, except references to their working locations and the fact that they were always experienced controllers. Furthermore, although it was often stated they received practice in new aspects of control operations preceding the experiment, the reports gave little or no detailed information about such practice.

It is also unclear from the reports how the pilot simulators were distributed in an experiment among controllers, experimental conditions, and successive runs. Presumably the same pilot simulator personnel functioned throughout an experiment and were subject to practice effects. The problems of their selection and training did receive some attention in a study by Courtney and Company, a human factors contractor at NAFEC (Danaher, Eberhard, and Colman 1959).

The study report noted that these personnel not only simulated pilots but also collected much of the data behind the experimental results. It emphasized the obvious importance of recruiting, screening, and training.

*Controller Activities.* An experiment to determine the effects of certain sector characteristics on the activities of en route air traffic controllers was undertaken in 1962 by another human factors contractor, the Matrix Corporation (Davis, Danaher, et al. 1963). By "sector" was meant a particular geographical area. The eastern half of Sector 4 controlled by the Great Falls, Montana, ARTCC was simulated with the Indianapolis apparatus; some small modifications were made to make the area more typical. The adjoining sectors, 3 and 5, were simulated as ghost positions within the ARTCC. Four three-man controller teams served as subjects, each team encountering each experimental condition. These conditions were the variations in sector characteristics, which were four levels of traffic density, three levels of traffic mixture (30%, 50%, and 70% of the traffic overflew the area instead of arriving and departing), and the number of terminals–one or two. Each of the twenty-four 90-minute runs contained two 35-minute data periods, conditions being counterbalanced within the runs for each team to control for the effects of learning. The activities of the controllers were categorized into four communication groupings: communication with pilots, co-ordination among controllers, unsuccessful co-ordination, and routine relaying of information. A fifth activities category was "manual operations." An observer who was a qualified air controller stood behind each controller subject and recorded manual actions, while vocal activity was registered on tape. Data were tested for significance with eleven analyses of variance.

*Experimental Methodology.* Experimental methodology* was the object of an experiment with the Model A simulator apparatus by the contractor that built it, Aircraft Armaments Inc., between February 14 and March 29, 1963 (Kidd et al. 1963*a*,*b*). The experiment had several aims within the over-all purpose of methodological improvement in the complex man-machine system experiments NAFEC was conducting.

One objective was to find ways to reduce the error variance, that is, the random fluctuations which cannot be isolated in an analysis of variance test. Such a test determines the statistical significance of differences between effects of variables in experimental results. As Chapter 2 noted, the larger the random variance in experimental results, the less likely it is that differences between the means for values of a variable will be significant. This is because the random variance is the denominator in a ratio (F), the size of which determines whether the differences are statistically significant. Nonchance differences between means can be obscured by extensive random fluctuations.

It was conjectured that one way to reduce the error variance was to make the traffic sample more homogeneous. Then diversity of aircraft would not diversify controller performance in a random manner. Diversity of aircraft ap-

*Another experiment investigating methodology (system and individual performance criteria) was reported by Buckley, E. P., O'Connor, W. F., and Beebe, T., in 1969 in FAA-NAFEC Report NA-69-40 (RD-69-50), *A comparative analysis of individual and system performance indices for the Air Traffic Control System.*

proach speeds resulting from different kinds of aircraft became an independent variable in the experiment. In one sample all speeds were the same. In a second the variance among speeds resembled what might be encountered at the airport and terminal area which was simulated (with some modifications) throughout the experiment—Logan Airport in Boston. A third sample had both more fast and more slow planes. The fourth, with the greatest diversity, included some vertical take-off and supersonic aircraft. In all the samples sixty-five aircraft arrived and sixty-five departed during a period of 90 minutes.

Another source of error variance might be a load accumulation effect, occurring if aircraft early during a run were handled more expeditiously than later ones. Such a difference in handling might result from a relatively slow build-up of inputs. Coupled to the load accumulation effect could be a loss of data toward the end of the session, because aircraft movements initiated late during a run would not be completed before it terminated. In the experiment a movements input variable had two states: a constant flow of traffic into the system, expected to result in increasing loads on the controllers; and a high entry rate at the start, tailing off to a low one at the end, expected to yield a rapid build-up of load followed by a constant load.

Still another experimental objective was to investigate the effects of various ways of composing crews. Chapter 2 pointed out that in experiments with multiman teams it is difficult and costly to procure a multiplicity of independent teams, yet the more there are, the more generalizable is team performance to the total population of teams. Experimental conditions can be also more validly compared when data for each condition come from a number of crews. This experiment had eight crews, each consisting of nine individuals. Four positions were regarded as critical: two approach controllers, a co-ordinator, and a local controller. The other positions were a departure controller, three en route controllers, and a flight-data man performing clerical duties. (The composition of pilot simulators was not mentioned in the study report; they all received the same instruction sheets containing rules and procedures.) Apparently the five noncritical subjects remained the same throughout the experiment. The composition of the four critical positions constituted a two-state (independent vs. mixed) independent variable. In four crews all the individuals filling the four critical positions in any crew differed from those in every other crew. In the four other crews two individuals were different in each crew and two were the same, but the latter rotated among the positions.

To keep all factors other than the independent variables as constant as possible, all runs had the same geography, weather, procedures, and, as already noted, number of aircraft and duration. Eight experimental conditions resulted factorially from the 4 X 2 design for the first two variables. The sequence of conditions was nonrandom. Instead, the method of precluding sequential effects was an 8 X 8 Latin square in which the experimental conditions were the interior letters, the rows represented successive weeks, and the columns represented runs during the week. The sequence of the two kinds of crew followed an A B B A order along the rows, two crews of the same type per letter. Although the crews did not have practice runs, eight shakedown sessions for the benefit of the experimenters and simulator pilots revealed several problem areas in equipment capacity and sample design.

The experiment also sought to evaluate various measures of system performance. It was hoped that some could be found which discriminated between experimental conditions, whereas some would not; and it was suggested that the latter then could be omitted from such experimentation—thereby terminating the practice of obtaining a considerable number of different measures in each experiment as insurance against missing some effect. It was found that measures of departure and arrival delay most clearly indicated the differential effects of the input rate during the session and that a measure of excessive (greater than 30 minutes) delay frequency was particularly sensitive in this respect. In addition, these measures were considered meaningful in air traffic control operations. Frequency of conflicts or conflictions (where two aircraft were headed toward the same air space) was also both a sensitive and a meaningful measure but conflicts did not occur very often. On the other hand, sensitivity seemed to be lacking in such measures as arrivals landed, departures completed, total aircraft processed, missed approaches, communications activity, and average arrival interval. It was suggested that to ascertain these might be a waste of time and energy.

At the end of every run the "critical" controllers answered 12-item questionnaires (mostly multiple-choice and fill-in items) and took part in a debriefing critique. After comparing the subjects' opinions with the objective data, the researchers (Kidd et al. 1963*b*) commented:

The results show that controllers' judgments of system effectiveness are not congruent with the quantitative data. In the present case, their opinions were that the speed spread between aircraft in the traffic samples was more important to performance than the input schedule. Quantitative findings indicated the reverse. Also, controllers felt that they were more effective when a homogeneous traffic sample was employed and this was not the case. . . . These results indicate that care should be exercised when controller opinions are used to evaluate experimental conditions. These opinions will not necessarily conform to objective measures of system performance.

In this experiment the error variances were more important than the means, since the effects on such variances were what were being investigated. Contrary to the researchers' conjecture, the greater the heterogeneity of the traffic sample (aircraft speeds), the smaller was the error variance, that is, the more consistent was the controllers' performance. Their mean scores showed no such trend.

Because in the design the crews were an independent variable, it was possible to determine the variance between crews and thus eliminate it from the error variance. It may have been expected that greater variance between crews would have resulted from the independent crew condition than from the mixed condition, in which some members were common to the set of critical positions. But just the opposite occurred.

Did the teams' performance change during the experiment because they were learning their tasks? On some measures, performance remained fairly stable during each sequence of eight runs per team among the independent teams, but it improved among the mixed teams. It could have been presumed that the independent teams, composed entirely of persons without previous experience in the experiment, would have been the ones to show more learning. On other measures all teams displayed considerable stability. The researchers attributed this to the fact that the subjects were highly skilled professionals. They concluded that to reduce costs, "critical crew sub-units can be formed by a balanced rotational scheme which would insure practical independence between crews but require only 60–70% of the number of controllers needed for strict independence. To insure representativeness of controller crews a minimum of eight controller crews should be tested in all major simulation experiments. In most cases, it would be preferable to use ten or twelve crews."

The experiment had one further methodological goal, to compare the dynamic simulation on the Model A apparatus with graphical simulation in an effort to determine how well the results with one method matched those with the other. (In addition, graphical simulation preceding the principal simulation helped the experimenters establish a maximum load level.) Since graphical simulation (described in Chapter 24) was much less expensive, if it produced valid results it could be a useful pre-test technique. Some of the dynamic simulation runs were matched in a graphical simulation in which aircraft represented by pins were moved across a map of the Boston terminal area. The results of the two simulation methods were highly correlated, verifying the value of graphical simulation as a relatively simple predictive and design method that could save time and money. The researchers concluded that if programming costs were justified, similar simulation might be performed with a digital computer.

The experiment also produced some serendipitous information concerning the management of such undertakings. For example, equipment malfunctions would temporarily halt a run, thereby giving the controllers a chance to rehearse their tactics at leisure. In this manner, malfunctions contributed to error variance. During the experiment some changes were made in the distribution of crew workload and in an aspect of the traffic sample; presumably they should have been done before the data runs began. Finally, since there was no fixed procedure for doing so, the crews were not always equally briefed before starting a run; as a result, some lacked necessary information. It was concluded that there should always be a formal period for briefing.

# 16

# Operational Applications Laboratory

In a research program initiated within the Air Force at Bedford, Massachusetts in 1959, equipment developed earlier for a particular system was exploited for experimentation to produce general knowledge about the ways military commanders make decisions. The equipment came from an experimental Tactical Air Control System (AN/TSQ-13) (see Chapter 6). Three experiments and a pilot study were performed by the Operational Applications Laboratory in the Electronic Systems Division of Air Force Systems Command, in conjunction with the Detection Physics Laboratory of Air Force Cambridge Research Laboratories (AFCRL). (The laboratories had other names during their careers. The Detection Physics Laboratory was called the Astrosurveillance Science Laboratory when the program started, and the Operational Applications Laboratory eventually changed its name to the Decision Sciences Laboratory. AFCRL and laboratories in the Electronic Systems Division earlier constituted Air Force Cambridge Research Center.)

The TSQ-13 equipment was a first-generation version developed by the Laboratory for Electronics for air defense in a tactical environment. It had undergone field testing at Shaw Air Force Base in South Carolina (described in Chapter 6) and was moved to the Katahdin Hill Site at L. G. Hanscom Field, in Bedford, for additional testing. Eventually it was converted into a research tool. It included an automatic analog tracking subsystem named Cartrac together with a plan position indicator (PPI) display and an interceptor guidance (vectoring) subsystem named Airmap, as well as a data assignment panel, a digital (geographical or "situation") display, and other ancillary units. When its testing was completed and it was no longer related to specific operational use, it presented an unusual opportunity. Parts of it could be employed in experiments as though these parts constituted a real system, yet this system could be regarded as representative rather than as operationally specific, with the additional advantage that research data could be unclassified. As stated in a report by the contractor which altered and maintained the equipment for experimental purposes (Stavid Engineering 1959), this TSQ-13 equipment lost its special identity and became simply a data processing system.

In the program begun in late 1959, it operated on a simulation of the air environment. This consisted of hostile attacking aircraft and missiles shown on the digital display, and defending aircraft and missiles shown only on status boards, that is, their pretended maneuvers were not displayed geographically. The research was aimed entirely at the problem-solving or decision-making of one individual, a commander who observed the tracks of the attacking aircraft or missiles on the digital situation display and ordered the dispatch of his own aircraft and missiles in defense. The sites from which the latter could originate were shown on the same situation display that tracked the attackers, while their various identities and characteristics were listed on the status board. The commander, assisted by a technician (talker) and status board keepers, was required to evaluate the threat and select appropriate actions. These experiments were sometimes labeled TEAS studies (threat evaluation and action selection).

On the situation display appeared various types of hostile aircraft and missiles with differing capabilities for destruction and with tracks which might or might not be considered threatening to the defended area, proceeding at various speeds and altitudes. As listed on the status boards, the defending forces which the commander could select and dispatch also varied in their types and capabilities, which could match, overmatch, or undermatch those of an attacker or attackers against which the commander might direct them. How quickly after its detection the commander committed one of his vehicles against an attacker could determine whether it scored a kill. The commander could dispatch more than one of his vehicles against a single target. He could conserve his strength or spend it lavishly. In other words, the commander had many choices in managing his resources to solve the problems of the enemy attack.

## EXPERIMENTAL OPERATIONS

The pilot experiment (Doughty 1960) demonstrated the feasibility of the simulation and data-gathering techniques. The first experiment (Fox and Vance 1961), which presented nine thousand hostile tracks in 120 experimental sessions to nine experienced Air Force officers, was intended to establish baseline performance as a function of task load (number of attackers) and extent of threat in terms of the capabilities of attackers. In the second experiment (Connolly, Fox, and McGoldrick 1961; Connolly, McGoldrick, and Fox 1961), an additional variable was the effectiveness of detection and tracking by the defender's simulated surveillance system. In contrast to the experimental results, idealized or machine solutions were also obtained in these two studies for the same input situations. The third experiment explored how the commanders evaluated and altered management decisions, when these were made by others. No report on this study was published.

The principal experiments varied load by including 60, 72, 84, or 96 attacking tracks. The run would start with five to ten "unassigned-against" hostiles already tracked. Then new tracks would be introduced at rates ranging from 3.5 to 5 per minute, depending on the total number to be entered, and the run would last 35–45 minutes. The commanders could query the display console to get "tag" information on the display concerning the characteristics of the aircraft being tracked.

Different input tapes were constructed for two extent-of-threat conditions. The tapes were generated in units of twelve tracks each by using 15-J-lc electromechanical target generators (described in earlier chapters), the Cartrac equipment, and the digital coder in the formerly operational equipment. First, by operating manual controls, operators inserted the scripted tracks into the 15-J-lc's, which transmitted signals to the Cartrac equipment as though inputs were being entered for a run. Operators at consoles manually assigned tracking gates, tag numbers, and auxiliary information. This information and the x-y analog position voltages of each tracking gate were converted to digital form in the coder and passed to a digital communications unit, from which the data were recorded on magnetic tape. Considerable checking for error was required against errors which could accrue from misalignment of the target generators, perturbations in Cartrac tracking, and noise elsewhere in the process. Nevertheless, the method was a workable one with available equipment for creating reproducible and heavy input loads which could not be introduced directly by the limited-capacity, imprecise 15-J-lc devices.

The simulated site is schematized in Figure 21, and the information flow is diagrammed in Figure 22. As already indicated, the commander interrogated a new track on the situation display (new targets "blinked") to ascertain its characteristics and then selected from listings on a status board a counterweapon (or weapons), the site from which it would proceed, and its type of armament. Before the run started he received an intelligence briefing on the estimated threat and its confidence level. He communicated his selections (decision) to his technician for transmission. A scramble clerk passed an assignment slip containing the commander's order to an inventory clerk for aircraft or another clerk for missiles. They deducted the assigned vehicles from the inventory and communicated the changes to the two keepers of the commander's status board, where the changes were registered; then they passed the assignments along to two referees, who evaluated the action selection. If these judged that a kill would ensue, their evaluation was passed to a closeout technician who, at the time the track should disappear from the commander's display, caused it to do so and notified the commander to this effect. On the other hand, if a kill were not to ensue, the commander was notified at outcome time that none had occurred. In addition, there was a damage clerk, who had to pass information to the commander's status board keepers about any damage by attackers which might close down a defensive site or destroy defensive aircraft or missiles. He made such determinations by comparing the commander's assignments and referee evaluations with a running record of all preprogrammed potential damage which would be inflicted by any hostile attacker.

The two referees were busy individuals. They divided the incoming tracks between themselves. For each track they had two prepared evaluation sheets, one for defending aircraft and one for defending missiles. The sheets were so constructed that a referee could find on them any type of counterweapon from any site and any number of weapons, with any of the available armaments, and with any of a set of assignment times related to the life of the hostile track. Kill probabilities and times were indicated on the sheets, taking into consideration

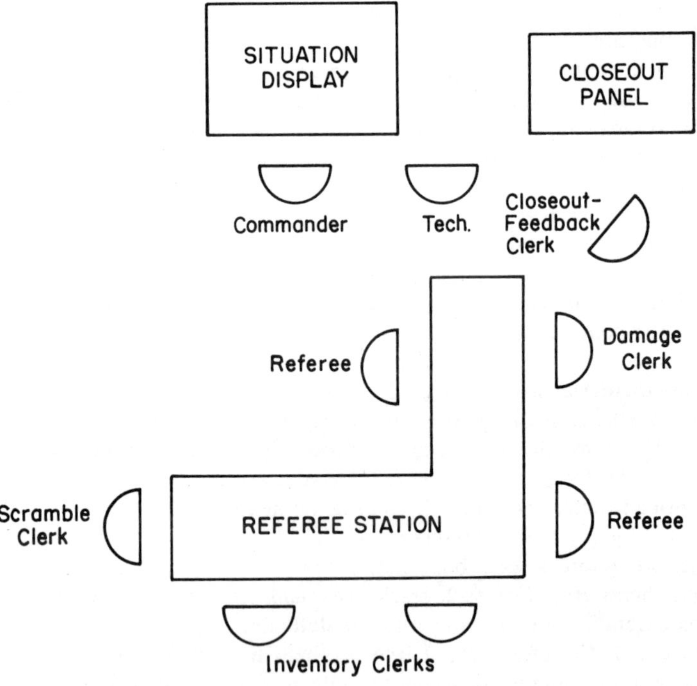


Fig. 21. Schematic Site Layout (Connolly, Fox, and McGoldrick 1961).

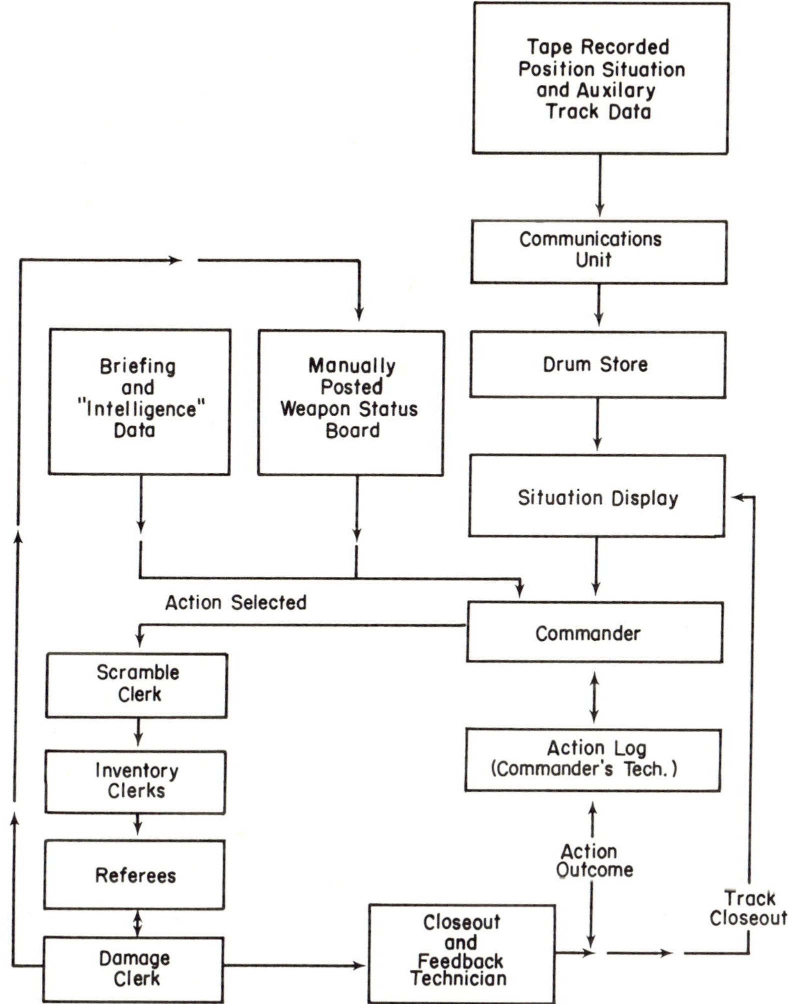


Fig. 22. Information Flow Diagram (Connolly, Fox, and McGoldrick 1961).

the time of assignment, site, type of weapon, number of weapons, and armament.

## THE THREE EXPERIMENTS

In the first main experiment nine Air Force captains and first lieutenants acted as commanders. All had had air defense experience, but their training before the experimental runs differed at two levels. At its maximum this training consisted of seven runs through a 96-track problem with defense weapons at two sites, followed by practice with problems of 24 to 72 tracks and eight weapon sites. Although the experimental design called for 168 sessions, only 120 were completed, distributed among the four levels of track load, two levels of threat complexity, two weapons siting configurations (eight and five sites), and the two amounts of training. The threat level and sequence of site configurations varied systematically within track loads, whereas track loads were increased progressively. This arrangement, according to Fox and Vance (1961), "was intended to permit both the subjects and the referees to function in a reasonable manner and to permit completion of the data collection trials in a minimal time. Time was a concern due to the difficulties of scheduling and retaining military subjects who had other duties to perform, because of cost factors and other problems associated with the real time simulation of complex problems . . . . To investigate fully all combinations of experimental variables in random order would have been prohibitive, both in terms of time and cost."

Postmission debriefings were conducted to determine whether the subject "was aware of his success or failure in coping with each mission, whether or not he could evaluate the adequacy of the advance-intelligence briefing, and his evaluation of the adequacy of his weapon inventory."

*Second Experiment*

Five of the more experienced subjects in the first experiment became the commander subjects in the second. Again there were two levels of threat complexity and four of track load. Data quality, the new variable, was varied in approximately the same fashion within each combination of these. According to Connolly, Fox, and McGoldrick (1961), "Each commander (except one) faced various combinations of 60 track loads from each tape four times. Each commander faced two different combinations of 72 and 84 track loads from each tape. Since there was only one combination of 96 tracks in each tape, this load was not replicated." There were eighty-six runs, and "Attempts were made to control learning effects by counter-balancing and/or mixing of the order of presentation of conditions within each subject." The experimenters stated that "the experiment was not set up as a statistical design," because time and cost would have been prohibitive, and also because "the interest was in the practical or indicative types of results rather than the inferential statistical type." No significance statistics were derived in this experimental program.

The authors stated further:

> While the main design was intended to be fully counterbalanced with respect to load and threat performance level, it was not always possible to adhere to this plan for logistic reasons. The experimenters were forced by circumstances to make use of the various commanders when their time could be made available rather than vice versa. Added to the factor of personnel availability, the vagaries of equipment malfunction literally forced a certain amount of catch-as-catch-can scheduling. Experience with and foreknowledge of the likelihood of such difficulties . . . were further, though not major, reasons why no attempt was made to fabricate or execute one of the more complex classical experimental designs.

The additional variable of this experiment, data degradation, was applied in ten steps of 10% each in two ways: tracks disappeared or stood still for various proportions of their track lives; and characteristics information, e.g., speed or altitude, was erroneous or missing. The first category of degradation, accounting for three-quarters of the total, might represent the fades which occur in radar surveillance or the concealments which may develop due to electronic countermeasures; the intelligence briefing said that such countermeasures might be expected. Thus, this experiment simulated more realistic conditions, whereas surveillance quality was perfect in the first experiment. It should be noted, however, that in this experiment temporary track disappearances or stoppages did not degrade the subsequent performance of the tracking system. When the track reappeared or moved again, the indicator for the hostile aircraft or missile appeared where the simulated aircraft or missile actually was moving in geographical space. Accordingly, unlike what might occur in such surveillance situations in reality, the track position data presented to the commanders was never faulty, merely unavailable.

As noted earlier, accompanying both the first and second experiments were analytic solutions to the problems which the subjects had faced. There were three types, according to Fox and Vance (1961): "(1) idealized 'human' solutions based on two rates of assigning actions, (2) idealized 'machine' solutions where delay in assigning action was zero, and (3) random 'machine' solutions where action delay was zero . . . . but action selections were drawn at random. These three solutions were derived from two levels of load (60 and 96 tracks), for threat level two, and for the five-site configuration."

### *Third Experiment*

In the third experiment the commander encountered action decisions which were already made—the weapon or weapons were already assigned against a particular hostile track. However, he could do any of several things to change this decision. He could countermand the action; reduce the magnitude of the counteraction by reducing the number of defending weapons or their mode of deployment; increase the magnitude of the counteraction by augmenting the number of his defending weapons or altering their types; or otherwise change the action. The prior decisions which he had the option of changing had one of several possible origins, each being a variable in the experiment. One source was an optimum strategy which would provide the maximum kill at minimum cost to defending weapons in the shortest time possible. Another was a poor strategy in all these respects. A third consisted simply of replays of the decisions which commanders had made in similar situations in the prior experiments. Since the subjects were still the same, in this third experiment the commander sometimes faced a decision which he himself had made. (There was no indication this was ever recognized by a subject.)

A variety of measurements marked these experiments, such as weapon assignment rate as a function of load, number of sites, and threat complexity level; weapons assignment against hostile aircraft and missiles which the commander regarded as threatening but were not, compared with assignment against those that were actually threatening (according to their paths); weapon selection delay times; intercept times; damage inflicted by the hostile tracks not processed; inappropriate application of defensive weapons, such as overmatching, under-matching, and applications resulting in out-of-range intercepts or tail chases; kills; percentage of weapons inventory assigned, etc.

## RESULTS AND THEIR IMPLICATIONS

According to Fox and Vance (1961), in the first experiment the commanders were able to prevent substantial damage from hostile weapons and maintain a capable posture with defensive weapons by devising successful strategies, but they were unable to verbalize these strategies very well. Track load did not degrade performance as expected; the more experienced commanders showed little evidence of leveling off with the higher loads. In fact, there appeared to be "a definite pacing effect," so that performance rate increased with load. The subjects were effective in dynamic sorting of categories in terms of threat, counterweapons, locations, etc., and in "ongoing weighting of alternatives selected in terms of the overall outcomes desired." It was clear that selected, highly trained individuals could make complex judgments appropriately under heavy loads; experience at high-load levels seemed to be the "sine qua non of maximum performance."

In the second experiment, while there was "a noticeable trend of deterioration in performance at the 84 and 96 track levels there was . . . no clear breakdown point," any more than in the first experiment, for the fully trained subjects (Connolly, Fox, and McGoldrick 1961). The limiting capacity for making the kinds of decisions required in this experiment seemed to be five to six and occasionally seven per minute, but even when the requirement exceeded this capacity the quality did not deteriorate seriously. The degradation of surveillance data in this study had no clearcut effect on the management of decision-making required. The subjects indicated they did not realize the extent of the degradation. Had there been indicators of data quality their behavior might have been different.

In spite of the subjects' level of performance in weapon assignment, the experimenters concluded that "the threat evaluation activities of the commanders . . . were of a rudimentary sort." They assigned defensive weapons against "real" threats and "apparent" threats "almost indiscriminately," because they failed to use all the cues in the display, relying simply on the category of the hostile weapon. "They were too pressed by the overall task to try to make fine, individual evaluations or discriminations." The commanders matched defensive weapons against offensive weapons fairly well, although mismatches tended to increase with heavier loads. But on many occasions they were guilty of assigning defenders against attackers when the latter were destined to be destroyed by defensive weapons already assigned. This occurred most flagrantly with the lighter input loads, so it "may be inferred that, when a substantial superfluity of weapons appears to exist, the human decision maker tends to let considerations of damage prevention and destruction of enemy forces override considerations of economy of weapons. . . . There was a definite tendency on the part of commanders to 'use up' available weapons and even squander them somewhat when the supply appeared to be much greater than the 'demand',"

and although this may be a perfectly natural tendency in such situations, it may not be good strategy.

An inclination to be prodigal with defending weapons also characterized the commanders in the third experiment. The subjects changed approximately 50% of the previously made decisions, often those they had made themselves. All tended to augment the commitment of defending weapons which had been selected either according to optimized strategies or from the precedent of the commanders themselves. The commanders achieved the same high decision-making rates as in the first two experiments. Unfortunately, the half-million items of data gathered in this third experiment received only limited processing, apparently because the person responsible for such processing left the project and the research program came to an end.

The experiments surely suggest some design considerations for command and control systems in which commanders solve, or try to solve, problems on the basis of surveillance data, an inventory of defending weapons, and some associations between these. Some implications have been advanced by Connolly, Fox, and McGoldrick (1961):

> At least some gross indications of the reliability of surveillance data should be provided for commanders. . . . Decision criteria for man-machine decision makers should be developed to include economic and logistic considerations even at the direct action or battle level. . . . Large amounts of intensive experience in reasonably realistic conditions and under high task loads should be provided for commanders at all levels. Such experience not only promotes maximum performance but may prevent overload from producing drastic effects.

## OBJECTS AND OBSTACLES

In the light of the contents and results of these experiments and the conclusions and recommendations which have been published, it is interesting to go back to some of the statements about their objectives. For example, according to Connolly, McGoldrick, and Fox (1961), "Obviously the main interest here is not to find out how men compare to automated systems, but to find out how to make the most sensible and productive use of the unique capabilities of men in automated and semi-automated systems." Along the same line, Connolly, Fox, and McGoldrick (1961) stated that "it has been the objective of these experiments to investigate man-machine capabilities in surveillance decision-making and, even more particularly, human performance of complex command and control types of activities. The value of knowledge of human capabilities and limitations in this type of behavior, if it can be obtained, and of the general and specific functions whereby such tasks are accomplished should be twofold. At the very least it should cast light on reasonable (if not optimum) means of aiding such human capabilities or of using them to best advantage in systems of the future. In addition, even general fundamental insights into human performance in this area are desirable if not essential to any future 'automation' of all or part of the decision-making task."

On the other hand, W. H. Vance, Jr., the chief of the Detection Physics Laboratory, suggested an even broader scope of objectives for the research (Fox and Vance 1961), namely to derive information about:

1) "The basic parameters of the decision process and how they are related": whether complex decisions can be reduced to definable parameters; methods of reducing complexity of decision-making alternatives; effects of type and quality of input data on decision processes.

2) "The critical aspects of man-machine interrelationships in the evaluation-decision process": effective criteria for optimum trade-offs between men and machines; essential parameters for man to serve effectively as a monitor and as a commander; effects of overloading; criteria for altering decision strategies; functional specifications for displays and controls.

3) "Significant parameters for human performance in evaluation and decision-making": procedures utilized by man in such processes as recognition, learning, reasoning, inference; methods of handling uncertainty, missing data, errors, delayed data; limitations imposed by human characteristics; identification of personality or other factors which could predict good or poor performance as a decision-maker; methods for testing mathematical models or automatic devices vs. human performance in decision-making; effects of overload, stress, high risk, etc., on human decision-making.

This program started, it would appear, with ambitions which were grand, if not grandiose. The foregoing outline gives an idea of the scope of a comprehensive investigation of military decision-making. There had been plans to go on to a fourth experiment in which there were to be three commanders interacting with each other: a principal and two subordinates. In addition, the Air Force Cambridge Research Laboratories, during 1961, had sponsored a study by the Planning Research Corporation of the requirement of a TEAS simulation research facility (Dodson et al. 1961). It was estimated that such a facility, to support the kind of system experiments the Operational Applications Laboratory had been conducting, but on a wider scale, would cost $250,000 to build and equip, not including any computer and associated equipment or personnel costs (Blanchard 1961). It was also estimated that fifteen laboratory personnel would be needed, to match fifteen experimental subjects.

But the program was phased out. One of the considerations was cost. The program had been costing about $250,000 per year for personnel, including $100,000 to a contractor to maintain and modify the equipment and $75,000 in contracted personnel—simulation operators, data-gatherers, evaluators. It was believed that personnel costs could be reduced greatly, perhaps by 50%, if the experimentation turned to a digital computer, but then there would be the computer and programming costs. It was felt that the facility and apparatus used for the three experiments would be inadequate for more complex experiments, such as the planned multioperator study. Also, the equipment was not reliable enough to produce correct inputs; a considerable percentage did not reflect precisely what was intended (although the subjects said they thought the data were realistic). And the data processing capability would be inadequate for larger-scale experiments; even the second experiment entailed 600,000 punched entries on cards.

Finally, uncertainty arose concerning the research potential of the available equipment if multioperator experiments were undertaken where there would be serial processing. In this, the inputs to one subject or subsystem depend on the outputs of a subject or subsystem earlier in the processing sequence. In such a situation one cannot control experimentally the inputs later in the sequence. Such loss of control over variables can make it difficult or impossible to learn with assurance what brings about differences in final measures of system performance. This is not a trivial problem in the technology of system experimentation.

# 17

# System Development Corporation Laboratory Experiments

In addition to its field experiments, described in Chapter 11, System Development Corporation (SDC) researchers conducted man-machine system experimentation in six laboratories between 1958 and 1966. Five were in Santa Monica: the Second IDC (Indoctrination Center), the Systems Laboratory (later called the Human Factors Laboratory), the Systems Simulation Research Laboratory, the Command Research Laboratory, and the Emergency Operations Research Center. The First IDC was operated by the RAND Corporation during the early days of the system training program. Its research, mostly on feedback and debriefing in system training, will not be described here because apparently no reports were published for circulation outside RAND and none could be identified; in any case the work seems to have been largely of an informal, exploratory nature. The laboratory not in Santa Monica was the Simulation Facility (SimFac), in Paramus, New Jersey. The Emergency Operations Research Center and its research are described in Chapter 22.

## SECOND IDC–AIR DEFENSE LABORATORY

The Second IDC was a location in which air defense personnel were indoctrinated in conducting system training in the pre-SAGE air defense system, much as they had been in the First IDC. The Second IDC was physically situated in an SDC building which also housed a SAGE AN/FSQ-7 computer used primarily for programming, since at this time the installation of SAGE was beginning; this computer was not otherwise associated with the IDC. The Second IDC, sometimes called the Air Defense Laboratory, was the scene of the WSEG–SDC ECM laboratory experiments.

### *WSEG–SDC ECM Experiments*

In 1959 a very large, four-experiment study was performed by R. H. Davis, R. A. Behan, and E. R. Pelta of the System Development Corporation for the Weapons Systems Evaluation Group associated with the Department of Defense. The objective was to determine, through simulation, some of the effects of electronic countermeasures on the Air Force's manual pre-SAGE air defense network of radar sites and interceptor bases in the United States. Although security constraints preclude describing the details of this study or its results, enough information is now unclassified to give an over-all picture.

Possibly the most interesting methodological aspects of this study were its four-part nature and (even for this kind of experimentation) its considerable scale. One experiment examined the operations, in a laboratory, of sixteen air defense crews manning two simulated sites in 192 40-minute runs. Another encompassed eight operational air defense sites where, in four runs, simulation inputs were generated by the same equipment that had been used in the laboratory. The other two experiments dealt with individual performance; in one instance eighty Air Force subjects took part in 640 20-minute runs, while in the other, engineering-oriented data were obtained from six SDC subjects.

The strategy behind this large study was to exploit the laboratory to present to crews a variety of conditions which could not have been covered in the field except at enormous expense and with degradation in experimental control; to study similar crews actually in the field, coping with the same kind of situation but without as much variation in conditions; to measure individual performance in the laboratory in a manner that could not be accomplished in a multioperator crew context; and to investigate with laboratory control the effects of some operational equipment which had been used only in the field experiment.

The geographical area which was simulated in the crew experiment in the laboratory was part of the setting for the field experiment. Similar simulation devices were used in these two experiments, although the tactical inputs differed and some of the operational equipment used in the field was absent from the laboratory. These variations permitted only approximate comparisons to be made between laboratory and field results; indeed, only general comparisons were ever intended or feasible. The data from the experiments on individual subjects could be related to the crew experiments also in a general way, despite differences in simulation inputs and devices. Still other comparisons, again approximate, could be made between the outcome of the field experiment and some Air Force tests with live aircraft, which were a major feature of a program of which the SDC experiments were a part.

For surveillance operations, aircraft radar signals and various types and intensities of electronic countermeasures were simulated by means of the AN/GPS-T2 (see Chapter 11) and its OA-1767 accessory, commonly known as ACTER (anti-countermeasures trainer). As indicated earlier, this equipment converted markings on 70-mm. film into radar and ECM signals which were fed into actual radar receivers. Each film stored a multiaircraft hostile air attack prepared according to a systematic design. This was the same simulation apparatus—the only apparatus available for experimentation—that was used in the System Development Corporation's System Training Program, in which it helped furnish the considerable amount of training and experimental data about that training described in Chapter 11. However, it could not present target and ECM parameters with the fidelity desirable in a test of the effectiveness of a man-machine system. In the laboratory experiments, other features were added in an attempt to increase realism, but an important question may still be raised. Under what circumstances—if any—do training simulators qualify for system testing?

In both team experiments, in the laboratory and in the field, the interception function of the air defense crews was carried out with the AN/GPA-23 analog tracking and computing equipment which figured in Chapter 7. Interceptor aircraft were simulated in the field experiment with the 15-J-lc devices described in earlier chapters and in the laboratory experiment with a newer simulator; in each case the equipment, like the AN/GPS-T2 and ACTER, was designed for training, not for testing.

In the laboratory team experiment, the two sites represented were associated with each other geographically and operated at the same time. The subjects, officers and enlisted personnel drawn from operating sites throughout the United States, were rotated between the two sites and through appropriate positions within each. In this fashion it was possible to create sixteen different crews from four groups of subjects, each group participating in the experiment for two weeks. In addition to the sixteen different crews, the independent variables included four aspects and four intensities of electronic countermeasures, and three tactical and three presentation parameters of aircraft signals. Four crews encountered each aspect of ECM, the ECM intensities varying randomly within each aspect; the aircraft signal parameters were assigned randomly to each intensity in all combinations of the two parameters in equal numbers. Analyses of variance could be performed to test the statistical significance of results attributable to differences within all variables. Various measures of crew performance were obtained for detection, tracking, interceptor assignment and direction, and communication between the two sites.

In the field experiment, some requirements were easier to satisfy than in the laboratory, some more difficult. Communications, consoles, and radar receivers were operational, so they did not have to be constituted for the purpose, as had been the case in the laboratory. But the simulated, co-ordinated air picture, with its attacking aircraft and electronic countermeasures, had to be created for, and generated at, eight sites linked together in joint runs instead of at merely two sites, a much more difficult problem in co-ordination and equipment maintenance. Further, in each run there were eight crews, one per site, whose adherence to rules of the experiment had to be enforced (if possible) and whose performance had to be measured.

Two crews were selected at each site. They were supposed to be approximately equal in performance and remained distinct from each other, retaining the same composition; the experimenters monitored this requirement. Four runs lasted approximately two and one-half hours each. Two different simulation input problems, equivalent in load difficulty, had been designed for the eight sites. Each crew experienced each problem and thus had two runs. One run included electronic countermeasures, the other did not. The order of these two conditions, which constituted the independent variable in the experiment, differed for the two crews at each site. (With this design, and with the admonition to commanders to provide two equivalent crews which remained distinct, it was hoped to avoid the problem of the "tiger crew"—a selection of superior performers. Commanders sometimes select such crews in official and important tests, when it seems advantageous to demonstrate a high level of unit performance.) One run occurred per day, on successive days. In addition to objective performance data, the experimenters obtained information from debriefings and questionnaires after the runs.

The two experiments involving individual subjects had no particular methodological interest in themselves, since their design and conduct were reasonably straightforward. There was one unusual aspect, however, about the larger one. The eighty subjects were assigned to the System Development Corporation's Air Defense Laboratory in groups of twenty, for four weeks each. During the last two weeks they were subjects in the laboratory team experiment. During the first two weeks they were subjects in the individual performance study. However, since in the latter only a few subjects could undergo the three-hour-and-twenty-five-minute runs at the same time, there was considerable free time. This the subjects spent profitably with other simulation apparatus, being trained to operate the AN/GPA-23 interception control equipment. Such occasions, where a simulation facility has been exploited for both experimental research and training, have been relatively rare.

## SYSTEMS LABORATORY–HUMAN FACTORS LABORATORY

A number of experiments took place in this laboratory, which was created in 1958 and in 1959 changed names, moved to another location, and saw its capabilities expanded. Even more experiments were projected. For example, one which never came to pass would have examined both debriefing and feedback in a variety of settings; it will be recalled from Chapters 8 and 11 that debriefing and feedback were cornerstones of SDC's system training program. It was proposed to examine as independent variables both the time of debriefing (immediate—that is, within 15 minutes; delayed to the next shift; and delayed to just prior to the next exercise) and the content of feedback (box score or general summary; a problem-oriented approach, seeking the source of error; and a component-oriented approach, providing information about individual performance). An experiment primarily aimed at another objective did include these variables partially and in a confounded manner; it is described further on. More extensive and better controlled research would have been an appropriate follow-on to the RAND experiments and SDC's field experiments.

Initially, the Systems Laboratory-Human Factors Laboratory was a relatively modest facility, consisting in its most elaborate form of a 26 by 14 foot area which could be compartmented into two or more smaller areas, with small platforms on two sides for observers, experimenters, and simulators. Equipment consisted of several plastic, manually inscribed wall-type displays, as well as clocks, desks, a number of telephones connecting subjects to each other and to the simulators, and equipment to record the telephone conversations. No computer was involved.

When the laboratory moved and was expanded, it acquired a total area (air-conditioned) of 57 by 28 feet in which the main, compartmentable experimental area was 24 by 14 feet. Surrounding three sides of this area and adjoining the briefing room was an observation deck. A substantial array of equipment was procured (Grant, O'Connell, and Stoker 1960). The sound system included microphone and speaker jacks at all subject and experimenter positions, microphones and speakers, white noise and tone generators, and a patching bay. The telephone system had fourteen stations for experimental operations and twelve for maintenance, plus a central patching bay. Six tape recorders and some disk recorders could register microphone and telephone inputs. A closed-circuit television system consisted of five cameras in the experimental area as well as nine monitors, each with a camera-selection panel, in the experimental area or on the observation deck. The monitors could serve as system displays for the subjects. There were several ways to generate stimulus material which the cameras registered and the monitors displayed. These were: (1) two "data generation tables," with a 35-mm. filmstrip projector modified for fine projection, a small screen, and a TV camera; (2) three "current marking tables," at which subjects could mark large Plexiglas slides which were alternately slid under a TV camera and photographed; and (3) a vu-graph projector with a vertical screen which a TV camera photographed. Additional equipment included 35 mm. cameras; push-button (5 × 6 matrix) panels with associated lamp panels; an eight-channel regular interval timer; a variable interval timer; and an operations recorder. Again, there was no computer. The laboratory was discontinued in 1961 when its functions were taken over by the Systems Simulation Research Laboratory (SSRL), which was computer-based. The closed-circuit TV system was adopted by SSRL, and the sound, recording, and telephone equipment were absorbed into those developed for the new laboratory.

### *Experiment on Interaction between Problem Load and Level of Training*

In this 1958 training experiment, described by Behan et al. (1959, 1961), each of twelve five-man crews had eight 40-minute training sessions on successive days, a 12-minute warm-up session before each training session, and daily 40-minute pre-test and post-test sessions. The task was an analogy of the surveillance function in a manual system Air Defense Direction Center. An operator viewed analogues of tracks on a display and reported them by telephone to a second operator. This operator plotted them on a second display. A third operator examined this display and reported the tracks to a fourth operator, who plotted them on a third display. A fifth operator viewed that display and compared its elements with those on cards which had been prepared previously. He announced matches and discrepancies. The displays were the vertical type, with plotting by china-marking (grease) pencils.

Dots were projected from prepared material on to the initial display, which bore a rectangular grid with identifying symbols for the cells in the grid. The first operator reported these identifiers to the second whenever three dots lined up in a sequence along any of the eight major points of the compass. The second operator plotted the reported sequence. (Noise was introduced in the form of dots which did not fit into a sequence.) The second display was marked by polar co-ordinates, as well as the grid co-ordinates, and the third operator read the dot positions to the fourth in polar co-ordinate terms. The display on which the fourth operator plotted the dots was marked only by polar co-ordinates.

Additional dots in a sequence had to be reported, as well as sequence direction and appearance time. The number of dot sequences to be relayed and plotted during a training session was 10 (lightest load), 20, 40, or 60 (heaviest load); intervals between appearances of successive sequences in these loads were 4 minutes, 2 minutes, 1 minute, and 40 seconds, respectively. In the experimental design, pairs of sessions had the same load, and the order of load was different for each crew. After each session a crew was informed of its performance and held an unstructured discussion. Performance was measured in total errors and in number of errors in each of eleven error types. Total errors were an approximately linear function of load. However, different kinds of errors yielded different functions. For example, errors in estimating position did not differ greatly among the three load levels. Missed sequences were rare among the three lower load levels, then rose dramatically with the heaviest load. The effects of practice were also interesting. With experience, total errors decreased significantly but not extensively, since the reduction in total errors was attributable only to the first operator in a set, coping with the two higher load problems. The second operator's errors actually increased with experience when he handled the higher load problems. Increasing experience neither improved nor degraded the subsequent operators.

The researchers concluded that the errors which could occur at the later positions depended on the performance at the start of the series. They believed that if practice had continued, the errors of the later operators would at first have increased, while those of the earlier operators would have decreased. Experimental difficulties posed when a system processes data serially in this fashion have been discussed in Chapters 11 and 16. The researchers further concluded that a criterion of a well-organized system might be the finding that "the interaction between system position and experience at a given load level no longer contributes significantly to variance. If we, in addition, specify the level of load at which this interaction term is not significant, we have an objective criterion of the level of training of a system."

### *Two Crew Turnover Experiments*

These 1958 experiments, reported by Rogers, Ford, and Tassone (1959, 1960, 1961), featured essentially the same tasks and equipment as the load-training experiment. However, crews expanded to seven persons (college students), with a supervisor and a second man at the front end of the serial processing. Now one man (the reader) identified the projected three-dot sequence and affixed a direction (heading), track number, and time alongside the dot sequence (track); a second man (teller) read this information and the position information to the next operator in the chain.

The second experiment was largely a replication of the first. The major independent variable in each was turnover within crews—the departure of members and their replacement. Another variable was the experience level of the crews in which turnover occurred: they were inexperienced or they had acquired experience only as subjects in the experiment. Still another variable was the level of experience of the replacements. Finally, the experiment compared (a) giving replacements pre-exercise component training, post-run debriefing, and knowledge of the results about system performance, with (b) abstaining from pre-exercise component training but giving them pre-run debriefing and knowledge of results about individual performance. This was the experiment mentioned earlier that confounded debriefing and feedback variables.

In the first experiment, four crews were formed from twenty-one subjects. Two inexperienced crews had eight exercises each. One was stable, the other had turnover. Then the stable crew became an experienced crew encountering turnover in seven exercises. It drew its replacements from those leaving the inexperienced turnover crew. Thereupon it became a stable crew with no further turnover for five exercises. In a second phase, two crews were formed from twenty-eight subjects, fourteen of whom had served in the preceding phase and thus were experienced. Both of these two crews had nine exercises, with turnover occurring after every exercise. Each was composed initially of subjects who had taken part in the preceding phase. Exercises lasted 40 minutes and consisted of four problems, two of which were simply the other two run backward. All problems had forty tracks (dot sequences). The loads in those run forward began slowly and ended at full load, so the loads in those run backward started at full load and tapered off.

The second experiment differed by incorporating (1) a gradual build-up in all four problems; (2) only thirty-three tracks; (3) pre-exercise component training to a performance criterion instead of a fixed amount of training; and (4) fourteen exercises for the crews in the second phase.

Crew turnover had no appreciable effect on inexperienced crews, the turnover crew doing about as well as the stable one, but it degraded performance in experienced crews by preventing improvement with practice. The researchers (Rogers, Ford, and Tassone 1961) concluded that system performance was degraded as a consequence of "a kind of turnover which produced a large amount of skill dilution" and that the "concept of 'skill dilution' could be used to account for the direction and relative magnitude of the effect of turnover on an information-processing system." They also interpreted their data analyses to conclude that turnover in one position could result in degraded performance in another position which was (1) closer to system input; (2) linked by telephone to the turnover position; and (3) overloaded by greater demands on memory storage. The researchers found in their results no sure indication of what kind of training would be best to counteract turnover. Component training to a criterion seemed better than just a fixed amount of such training, but this variable was confounded with others. The researchers noted that still unexplored were the effects of turnover on "crew motivation, crew cohesion, crew standards of acceptable performance . . . "

### *Crew Development Studies*

These studies were more dream than reality. Actually only one study came about, at the end of 1959, and it should probably be called a game rather than an experiment, since no variables were systematically manipulated. It has been reported by Ellis, Jensen, Jordan, and Terebinski (1960), Ellis, Jensen, and

Terebinski (1960), and Jordan, Jensen, and Terebinski (1963). Five crews of three persons each carried out a logistics-type task, four crews in six one-hour periods and one crew in twelve periods. The crews were variously composed of research staff, other professionals, college students, and secretaries.

This study was of less interest than the program of which it was a part. The program's notable features were: (1) the discrepancy between aspiration and accomplishment; (2) the simulated task; (3) simulation of a computer; and (4) methods of recording and interpreting subjects' behavior.

The program was proposed in 1959 as a two-year, four-phase investigation of crew development and problem solving. At one point (Ellis, Jensen, Jordan, and Terebinski 1960) this program was to include examination of the variables of timing and content of feedback information in debriefings, but this objective faded from view. As subsequently stated by Ellis, Jensen, and Terebinski (1960), the program was addressed to two questions: "How does an aggregate of individuals develop into a smoothly-functioning team? By what characteristics and dimensions can the change be described?" Thus, one objective was to determine whether discrete stages of crew development could be identified in behavioral terms, the hope being that such stages could constitute training goals and evaluation criteria. The researchers also wanted to learn how to observe and record the behavior of crew members in man-machine systems. These themes arose from the researchers' own experience with operational teams in field situations in the system training program.

The inquiry's four phases were to be the development and test of data-collection techniques, the collection of data on crews in different systems, the analysis of those data and the formulation of hypotheses, and the "experimental investigation of the more important hypotheses." Only the first phase was completed, perhaps without entire success. Experiments in "communication restriction" were proposed in 1960 but not undertaken.

The logistics-type task was a simulation called "the railroad game," described in the reports already cited and by Jensen (1961). It was designed to include the selection, decoding, and sorting of information, the co-ordination of work among several persons leading to their making decisions, the coding and transmission of those decisions, and the interpretation of information about changes that resulted from crew actions. It also provided a minimum of simulation problems, flexibility for variation, and opportunity for easy observation.

The three-person team operated a railroad's freight car dispatching office. An experimenter represented a central information office and stationmaster, while two simulators represented recording and accounting functions which might be carried out by a computer. The simulated environment consisted of either ten stations connected by a railroad, or ten cities connected by railroads, with five hundred cars of six kinds available for freight shipments. Weather and track conditions could be varied. Except for a seaport, each station or city dealt with a unique product. Shipping orders on a filmstrip were presented to the crew via a TV screen, with variations in load. The railroad was paid for hauling freight on the basis of the shortest distance between points, and it was charged for hauling empties, for maintenance, and for unfilled orders. The crew had to decode the orders on the TV screen, select the appropriate freight car for shipment, determine the routing, encode the order into a binary code, relay the order to the computer by means of the 5 X 6 pushbutton panel mentioned earlier in the description of the laboratory, and maintain a status display. Status data were furnished by the computer. The goal was to make as much money as possible through optimal use of resources.

To simulate a computer, two individuals at a marking table received binary coded car orders, checked them for completeness and for car availability; rejected them if an order was incomplete or the requested car was unavailable, and transmitted revised status data to the crew. To accomplish this last step, they penciled the data on large slides which a TV camera photographed for display on a TV monitor in the dispatching office.

As indicated earlier, the researchers were primarily interested in methods of recording and interpreting the subjects' behavior. Experimenters dictated their visual and auditory observations of what the subjects were doing and saying into a tape recorder, and the subjects' speech was also recorded directly. The tapes were transcribed, and the experimenters, their recall stimulated by the transcriptions, reconstructed the events and produced written or dictated protocols and interpretations. At first there were four observers, then two, finally one, the reduction resulting from increasing duplication of data and facility in recording. Protocols avoided ascribing motives or causes except when these were explicitly requested. With experience, observers could produce a coherent description or summary, rather than isolated items, and do it during the observation period itself. Actions were tallied as specific items, not as instances of general categories. At first there were no rules for defining what should be reported, but eventually some rules of allowable omission were adopted to cover tics, certain repetitive acts, and actions which could be clearly inferred from other actions reported.

As was to be expected, such procedures were extremely time-consuming. So were those for preparing the data for analysis, such as (1) "post-categorization," that is, fitting component actions into behavioral categories established after the fact; (2) interpretations of actions in terms of (a) the effects of one person's acts on another and (b) an individual's goals; and (3) characterizing behavioral elements by means of special graphic symbols. Although these approaches constituted a brave attempt to describe the streams of interperson behavior in a work-oriented, problem-solving situation, apparently the magnitude of the required effort overwhelmed the researchers. They documented speculative and impressionistic notions but no analyses of the data.

### *Composition of Debriefing Participants*

In the late summer of 1959, the intercept function of the SAGE system was simulated in the Human Factors Laboratory with girls as subjects. The girls were drawn from southern California colleges, junior colleges, and high schools. According to Burwen et al. (1960), a survey had indicated that females were more available than males for work during the summer months; in addition there was "a general bias in their favor with respect to reliability and ease of management." It was presumed that any difference of temperament between men and women would affect experimental conditions equally. After the experiment, the researchers concluded that "the similarities in the behavior of the crews and military crews observed in real operational environments supported the choice of female subjects."

There were four crews of ten individuals each; each crew had eight sessions, two per day, after one day of orientation and training. The ten individuals in a crew consisted of a weapons director and three teams, each composed of an intercept director and two pilot simulators whom the intercept director guided in making simulated intercepts. The pilot simulators in two of the crews were present (but did not otherwise participate) in debriefings after each exercise; those in the other two teams were absent. This was the independent variable. The results indicated it did not affect crew performance.

The simulation of SAGE was no more realistic than the girls' simulation of Air Force officers and airmen. Target and interceptor tracks were shown on TV monitors to the intercept directors, who gave heading directions to the pilot simulators. The latter sat at marking tables and plotted the required interceptor tracks; these were sent via TV camera to the monitors. The input of target tracks came from stripfilm projection and camera pick-up at a data generation table.

In such experiments it has often been deemed important that the experimenters be able to keep the subjects under observation, partly to make sure the subjects do not discuss the experiment among themselves. But in this study, when the pilot simulators went in a group to the rest room this control was lost. On another occasion, however, when the subjects were assembled in a projection room, a microphone enabled the experimenters to listen to their conversations. It was reassuring to discover that the topics the girls discussed, while interesting to the experimenters, were harmless to the experiment.

### *Other Studies*

Several other studies were undertaken in the Human Factors Laboratory. One was another SAGE simulation, this time of two-man teams—the weapons director and his assistant—with thirteen other positions being represented by simulator personnel. Each team had three one-hour runs. Some teams had had extensive experience, others none. Various co-ordination and planning activities were observed and tallied in an effort to show how co-operation develops in a dyad with decision-making functions (Ford and Katter 1960*b*). Co-ordination and planning were also the theme of an experiment involving thirty five-man teams. Each team sat at a vertically partitioned table where members silently passed cards through slots to each other until all reported knowledge of a given symbol. The independent variable was the extent to which extra messages could be exchanged (Shure, Rogers, and Meeker 1961). In a third experiment, four three-man teams confronted thirty-four decision situations distributed over five one-hour sessions in a simulation of a SAGE battle staff (Rogers, Shure, and Meeker 1962; Shure, Rogers, and Meeker 1963). This experiment is described in Chapter 21.

## SYSTEMS SIMULATION RESEARCH LABORATORY

The System Development Corporation's most elaborate facility for man-machine system experiments was the Systems Simulation Research Laboratory (SSRL). It was created for the express purpose of performing this kind of research, and its history is illuminating in many ways. Costs have not been ascertainable, but they undoubtedly were high, as a description of the facility will indicate. Funding came from fees charged in contracts with the Air Force.

A new two-story building was constructed in Santa Monica in 1961, attached to a previous building which housed a SAGE FSQ-7 computer and the Second IDC. The new structure was built to contain not only SSRL but also offices for SDC's research staff and the AN/FSQ-32V computer, on which the Command Research Laboratory came to be based. (That laboratory is described shortly.)

The facility occupied between 13,000 (Harman 1963*a*) and 20,000 (Harman 1963*b*) square feet. The principal area of experimental operations was a two-story room (20 feet high, to admit a 20-by-20-foot display) occupying about 2,000 square feet. It was surrounded on three sides by smaller, one-story spaces for a debriefing room and for experimental operations on a smaller scale. These spaces also totaled about 2,000 square feet. Above these areas and surrounding the main room were locations for observers, human simulators, managers, and visitors, as well as a visitors' briefing room. One-way vision windows permitted observation of the activities in the room below.

Along one side of the building, on the first floor, were rooms for a large, solid-state digital computer, for buffer, communications, and control equipment, and for computer maintenance and data preparation (including EAM equipment). These occupied about 3,800 square feet.

### *Equipment*

The computer was a Philco 2000. As a buffer and coupler between the computer and the input devices in the experimental areas, SDC itself built an "RL 101-Real Time Input-Output Transducer." This device could receive electrical signals from the experimental area's subjects in real time through many channels, store them briefly, then transmit them in fast time over a few channels to the computer. It could also accept signals from the computer, store them, and distribute them to displays. Plugboards made it possible to connect the input and output devices in the experimental areas to various locations in the RL 101 storage.

As an indication of the difficulty in planning such a device, experience in using the RL 101 led to modifications and additions. For example, its output capacity was expanded after it became necessary to delete some functions from an experiment because the capacity was inadequate. Plugboard arrangements were amplified and altered so that it was no longer necessary to interchange cable connectors after an experimental run to adapt to another experiment; due to continual insertion and removal of cable connectors, both cables and connectors were breaking, the interchanges took a great deal of time, and they sometimes resulted in wrong connections. Other improvements of the RL 101 increased its reliability.

Because it was presumed that the design of the facility would have to conform to the particular computer and because long lead times were involved, the digital computer for SSRL was selected and bought before the first experiment was planned. Lack of ready-made programming packages from the manufacturer and limitations in core storage somewhat restricted the computer's use, especially when the AN/FSQ-32V became available nearby and was favored by some researchers. However, during its first two years of daily operation the Philco 2000's usage at one point almost reached two full shifts of operation, before leveling off to slightly more than a single shift.

*Displays and Entry Devices*

The displays by which computer outputs were presented to experimental subjects were designed as general purpose displays, to be adapted in consoles of various kinds for particular experiments. One type, of which there were eight units, was a "situation" display, showing moving objects on a map-like picture of an area. A raster-scan, TV-type, cathode ray tube display was designed by SDC itself. It had 293 lines on a 8 1/2 X 11 inch raster surface. Each line was divided into computer words which in turn were divided into computer bits. The result was 89,000 dots which were individually controllable by the computer through its program to be either on (white) or off (black). Initial planning called for 196,658 dots, which would have provided greater resolution; von Buelow (1962) explained the development as follows:

Many of the systems upon which research in SSRL seemed imminent required a dynamic maplike display with identified geographic locations and moving objects. Displays of this type which are commercially available, or in some cases still in the research state, were surveyed, and the feasibility of their use in the laboratory was evaluated. While displays with many highly desirable characteristics were found to be in evidence or under development, budget limitations made most of them prohibitive. SDC, therefore, decided to design its own system, sacrificing some of the extra features such as expansion, offcentering, and lightgun capabilities. It was further decided to put the burden of generating and changing displays on the program rather than on the hardware. This meant either imposing some limitations on the size of systems being simulated or slowing down the real time operation. For the purpose of starting the Laboratory, this seemed to be an expedient compromise.

A second display type, also SDC-designed, was tabular. There were twenty-four of these. Each could show ten symbols on each of ten horizontal rows. A raster was generated by a vertical sweep, and a symbol–letter, numeral, or special symbol–was created by program-controlled selection of spots in 5 X 7 matrices. In this fashion any symbol which could be formed from such a matrix could be produced by means of the computer program. Concerning this development, von Buelow (1962) commented: " . . . efforts were made to utilize commercially available CRTs with built-in character matrices or with external character-generating hardware. Here again, to get the variety of characters required by a general-purpose laboratory too expensive a system would be required."

The same author noted that by changing the amount of space on the computer storage drum devoted to writing symbols on the tabular display, it was possible to have 10 rows of 20 symbols each, 19 rows of 10 each, 19 rows of 20 each, and other combinations. Changes from one format to another could be accomplished by rearranging connections on a small plugboard.

Of particular pertinence is the point that these displays did not match those in any real system. They had less capacity than those in some systems, perhaps greater than in others, and in all cases were qualitatively different. The same observation could be made concerning the entry devices whereby experimental subjects communicated with the computer. These were various kinds of switches, most of them sets of push buttons.

How closely should the display and entry devices, by means of which a subject interacts with a computer in a simulated system, resemble those of the system being simulated? In what way does lack of realism at this interface affect the applicability of experimental results? The question is a critical one. The dilemma becomes acute when the laboratory, as in the case of this one, is oriented toward future systems for which there exists no hardware to duplicate or simulate.

It may be possible to program a laboratory's general-purpose computer to match the programs in the real system's computer, and also through programming to match ingredients of displays. But some of the display characteristics are fixed—frozen into hardware or limited by computer storage capacity—and the hardware constraints of entry devices are even more restrictive. Since all real systems do not have the same display and entry devices, it may not be feasible to simulate such devices realistically at the man-computer interface by either general-purpose or off-the-shelf equipment; further, interface devices which realistically simulate those in one system may not do the same for another.

As a case in point, in the Terminal Air Traffic Control System experiment to be described shortly, the SSRL displays and entry devices simulated those in a future computer-based system for terminal air traffic control. But since such a system had not yet been designed and built, there was no way to be sure that these displays and entry devices typified those that would be a future reality. In all probability, they did not.

### *Communications and Recording*

Another set of equipment in the Systems Simulation Research Laboratory (von Buelow et al. 1961) handled communications, monitoring, and recording. To provide telephone and simulated radio communications between participants in an experiment, there were fourteen nine-line and twenty five-line two-way units in the experimental spaces, while twenty additional five-line units enabled observers to monitor intersubject telephoning. Thirty-four conference circuits could be established. Maintenance personnel had an independent communications system between laboratory locations and maintenance areas. All single lines, conference circuits, and monitor lines were set up by patch cords at a central patch bay, so that the system could be completely and rapidly reconfigured between experiments. A communications control console was situated on the left wing of the console for RL 101 control. The patch bay was also in the RL 101 room.

A public address system had speakers in the main experimental room, on the observation "decks," in the computer room, in the RL 101 room, and elsewhere. Microphones were placed in most of the same areas. Those in the main experimental room could pick up nontelephone conversations or exchanges among subjects. Four tape recorders could be switched to record what was being picked up by these microphones or spoken over any part of the telephone system. They could operate continuously during an experiment or be turned on and off at predetermined times. In the experimental room or on the observation decks, the five portable television cameras originally procured for the Human Factors Laboratory could record what subjects were doing as well as the displays they were viewing; these were usually too small and distant to be visible to observers, who instead could watch eleven display monitors in this closed-circuit TV system.

*Establishment*

There is some value in examining how SSRL was established. The management of the System Development Corporation, with the concurrence of the board of trustees, decided in October, 1959, to "create a general purpose computer-based simulation facility for research in systems" (Harman 1960). Early documentation indicates that the driving force was an interest jointly in systems and in simulation, but that ideas as to just what research would be conducted and what problems attacked were nebulous. Nevertheless, the future of the facility was viewed optimistically. H. H. Harman, the director of SSRL, wrote (1960):

> The initial planning and building of the Laboratory is entirely at Company expense but it is expected that after a couple of years of operation a large part of its cost will be underwritten by specific projects. The first order of priority will be self-sponsored basic research activity. Secondly, the Laboratory will be available in support of research and development in connection with the ongoing Company activities. In the third priority order, specific contracts might be written for work in SSRL. Typically such contracts might entail testing and evaluation of proposed new control systems through simulation in the Laboratory. Finally, on a lower priority basis, the Laboratory will be made available to outside researchers, that is to people in universities or research institutions. The cost of such research might be underwritten by such an agency or by SDC.

Harman also noted that a multidisciplinary staff of "six senior research people" in SDC had been formed to formulate requirements for the laboratory and design it, to "take a leading role in suggesting and generating appropriate studies for the Laboratory," and, when the latter would be in full operation, to "serve principally as Research Coordinators working closely with research teams throughout the Company in the design, planning and executing of experiments in the Laboratory and in the subsequent data reduction, analysis and interpretation of results." By no means all of this staff had significant prior experience in generating large-scale man-machine system experiments.

The design of the laboratory was completed in December 1960, and construction of the physical plant was finished four months later. The computer arrived shortly thereafter. The laboratory was dedicated on September 28, 1961. at a ceremony attended by five generals, various government officials, scientists from universities and other nonprofit organizations, and the press. Dr. John W. Gardner, then president of the Carnegie Corporation, gave the dedication address.

There had been a two-month lag in selecting a computer, but computer delivery was ahead of the original plan; the lead time was only eight months instead of the expected twelve to fifteen or eighteen. It was hoped that "full operation will begin before the end of 1961" (Harman 1960). Another planning document expressed hope that a pilot run of the first major experiment would start at the beginning of December 1961, with the first experiment beginning early in January 1962. Among the plans prepared were a computer plan, a facility plan, a development plan for special equipment, an equipment purchase and installation plan, and a development plan for the programming system.

The first experimental run of the first full-scale experiment actually took place on July 11, 1962. Programming and, in particular, system testing and debugging took longer than expected. The last run of the same experiment occurred on August 10, 1962. SSRL produced no further man-machine system experiments of the scale and complexity for which the laboratory was especially created, although one set of studies (Leviathan) approached these. The experiment which did get done was the Terminal Air Traffic Control System Study (TATCS), to be described shortly. The laboratory was not idle, however, as the following data indicate.

*Utilization*

During fiscal year 1963, the computer was used 2,566 hours for computer program development, that is, by itself for developing compiler, executive, diagnostic, and statistical programs and for all-computer studies. It was used 581 hours during the conduct of experiments (including TATCS) and in checking experimental arrangements and programs as well as processing the data. The laboratory research in addition to TATCS involved an investigation of computer-assisted instruction for a classroom; studies of individual pattern recognition and human data processing; the Leviathan program (described later in this chapter); bargaining and negotiation studies (described in Chapter 23); and some of the work subsequently transferred to the Command Research Laboratory (described later in this chapter). One of the serendipitous products of SSRL was increased exploitation of a digital computer on the part of some of the behavioral researchers, notably for presenting stimuli to individual subjects, directly registering their outputs, and immediately reducing the data from them.

However, eventually computer usage declined. Those who managed the laboratory acknowledged that the projects were not of the large-scale man-machine variety which they had expected to materialize. In addition, almost all the work fell in the category of SDC-sponsored research, rather than in the categories of supporting SDC's contractual commitments, or use by non-SDC consumers as a marketed resource, or use by outside scientists as a public service facility. For one reason or another, other parts of the company did not fulfill the hopes of the originators that they would use the laboratory for system development and checkout; there seemed little demand for running large-scale experiments of any kind in this laboratory. Of those interested in such experimentation, some pinned their hopes on the newer Command Research Laboratory (CRL) and its larger, more versatile computer. Others, who also were in a different department, created still another laboratory (without a computer) for studies of civil defense and response to disaster (described in Chapter 22).

The main experimental area was eventually partitioned into smaller, single-story spaces, the Command Research Laboratory experimental area being established on the second floor. Finally, in April 1965, both of the laboratories were discontinued as separate entities, merging into a Research and Technology Laboratory on the second floor in the old CRL area. The Philco 2000 computer was tied into that location by recabling. Subsequently it was disposed of and its functions were transferred to a new computer complex based on an IBM 360 system.

## TERMINAL AIR TRAFFIC CONTROL SYSTEM STUDY (TATCS)

After it had been decided to build and equip the Systems Simulation Research Laboratory, the SDC management debated what kind of system should be investigated in SSRL's first man-machine system experiment. Among those considered were systems for missile defense, for waging limited war, and for transportation, but the decision settled on future air traffic control—in the time period subsequent to 1975. It was assumed that by then computers would be widely used in air traffic control and that increases in air travel would present even more serious problems of system co-ordination. The SSRL staff was asked in the fall of 1960 to design and build a laboratory model of a post-1975 terminal air traffic control system as the first major simulation project in the new facility. The emphasis at this point was placed on simulation and the system to be simulated rather than on particular themes or questions to be confronted through experimental inquiry.

*Objectives of the Experiment.* Published statements of the experiment's objectives failed to mention SDC aspirations to contract with the Federal Aviation Agency to help design a future computer-based air traffic control system. One statement noted that "the immediate objective of the study is not to improve operations at the particular airport complex" (San Francisco) simulated in the experiment but rather "to provide a vehicle for study of systems of this general type" (Harman 1963*b*). Later this author, in briefing the board of trustees, said that the study "was developed as a test and evaluation vehicle for SSRL planning and implementation, and as a demonstration of SDC capability and effectiveness in the use of simulation technology." Alexander (1962) wrote: "Useful information for designing a control system of the future can be obtained by operating the system in this environment and by manipulating and testing its various parts. In addition, certain underlying relationships and principles of systems in general can be identified and the technology for simulating systems in the laboratory improved." He added, as reasons for the experiment, that air traffic control was recognized as an important public problem; that the study would not be subject to security restrictions; that SDC's interests and experience were suitable (in view, for example, of its other air traffic control projects); that an "air traffic control system evidences most of the operational functions and relationships common to many information-processing, command and control systems"; and that terminal air traffic control embodied both tactical and strategic functions. Earlier, Alexander and Cooperband (1961) wrote that the study goal was not to affect the design of a system which was to be implemented but, "rather we expect that the systematic experimental manipulation and intensive study of the operation of a laboratory model will produce results which will contribute to general system theory." Later in the same report they said:

The research project for which the TATC Experimental System was designed has three purposes. The first is to learn more about the factors which influence the effective working relationship between tactical control and strategic planning functions of systems; the second is to advance the development of laboratory research techniques for simulating man-machine systems in real-time; the third is to make some contribution of improved methods of handling terminal air traffic.

And subsequently, Alexander and Porter (1963) wrote:

SDC chose the problem of research on terminal air traffic control in the post-1975 period for a number of reasons. First, terminal air traffic control is a problem of increasing public importance and any contribution to an understanding of the problem is in the public interest. Second, there was the need to establish or deny the hypothesis that it is possible to simulate in the laboratory a complex system and from its operations to gain insight into how the system should be designed in real life. And, third, the operation of a simulated terminal air traffic control system held promise as a vehicle for the study of a number of basic problems in man-machine system design.

The experiment was definitely envisioned as a way to "shake down" the new Systems Simulation Research Laboratory. Much of the laboratory's display and entry equipment was designed with the experiment in mind. The experiment was characterized in several documents as just the first in a sequence which would investigate terminal air traffic control. Planning documents listed a number of air traffic control problems, experimental investigation of which would be expected to yield generalizable results. The first—and only—experiment did not attack these. Instead, it was restricted to a minimal system and incorporated the following five independent variables:

1. Uniformity of the rate at which inbound aircraft penetrated the system. The rate was or was not uniform.

2. Geographical distribution of the inbound aircraft. Equal numbers came from north and south, or twice as many came from north as from south, or the opposite.

3. Homogeneity-heterogeneity of the aircraft. In the homogeneous case, all were high performance vehicles such as commercial jet aircraft; in the heterogeneous case they were divided between supersonic and medium performance aircraft.

4. Two teams of subjects. These had received different kinds of training in an associated experiment (described later) and had been found to be markedly different in "cohesiveness."

5. Procedural flexibility. In an open-scheduled configuration, flight plans contained only initial and final fixes (routing points) for inbound and outbound aircraft, whereas in a prescheduled configuration there were at least three fixes which the controller had to make sure an aircraft reached. Separation standards in the local control area also differed in the two configurations. Controllers operating in the open-scheduled configuration could exercise a wider range of discretion. This variable was also called "rules flexibility." Since it produced significant results and led to a number of concepts about system functioning, it was the focus of the account of the experiment which reached the open literature (Alexander and Cooperband 1964*b*).

*The Simulated System.* The minimal system, consisting of portions of San Francisco terminal air traffic, was abstracted from the experimenters' design of a hypothetical post-1975 air traffic control system for the entire San Francisco–Oakland area. This system was put together by May 1961, after two extensive surveys, one of current terminal air traffic control, another of forecasts. It was expected that the larger system would be progressively approached in laboratory experiments, each of which would contribute to its redesign (Alexander and Ash 1962).

In the minimal system were two local air traffic controllers, two conversion controllers, a traffic co-ordinator, a supervisor, and an assistant to the supervisor. The supervisor issued directives, flight plans, weather reports, schedule changes, etc. He was one of the experimenters. The other six individuals were experimental subjects. Each of the controllers sat at a console which included a situation display, a tabular display, pushbutton entry panels, and telephone and simulated radio communications. The assistant to the supervisor was the contact between the supervisor and the subjects.

There was also an embedding system, representing air traffic control functions with which the minimal system had to interact. It consisted of two ground controllers, two sector controllers, and seven pilots. These were what have been called in other chapters quasi subjects. Each pilot sat at a console by means of which he could handle up to five flights. His switch inputs at the console were interpreted by the computer in a manner analogous to the way in which an aircraft would respond to its pilot's control actions. Outbound flights passed from ground control via the traffic co-ordinator to local control, then to conversion control, then to sector control. Incoming flights went through an approximately reciprocal process. All controllers communicated with the pilots by voice radio, and controllers could communicate with each other by telephone.

In this minimal system the computer did not play an extensive operational role. Its principal operational function was to organize required displays, such as those of a take-off queue; but it did not organize the queue itself, otherwise exert direct control, present alternatives, or calculate probable collisions or clearances, except to generate conflict-free aircraft altitudes. As a matter of fact, an analysis showed that conversion controllers had to insert four times more information into the computer than they received from the computer. The computer accepted switch actions from controllers and pilots and translated pilots' switch actions into displayed movements of aircraft on the controllers' situation displays. It recorded all switch actions and processed and reduced the data from them as the experiment progressed.

*Experimental Arrangements.* The subjects and quasi subjects were male college undergraduates. The criteria for choosing the two six-man crews in the minimal system were reasonably clear speech, normal hearing and vision, an intelligence test score within the range of plus or minus one standard deviation from the mean score for college freshmen, and lack of knowledge about air traffic control. The quasi subjects were selected without reference to any particular criteria. Probability of attendance throughout the study was encouraged by dividing hourly pay into two parts—base pay and a bonus which required perfect attendance.

The quasi subjects were trained over a three-month period, while the training of the regular subjects lasted five weeks, starting with a one-week orientation course and continuing with one one-hour and eight two-hour operational problems "on the job." The training problems were divided into thirty-minute segments, each of which contained at least two or three critical air situations requiring important control procedures, such as an emergency procedure, a change in flight plan procedure, or a handover procedure. The subjects trained as crews, each member retaining his position throughout. Subjects had been assigned to crews by matching them through centile scores on American Council on Education language (ACE-L) tests, and to positions within crews through scores on an achievement test measuring knowledge gained in the orientation course supplemented by opinions of experimenter-instructors.

The staff required to conduct the experimental sessions varied between six and ten persons. Each of twenty-four sessions, twelve per crew, lasted two hours and consisted of four thirty-minute problem periods. Each of the ninety-six problem periods scheduled six inbound and four outbound flights. A set of twelve different problems was presented four times for each crew. The design and production of these problems were described by Cooperband, Alexander, and Schmitz (1963) as follows:

> Twelve standard sets of flight paths were generated, one for each of the unique combinations of traffic variables. In preparing the schedule for a problem period, these flight paths were translated into flight plans according to the starting time of the particular problem period. Since part of the system design was based on the assumption of a central scheduling function which guaranteed flight plans to be free of conflict at fixes, the standard flight paths were processed by a computer program which resolved such conflicts by revising altitude assignments. The output from this conflict resolution program was examined manually and adjusted further where necessary to conform as closely as possible to the standard paths. Then flight-plan strips were printed automatically for the subjects and for simulators, observers, and experimenters. Certain parameters of these flight plans were used by another computer program to produce a control deck of punched cards which supplied all the necessary information which the com-

puter programs in the test and embedding systems needed to "create" these flights.

Thus the rate of penetration, geographical distribution, and aircraft homogeneity-heterogeneity variables were distributed within each set of twelve problems by a factorial design. For both crews, the first twenty-four problem periods were devoted to the inflexible controller procedures, the second twenty-four to the flexible ones. The experimental sessions alternated between subject crews. The quasi subjects remained the same. Both subjects and quasi subjects had briefings before each session and debriefings afterward. The subjects' debriefings consisted first of an unsupervised discussion of their performance and then an experimenter-led discussion following a report of errors committed. The experimenter discussed procedures with the subjects and tried to get them to evaluate system design and make suggestions. Procedural matters which came up in one crew's debriefing were introduced in the other crew's debriefing.

The subjects were monitored visually at all times and their telephone and simulated radio conversations were monitored as well. Data concerning individual controller performance were obtained from voice tapes. But the data for the analysis of system performance came from neither films nor voice tapes. Instead, it was based on the computer-collected, magnetic tape records of switch actions and flight histories of all aircraft. These records made it possible to re-create any problem period in its entirety.

System measures included probability of collision; safety violations (closer than specified separation limits); per cent of time aircraft spent "holding"; per cent of aircraft that were held at least once; difference between actual flight time and the time required to fly the shortest available path; the ratio of these two times; mean time spacing between successive aircraft at designated points; aircraft waiting time between conversion control and local control; aircraft time in conversion control but not holding; fuel consumption; and variability in aircraft arrival times. These measures were aggregated into three groupings: safety, expeditiousness, and orderliness. Measures of individual controller performance included average number of radio communications per aircraft; average length of each control message; total controller talking time; total time of controller switch actions in computer entry; and mean number of clearance points through which aircraft were routed.

*Outcomes.* Results, tested in analyses of variance, showed that the three independent variables concerned with the composition and distribution of traffic had little impact. As for differences between crews, one crew was significantly more expeditious than the other, due to better local control, but was not otherwise superior. The variable of procedural flexibility had significant effects. When controllers could be flexible, aircraft spent only 1% of the time holding instead of 4%; and 12% of the aircraft were held instead of 36% under the inflexible conditions. Safety was not compromised. However, flexible scheduling added 11 seconds per aircraft to the minimum schedule and nine seconds to the average transition time. "To maintain adequate separation between aircraft and still impose little delay on their progress," wrote Cooperband, Alexander, and

Schmitz (1963), "the controllers seem to have assigned them to routes which were slightly longer than the most direct path. In other words, to accommodate the exigencies of the traffic environment, the controllers were trading space for time."

In evaluating the outcome of procedural flexibility, Alexander and Cooperband (1964*b*) drew attention to this experiment's indication "that adaptation occurs under discretionary conditions even when load is held constant." They recalled that the RAND air defense experiments (Chapter 8) had shown that such adaptation occurred when load progressively increased. They emphasized that the superiority of rules flexibility characterized the crew which was "cohesive" far more than it did the other crew. But they noted that the conversion controllers in the less cohesive crew did better on the individual performance measures when the rules were flexible than when they were inflexible. Their findings led them to formulate concepts of stress index and discretionality in system behavior.

The researchers noted a number of by-products (Cooperband, Alexander, and Schmitz 1963). One was the ability, through the computer, to acquire measures of operational effectiveness as the experiment progressed. Another was the development of a data regeneration technique for recording on magnetic tape all switch actions taken by operators for later replay of what happened. (See Chapter 11 for a similar development.) The experimenters learned it was very difficult to standardize and control the activities of the quasi subjects in the embedding system. They exhibited too much unpredictable variety. It has also been suggested that the development of simple interrupt and swapping features to interleave the experimental use of the computer with other uses at the same time was an early instance of time sharing.

The difficulty of communication among engineers, programmers, and human factors personnel while the TATC system and experiment were being designed and developed led to a "schematic simulation" method of simulating the system's operation (Alexander and Cooperband 1964*a*). ("Schematic simulation" is described in Chapter 24.) A number of future research questions were thereby indicated, particularly the degree of automation which should be introduced into the scheduling and routing of aircraft and into conflict detection and conflict resolution.

Some follow-on research addressed to these questions was proposed by Alexander (1963), but it was not oriented toward large-scale experimentations; and some was done, namely, the development of a terminal air traffic scheduling model and an experiment on individual detection of compound motion in the behavior of predicting two-target collisions (Cooperband and Alexander 1965). No further TATCS experiments were conducted.

The history of the one TATCS experiment not only suggests the difficulty of keeping a project of this nature in operation but also illustrates some of the time requirements for such an experiment. The following estimates came from H. H. Harman: designing the reference system, about four months; programming, fifteen months; design, construction, installation, and component test of hardware, nine months; writing and modifying the system procedures, three months; system test, three months; indoctrination and training of subjects, somewhat more than one month; experimental runs, one month; analysis and publication of results, about twelve months. Some of these items, of course, overlapped. In brief, preparation required about a year and a half; training and running, about two months; analysis and publication, another year.

*The Training Experiment.* As noted earlier, the two six-man crews for the TATCS were subjects in a training experiment which also provided their training for the TATCS experiment and preceded the TATCS experimental runs. This training experiment has been reported by Rundquist (1963) as only an exploratory study, with results being no more than suggestive. It compared two methods of training. In each the subjects processed flights as a team. In one the subjects used the TATCS consoles and communications; in the other they sat around a table and worked with schematic diagrams (those used in schematic simulation in system design) of those consoles and communications, stating their actions verbally; instructors presented the air picture on a magnetic blackboard. With both methods the subjects, at the end of a session, received knowledge of results and conducted a discussion in a debriefing.

In test-retest comparisons, the crew which had been trained on the actual equipment performed better, but various factors prevented stating this result as a firm conclusion. For one thing, the crews differed markedly from the start in ratings for cohesiveness, the crew working with the schematics being much less cohesive. Its members were inclined to be noisier, more argumentative, more hostile toward each other, and less inclined to accept leadership; they seemed to have more sharply defined and individualistic personalities; and they had less effective leadership.

"The schematic training proved very difficult to operate," Rundquist (1963) reported. "It was impossible to process complete training problems . . . . The schematic training was not a happy experience for either instructors or students. The instructors had to bear down to get proper attention and effort from the students . . . . It was learned after the completion of the study that the (schematics) crew initially thought they had been chosen for the schematic training because of poor performance on pencil-and-paper tests of system knowledge." These tests were used for position assignment within crews, but not assignment to crews, which, as pointed out earlier, was based on matching through ACE (L) scores.

## LEVIATHAN STUDIES

The Leviathan Studies consisted of ten simulations of a large intelligence-gathering organization, five in 1963 and five in 1964 (Rome and Rome 1964*a*, *b*; 1965, 1967). In these, human subjects played the roles of control personnel in the organization, while the personnel who did the productive work—processing the intelligence data—were represented in a computer. The computer displayed the stimulus information to the live subjects, registered their responses, and reduced the data from these responses. Earlier, the simulation had been entirely within the computer (Rome and Rome 1961, 1962), but no actual simulation runs were reported. In the earlier work the main effort was to conceptualize the computer simulation of a large organization and produce the computer programs for that simulation; programs then had to be redone to fit the Philco 2000 in SSRL.

It is difficult to say to what extent the Leviathan studies should be construed as man-machine system experiments. They were oriented primarily to contexts of organizations, rather than man-machine systems. In addition, they lacked many of the aspects of an experimental approach, even though they were called experiments. However, they will be described briefly, since they had some features of interest to the theme of this book. There is another reason to keep the description brief. Although a number of Leviathan reports have been published, these omitted much of the methodology employed, particularly concerning the 1964 series. Further, as of 1967, only rather general results had been published and almost none for the 1964 series. Rome and Rome said in 1967 that "Data analysis of both series of experiments is still in progress."

The Romes simulated a six-level hierarchical organization said to be representative of large bureaucracies of many varieties, such as a school system, an industrial plant, or a United Nations agency. The two lowest levels existed entirely within the computer and consisted of 704 "robots" in sixty-four squads. The next higher levels were sixteen group leaders with staff assistants, four branch heads, and a single command figure. These were all played by graduate students. The very top level was a still higher organizational entity represented by the experimenters. The computer programs embodied "seven basic elements essential to all large hierarchical organizations": a formal authority structure, a technological or productive system, a production task, continual interaction with competing external environments, performance feedback reports, communication media, and policy formation. The echelons played by actual people exerted control over four functions: traffic, manpower resources, priorities, and production. One of the questions to which the researchers addressed themselves was the method of distributing control of functions among components of the hierarchical levels. All actual production was accomplished at the lowest level, by the robots in the computer.

In the two series of studies, the simulated organization was a hypothetical intelligence communications control center within a national intelligence agency. The center's production consisted of processing communiqués (raw material) which it received from all parts of the world (sources), its output being transmitted to various government agencies (consumers). This kind of production was advantageous for computer-based simulation. It dealt entirely with data—verbal materials—rather than with physical objects which would have to be transformed into verbal descriptions in order to be computer-processed in a simulation laboratory.

Although the researchers have described their studies as "experiments," they have qualified them as "open ended" (Rome and Rome 1967). Under experimenter control were such things as the size and configuration of the organization; the allocation of productive energy and its amounts and costs; the kinds, amounts, timing, and distribution of information feedback reports; the channels, methods, and idiom (when it was computer-mediated) of communication among the live participants; and the composition and timing of communiqués. The extent of this kind of control would justify viewing the simulation exercises as controlled observation.

The experimenters have asserted there were purposeful limits on their control. "Subjects and experimenters act as two distinct social forces that mutually influence one another during the progress of an experiment," Rome and Rome (1967) commented. However, subjects and experimenters never met face to face during an experiment–a precaution to preclude irrelevant conduct on the part of either. The Romes added:

> As an experiment progresses, we continually seek evidence that the subjects are moving both to more authentic action and to more effective performance. We attempt to assess which tendencies are likely to inhibit the subjects' two-fold development, which to further it; and thereon we either resist or encourage the directions in which the subjects are developing. This we do by enacting the roles of suppliers, consumers, supervening bureaucratic authority, and impinging cultural influences. Thus an ever-progressing spiral is established between subjects and experimenters, in which a mutual monitoring or balancing-counterbalancing interaction takes place. Each thrust attempted by the experimenters can be assessed by the subsequent conduct and achievement of the subjects . . . we use no a priori formula, prescription, or preconceived model concerning how an organization should structure itself, how it should operate . . . . Nor do we follow fixed, pre-set schedules. Instead, we design and time our intervention in tune with the emergency . . . of specific, organically developing patterns of interaction or specific, ever-rising levels of technological accomplishment.

*1963 Series*. The five studies or "runs" in the first series involved twenty-one student subjects who were hired to serve throughout the total of 120 laboratory hours; attrition figures have not been published. Each "simulated day of operations" was called an "epoch"; this time unit has not been further defined in available documentation, nor is it clear how long each laboratory session lasted.

The first study began with five epochs entirely in the computer, followed by fourteen in which the subjects participated, whereupon more computer-only epochs followed. The subjects had initially been given relatively little information or instruction about what they were supposed to do, and on-going feedback did not provide much more. Not too surprisingly, production steadily dropped. After the fifteenth epoch (tenth with subjects), the subjects received feedback (in a debriefing) which enabled them to understand much better who was who, who was responsible for what, and what their mission was. Apparently the feedback emphasized "trend" information, and the briefing officer exhorted them to "emphasize the system point of view."

Of methodological interest were two comparisons made by the researchers. In one, they simulated in the computer a projection of subjects' performance based on what it had been prior to this debriefing, and they compared this projection with what actual performance became as a consequence of that event. In another, they predicted the performance of the subjects during the next four epochs as a consequence of that event and compared this all-computer projection with the computer record of what actually occurred during those four periods. The organization's processing delays decreased and productivity rose, they found, after the people who controlled it began to discover more about the organization and how it might function.

In the first study, each branch head was responsible for controlling all of the four functions mentioned earlier, but each of his group heads specialized in one of those functions, controlling four squads of the same specialty. In the second study, each group leader was responsible for all of the four functions, with one squad for each speciality; his cognizance extended over one-fourth of his previous territory and his feedback pertained only to his more confined domain. The additional co-ordination requirements were subjected to another change. The scenario now introduced crises from epoch to epoch. This study terminated after the fourteenth epoch because "a technical mistake was made by one of the officers that resulted in disaster to the center in the next epoch, and we had to stop the experiment. Our programs are now coded in such a way that this cannot happen again" (Rome and Rome 1964*a*).

The third study, containing fifteen epochs, maintained the same organizational structure and crises environment but reduced the voluminous feedback given to the subjects. Now they received reports "by exception" and all quantitative information at the component level in only one epoch out of every five. "What they did receive every epoch were system performance reports and component failure reports," the researchers wrote. They made comparisons with the subjects' performance during the second study. Production was greater and delays were smaller. The researchers did not mention the possibility that the improvement was due to practice, for which there appeared to be no control in any study.

In the fourth study, a crisis environment continued but organizational structure changed. The functional responsibility of the four branch heads was differentiated, so each head had a unique responsibility. However, only one of his four group heads had the same functional responsibility, all of them continuing to exercise their specialties as they did in the third study. With this unusual arrangement, organizational productivity paralleled that of the third study. It was constant for twenty-three epochs and then accelerated until the twenty-eighth and last.

In the Leviathan research, the subjects operated twenty-four small consoles which were on-line with the Philco 2000 computer. They sent messages to the computer, i.e., to robots, by means of alphanumeric messages on a CRT display. The subjects could generate messages composed in many steps, each following a particular display. For example, the display might present a set of possible transactions, each listed with a number. By pressing the pushbutton bearing that number, the subject would initiate that transaction, which he might amplify with letter pushbuttons—thus identifying, for instance, some robot. This "language tree" arrangement provided a large repertoire of transactions stated in "natural" English, despite the requirement for unambiguous and rigorous formulation for computer processing. In the fifth study, in 1963, this computer-subject communication capability was exploited to provide to the subjects numerous exhortations and much advice as to how to increase production.

Although the researchers did not make the point, apparently such instructional inputs to the subjects were more effective than feedback.

*1964 Series*. This series was approximately as long as the 1963 series. The twenty-eight subjects were a new set of students. In the first study, which ran for forty-two epochs, the subjects were instructed about their organization and its functioning through their console displays in six four-hour sessions. In other words, the computer taught them. There was also a human briefing officer who furnished, the experimenters wrote, "intensive technical guidance." Although the cost of operations was raised about 30% over the levels in the 1963 series, the productivity of this fresh set of subjects surpassed that of the first three 1963 studies, an effect the researchers attributed to better planning and exploration of future contingencies as a result of the early computer-generated instruction.

The second study in the 1964 series was apparently the later portion of the first run (Rome and Rome 1964*a*). A procedure of reporting by exception was introduced into its thirty-seventh epoch. Productivity remained high and a large backlog was reduced, but the subjects complained they were getting insufficient information. In the third study, the subjects could share their feedback information with other officers in the command hierarchy. Although the researchers wrote that this 1964 series consisted of five experiments, no description of a fourth and fifth had been published by mid-1967. The program was discontinued at the System Development Corporation in that year.

Although the programs and studies lacked experimental design, the researchers in their reports have offered various comparisons between the results of studies to account for the influence of a number of variables. However, they have not discussed the question of confounding of variables which is obvious even in the summary reports. It is hardly surprising that no significance statistics were applied in the studies. The funding for these interesting explorations of a new territory and for the associated computer programming, on-line computer time, subjects' consoles, and subjects' fees came mostly from SDC's own resources, with support in later years for "development of the theoretical aspects of the research" from the Air Force Office of Scientific Research of the Office of Aerospace Research.

## COMMAND RESEARCH LABORATORY

The Command Research Laboratory (initially called the Command Systems Laboratory) was a major feature of a project funded by the Advanced Research Projects Agency (ARPA) of the Department of Defense and initiated toward the end of 1961. As a facility for experiments, the laboratory went into operation early in 1964. The project continued after 1964 but not the laboratory; as mentioned earlier, in 1965 it was merged with the Systems Simulation Research Laboratory (SSRL) to form the Research and Technology Laboratory, which was not the locus of any man-machine system experiments. The ARPA project's research and the Command Research Laboratory have been described by Cooney (1964).

Planning for the laboratory began in February 1962, and construction started in the spring of 1963. The laboratory was based on the very large AN/FSQ-32V computer, a machine which International Business Machines had built for the Air Force for another purpose. The central processor of the Q-32 was shipped to the System Development Corporation in Santa Monica in 1961 and installed in the same newly constructed building with the SSRL. Although the central processor was thus provided free, this was not true of the necessary peripheral equipment (e.g., input-output), much of the needed storage, the computer programs, the displays and entry devices, or the buffer and transducing equipment. These had to be obtained with ARPA funds. In other words, the laboratory was a costly investment, even though it was unnecessary to buy or rent the central processor (main frame). The buffer-transducer was another computer, the PDP-1. It not only accomplished the same kinds of tasks that the RL 101 did with the Philco 2000 computer in SSRL, it was essential for implementing the time-sharing mode which was developed by SDC data processing specialists for the Q-32. Still another computer was used with the Q-32, an IBM 1401, for off-line support operations.

*Equipment*

It can be difficult at times to distinguish between a laboratory as an experimental facility, a laboratory as an organizational entity, a laboratory as a project, and a laboratory as a label for some arbitrarily bounded set of activities in search of scientific dignity. The Command Research Laboratory was no exception. In this report it is construed as meaning the experimental area with component equipment and programs when these were being used for experimentation. The laboratory was not synonymous with the computer and its associated equipment. The principal entry devices associated with the Q-32 were more than two dozen teletypewriters and typewriters, but only four of these were used for experimentation in the laboratory as it is defined here. The computer was in almost constant use—nineteen hours per day during much of 1964, for example. But very little of this use was devoted to laboratory experimentation.

The dimensions of the experimentation area on the floor above the computer were approximately 38 by 30 feet. Subjects performed in half of the space. The remainder was allocated to a simulation, observation, recording, and control area separated from the subjects' area by one-way vision glass; and, on another side, to an area for cathode ray tube (CRT) consoles and photographic and projection equipment. In its only configuration for large-scale experimentation, the subjects' area had three CRT-equipped consoles; three teletypewriters next to these; three telephones at the consoles; and a wall screen for rear projection. Six simulation telephones, manned by two persons, were connected to the console telephones. Two voice recorders could record telephone communications. The simulation personnel had a CRT console for monitoring and a teletypewriter.

The CRT displays could be both tabular (alphanumeric) and pictorial or map-like (formed by vectors); as many as two thousand characters or vectors could be presented in one display. The consoles were built especially for the laboratory by a contractor. The displays were formed by stroke-type character generators driven by the computer. A capability was developed to use the displays also for data entry in conjunction with a light-pen (Burnaugh and Moore 1964), but this was unavailable in time for experimentation so the teletypewriter keyboard constituted the only entry device. The teletypewriters furnished a permanent record of what had been entered. Cables connected the teletypewriters and CRTs to the PDP-1 in the computer area a floor below. Plans to create more advanced display systems with keyboards integral to the display consoles were never realized.

Large displays were presented on the vertical rear-projection screen in a manner arising from "budgetary considerations" (Cusack 1964*b*). A Polaroid camera photographed the face of the CRT on the console in the projection area. Then a transparency made from the photograph was projected on to the rear of the screen, the entire process requiring about 35 seconds. These displays tended to lack sharp definition, precise registration, and desirable co-ordination with fixed reference data.

*Program*

The experimentation area and its equipment, it is clear, represented a relatively modest effort, which was perhaps just as well in view of the limited use to which the laboratory was put. Only one man-machine system experiment was conducted in it, and this one, the Display I experiment (Cusack 1964*b*) was unpretentious and aimed largely at shaking down the laboratory. Not only had more display experiments been planned (as implied by the Roman numeral "I") but a 1962 planning document forecast a considerable array of studies. The range extended from single operator studies to multiple node-multiple side studies with more than two sets of teams representing several networks of nodes in a many-sided conflict situation. Intermediate arrangements were to be single-node in one side (a small team representing critical functions of one command node), multiple nodes in one side, and one node in each of multiple sides. It was hoped that by July 1963 studies would begin that involved extensive competitive situations with as many as twelve participants. It was forecast that after October 1963 the laboratory would be the scene of multiple simultaneous operations, and simulations even of multisided, multinodal competitive situations.

The over-all aim of the ARPA project was to gain new knowledge concerning command processes and increased understanding about the management of forces, and to improve the application of computer technology to command and control. But the translation of these goals to experimental objectives proved difficult for the staff of more than two dozen researchers in the project. From the start it was unclear in what direction the project was heading; and emphasis was placed not so much on experimental inquiry as on developing a facility and creating data bases, models, and scenarios (Bayless 1962).

In addition to the Display I experiment, the ARPA project gave rise to three sets of smaller-scale experiments, all on decision-making by individual subjects. Four of these experiments were performed in the Command Research Laboratory. These three sets are described briefly in Chapter 21. One set of three experiments dealt with "probabilistic information processing" (PIP) based on the Bayes theorem (Kaplan and Newman 1964*b*). Only the last of these was done in the Command Research Laboratory. A pair of experiments on "HEMP target analysis" (Merrifield and Erickson 1964*a*, *b*, *c*) used the Command Research Laboratory area; subjects sat at desks in paper-and-pencil situations. The third set consisted of three studies of "Force Allocation" (Wood and Friedman 1964); they also employed paper-and-pencil methods. The last of these was conducted in the Command Research Laboratory area. A fourth research program was called "Multivariate Threat Analysis" (Bayless, Erickson, Grant, and Horst 1963; Bayless, 1964). Although one of the reports was entitled "Multivariate Threat Analysis: Experimental Results," the program included no experiments with human subjects; and that same report, in explaining why no experiment was described, stated that "The data that are reported here are the outcome of these exploratory sessions of the project, rather than the result of an experiment with specific hypotheses and a formal design." The exploratory sessions, involving individual behavior, occurred before the Command Research Laboratory became operational. Finally, a series of experimental studies of human dyads, engaged in bargaining and negotiation behavior, was sponsored by ARPA but conducted in the Systems Simulation Research Laboratory, as mentioned earlier; these studies are described briefly in Chapter 23.

### *The Display I Experiment*

This experiment investigated the differential effects of four display arrangements on the performance of one of three simulated military officers in an alternate command post. These officers diagnosed the outcome of a nuclear war's first exchange and planned the restrike. The four arrangements were a large screen display, a console display, a combined console and large display, and no display. How this experiment came about has been summarized by Cusack (1963), the principal investigator: "Some months ago it was decided by management that the ARPA laboratory should be modeled after some real-world counterpart to add realism, content, and applicability to our programs. An Alternate, Mobile Command Post (AMCP) was chosen as the laboratory vehicle, to be physically and conceptually similar to such a facility at the CINC level within a military organization."

Cusack further explained that "The size of such a command post (probably housed in an airplane) matches well with the amount of equipment available for the ARPA laboratory"; and that the selection was topical and could yield applicable results. The next planning step, he said, "was the selection of an environment in which to operate and which would provide inputs to the AMCP." It was decided to use portions of a model of interconnected command posts which had already been developed in the ARPA project; this COMMAND model included a scenario, an engagement analyzer, a situation recognizer, an action selector, and a data base. Since presumably this elaborate model had been developed to support project research, here was a chance to exploit it. (Project personnel appear to have made relatively little other use of it.)

The third step was "the selection of a decision area with which to challenge the experimental subjects." The choice lay between four episode periods: a considerable time before hostilities, a short time before, a short time after their onset, and a longer time afterward when damage could be assessed. The fourth was selected.

The last step was to determine what to experiment about, within the foregoing contexts. Cusack (1963) wrote:

> A list of candidates for laboratory research was then solicited and evaluated . . . . Out of those submitted, a set of six display studies were the only ones which met the objectives and were, therefore, selected as the core concept of the initial laboratory plan . . . . We are concerned with the first such display study: individual versus group displays. The hypothesis to be tested is: Does the inclusion of a large-screen (wall-type) display—capable of multiple viewing within the operating system—assist, detract from, or have no effect on the speed and accuracy of reaching damage assessment decisions?

*Experimental Operations.* The situation required three subjects in each run: a damage assessor who tried to find out what U.S. facilities (e.g., ICBM launch sites, cities) had been put out of action or destroyed by the enemy's strike; a strike assessor who tried to ascertain what U.S. missiles and aircraft had done to the enemy's territory; and a planner who "was charged with finding out which enemy targets that had been selected for destruction were not destroyed, and what missiles he had available that could be paired with those targets." The planner received information from the assessors directly, by telephone, or by display via the computer's processing of their inputs. The assessors received messages from simulator personnel representing air bases, missile sites, cities, photo-reconnaissance centers, etc. One simulator phone line also connected the planner with a simulated commander in chief who kept requesting information. The equipment with which the subjects and simulators worked was that listed in the description of the laboratory.

The twelve subjects in four three-man teams were selected by performance scores from a larger group of U.C.L.A. ROTC students who had been participating in other ARPA project studies; the top performers in these studies were assigned as planners. Each team encountered each of four display conditions once, but these conditions directly affected only the planner. He could view a condition which consisted of the large display, console display, both, or neither. Each assessor always had his own console. Five different "wars" were scripted, one for training and the others to provide different inputs to a team in successive sessions. The scripts were originally designed as approximately equivalent in load, with thirty-nine bomb bursts on each side.

The sixteen sessions each lasted one hour. Two-hour sessions had been planned, but pilot runs indicated clearly that "sessions of this length were almost sure to be either seriously interrupted or terminated by computer or program malfunction" (Cusack 1964*a*). The session duration determined the input load. The computer and program malfunctions also led the researchers to convert the first week of sessions into training sessions exclusively, in place of alternating training and data-taking sessions, and to postpone the latter a couple of months. Even so, or possibly because of the interim period, the performance of the subjects continued to improve during the experimental sessions—except between the second and third sessions, when there was a seven-day layoff.

The experimental design consisted of two 4 X 4 Latin squares superimposed, with the effect that the display conditions and the war scripts were confounded; a particular war script was associated (twice) with a particular display condition and only that condition. This lack of orthogonality would probably have made little difference if the war scripts had remained equivalent in load. However, during the training sessions it was found that the planner could not complete his work during a run, so the rules of the game were changed to make things easier for him. But "by changing the rules, we unknowingly changed the difficulty of the problems and upset the principles of the design" (Cusack 1964*b*).

The researchers realized that the four-condition design, in any case, did not fully compare the display situations. For example, the large screen display could be viewed by all three team members, but the display at the planner's console with which it was compared could be viewed only by the planner. Further, for the large display there was a 35-second delay in updating the displayed data, whereas the console had no delay. However, the researchers felt it would be unwise in a precarious shakedown experiment to take the time to add the conditions needed to assure full comparability. Cusack (1964*a*) commented that "it is apparent that interpretation of the results for this curtailed investigation will have to be extremely conservative." There was no attempt to evaluate the results for statistical significance.

*Results.* Summarizing the results, Cusack (1964*b*) said that the console by itself proved to be the best display condition. For some reason the large display degraded team performance when the planner also had his console, even though under this condition the planners virtually disregarded the large display. A number of other results hold interest for system design.

1. Planners made many errors in asking the computer for displays by operating their teletypewriter keyboards. Errors occurred in 25% of the display composition messages related to the large screen display. The percentage was halved when the planner was using a console, apparently due to the rapid feedback of information made possible by the console. It was thought that better human engineering of message formats would have lowered the error totals in either case. It seemed questionable whether a teletypewriter should be used as an input device for the displays.

2. Although all the damage assessors happened to be experienced typists, their error frequency too was high, ranging from 6.5 to 10.6%. Two of the strike assessors did not know how to type—an acknowledged oversight in selecting the subjects. They averaged more than twice as many errors as the other two. Cusack (1964*b*) suggested that in military computer-based systems of this nature, the officer operators should be good typists, or each should have a highly trained clerk, or display requests should be made by means of pushbuttons.

3. Presumably because they had fewer and less complicated messages to transmit, the strike assessors had about the same range of error frequencies as

damage assessors. Planners had the most messages and the most complex ones to handle, and they were the most susceptible to error. Accordingly, error frequency seemed to be, at least in part, a function of the complexity of the messages.

4. More than twice as many information requests were made by the planner when he used the console than when he could use only the large screen, indicating that "the ease of getting information . . . has more influence on whether the information will be requested than any other consideration." But results suggested that the larger number of display requests was not the factor that made the planner's performance (in pairing missiles with targets) better with the console.

5. The subjects were highly motivated, but they needed more training than had been estimated as necessary. Also, there were major differences between the performances of different teams. Thus, it appeared that the selection and training of officers manning an alternate command post in a nuclear war could profoundly affect its outcome.

Cusack (1964*b*) observed that selection, training, and human engineering presumably interacted with each other, in that the contributions of any one of these could reduce the need for either of the others. He said the Command Research Laboratory constituted an excellent facility for investigating this interplay and its implications for total system design.

## SIMFAC–THE SIMULATION FACILITY IN PARAMUS, NEW JERSEY

In 1959 the System Development Corporation subcontracted with the International Electric Corporation (IEC)–a subsidiary of the International Telephone and Telegraph Corporation–to provide system analysis and develop computer programs for a computer-based command and control system projected for the Strategic Air Command (SAC); later SDC became an associate contractor. The system is known as 465L, described by Parsons and Perry (1965) and Shaw (1966).

From the outset of this work, which was centered first in Paramus, New Jersey, and later in nearby Lodi, SDC envisioned operating a simulation facility (generally called SimFac) in which to conduct man-machine system experiments. Such a facility initially had such objectives as: "1. To analyze the human operators' abilities relevant to the use of visual and auditory displays and workspace layout. 2. To try out and analyze system and subsystem operational configurations and decision-making capabilities" (Jaffe 1959). The facility was supposed to contain a partial physical model of the SAC Control Center in Omaha, Nebraska, as it would exist after the 465L system was produced and installed there (Redgrave 1962).

SDC researchers hoped that SimFac construction could begin by February 1960 and that the first multicrew experiment could start in August or September 1960 (Jaffe and Adamson 1959). However, the project was delayed, partly because the prime contractor, IEC, was less than enthusiastic. SDC had to present a briefing in mid-1960 to indicate SimFac's needs, uses, plans, advantages, and financing (Rhine 1960) before IEC agreed to SimFac's's creation.

One reason for the prime contractor's hesitation was the fact that it was establishing a test facility of its own to check out system equipment and programs and to demonstrate system feasibility. SimFac was accepted on the basis that it was aimed, far more modestly, at the empirical investigation of procedures, training methods, aids, forms, and displays–in other words, at the system's operators. However, since it had its own test facility, the prime contractor was never too eager to use SimFac itself; neither was the ultimate recipient of the system, although SAC did provide subjects for an experiment.

SimFac was finally completed and became operational in the spring of 1961. It occupied 4,300 square feet in an IEC building at some distance from the SDC offices. It comprised two laboratories, the Simulation Laboratory for conducting complex crew studies, and the smaller Human Factors Laboratory for component research. The Simulation Laboratory included an operations room, consisting of two elevated daises facing a large wall projection screen; an observation room for monitoring, control, and "embedding" simulation; and a debriefing room. In addition the facility contained a data production and processing room, an equipment and maintenance room, a photographic laboratory, and a projection room.

Preceding SimFac were three small experiments on display design, all on various aspects of symbol legibility. The Human Factors Laboratory was the scene during 1961–63 of five more small experiments. One concerned the need for vertical lines in tabular displays. Two dealt with item tags in messages, and two more investigated message processing.

Four experimental investigations of displays took place in the Simulation Laboratory in 1962 and 1963. One measured how fast ten individuals, who were SDC employees, individually read a variety of tabular displays; the purpose was to determine how long the displays should be held for viewing (Foster 1963). In another, seven individual subjects (supplemented in one part by six more) requested displays by means of a display request panel and a "Prototype Display Request Message Handbook"; the objective was to find out how long it took to request messages, how frequently errors occurred in doing so, and what kinds of errors they were. A third study examined the relative advantages of wall displays and simulated computer printouts (Schwartz et al. 1963). Six subjects (SDC employees) individually viewed a large number of displays twice in each of three situations. In one, displays were presented on a large screen instead of in printouts if, operationally, there was an option. In another, displays were presented only in printouts. The third condition used a mix of displays specified in the system design. The subjects, simulating a single SAC controller during a peacetime force exercise, had to detect and take action in response to mission deviations and emergencies.

### *Simulation Study I*

None of the foregoing projects really had sufficient scope and complexity to justify construing it a man-machine system experiment. However, one experiment in the SimFac Simulation Laboratory came closer to qualifying as that kind of study. It was the only one that did so–although such experiments had been the principal objective of SimFac. This was Simulation Study I, reported by Wakeman (1962) and conducted in the Simulation Laboratory during February of 1962; in conjunction with it, a display experiment was done in the Human Factors Laboratory using the same subjects, who were SAC officers. The Roman numeral "I" correctly implies that additional studies of this nature were contemplated. Simulation Study II reached the planning stage, and the objectives of four more were laid out in 1961 planning documents. Some of these, as well as earlier aspirations, will be noted shortly.

Simulation Study I investigated 465L's "data presentation subsystem from the view of the Operations Controller. The objective, broadly stated, was to determine the information requirement for SAC operational control, and to see how well these were met by the current display package. The method, also broadly stated, was to present to experienced SAC Control personnel a representative group of control situations, and to observe and record their relevant behavior, and comments, for analysis" (Wakeman 1962).

The study had been planned to consist of twenty-four one-week experimental sessions (Scott 1961), although it was realized that the scope might have to be reduced if enough military subjects were not provided. Actually, the study lasted a single week. Four two-man teams of experienced SAC Operations Control personnel were the subjects, each team taking part in a two-hour exercise, followed by a debriefing session. Monday was devoted to orientation and training, Tuesday to the Human Factors Laboratory displays experiment (which will not be described here because the report is classified), Wednesday and Thursday to the four experimental sessions, and Friday to a workshop at which the officers critiqued the system's displays.

A shakedown run of the study had been held in January with SDC personnel stationed at SAC headquarters as subjects. A pilot study four months earlier had revealed inadequacies in the experimental vehicle and in some of the laboratory operation procedures (Scott 1961). This pilot study used SAC officers as subjects, and their experience during its somewhat frustrating course was one of the factors that dimmed the military enthusiasm for the research project.

In Simulation Study I a single scenario detailed the operations of a hypothetical SAC force that, in contrast to a real one, was altered and reduced somewhat in size and structure to meet both security and laboratory limitations. In the scenario, numerous operational problems developed and the SAC controllers had to cope with them by using displays. The force consisted of fifteen SAC bases in the United States and 450 aircraft. Two simulator personnel, communicating with the controllers by telephone or simulated radio, represented these bases, the three numbered Air Forces to which they belonged, bomber and tanker crews, internal headquarters elements, the North American Air Defense Command, and the Joint Chiefs of Staff. The experimental staff also included six persons to manage and operate the presentation of wall displays and printouts; three to make observations; and one to act as the co-ordinator. In addition a staff member acted as an airman, assisting the subjects by transmitting display requests to display control.

To simulate the wall displays, which were to be a critical computer-generated feature of 465L, random-access slide projectors using a rear-projection technique presented preproduced slides indicating changes in various categories of data. A button-operated request device selected slides for display; it was situated in display control on the observation deck. Computer-generated printer outputs (printouts, i.e., hard copy) were simulated by pretyped and photographically reproduced pages of information which were dropped to the subjects from the observation deck through a slot; a bell rang each time one was delivered.

The rear projection screen was 10.5 feet high by 34 feet long. Twelve projectors (out of the laboratory's actual complement of fifteen) could display up to thirty thousand alphanumeric characters of information on the tabular displays, which occupied most of this screen in matrices of 48 rows by 72 columns and 16 rows by 72 columns. Between the matrices was a map display 71 inches high by 89 inches wide. These various displays were somewhat smaller than those being developed for 465L, due to laboratory dimensions. The laboratory was equipped with four individual request units to select which projector to use. This meant selecting some portion of the total screen surface and its particular repertoire of displays. Through a request unit it was also possible to select individual slides, or an entire sequenced set of ten, among the forty slides which were associated with each random-access projector. A timing control system could sequence the presentations of slides automatically. A method developed for producing new slides during an exercise was not exploited during this study. Slides were prepared by photographing carefully typed McBee Keysort cards. Photographic equipment included a 2 × 2 glass plate camera, a 4 × 5 view camera, a 35 mm. microfilm camera, and a standard 35 mm. camera. SimFac and its equipment have been described by Martin (1961*a*, *b*).

A telephone system consisted of twenty 9-line talking stations and ten 18-line monitor units, with three patch panels to reconfigure connections. The SimFac staff used a separate intercom system composed of four master units and eight slave units. Microphones and speakers were included in the sound system. Voice communications could be recorded on twelve tape channels, and subject activity on a 40-channel event recorder. Illumination could be controlled throughout SimFac. For example, the operations room of the Simulation Laboratory was divided into four illumination level control zones, three over the two daises where subjects were located and one over the display screen.

Simulation Study I addressed itself to four questions about 465L displays (Wakeman 1962). One asked how SAC controllers used them. The primary data source was a record of the displays which were selected by the subjects during the experiment; additional sources were a record of the sequence of controller actions, a record of intersubject communications, and subjects' comments in the debriefings. The second question asked what portions of the display system should be redesigned to be more useful, and how. Here the data source was the workshop, supplemented by debriefing comments and the intersubject record. The third question was: What information necessary for SAC control is not provided by the displays? A telephone request record was the principal data source, supported by the intersubject record, debriefing comments, and the workshop. Finally, what would be the most useful physical arrangement of displays, and the procedures for accessing them? The answers to this question came from analysis of the display record, the intersubject record, and debriefing comments, and the question was discussed in the workshop.

*Display Design Findings.* Although the available report on Simulation Study I included no quantitative data among the results, it stated numerous conclusions concerning display design (Wakeman 1962):

> . . . 465L displays require the responsible controllers to go from display to display to pull together the elements of the problem. This was one of the most frequent complaints of the subjects. They felt that it should not be necessary to look in several places to get the "complete story." Fewer displays containing more complete data would give them what they felt was the required information.
>
> . . . when a number of problems come up simultaneously, the screens can be flooded. This problem of lack of priority exists in 465L as it is currently conceived, and the whole area must be reconsidered. . . .
>
> [A more complete mission profile] is required in order for the controller to review the seriousness of the problem . . . if the situation is an emergency, all available information must be available immediately. It was strongly recommended to have problem-related information forced with the profile information . . . .
>
> The subjects pointed out that the bomber-tanker operation is most critical in the refueling phase. When this is a part of the mission being monitored, it is important to have both sets of information co-located. Many of the refueling difficulties originate in a deviation of one or the other aircraft and the timing information for each is needed in order to determine quickly the rescheduling requirements . . . .
>
> While the general sequence of the mission was depicted by the order of the various columns, the deviations were grouped out of order: review would be more easily accomplished if each deviation were presented next to the related category . . . .
>
> Comments indicate dissatisfaction with the treatment of deviations in terms of the method of data presentation. The present technique suggests that all deviations are the same, and that the circumstances surrounding them do not vary. We know, of course, that the designers did not intend this, but we also know they intended to place the burden of differentiating among deviations and circumstances on the operator . . . .
>
> The controllers' long standing experience with weather symbology and descriptions led many of them to question the decision to eliminate these symbols from the group display generator inventory. While granting that training could remove the problem, they were nevertheless concerned with the change . . . .
>
> Too many clock times must be used in SAC today during emergency conditions. The situation is complicated by the displays if they use variations on this set . . . .
>
> For weather data the technique of signalling when a problem exists must be abandoned for an indication of a trend toward weather deterioration . . . .
>
> When presented, the data should be grouped so as to clearly define the scope and characteristics of the problem, and the impact of the problem on other areas. As presently designed, rather than relieving the system operator from extensive scanning (easing his problem definition task), the operator is forced to jump from display to display to tie together related data . . . .
>
> To recapitulate, the design indicates a lack of understanding of the need to eliminate uninformative displays, increase operator confidence in the accuracy, completeness, and pertinence of the displays, and expand the scope of the descriptive capabilities to reflect the circumstances underlying the events.

Hopefully, the final design of displays for 465L incorporated the suggestions which emanated from Simulation Study I. It has not been possible to determine whether this occurred.

*Other Plans*

In October of 1961 (Scott 1961) and again in March of 1962 (Scott 1962), Simulation Study II was being formulated to examine human factors problems in emergency war operations (EWO) in the 465L era. But, in December 1961, its purpose was stated as examining the effects of data availability and update frequency on decision time and the relative advantages of printouts and wall displays for presenting tabular information. Simulation Study III, requiring ten crews, was to be aimed at determining minimum data requirements under conditions of hurried decision-making. Study IV was supposed to find out what happened when controllers lacked non–465L communications; studies V and VI were aimed at problem solving.

In earlier planning (Jaffe and Adamson 1959), SimFac research was to include "directness of communication"–e.g., what types of data are best communicated from the original source direct to the user; number of communication links; and time density of transmission. The study of this area was to take three months, either as a single experiment with a relatively small number of crews in a single analysis of variance design, or as several experiments involving smaller crews. Another project of the same magnitude and time duration was planned to study "The requirements for updating and accuracy of data displays." A third would have investigated 465L to find out which portions of the system most needed training. In contrast, the briefing already mentioned at which SDC management justified SimFac to IEC management (Rhine 1960) emphasized its simulation methods and equipment and how SimFac would differ from the test facility. No experimental program was outlined.

In the story of SimFac, as in the accounts of other SDC laboratories, events and non-events speak for themselves. Some aspects of man-machine system experiments received more emphasis than others. Execution failed to match aspiration. Possibly the aspiration was the legacy of the RAND SRL experiments, described in Chapter 8, that were responsible for SDC's existence and the laboratories it spawned. But certainly these laboratories never rivaled in production and impact the RAND SRL and its research.

# 18

# Applied Physics Laboratory

Man-computer interactions in making military decisions were extensively investigated by R. M. Hanes and J. W. Gebhard at the Applied Physics Laboratory (APL) of The Johns Hopkins University over a four-year period beginning in 1962. Their five experiments dealt with task force anti-air warfare (air defense) centered in a simulated shipboard combat information center (CIC). (For accounts of other CIC studies, see Chapters 3, 4, 5, and 12). Although Navy officers were the subjects, the research was not a contracted project.

One of the critical features of all air defense operations is the attempt by the enemy to cripple the effectiveness of the defender's radars by jamming them, that is, by sending back electronic signals which prevent the defender from ascertaining the location of the jamming aircraft and which also can hide other aircraft from radar detection. Very properly, Hanes and Gebhard incorporated electronic warfare into their simulation of a hostile attack. But since electronic warfare (see Chapters 11 and 17) is a highly sensitive subject, most of the contents of the individual reports describing their experiments have remained classified (Gebhard and Hanes 1963, 1964; Hanes, Gebhard, and Emch 1962, 1963; Hanes and Gebhard 1963, 1964, 1965*a*, *b*). Fortunately, a review of the program has appeared in the U.S. Naval Institute Proceedings (Hanes and Gebhard 1966) and an unclassified report has been published describing the simulation techniques (Fagan 1963). What follows has been drawn from these two documents, unclassified portions of the other reports, and unclassified information from the experimenters.

The aspect of man-computer interaction Hanes and Gebhard examined was the extent to which commanders would accept or reject solutions to problems when these solutions emanated from a computer. The problems were those which a military commander might face in managing his weapons, particularly in the selection of weapons (aircraft and missiles) and in the choice of orders as to how they should proceed to repel hostile aircraft. Thus the experimental subjects, like those in the Operational Applications Laboratory experiments described in Chapter 16, were engaged in a series of management decisions, rather than in surveillance or direct control of interceptor aircraft. The dependent behavior in the experiments was essentially that of a single subject, called the commander. Assisting him was a team of four other operators, a talker who relayed voice messages and three plotters who manually maintained displays on which the commander largely based his decisions. All other personnel one would expect in a CIC were simulated by laboratory staff in the simulation center from which the exercises were run.

One of the displays was a computer-recommended action display, which showed computer recommendations as coming automatically from outside the simulated CIC. By doing nothing the commander could accept an action recommended on this display, in which case the action was executed. Or he could, within a fixed time period, operate a keyset to reject the recommendation, in which case the computer-recommended action did not take effect. In this case the commander could order a substitute action based on his own judgment. (See Chapter 12 for an earlier investigation of the extent to which human operators might intervene in a computer-aided CIC.)

In actuality, no computer existed, but the experimental subjects did not know this. The computer was simulated. Problem solutions were produced by a human simulator who was schooled in a program so he could produce the same recommendations which would have come from the program if it had been processed in a computer.

## TWO EXPERIMENTS WITH SOP

While the foregoing arrangement, described in more detail later, characterized the last three experiments of the program, the first two experiments did not pretend to incorporate a computer. Instead they were based on an interesting analogue of the computer-recommendation context. Before the experimental runs, the subjects (commanders) studied a comprehensive set of standing operating procedures (SOPs) prepared by the experimenters with expert consultation. When the simulation center staff adhered to these during a run, the resulting actions were analogous to the action recommendations from the computer in the later experiments. SOP solutions were reported to the commander during an experimental run; the commander could follow the SOP actions or not. One of the points of qualitative interest in the first two experiments was the extent to which a commander first stated his agreement with the written SOP and then put it into practice.

The first experiment was also intended to explore the effectiveness of the simulation technology and to obtain information about commanders' decision-making behavior in various tactical and mission situations. The ten Navy officer subjects had had CIC experience. During five days of training and experiment for each, one group got four practice runs and sixteen experimental runs of 30 to 60 minutes each; another had three practice sessions and sixteen experimental runs, some of the latter lasting only 15 minutes. A debriefing was held for each officer at the end of his experimental participation.

One reason for this procedure was that after the first set of runs, changes were made in one of the basic inputs as a result of experience with that input; further, one of the tactical parameters was changed to acquire information with which to design a subsequent experiment. Various attack tactics and parameters in the simulation were systematically introduced within each set of runs, in such a manner that comparisons could legitimately be made among the results. Measurements were obtained concerning the number and latencies of a commander's orders for his own aircraft, in addition to data about the distances they flew and the extent of hostile penetrations.

The second experiment involved only four officer subjects and followed much the same procedure as the first; it was regarded as complementary to it. In some of the eighteen experimental runs per subject, following three practice sessions, the commander was required to stick to the SOPs; in some he could deviate from these and adopt his own options. (See Chapter 16 for a related type of situation in one of the Operational Applications Laboratory studies.) Again, tactics and parameters were systematically varied within and between runs. Thanks to experience gained in the first experiment, some of the simulation technology was altered in the second. Although several types of measurement were added, measures again reflected both system performance and the particular management actions of the commanders.

Throughout this research program the data reported were both quantitative and qualitative. Experimental design helped assure that differences in results were attributable to different states of independent variables among inputs, procedures, and subjects. Quantitative data were subjected to significance statistics using the 5% level in most cases. Generally subjects served as their own controls, but in two experiments they were confounded with other variables. Qualitative data came from subjects' comments during the experiments and in debriefings; and some were impressions which the experimenters themselves acquired.

## THREE EXPERIMENTS WITH A SIMULATED COMPUTER

As already indicated, the third experiment initiated the simulation of a computer and interaction between it and the commander. Each of eight CIC-experienced Navy officers went through four practice and twelve experimental runs of 30 to 60 minutes each. The simulation inputs for the practice runs were drawn from the first experiment, those for the experimental runs from the second experiment. Instead of studying an experimenter-written SOP, each subject had to write down his own SOP before the runs began, for later comparison with the procedures which he actually employed. As in the earlier studies he also received a handbook describing various aspects of the operational situation in which his actions as a commander would be embedded. In addition, in this experiment each subject received extensive indoctrination not only in the manner in which he would receive computer recommendations and could reject them, but also in the program logic which would be responsible for those recommendations. Half the subjects received the latter instruction only when they had completed half their runs. Four computer demonstration runs were included. The commanders were also given data which purported to show how well the computer program had performed in comparison with human management.

The principal results from this experiment concerned the over-ride of computer recommendations, through rejection, anticipation, or substitution. In addition to frequency of types of over-ride, it was possible to establish how the frequency of over-rides was affected by various attack tactics and parameters, individual differences among subjects, and delaying the indoctrination in computer logic. Because the simulation inputs were the same, it was possible to make certain comparisons between the results of the computer-less second experiment and this one, in terms of system and manager performance, although the subjects were different. Guarded comparisons could also be made between the extent to which the subjects in this experiment over-rode the computer and the extent to which those in the second experiment over-rode (disregarded) the SOP to which they had been exposed; to a considerable extent the computer program in this third experiment coincided with the SOP of the second.

The fourth and fifth experiments elaborated on the third by giving the commander the option of selecting among a number of different computer programs with which to manage his forces in dealing with several different kinds of attack; he could still over-ride the computer as in the third experiment. In the fourth experiment he made his selection on the basis of advance intelligence reports, which predicted the kind of attack with or without error, and he could not change programs during the battle. In this, as well as the fifth study, some of the intelligence reports could mislead the commander by indicating a different type of attack from the one which he would encounter, some gave a correct prediction, and some provided no basis for prediction. In addition, the fifth study allowed the commander to change his selected computer program during the attack to a program which seemed more appropriate to the unfolding tactical situation. In a sense, this option was another form of over-ride, applied to an entire program instead of to individual program-produced recommendations.

As Hanes and Gebhard (1966) described the situations, "three defense logics were made available to the commander for comparison with the case where there was only one. We wrote one to be especially efficient against a massed raid from a single direction, another was for a broad-front attack, and the third was a generally conservative logic for a deceptive situation. The result was clear: commanders over-rode less when they were able to select the computer's defense logic to match their estimate of expected enemy tactics" and even less when they were able to "change the logic during an engagement, e.g., to switch from single-prong defense logic to general if the raid did not occur in the manner anticipated."

Various measurable behaviors on the part of commanders resulted from the various degrees of uncertainty created by the advance intelligence and subsequent attack. It was possible to compare the frequency of human over-riding of the computer recommendations when the computer program was appropriate to the kind of attack with the frequency when it was not. In the last study it could also be seen how often the commander changed the computer program when it needed changing and how often when it did not. Still another kind of measure resulted from a requirement preceding the experimental runs in both experiments. The subjects had to estimate the probability of each kind of attack, relying on the intelligence summary. Subsequent analyses looked at the relations between their estimates and their selections of defense strategy.

One of the management strategies programmed for the fourth experiment resembled what had been used in the third, permitting some interexperiment comparison. It was also possible to make extensive comparisons between the results of the fourth and fifth experiments, because not only were the various management programs the same in the two studies but so were the attacks and scripts. However, the experimental approach differed with regard to subjects and sessions. In the fourth study the subjects consisted of twelve Navy officers who had had CIC experience. Each first went through four manual practice runs (without the computer), four computer demonstration runs, and a single additional practice run with over-ride. Twelve 22-minute runs per subject followed. Different sets of four subjects each were assigned to the three different conditions of advance intelligence. In the fifth experiment, on the other hand, each of forty-four officer subjects spent only one day at the laboratory, in contrast to a full week for all in the previous experiment. Relatively few subjects in the fifth experiment had had any extensive CIC experience, but this variation permitted some assessment of experimental results in terms of operational experience. In his one day, each subject experienced three computer demonstrations, a single over-ride practice run, and four experimental runs.

## SUBJECTS

All told, eighty-one naval officers took part in this experimental program as commander subjects, ranging in rank from captain to lieutenant. Those in the first four experiments came from the Atlantic Fleet, those in the fifth experiment from various Washington bureaus. All were made available at no cost to the laboratory, and in most of the studies almost all had had relevant experience. That is not to say, however, that their experience by any means matched some of the simulated warfare which they encountered in the laboratory. As a matter of fact, in general they regarded their laboratory exposure to various attack tactics and aspects and their experience in counter-management as extremely educational and helpful to their operational activities. The experiments taught them how to fight. The simulated computer programs "challenged the commanders who came to APL to give more attention to AAW tactics and doctrine than they are usually able to do" under the burden of administrative duties (Hanes and Gebhard 1966). It is also believed that many brought back with them to the fleet some of the SOP or program procedures which they learned in the experiments and through the handbooks which accompanied these.

One of the questions raised by this serendipitous result of the research program is whether the procedures the officers acquired would be valid in the operational setting, a setting which can be construed in two ways. One is that of actual hostilities. It is difficult to evaluate the degree of transfer from the simulation to wartime operations, but the experimenters attempted to do so in the only way available: they recorded the running comments of the subjects on tape and also queried them about the realism of the simulation (Gebhard and Hanes 1964). The other kind of operational setting is that of peacetime shipboard exercises, performance in which may be important for personal advancement. It is not clear how appropriate the special experience gained in well-simulated warfare would be for such achievement.

## COMMAND AND SIMULATION CENTERS

The physical arrangement for the simulation comprised a command center (a sort of stripped down CIC containing Navy equipment) and a simulation center. As noted earlier, the commander made his decisions in the command center while watching several displays; and he communicated them by pressing a button on a keyset to reject the computer recommendations or, alternatively, expressed his own decisions for vocal transmission by his talker. The command center is illustrated in Figure 23. One might regard it as a more or less contrived or hypothetical system in the simulation of a real context.

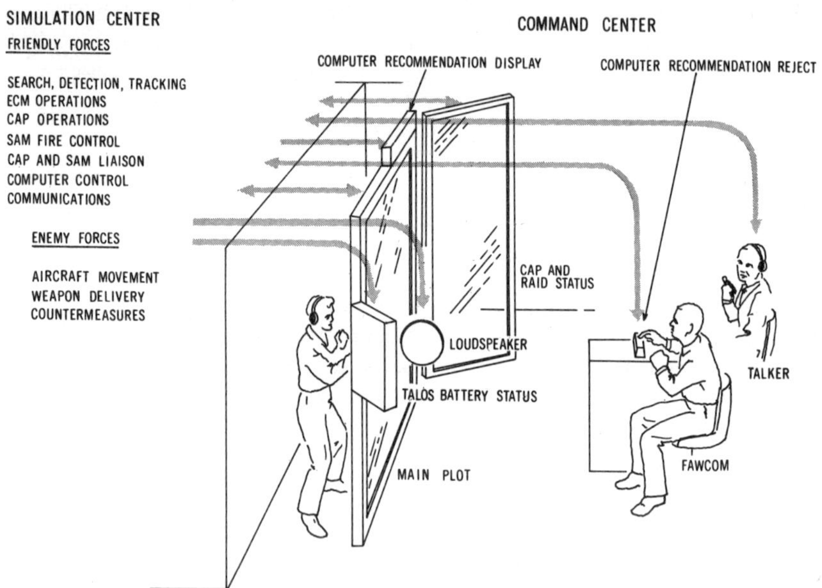


Fig. 23. The Simulated Command Center (Hanes and Gebhard 1966).

The recommendations display consisted of four variable indicators which communicated messages, and a fifth indicator which, when it appeared, always recommended launching more aircraft from the carrier. In each of the variable indicators the assignment of aircraft or missiles to specific hostile aircraft could be shown by a selection of letters and symbols. These recommendations included the stationing of aircraft, their return to the carrier, and their potential loss due to inability to return to the carrier upon completing an assignment. A color code designated aircraft speed. The number of elements increased between the third and fifth experiment, suggesting a growing sophistication in the simulation.

Whenever a new recommendation appeared, a green light appeared on the commander's control keyset for 20 seconds, followed by a yellow light for 10 seconds; the commander had 30 seconds in which he could reject any or all of the recommendations, and if he did so a red light appeared.

In addition to the computer recommendations display, the command center contained a geographical display which showed positions of attacking and defending aircraft, a status display for both of these, and a status display for missiles. The plotters at these displays received the data they placed on the displays either over a telephone line or from a loudspeaker.

The simulation center, diagrammed in Figure 24, demonstrated what can be accomplished with ingenious manual operations when neither a computer nor a

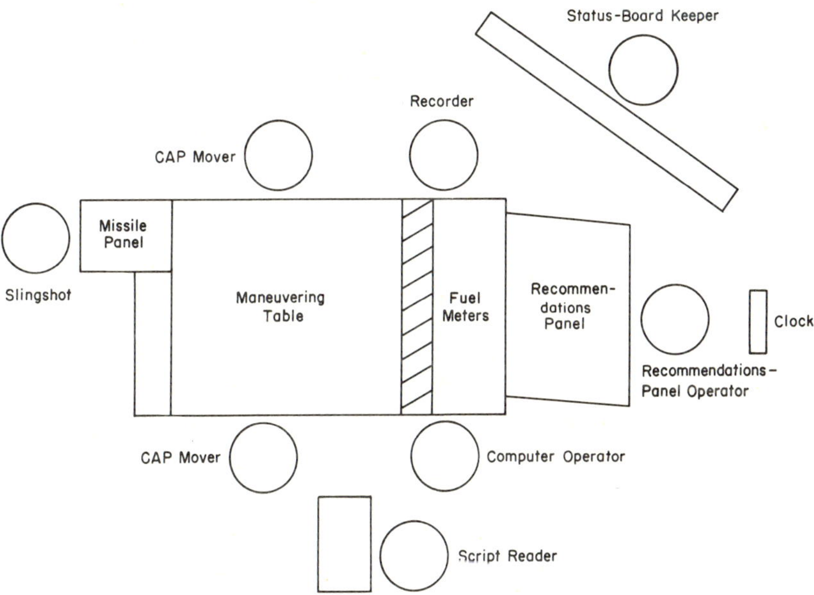


Fig. 24. Diagram of the Simulation in the Third Experiment (Fagan 1963).

complex of electromechanical target generators is on hand to create the air environment in an air defense simulation. The geographical picture was maintained on a horizontal maneuvering table which had a transparent top. Paths of attacking aircraft were represented by sequences of lights under the surface; a new lamp lit up every minute to show the progress of the attacker toward the center of a carrier force, and the progression occurred automatically once it had been switched on. In coincidence, a prerecorded tape played over the loudspeaker conveyed hostile aircraft positions to the command center. As soon as the prerecorded tape was invalidated as a result of command center action, a human substituted on the loudspeaker and provided both preplanned and new information, such as the destruction of attackers and changes in their designations. The tape and human sources also produced information about hostile electronic countermeasures.

Interceptor positions were represented on the maneuvering table by colored tokens which were moved by hand by operators from one grid position to another in accord with movement orders observed on an aircraft status board in the simulation center. An operator maintained this status board by listening in on the telephone line from the commander's talker in the command center, and also by monitoring the instructions given by the human who was simulating a computer. The latter did not himself operate the switches which activated the computer recommendations panel, but instructed another person in the simulation center.

Other personnel in the simulation center included Slingshot, who operated a missile control panel, and a recorder, who kept a manual record of information exchanged during the run and also gave fuel, ammunition, and "splash" (kill) reports. Fuel meters were simulated for the interceptor aircraft, since jet fuel is a critical factor in interceptions and its consumption per mile varies with interceptor speed and altitude. Tape machines recorded the various voice communications on telephone lines. The command center displays were systematically photographed. At least eight people were required in the simulation center during a run, in addition to the four who assisted the commander in the command center. Some changes in instrumentation came about with experience. A commander's plan position indicator (PPI) display was eliminated, as was television transmission between the two centers.

The operator who simulated a computer filled the most difficult role. He had not only to know the prepared computer program but he also had to adapt it around orders which the commander himself issued and derive recommendations from the adapted program. Fortunately for him, and for the entire simulation effort, these experiments were one-sided rather than two-sided simulations. The latter would have demanded a degree of reactivity in this simulation which would have been most difficult to execute. As a matter of fact, initially considerable thought was given to basing the research program on two-sided games, but the idea was abandoned as unfeasible.

## SYSTEM DESIGN OPTIONS

Hanes and Gebhard (1966) have pointed out that in designing the research program, choices had to be made concerning the role of the simulated computer. For example, should it present recommendations automatically or only on demand? If the former, how frequently? Should the commander be required to take some positive action for the recommendation to be accepted or should this take effect automatically if he failed to reject it? Should the computer repeat a rejected recommendation?

The answers to these questions came partly out of the experimental research itself. Because in the precomputer experiments "some conservative commanders delayed joining battle for an obviously excessive amount of time," the automatic-recommendation mode was adopted "because it overcomes a demonstrated inertia in the manual system." It was also found that the experimental commander apparently could comprehend and evaluate as many as five computer recommendations in 30 seconds, a 60-second period being considered by most commanders as too long. A few officers disliked the idea of commanding by default but most approved of the reject method; a positive acceptance approach encountered the objection that "an indecisive commander can delay indefinitely by doing nothing," and the experimenters felt a computer-aided control system should minimize delay. Some commanders would have preferred the computer to repeat its recommendations, but it seemed better to the experimenters "to consider that the commander has an alternative in mind when he rejects, and consequently should not be dunned with the same recommendation." The experimenters also considered whether to design the simulated computer program so it would explain its recommendations, but they decided it was sufficient to display clearly "all the data on which the recommended actions are based." These various problems and data pertaining to them have been also discussed in another report (Gebhard and Hanes 1963).

Hanes and Gebhard (1966) expressed the opinion that many of these options among which they had to make selections should be put to experimental test in further research. Their experimentation simply initiated an attempt to show "how the commander interacts with the machine that is presumed to help him." There was a presumption that "to bring automation effectively to the aid of the commander, it is first necessary to discover the conditions under which he accepts it." It was noted that while commanders welcomed computer-aiding in principle, their actual "acceptance of tactical action recommendations, as revealed by our experimental data, varies from 0 to 100 percent," and was determined by a number of factors which included "the personality and background of the commander, a matter that the machine designer often overlooks." The researchers added:

> Some commanders overrode the computer program extensively. . . . Anticipations of recommendations show that the commanders often think the computer is too slow on the up-take. Conversely, rejections are often made because they think the computer is "jumping the gun." Variable computer assessment time under the commander's control is a matter that should be investigated . . . . Override often occurred simply because the commander felt he had too little time in our situation to evaluate the recommendations. Override here meant stopping the action to gain time to consider. It did not indicate necessarily that he disapproved of specific orders. Controls for selecting the number of recommendations and length of time they are displayed might also be useful things to give the commander, but we predict that they would result in additional delays. The quality of the data on which the program bases decisions is frequently questioned. This suggests that a weighting factor for reliability should be attached to input data and taken into account in the computer's deliberations . . . .
>
> For the most part, commanders feel challenged by an automated system that keeps them on their toes—that forces them to evaluate and control a combat situation at a lively pace. The effect of this challenge is not just one of competition with the machine, but rather an impressed requirement for cooperation—to work effectively with the machine lest the situation get out of hand.

It can be readily appreciated that although rigorous design criteria as such have not emanated from this research, this conclusion and the others expressed by these authors have important implications for the design of computer-augmented systems. However, although some of the recommendations of the officer subjects concerning the automation or nonautomation of certain displays may have been adopted, there seems to be no evidence as to whether system developers have as yet exploited the findings of this research program.

# 19

# MITRE Corporation Simulation-based Testing

In conjunction with some of its developmental work, the MITRE Corporation has carried out a number of simulation-based testing programs that should be described in a review of man-machine system experiments, even though in some instances the MITRE researchers made no ardent claim that their studies were experiments, and in others this label may be debatable. The MITRE projects will be described under four headings: the Systems Design Laboratory, the AESOP program, command and control systems, and air traffic control studies (which definitely included large-scale experiments).

## SYSTEMS DESIGN LABORATORY

As part of a joint program of the MITRE Corporation and the Electronic Systems Division of the Air Force Systems Command, the Systems Design Laboratory was built in 1963 at L. G. Hanscom Field, near Boston, Massachusetts, and was formally opened in December of that year. Major General C. H. Terhune, Jr., commander of ESD, indicated on that occasion that one major purpose of the laboratory was to simulate and test new command and control systems. He said (Bamford 1964): "In order to test new systems properly, we must be able to make precise measurements, simulate actual operating conditions, and handle large masses of data . . . . This requirement points up the need for computerized systems that will enable commanders to virtually tailor-make their weapons' response to suit any situation at a moment's notice . . . . This is the facility that will do that."

The Systems Design Laboratory was a new, windowless, air-conditioned, two-story building with 45,000 square feet of floor space. On the bottom floor was its principal component, the very large and powerful IBM 7030 computer, called STRETCH. On the second floor were large areas to accommodate the extensive sets of consoles, wall displays, and operator teams that characterize command centers in the major command and control system of the Air Force—such as SAGE, sketched in Chapter 11; the Strategic Air Command Control System (465L), brought to view in Chapter 17; and the North American Air (subsequently Aerospace) Defense Command's Combat Operations Center (425L), known as NORAD COC.

By the end of 1966 the large experimental space had shrunk to two small areas, one containing four input/output stations (each with a cathode ray tube display, light pencil, typewriter, and printer), the other consisting of three cubicles, each for an individual operator equipped with manual aids. This was the set-up for the second of the AESOP studies, the only program approximating the nature of man-machine system experimentation undertaken in the laboratory. Much of the former experimental space had been converted into offices for programmers. The experimental area had seen little activity. Some evaluation of operations at prototype BUIC (back-up interceptor control) System consoles constituted the only operations-centered work on command and control systems. The laboratory was not mentioned in a comprehensive review of MITRE's efforts in the evaluation of such systems (Jacobs 1965). But the computer was not idle. It was a much-used programming tool. A number of computer programs were developed and tested with it, some of them with applicability to command and control. Estimates of the cost of the Systems Design Laboratory have not been published.

## AESOP PROGRAM

AESOP is an acronym standing for "an evolutionary system for on-line processing." It consisted of a set of computer programs and the input/output stations mentioned above. Summers and Hazle (1965) described it as "an on-line program system providing capability for operator controlled retrieval from and modification of a data base and for real time construction and execution of logical and mathematical procedures." It was also characterized as "an experimental on-line information control system . . . a prototype for a class of management or command information systems capable of giving the user as much on-line control over system performance as possible" (Bennett, Haines, and Summers 1965; Spiegel, Summers, and Bennett 1966; Summers and Bennett 1967).

In AESOP, the computer presented displays on the surface of a cathode ray tube (CRT), in tabular or "tree" format, and also on a printer. An operator gave orders, new data, and data retrieval requests to the computer by signaling dots on the CRT with a light pencil or by typing on a standard keyboard. The computer responded fairly quickly. Thus the operator and his input/output devices were on-line, operating in real time. The AESOP programs and input/output devices time-shared the computer, that is, they were interleaved with other programs and input/output devices which the computer serviced during the same time period. Most of the AESOP programs were basically general-purpose for dealing with any large data base. Others were problem programs concerned with particular data bases associated with particular systems. The first system to which AESOP was addressed, and the one which figured in the Systems Design Laboratory program, was a tactical air control center (TACC).

MITRE's interest in the management of tactical aircraft initially centered on the modeling of higher headquarters' planning of aircraft deployment; but when acceptance of this approach seemed difficult to achieve, it shifted to "next day" planning at a tactical air control center. Such a center might be a relatively small, perhaps mobile unit, close to the scene of such operations as interdiction missions. A laboratory replica of the current planning division of a current, nonautomated (manual) center was created in the Systems Design Laboratory late in 1964 and was exercised in a simulated environment. Then a parallel, computer-supported version was designed with AESOP. The two versions were operated in parallel in a comparison study in February and March 1965 (Doughty, Schwartz, and Cohen 1965). Subsequently, the structure and operations of the planning function at the tactical air control center were modified, the AESOP programs were developed further, and a second test series was conducted between October 13 and November 2, 1966 (Doughty 1967). Again, the manual system and the computer-supported system were operated at the same time and compared.

The MITRE researchers conceived of the laboratory representation of a tactical air control center as a "testbed" for AESOP. Their interests, in other words, seemed to lie more in developing a computer-based technology for querying and manipulating data bases than in tactical air control. It would appear, however, that tactical air control was chosen as the testbed because of the MITRE Corporation's role in engineering and co-ordinating the development of computer-based command and control systems for the Air Force; it was hoped that the AESOP-type technology might actually be adopted for tactical air control. In explaining their testbed philosophy, the MITRE researchers (Bennett et al. 1966) said that "experimental applications of on-line digital aids to planning can be exercised in order to systematically evaluate their utility. Such a procedure helps to insure the operational relevance and utility of the prototypes. Successful development and test then lead directly to field application and/or to the establishment of specific new requirements for computer assistance."

As additional uses of a testbed, the researchers specified the elaboration of planning requirements for the particular system, "experimenting with the planning process itself," and "demonstration." Visitors to the Systems Design Laboratory who spent a little time learning how to interact with the computer through the AESOP programs received a personally addressed diploma in a print-out from the STRETCH computer.

### *First Test Series*

The first series of test or experimental sessions comparing the manual and AESOP methods of next-day planning of tactical air strikes involved a three-man crew in each system. The planning resulted from interactions within this crew. Two crews of planners were composed of members of the MITRE technical staff and technical support staff. Each team operated both methods, manual and computer-aided, in alternate sessions; and in any one session both methods and teams functioned at the same time, with the same inputs. The inputs consisted of a scenario and associated messages, including telephoned messages. The scenario differed from session to session.

Each team had to allocate available aircraft, crews, and ordnance resources to satisfy requests for tactical air missions. The scene was laid in Iran. Both methods incorporated table and wall maps, work forms, a working assumptions manual, telephone communications to the simulated environment, and simulated teletype inputs. They differed mostly in the location of files (e.g., aircraft status, aircraft characteristics, ordnance status, planned missions, base-to-target distances). In the manual system these were conventional "paper" files; in AESOP they were stored in the computer. Actually, AESOP "did not provide adequate computer assistance in any of the more time-consuming and complex activities of the planning process; namely calculations, decisions, and clerical tasks" (Doughty, Schwartz, and Cohen 1965).

It had been planned to have twelve test sessions following two familiarization and twelve training sessions, the load progressively increasing in both training and test. Two of the training sessions were aborted, one was canceled, and in another the teams were not fully manned. At the end of the training sessions it was decided to terminate the series, so the only data available were from the training sessions. These data led the researchers to observe (Doughty, Schwartz, and Cohen 1965):

> Both the manual and computer-based versions of the TACC planning operations were allowed to evolve in the laboratory throughout this period. The test environment, data gathering techniques, and test procedures, under development during this period, were also changing constantly . . . . Therefore, the data reported here merely indicate trends and the conclusions drawn are of a tentative nature, pending verification in forthcoming, more rigorously controlled test series.

The data from six training sessions indicated a superiority for the manual system, but this characterized only one of the two crews. The manual system responded to more mission requests and required fewer sorties, fewer aircraft, and fewer bombs to meet desired kill probabilities. Also, the manual planners needed considerably less time to plan each mission. The two crews differed considerably in most measures, including frequency of error in typing inputs for AESOP. Although these dropped from initial highs of 40% for each team, the error rate in the final session was 24%.

### *Second Test Series*

The simulation of procedures and environment, described by Schwartz et al. (1965), had been based on considerable investigation of tactical air planning; and after the series there was more exploration, including a field study at the Air University to find out what criteria planners would use in formulating a particular plan or in comparing alternative plans (Edwards and Morrill 1965). MITRE researchers gave eighteen Air University personnel a tactical air control problem to solve, followed by a questionnaire probing how the solutions were achieved. For example, a respondent was asked whether he would send a single aircraft on a mission to attack a target, and what proportion of time he spent in decision-making, clerical tasks, organizing his job, and other aspects of the planning task. As a result of this study the testbed was revised and elaborated for the second series (Miller 1965).

The second test series again compared the manual and AESOP methods, three subjects operating each method during the same session and alternating between methods. But they performed as individual operators, not as members of a team. Each represented the planner in the fighter section of the current plans division of a tactical air control center. The manual planners occupied separate cubicles. The AESOP planners worked at input/output stations somewhat separated from each other. In this series, more than in the prior one, AESOP included automatic computation, copying, detail-generation, and bookkeeping functions which the operators themselves had to perform in the manual system.

Again MITRE employees were the six subjects; "last minute pressures to complete the test as soon as possible" allowed insufficient time to select, assign, and train Air Force officers (Doughty 1967). The MITRE subjects were trained in the two methods for four weeks to the point where they could produce acceptable plans with either method. Ten four-hour data sessions followed. Occasional interruptions resulted from computer or equipment failure in the AESOP method. Two staff members observed the subjects using that method; one observed those performing manually. All AESOP actions and their times were automatically recorded by the computer; intervals between operator action and computer response were recorded by the observers with stop-watches. The manual planners time-stamped each work form as it was used, and a camera photographed the wall-mounted mission board. Pending statistical analysis of the test data, Doughty (1967) reported that AESOP surpassed the manual method "in all measures of excellence and desirability: Total time to prepare a plan (process desirability), plan quality (a combination of output excellence and desirability), and planner preference (process excellence). The extent and significance of the differences, however, remain to be determined." Much emphasis was placed in this study on the selection of criterion measures, with guidance from Jacobs (1965)—see below. There were two kinds. Output measures were combined in a single criterion score. Process measures consisted of the times required to accomplish tasks. The subjects critiqued the methods at the end of each session and when the series was completed.

## COMMAND AND CONTROL

Although command and control systems were not investigated extensively in the MITRE–ESD Systems Design Laboratory, MITRE personnel conducted simulation-based investigations of two of them near or at the site of the operational system itself. These two, 425L and 473L, were the NORAD Combat Operations Center System in Colorado and the Air Force Command and Control System at Air Force Headquarters. Both are the command variety of command and control systems and have been described in reviews of a spectrum of command and control systems by Parsons and Perry (1965) and Shaw (1966). In addition MITRE developed a national command post simulation (Doughty 1963) and proposed simulation for the national military command system (NMCS). MITRE (and before it, Lincoln Laboratory) was also active in live testing of SAGE (Jacobs 1965).

New command systems like 425L and 473L, supplanting manual systems, present difficulties in simulation and measurement. They are one-of-a-kind systems, that is, in each only one operating center exists, unlike multiple-center systems such as SAGE. Simulation may interfere with on-going operations to a serious extent. Senior military personnel staff much of a command system, especially critical positions. The systems manage the deployment and redeployment of resources; it is not easy to measure the success of the actions taken.

For a number of reasons, MITRE's simulation-based work with 425L and 473L might not be regarded as experimental—although the term was at times adopted—so much as exploratory observation. MITRE researchers described themselves as oriented toward "flexible" and "non-rigorous" evaluation techniques. For example, displays might be "played around with" to investigate the many ways of going from different inputs to different outputs.

They took this approach partly because the systems were undergoing development at the same time they were being evaluated, and changes in them occurred continuously or in sizable steps; evaluation plans also encountered unanticipated alterations. It was difficult to adhere to a systematic program. As Jacobs (1965) phrased it, it was difficult to "hold the system still." Another reason was present in the very nature of the systems. As just noted, in command-type systems it is hard to quantify much of the output, which may consist of decisions concerning the allocation of resources over a period of time. What, then, should be measured? And without measurement, how rigorously can one test, or experiment? Further, as Jacobs emphasized: "Realism of the environment is particularly difficult to achieve in command and control systems and it becomes more difficult at the higher levels because of the tremendous number of possible war situations that a command may be called upon to engage in. Therefore, in almost any evaluation, the external environment assumed is subject to serious question."

The same author drew attention to dangers in the selection of evaluation criteria. One danger was "choosing a single simple measurable criterion which is representative of the principal goals of the system but neglects other important goals." Another was "choosing sets of performance measures whose relationships to the main goals cannot be established and for which the relative contributions of the individual performances cannot be established." The third was "neglecting important costs of the system in the criteria, particularly the costs of maintenance, training, quality control, etc." Jacobs distinguished among the following system "qualities" from which criteria could be constructed: (1) capacities in performance ("Can the system perform?"); (2) operability effectiveness ("Can the system do the job?"); (3) maintainability ("Can you support the system?"); (4) reliability ("Can you count on it?"); (5) survivability ("Can it survive damage?"); (6) integratability ("Can the system be integrated?"); (7) creativity ("Can the system decide how to change?"); (8) adaptability ("Can the system be changed?"); and (9) intellectuality ("Do you have the context required to decide how to change?").

Jacobs also described four orientations toward the value of a command and control system: (1) excellence, indicated by capacity measures ("Is this system better than that system?"); (2) utility ("Does the system do the job—meet the need?"); (3) desirability, indicated by performance/cost ratios and resources ("Do I want this system doing this job more than that system doing that job?");

and (4) formality, involved in system acquisition ("Does it pass all the tests?"). These orientations, Jacobs said, were appropriate to different sets of people—the first to techniques researchers; the second to system researchers, designers, and developers; the third to managers and commanders; and the fourth to administrators and system acquisition managers.

*425L System Testing*

Jacobs (1965) characterized most of the testing of the 425L system (NORAD COC) as having the excellence orientation, with the objectives of design improvement, establishing performance criteria, developing operating procedures, developing techniques for category testing, and developing exercise and evaluation techniques. For a typical test there were no quantitative measures or criteria; instead the evaluation took the form of judgments by observers in response to various questions.

The 425L test program has been described at length by Lesiw (1967). It occurred, for the most part, neither in the operating system being replaced nor in the location of the new system, a hardened site in Cheyenne Mountain near Colorado Springs, but rather in what was called an "experimental facility" not far from these; however, some Category II testing did take place at the final site. (The Air Force has three categories of testing, which differ in scope and content.) The program suffered from equipment changes and deviations from final design. For example, a different computer was introduced at mid-point; the main wall display and the closed circuit television were unavailable much of the time; and the consoles were fewer than had been planned. One of the phases of computer programming was canceled, necessitating the modification of the prior set of programs and their testing, which was interrupted by a decision to make the final set of programs more austere and responsive to input demands. Testing had to be reoriented, and much of it had to omit coverage of man-computer interactions.

In addition the personnel configuration defining operator positions and console allocations underwent several changes, some as a result of test experience, some by directives of the NORAD command. Considerable turnover occurred among the crews of military personnel assigned to man the fifteen consoles. Test planning reflected the discontinuity of the developmental program. Nine test orders incorporating plans and objectives proved to be too ambitious to be realized, owing to low reliability of programs, computer, displays, personnel, and procedures. The terms "experiments," "experimentation," and "experimental" can be found scattered through the test program's documentation, but no particular experiment was described by Lesiw (1967). As a matter of fact, he commented: "The term experimentation has not been used in the classical sense which would imply discretely manipulable variables and laboratory conditions of control."

It is difficult to discover how many test sessions were held. However, in one "demonstration" portion of the programs there were fifty "runs," successful or partly successful, to develop and document operating procedures and to familiarize NORAD COC personnel with system performance. Two different "command post demonstrations" were also conducted a number of times.

They and the tests were based on simulation of the system environment through fourteen "problem packages," each of which consisted of an input message card deck, a voice script card deck, event listings, reference maps, voice scripts, background material, and operational briefings. The packages were developed from scenarios representing hypothetical war conditions. The card decks were computer-produced and, in turn, were the sources of magnetic tapes that were run through the computer to produce simulated displays on the consoles. Input messages, perhaps the major components, represented messages to the NORAD COC from military commands and government agencies describing hostile attacks (missile and aircraft), battle damage, resources, and the actions of U.S. forces. Data concerning console actions could be recorded by the 425L computer itself, reduced, and printed out.

Lesiw (1967) concluded that the test program "facilitated an orderly and economical evaluation of a system demonstrably capable of performing its specified mission . . . . The experimental facility . . . did in fact provide a relatively economical and effective vehicle for establishing, exploring, assessing and improving initial system concepts, design and operations." He also observed that such a conclusion "does not imply that the program was ideal nor that methodological, administrative, and management problems were entirely absent." For one thing, he said, the problem packages initially lacked realism because the military participants did not take part in their design, and when these people did become involved, disruptive changes occurred in problem designs and simulation techniques. For another, personnel assignments were unreliable. Operating procedures were inadequately documented, so it was difficult to pass them on to new position incumbents. Finally: "Experimentation as a phase of development was not fully understood or accepted as an important step in system evaluation; consequently, support requirements were not satisfied as readily as might have been expected. At the same time, the experimental program did not enjoy a high priority in budgetary and developmental schedule considerations."

Observers at the time of the test program expressed doubts about the value of developing operating procedures for the final 425L system so far ahead of time, in a test facility in which displays, consoles, and communications differed extensively from those that would be employed in the final system. Many aspects of the interfaces with internal and external agencies had not been established, although procedures must deal with them. What could be tested, it was asked, while procedures and measures were being developed, since neither procedures nor measures were stable? In addition, the simulation was regarded as insufficiently realistic. It was suggested that of the three objectives, namely, development of procedures, test of procedures, and development of test measures, only one could be pursued at a time with validity; but if one wanted to go around in circles, the runs should be called "exploratory," not "tests."

Other suggestions were that (1) the test conductors should clearly delineate the permissible activities of all participants; (2) subjects should be isolated or at least not interfered with while they were performing; (3) subjects should not talk to test personnel during a run; (4) the scenario should not be revealed to the subjects before the run; (5) guests should not be permitted to interact with subjects at all, and with test personnel as little as possible; and (6) if equipment problems arising during a run could not be ignored, either the malfunctions should be handled in the same way they would be under operational conditions or the run should stop until the failure was corrected.

The 425L testing program provided many lessons for the conduct of man-machine system experiments.

*473L System Exercises*

A method of "normative exercises" for evaluating performance and facilitating design in the Air Force's Headquarters Command Post (473L) has been described by Proctor (1963), also Porter and Proctor (1962), who took note of several system exercises during a period of eighteen months.

The first problem in the user's facility ran for eight hours and involved some 60 players in various jobs, teams and functional areas of the manual system. (This problem was repeated some months later in an alternate site with less than two dozen players in a relatively austere physical environment.) The second problem was presented in a twelve-hour exercise, again involving over sixty players. The system by this time had acquired a minimal automated data storage and retrieval capability. In all three exercises, over thirty commands and agencies were simulated in SIMCON.

SIMCON was a simulation control team which provided inputs to the system being exercised and responses to the system's outputs. The team could number as many as twelve persons, including a message composer and a deliverer of hard copy to the exercise participants.

The SIMCON personnel performed three functions. One was to send scheduled messages to the command post team, according to a script called an integrated operations schedule, and to originate unscheduled messages in response to participants' inquiries. The second was to observe "activity within particular system areas and time periods" and summarize observations on recording sheets. In addition, voice messages were taped, and both the SIMCON and command post personnel kept logs of messages. The opinions of both sets of people were obtained after an exercise through a structured questionnaire.

It was the SIMCON team's third function which furnished this exercising method's special character and its name. A "normative solution path," that is, a preferred problem solution ("revised as necessary in light of exercise findings") was developed prior to an exercise, but the command post personnel did not know what it was. Rather, the SIMCON team, which did know it, was responsible for keeping the command post personnel on the right path and getting them back on it when they deviated. If these personnel took a wrong step, the SIMCON team redirected them by inserting "information to the system calculated to make players rethink their position and come to a more appropriate and timely response." To assist in detecting deviations, "outcome checkpoints were established by deciding what observable characteristics of system activity would certify that critical Descriptive Problem solution sets are being accomplished in the exercise."

Performance evaluation was apparently based more on judgments as to whether the command post personnel achieved "the right solution" or at least "an acceptable solution" than on quantitative data, although the latter were characterized as a goal. The limitations and advantages of normative exercising were presented thus (Porter and Proctor 1962; Proctor 1963):

Normative Exercising appears inappropriate if system activity is well defined in terms of outcomes and performance criteria. However, when criteria are lacking with which to test outcomes of system functioning, when there are many conflicting views as to how to arrive at a problem solution and where the current environment of the system does not contain essential characteristics of the environment in which the system is expected to function, Normative Exercising of the variety discussed in this paper is proving to be quite useful as an analytical and evaluative aid to system design.

## AIR TRAFFIC CONTROL STUDIES

Between December 10, 1961, and November 13, 1963, the MITRE Corporation conducted six simulation-based man-machine studies of en route air traffic control in its Boston Air Traffic Control Test Bed, so named because the air environment simulated was a large part of the Northeast, with Boston as the principal air terminal. The facility for the studies was at the Air Force's L. G. Hanscom Field, near Boston. The studies were sponsored by the Federal Aviation Agency, with support from the Electronic Systems Division of the Air Force Systems Command.

They originated in a program which investigated how SAGE air defense equipments and techniques might be adapted for control of en route (between airports) air traffic by the the FAA. That project, begun in 1959, was called SATIN, an acronym for SAGE air traffic integration. But the program's objectives shifted to a broad examination of computerized automation support in en route air traffic control using a computer (AN/FSQ-7, SC-1) and display/entry consoles which had been built for a SAGE development that was canceled.

The six studies varied in extent between 7 and 27 simulation sessions; the evaluation session total was 104, the hour total about 206. The name of each study was an acronym. In chronological order, the studies were: STEEL (simulated test environment to evaluate load); DAMDOT (package D air movements data only testing); TRICOM (triple comparison); DIAL (display alleviation); APEX (area planning experiment); and THOT (terminal handover testing).

Although the program explored a range of design questions in air traffic control, it ended before some of the projected areas were investigated. One of the most interesting would have been a comparison between the first and last hours in the eight-hour work shift of a control team. Another would have been a comparison of varying degrees of automation, from fully manual to fully automated control with variations between, through successive "peeling back" or "ablation" of the extent of automation.

The studies had a number of common characteristics. One was the use of experienced FAA air traffic controllers as the controllers in the simulation missions, while the simulator positions were manned by Air Force military personnel. There was considerable personnel overlap from study to study.

Signals of aircraft on the console displays consisted of computer-processed search or beacon radar data originating from magnetic tapes which contained the mission's problem (air traffic sample). Problems were largely computer-generated from a library of flight plans (paths and times). A "passive active dynamic simulator" computer program was developed to display controlled traffic, background traffic, and various kinds of noise, including weather, but due to difficulties an alternative program created for program testing in SAGE was widely used at first.

Another common feature of the studies was the introduction of a "shadow flight plan" into the simulation of each controlled flight. It was displayed only at simulators' consoles. Through this shadow input the simulator personnel, playing the roles of pilots, could report aircraft positions which diverged from those projected in the actual flight plan. Divergences usually resulted from deviations in aircraft ground speed. The simulator personnel could amend the shadow flight plan when the controller made some alteration in the actual flight plan.

The studies varied somewhat in the emphasis placed on experimental control; STEEL and THOT received less than the others. The reports of DAMDOT and APEX are particularly complete. Much documentary material was prepared for some of the studies, such as handbooks for operators, training manuals, handbooks of air traffic control information, and procedural handbooks for experimenters and simulators, concerned with inputs, recording, and control.

*Evaluation Methods*

Much attention was devoted to methods of evaluation, some of which will be reviewed here before the individual studies are described. In both STEEL and TRICOM a major criterion was the saturation load at which system breakdown occurred. Held and Wolff (1962) described the STEEL method as follows:

> A multiple observer technique was used. Four observers with varying professional backgrounds were used for each team. These observers were each provided with a set of performance criteria to watch during a test. They kept notes on time-ordered logs during each test. Following the test they would make judgments on whether the team had reached a capacity situation. The observers wrote short reports on each control problem following the test. These reports defended and explained their judgments. The control team personnel (including simulation) also wrote reports on each control problem. These reports discussed any capacity conditions the controllers felt existed during the control problem and the reasons why.
>
> A team (controller and observers) debriefing was held after each test was over and after the individual reports were written. A discussion leader attempted to ascertain if a collective opinion existed within the group on each control problem and to find the reasons why or why not such a collective opinion existed. He then wrote a short report summarizing the results of the debriefing discussion. At a later time an analysis committee consisting of at least one MITRE and one FAA member went over all the reports and logs for each problem. This committee made the final judgments on capacity conditions and the reasons why . . . . There was a highly satisfactory consistency in the reports and logs from the personnel involved in each test. Although an individual's judgment was frequently over or under-critical (usually due to incomplete information), the complete picture of events presented by all the reports and logs generally focused to a remarkably collective opinion.

The fact that opinions of controllers and observers often disagreed with objective data is noted in the above summary, although it does not receive the emphasis that actual instances indicate it should. For example, the same report said elsewhere: "Three of the seven times the three man team had a breakdown load judged for them, the judgments were extrapolated. That is, the team was not quite at a breakdown point but almost."

In the TRICOM study, the saturation load, described as "the number of aircraft handled by the control team during the ten-minute interval just prior to breakdown" instead of an instantaneous airborne count at the time of breakdown (which was variable), was determined in a different fashion (Lee 1963):

> To aid in the determination of the breakdown time, time-sequence charts were made depicting all hazardous events in each problem. A chart was made for each sector problem. The information on the charts was gathered from data reduction and observer logs.
>
> Controllers directly involved in the TRICOM test series took part in the determination of the breakdown time for each sector problem. The controllers were shown the charts one at a time, not knowing which problem it represented or what the traffic loads were. For each chart, each controller independently made and wrote down his judgment of the breakdown time. From the times associated with each chart, the median was chosen as the breakdown time. This median time was used to determine the Controller Judged Saturation Load.

*Hazard and User Convenience Sources.* Controller judgments were also invoked in arriving at measures of hazard and user convenience in TRICOM, DIAL, and APEX. The derivations differed among the studies.

In TRICOM (Lee 1963), ten controllers ranked for seriousness and danger seven general situations into which aircraft might enter. They arranged in order slips of paper with the situations printed on them. A frequency distribution of the ranks indicated that the seven situations could be grouped into four categories. The categories were given weights from one, for the least hazardous, to four, for the most hazardous. For each mission, the instances of each of the seven situations received the weights of the categories to which they belonged. The sum of all the weighted instances (normalized to eliminate differences due to different numbers of aircraft in the mission) constituted the objective hazard score of each mission. Then the objective hazard scores of all missions were rank-ordered. Differences between scores for missions with different experimental conditions were tested for statistical significance by a sum-of-ranks test to determine whether the experimental conditions resulted in differential hazard.

DIAL (Truesdell 1963) distinguished among eleven situations. One was "no hazard." Six levels of aircraft traffic conflict could be associated with four levels of aircraft nonconformance (deviation from flight plan position). The twenty-five combinations were written on slips of paper which controllers arranged along a line, the order and spacings between slips indicating relative hazardousness. The situations were given interval-scale values according to their positions. The instances of each situation during a mission as derived in data reduction acquired the situation value. The summed values constituted a mission score called "summated controller-judged hazard values." The statistical significance of differences between mission scores was determined with the Mann–Whitney U Test.

In APEX (Hazle and Lee 1964), thirteen situations including "no hazard" were combined to make twenty-six trios, which sixteen controllers scaled for relative hazard. The twenty-six combinations were the outcome of an incomplete block design. A value for each of the thirteen situations was given a score consisting of its value multiplied by the number of minutes of its duration. The total score was the system safety margin score of the mission.

A second hazard evaluation method was evolved for DIAL (Truesdell 1963). Time plots of five hazardous situations in each mission were constructed from the data for that mission. Fifteen controllers compared the plots of each mission with those of every other mission in a paired comparison procedure which rank-ordered the missions according to their hazard. This "mission comparison of hazardous event time plots" method was combined with the "summated controller-judged hazard values" method to create a third method, consisting of a rank-ordering based on the two together.

Still another approach, in TRICOM and DIAL, was to derive an "objective surveillance trouble score" based on two of the situations in the objective hazard score, and an objective violation score, based on three other situations in it, all having the same weights as the objective hazard score. The first produced a score for surveillance, the second one for control.

The user convenience score was constructed for TRICOM in much the same way as the objective hazard score. The same ten controllers ranked eleven situations inconvenient to the pilot or system user, such as holds, departure delays, and route changes. These were placed in four categories, which received weights of one to four. In DIAL the four categories differed somewhat from the TRICOM groupings. In APEX, fourteen controllers arranged twenty-nine slips describing inconveniences on an interval scale, for each of five types of aircraft. Although it is really the controllers' view of inconvenience that influences the operating system, some pilots' judgments of inconvenience events were ascertained as well. They diverged considerably from the controllers' views, and two DC-3 pilots differed substantially between themselves.

An interesting treatment of data in DAMDOT (Hett et al. 1962) was the construction and scoring of task-flow diagrams of conflict resolution and flight plan ordering by controllers. Procedures were represented as a series of yes-no decisions, the decision points being numbered. Then the percentage of affirmative choice was derived for each decision point and shown in a table along with the number of occasions on which the decision was made.

In APEX an attempt was made to find out whether hazard, as indicated by the system safety margin scores, could be predicted by measured system load, by predicted system load, or by controller workload. Measured system load was based on the values of eleven elements that actually happened—average number of flights, number of arrivals, number of departures, etc. Predicted system load was based on ten of the same elements with values reflecting what could have been predicted to happen during a given time interval, if a prediction had been made before that time interval began. Controller workload had six elements. Multiple regression analysis failed to show consistent predictability of hazard for any of the load measures.

*STEEL*

This investigation (Held and Wolff 1962) of high altitude en route traffic varied the number of controllers in a team between two and three to find out how many flights each team could keep under surveillance (not control) without breakdown. It concluded that 15.3 was the average breakdown load. There were seven simulation sessions: two for training, each lasting two hours, and five for evaluation, each lasting three hours; data were obtained from four of these evaluation sessions. Three one-hour missions took place in each of the evaluation sessions. In six different one-hour problems, the number of flights per problem varied between forty-one and sixty-three and the maximum number of simultaneous flights between twelve and twenty-one. Conflict rates also varied. An area about 400 by 400 miles was simulated for control by three teams, one per sector, and two supervisors; a fourth team was on stand-by. The surveillance controller position was unfilled in none, one, two, or all three teams. The teams, which remained the same, varied greatly among themselves in proficiency, which increased during the test series.

Preceding the simulation study were three investigations of the actual air environment (Members of the D–16 Staff 1962). One simply monitored the Boston area. A second, SCOOT (SAC co-operative testing), consisted of sixteen four-hour missions with seventy-eight flights of Strategic Air Command aircraft. Traffic was too light to provide more than engineering data—one reason why the MITRE researchers turned to simulation, which they had employed in seven exercises preceding the SCOOT sessions for planning and developing procedures. In addition, the SATIN controllers exerted little control over the SAC aircraft. The third investigation was STAM (SAC test aircraft missions), in which SAC flew two of three scheduled four-hour missions (nineteen flights). Although more control could be exerted, ultimate responsibility still resided with the Boston air route traffic control center. (ARTCC). More engineering data were collected.

*DAMDOT*

This study had two parts, DAMDOT and DAMDOT Extension. It concentrated on high-altitude aircraft control (rather than surveillance) and incorporated automatic (computer-detected) predictions of conflicts (occupying the same airspace) between aircraft (Hett et al. 1962). It also investigated methods of intrateam resolution of predicted conflicts and of intrateam control. Along with a supervisor, two-man teams similar to those in STEEL each controlled one or three sectors. For each controller team there was a two-man simulation team and a pair of observers. A two-month training program included classroom instruction, fifteen two-hour training missions, and three weeks' instruction in touch typing (so the controllers could operate teletype keyboards properly).

Eight two-hour sessions with automatic conflict prediction and display were followed by three without it (DAMDOT Extension). The four teams (including a standby during each mission) rotated through the sectors, but the same individuals remained in the same positions within each team. Each team performed in six missions, twice in each sector, in almost all cases with different sets of inputs on those two occasions. There were two sets of inputs–flight plans–with fifty-nine or sixty flights per mission and a maximum of ten simultaneous flights. Deviations from flight plan and planned conflicts were equivalent in the two sets.

The computer obtained data automatically from its own operations. Data also came from voice tapes, logs, and teletype messages. Among the measures were traffic load, controller usage of teletypes and consoles, frequency of various displays, flight plan interruptions, number of conflicts, and time in conflict. These last two were greater when the controllers lacked automatic conflict prediction. Teams differed significantly among themselves on some measures, not on others. The statistical significance of differences was tested by chi squares and Kruskal and Wallis analyses of variance by ranks (nonparametric). Interestingly, the significance tests were explained at some length in the study report, since they apparently were an innovation in MITRE testing. Concerning their use, the report (Hett et al. 1962) commented:

> In future test series, especially when statistical tests are to be performed on resulting data, the importance of complete development of analysis plans *before the test series begins* cannot be too highly stressed. Measures must be defined, major hypotheses to be tested must be postulated, and appropriate statistical tests and methods of presenting the data must also be chosen. When this is done, specifications for data reduction programs can be precise and in many cases, no manual transformation of data needed . . . .
>
> Of course, during the test series, unforeseen circumstances will appear, and all expected results will not. Such events will almost always cause some change in analysis techniques; i.e., additional needs for data reduction or the defining of entirely new measures. However, with a sufficiently flexible analysis plan, such events will represent minor changes, not complete revisions.

The report also presented a plea that new, complex data reduction programs "be designed, coded and checked out prior to the start of the test series with which they are to be used" so all the data required could be recorded and it would be unnecessary to undertake "the tedious, time-consuming manual correc tion" which was needed in the DAMDOT analysis.

For the benefit of future researchers, the study report also included an appendix describing problem areas as recorded in test managers' reports. In addition to the data-reduction difficulties noted above, problems included equipment malfunctions, failures of computer programs to perform as expected, and unexpected performance which was operationally undesirable, as "when a maintenance man dropped six rolls of TTY paper on the console," and it lost its power.

Controllers expressed their views in debriefings and in a "Summary of the Operational Viewpoint" included in the study report. They advised greater sophistication in the methods of presenting and dealing with conflicts. For example, they felt that to resolve a conflict by changing an aircraft's altitude, a controller should be given a display of available altitudes instead of being required to try other altitudes one by one, getting a lengthy printout on each trial. They urged further that planning data for a controller on six different displays be consolidated on a single display.

### *TRICOM*

Two innovations widely discussed in 1962 (e.g., in the Beacon Report of a national commission) for the control of en route air traffic were the installation of beacons in aircraft and automatic tracking of search radar or beacon signals by means of a computer. The stronger signals of the beacons (compared with ordinary radar echoes) improved discrimination by controllers when their displays also showed noise; and the beacon signals could be coded so the displayed signal identified the aircraft, or designated its altimeter altitude, or both. In automatic tracking the computer reassociated the displayed signal and the computer-generated track when these diverged; in nonautomatic tracking an operator brought the track back over the signal with a light-gun action at his display. The TRICOM study investigated these two innovations in a low-altitude (below 14,500 feet), high density, simulated air environment in which en route controllers exercised both surveillance and control. As additional variables, a sector control team consisted of either two or three men, the third working without a pictorial display; and the total area being controlled by one team was composed of either two sectors or one which was the size of the two combined.

Like its predecessor, this study (Lee 1963) had two parts, TRICOM in June 1962, and TRICOM Extension in July and August 1962. In the first were fourteen 90-minute problems (missions), and in the second eight of the same type. During a problem, data for 75 minutes were recorded and analyzed, and fifty aircraft were introduced. Traffic load progressively increased to expected saturation by adding thirty of the aircraft during the last 30 minutes, so the number requiring control at the same time reached twenty. The second part examined four modes of surveillance: all aircraft beacon-equipped and tracked automatically; all so equipped and tracked nonautomatically; none so equipped and tracking was automatic; and none so equipped and tracking was nonautomatic. The team size and sector size variables occurred only in the first part of the study, along with three surveillance modes; all aircraft were beacon-equipped with nonautomatic tracking; 60% were so equipped, with nonautomatic tracking of all; and 60% were so equipped, with automatic tracking of these and nonautomatic tracking of the remainder.

According to the study report, the experimental design balanced the primary variables across practice, input sets, sectors, and control teams, none of which had significant effects. Exercises were introduced between the two parts; proficiency was greater in the second. The two-man and three-man teams were equally effective in that they "saturated" at the same level of sixteen aircraft under control at the same time; it should be remembered that the third man had no pictorial display. The larger sector saturated at twenty-three aircraft, in contrast to sixteen aircraft in the double-sector configuration; however, according to the way the inputs were arranged, one-half as many aircraft were under control per unit area in the large single-sector condition as in the double-sector one. Data concerning automatic tracking in the first part of the study had to be disregarded because of extraneous problems. In the second part, the nonbeacon, nonautomatic tracking combination had a lower capacity than the other conditons. When all aircraft were beacon-equipped, automatic and nonautomatic tracking were equally effective.

*DIAL*

Operationally, air traffic control in DIAL (Truesdell 1963) resembled certain conditions of TRICOM's second part. Two control teams, one per sector, performed in each mission. The computer tracked aircraft signals automatically. Some of the signals were beacon, some search radar. Each team had two men; however, only one of these had a display console, while the other operated a teletype. The independent variable was the method of reducing clutter on the pictorial display. By clutter was meant information coded in letters and numerals (alphanumeric symbology) associated with a track in such quantity in a mapped area that confusion resulted. One method was to relegate much of the symbology to the margins or periphery of the pictorial display on the console's CRT. Another was to divide the air space into two vertical instead of horizontal sectors, assign each altitude sector to a control team, and display to a team only the symbology for the altitude sector assigned to it; this method halved the amount of symbology per unit area. The third method was simply the display arrangement (and horizontal sectorization) used in TRICOM.

Seventeen missions were conducted between November 5, 1962 and January 23, 1963, and ten of these were analyzed, six being disregarded because the subjects reached saturation too early; in another there was a computer halt. In 90 minutes following 10 minutes of build-up, 118 or 111 flights (below 15,500 feet) were introduced in the two input sets (problems). One set had instantaneous traffic loads ranging from ten to thirty-two aircraft, the other from eight to twenty-seven. The two teams rotated through experimental conditions and through both of the horizontal and vertical sectors. According to the safety criterion, the TRICOM arrangement surpassed the vertical sectorization, which in turn was better than the new kind of display which placed tabular material in its periphery, but the differences were not quite significant statistically. Both the TRICOM display and vertical sectorization were superior to the new display in terms of user convenience. But the subjects said they preferred the new display. They offered numerous suggestions for improving its design, which the researchers conceded had been done rather hurriedly. Human engineering as an aid to display design or system design was not mentioned in the report of this or any of the MITRE air traffic control studies.

*APEX*

This study, the largest and probably the best reported of the MITRE air traffic control investigations (Hazle and Lee 1964; Morey and Yntema 1964), stemmed from an FAA concept that aimed to improve en route traffic control in a multisector area in the 1970 time period. It was thought that a common planning team which arranged departure clearances could reduce the work loads of sector teams by minimizing the frequency of conflict situations requiring resolution. The computer programs in APEX included more sophisticated procedures of conflict detection, increased flight plan capacity, and improvements in simulation. Before the data sessions began, "various planning positions, functions, and procedures were tried and assessed" in thirty runs with FAA controllers as participants; displays were also examined. During the same four-month period there were twenty training sessions.

Thus in this study the simulation exercises or sessions devoted to developing the system (and to training the subjects) were kept distinct from those conducted to evaluate the system that had been devleoped. However, this approach did not deter the researchers from introducing changes in conditions during the series of twenty-two evaluation runs as a consequence of experience during earlier runs. Because the controllers' tasks seemed so easy up to that point, starting with the eleventh mission the inputs were extended to include controlled combinations of severe weather, flow restrictions, and military refueling flights.

Seven teams of controllers could interact: a three-man planning team; a two-man team handling a high altitude (above 17,500 feet) sector; three two-man teams for three sectors of lower altitude traffic together with a supervisor; a three-man team representing en route control teams for sectors adjacent to the high altitude sector and the three lower altitude sectors (which were the object of most of the study); and a two-man team representing teams controlling approaches to and departures from airports. In the last six sessions the high altitude sector was eliminated and traffic was increased in the three lower altitude sectors; these six sessions were not included in the main data analysis.

In one-half the sessions the three-man planning team was omitted and its job of planning aircraft departures was handled by one of the controllers in two of the three lower altitude sector teams. This variation constituted the principal independent variable. As a result in eight data sessions there were seventeen controllers in seven interacting teams; eight other sessions had fourteen controllers in six teams. In addition there was a two-man simulation team for each controller team and another two-man simulation team for entering flight plan inputs. Eleven input sets of controlled traffic (problems) were assembled from a library of 1970-era flights. Each input set figured in one planning and one nonplanning session. The sets contained, on an average, 178 flights for a period of two and one-half hours—commercial carriers, general aviation, SAC aircraft, other military aircraft, and 97 background (uncontrolled) flights, on and off airways. Of the controlled flights, 85% were beacon-equipped and 70% of these were identity-coded.

The study report (Hazle and Lee 1964) described the experimental design further:

> Missions were scheduled in sets of four in a PNNP or NPPN order so that experience was balanced across the two operational modes. This modular design also allowed for minor changes in operating procedures, computer program, equipments, or manning after each fourth mission without upsetting the balance of the experiment.
>
> Manning was held constant throughout the series. Except for minor changes because of illnesses, each control sector was operated by the same two people during all missions; the same Division Supervisor served throughout; and the planning team personnel remained constant. Absenteeism occurred in approximately the same degree for both modes of operation, so any effects of substitution were balanced . . . .
>
> The factor which forced the use of only one planning team was the limited preparatory and test time. (This time limit resulted in part from the scheduled release of APEX personnel—both MITRE and FAA—for other ATC commitments.) Within this limited time it was necessary both to train the controllers in

the operation of the Model 200 system and to develop the APEX procedures . . . .

It did not appear that the time available would permit the development of good procedures and the training of controllers for competent operation of multiple positions. Because good procedures and high quality performance were deemed more critical than position rotation to the validity of the APEX series, it was decided that manning should be held constant over all tests . . . .

The choice between rotating positions and good procedures and training was influenced by several considerations. First, the variability in performance to be expected from inadequately trained operators would have prevented the drawing of conclusions based on the statistical analysis of results. (This variability would have increased the size of the error variance, by means of which the effects of such variables as planning mode are assessed.) The expected reduction in the statistical reliability of the results was judged to be more deleterious to the validity of the test series than the bias introduced by constant manning. Furthermore, although it is known that controllers differ in their approach to a position or function and that these differences can affect the system's performance, it is believed that the procedures which define the way a function is performed have a much greater effect on the impact of that function than do the individuals who follow those procedures . . . .

We do not believe that the differences shown between the planning and non-planning modes were due to the particular APEX manning.

Experimental results showed that under routine traffic conditions the planning team did not improve system performance; but there was greater system safety—and less user convenience—with the planning team when simulation inputs included severe weather, flow restrictions, and military refueling flights. Five data sessions (not included among the twenty-two mentioned) were "discounted because of severe program, procedure, or manning problems."

### *THOT*

The STARE project at NAFEC has been outlined in Chapter 15. STARE meant "single terminal and runway experimentation." It embodied the use of a large terminal area (50 miles in radius from the airport) to regulate, with computer aid, the arrival time of aircraft on the ground by path selection, with a complex pattern of entry fixes (points), feeder fixes, approach fixes, and missed-approach fixes. The THOT study (Moros 1963) examined methods of regulating en route traffic and handing aircraft over to the STARE controllers as the aircraft approached the entry fixes. To hand aircraft over at particular times, controllers had to introduce delays during the en route portions of an aircraft's flight. Methods for doing so included speed changes, path stretching, and holding. The ability of a computer-based system to effect each of these singly or in combinations was one of the investigative objectives of THOT.

It was assumed that Boston was the automated STARE terminal. An area of 15,000 square miles in the Northeast was the simulated air space. It comprised three lower altitude sectors and one higher altitude sector covering the same geographical space as in previous MITRE studies of all-altitude en route control. Eleven positions in a simulated air route traffic control center were manned by two-man teams for each of the four sectors, a division co-ordinator, and one individual for each of two adjacent sectors. In addition, two STARE control positions were represented: an arrival planner and a sequence controller. Complementing the thirteen subjects were twelve simulator personnel: six representing pilots (with radio communication to the controllers), four handling teletype inputs and representing airport tower control teams, and two inserting flight plan data.

Simulated traffic was supposed to resemble that of the 1970 time period. The average arrival rate was twenty-five aircraft per hour. Instantaneous loads in any one sector averaged eight. Divergences from flight plans resulted from speed deviations, lateral (path) deviations, and radar errors. Initially, sessions lasted two and one-half hours and were conducted once a week; later they came twice a week and lasted three and one-half hours. Twenty missions took place during a two-month period. Data covering twenty-four hours of testing were obtained manually and from teletype printouts and controllers' postsession critiques. Quantitative results included conformance data (e.g., amount of delay and deviation from flight plan); arrival times; radar violation counts; and frequency with which the various methods were employed to sequence arrivals at the terminal area. Although some chi square tests examined the statistical significance of the frequency data, "formal statistical analysis was extremely limited" as "THOT was a developmental test series with a constantly changing program and procedural base" (Moros 1963).

# 20

# Institute for Defense Analyses Communication Studies

Intercommunication among individuals has been investigated experimentally over the years, and it would seem that those investigations which concentrated on various kinds of communication nets might be relevant to operational problems in real systems. Much of this work has been reviewed by Glanzer and Glaser (1961). This account of man-machine system experiments will not survey these studies. Not only have they been well reported elsewhere but also there is little evidence they have helped solve key problems of modern intercommunication, especially among individuals physically separated from each other.

These problems have been various. What are the relative advantages and differential effects of various intercommunication media, such as telephone, teletype, and television? What happens when intercommunication must undergo transitions between languages? When a number of individuals have to intercommunicate about a joint matter, how does a party-line conference hook-up compare with a point-to-point network? In a party-line arrangement, should the conference participants all have access to speak on the net at the same time or should such access be limited by some serial constraint? And how may a chairman's control be implemented?

Such questions become serious when it is necessary to link together statesmen or military leaders in different countries, and civilian or military leaders within the same government. After the Cuban missile crisis it seemed especially important to know how, if at all, fourteen heads of state, speaking different languages, could discuss the rapid establishment of a multilateral force. Accordingly, in 1963–65, the Office of the President's Scientific Advisor and the Office of the Director of Defense Research and Engineering (DDR&E) turned to the Institute for Defense Analyses (IDA) for guidance on how to design such linkages; and H. W. Sinaiko, J. Orlansky, and T. G. Belden of IDA began to ask such questions. Since the professional literature held no directly relevant answers, they established two research programs with the help of several subcontractors: Human Sciences Research; Aircraft Armaments, Inc.; and Stanford University consultants. The first program ran in 1963, the second in 1964–65.

Each program was characterized by what the researchers called "indelicate experiments." Summaries have been published by Sinaiko (1963, 1964*a*, 1964*b*) and Sinaiko and Belden (1965), who explained that "the Indelicate Experiment is characterized by simplicity, flexibility, smallness of staff, and rather gross measurements of performance." In describing one of the studies of the second program, J. S. Kidd (1965*b*) of Aircraft Armaments observed:

An indelicate experiment may be characterized as a method which provides for the orderly observation of some phenomenon but which does not have the usual emphasis on statistical analysis. The logic of experimental design is represented, as are the principal controls evoked for research involving human subjects; however, there is no emphasis on the prospect of conducting tests of statistical significance. The method is, therefore, analogous to naturalistic observation but affords the advantage of allowing the researcher to set the time, place, and conditions under which the phenomena of interest will occur and under which the observations will be carried out. Since the data cannot be processed by the impersonal techniques of conventional statistics, an inferential outcome depends very heavily on the insight of the observers.

Sinaiko and Belden further noted that their methodology represented a trade-off between urgency of getting useful information and precision achievable through experimentation, a methodology "somewhere between the quick-fix area and the usual time-consuming precise experimental solutions." It may be noted in passing that pending further analysis the potentials and constraints of this low-budget approach are uncertain for fields other than communications.

The facilities for this research came about with great speed—in each program in about a month—and at very low cost. This was possible in part because the space consisted of rooms in the IDA buildings (and elsewhere) and the equipment was mostly leased, off-the-shelf communication equipment (and to a limited extent telephones already in conventional use) and an established television facility. If some of the rooms were needed for normal purposes during the day, experimental sessions occurred at night. The local telephone company installed the communications equipment and provided special switchboards.

In the first program, each of five conference rooms contained an automatic send-receive teletype machine (ASR, Model 28), a "page receiving only" teleprinter, and a standard dial telephone with additional speaker phone. A five-station telephone net was isolated from the internal IDA telephone service. Two plug-in tape reperforator units were available. One of the conference rooms was the experimenter's control center. In addition to the telephone equipment, it contained an ASR 28; four "receiving-only" teleprinter units, each with a tape reperforator unit; and a patch panel to modify the teletype network. Dialing could automatically change the telephone net. Everything extraneous to normal IDA operations was removed when the experiments ended. The equipment in the second program consisted of more extensive telephone equipment and a prototype console from a projected communication system being investigated.

## FIRST PROGRAM

In addition to multioperator experiments, the first program included surveys of current teletype and other conferencing methods in the United States and analyses of variables for experimental inquiry (Bailey, Nordlie, and Sistrunk

1963). Preceding it was an investigation of "mood ambiguity" in command and control messages (Belden and Sinaiko 1963): eighty-eight subjects had to categorize messages derived from Navy CIC research at the Applied Physics Laboratory (see Chapter 18). This study supported an effort to identify unambiguous or "coherent" command language.

Next came a set of six language-oriented studies in which a number of subjects interacted, although the emphasis was primarily still on individual behavior. The first two involved an interpreter and two teletype (TT) operators. One TT operator sent text which the interpreter translated as it appeared on his receiver, dictating the translation to another TT operator whose transmission was recorded on tape. The sequence may be diagrammed as follows:

*Text* → TT Op → TT → *Interpreter* → *TT Op* → *TT* → *Tape*

In all six studies the text was the minutes of the 921st meeting of the United Nations Security Council. In the first study a French interpreter translated from English into French and an American interpreter from French into English. In the second study the direction of translation was reversed for the subjects, so that an American interpreter now translated from English into French, and a native of France translated French to English. In the third study the interpreters retranslated what had been recorded earlier back into the original language; the inputs came from the tapes rather than a TT operator, and the interpreters dictated their translations of the TT output to a typist.

The first three experiments emphasized sight interpretation, that is, how well an interpreter could keep up a running translation as the machine produced text. The next three studies put their emphasis on conventional translation and review of completed text. Again there was reversal of the "target" languages and retranslation back into the original language. The communication flow in the fourth and fifth experiments may be diagrammed as follows, the long lines indicating review of one translator's dictated text by the other:

*Text (1)* → *TT Op(1)* → *TT (1)* → *Translator (1)* → *Typist* → *Text*

*Text (2)* → *TT Op (2)* → *TT (2)* → *Translator (2)* → *Typist* → *Text*

In the sixth experiment the translator worked directly from translated text rather than from teletype outputs.

It was found that sight interpretations, at an average of thirty words per minute, required five times as much time as the original council meeting which produced the text. It was slower than the sixty-six-words-per-minute capacity of the teletype machines. The interpreters made so few errors that this method of translation appeared feasible; the American's performance in turning English into French was marginally acceptable. The conventional translation process took four times longer than the sight interpretation; again errors were few and the American's performance in the reversal mode was marginal or substandard, according to State Department requirements. According to the researchers, "To our knowledge, our studies are the only investigations in which the procedures of professional translators and interpreters have been tested under experimental control."

*"Miniature Experiments."*

The next investigation dealt with conference techniques in fourteen conditions the researchers called "miniature experiments," all having a five-station configuration. Two studies involved telephone communication only, in English, one with a chairman and one without, and both with an open party-line network. Teletype alone was the medium in nine, four in English only and the rest in both English and French with simultaneous two-way translation, except in one instance of consecutive two-way translation. The party-line circuit was used in five, while in the others the circuit was a Y-type in which all conferees had to transmit to a nonparticipant chairman, who rebroadcast each message to all stations via tape relay. Four of the common-circuit experiments had a chairman, in one case acting as a participant, in another with the prerogative of modifying and editing. This variation was also brought into the Y-type circuit studies. Other variations within the party-line circuit studies were subdivision into two subnets; transmission by each station to all others in turn; and physical separation of the principal at a station from his interpreter, so that they had to communicate with each other by telephone.

The last three miniature experiments, all in English and without a chairman, combined party-line teletype and telephone media. In one there were two private phone lines; in another the participants could telephone each other selectively; and the third had a teletype subnet of three and a telephone subnet of three, with one participant belonging to each subnet.

The fourteen conditions occupied four nonconsecutive days. Each occurred only once, that is, had a single replication. On two occasions the subjects shifted from face-to-face to telephone and later to teletype conditions during the session, momentarily interrupting the on-going task to make the shift. Teletype operators were IDA secretaries. The four principals in each of the conferences had backgrounds in economics, political science, medicine, the physical sciences, military operations research, and the behavioral sciences; they were drawn from the professional staffs of IDA and the Advanced Research Projects Agency. The number of conferences in which any one took part varied from two to eight. In addition, a native Frenchwoman was a principal in some conferences, and an experienced interpreter participated in the bilingual sessions. Each subject was first trained for about two hours, first with orientations, then a practice conference. Debriefings followed each session. Subjects said they preferred the telephone to the face-to-face condition for negotiation, but if they wished to take or maintain a firm position they preferred the teletype.

Before the actual experimental sessions began, the researchers conducted nine shakedown conferences with IDA secretaries as subjects. One of the purposes was to check the conference task, SUMMIT–II, described by Kidd (Aircraft Armaments, Inc. 1963). Each participant represented one of four countries in the "United Confederation" allied against the aggressive tactics of a common enemy. Participants' actions during fixed time cycles consisted mostly of contributing military units, in response to a request for assistance, levied against the Confederation. Each had a set of contingency plans and costs and each sought to meet the group demand at minimal group cost, at the same time keeping his own cost as low as possible.

SUMMIT–II developed from a number of efforts to create a suitable task, starting with an ancient Chinese mathematical puzzle and a "traveling-salesman" game, and extending through three versions of a NATO-oriented game and a first version of SUMMIT. Among the reasons for rejecting some of these were lack of relevance and meaningfulness (uninteresting to subjects); predominance of individual behavior rather than group communication; superfluous role-playing resulting in irrelevant factors and more information than could be used; excessive complexity and too great an accounting burden on the players; and undesirable favoring of subjects with military experience in a detailed war game. It was also apparent during this work that some tasks could make different demands than others did on subjects' behaviors, such as visual activity, and on control by the chairman. Some of the tasks were tried out on groups of college students.

SUMMIT–II had "the advantage that the detailed setting and rules presentation can be quickly adjusted to accommodate differences in the caliber of the subjects." By incorporating greater realism it was thought to hold player interest better than SUMMIT–I.

In noting that this developmental account "can provide guidance to others who may attempt to develop other such games in the future," the Aircraft Armaments report suggested a number of criteria for a task in this type of experimentation. It should be relevant, meaningful, provocative, sufficiently simple and easy to learn, scorable, playable regardless of the communication medium and with controllable duration, replicable (permitting repeated use of the same rules and implementation with the same subjects), expandable (no restriction on the number of players), and controllable (letting experimenters readily adjust the conditions of play). SUMMIT–II was believed to satisfy these criteria.

The results of these fourteen miniature experiments were never expressed quantitatively. "Performance measurement was not easy, nor was it ever solved satisfactorily," Sinaiko and Belden (1965) commented. "Although we made time-and-motion type records of who said what and to whom, we feel that our best data came in informal debriefings of subjects. Not to be overlooked, too, was the value to us as experimenters in being able to observe directly many conference arrangements." A representative conclusion was the value of hard-copy records made available by teletype. Among other qualitative findings were occasional negative reactions to the chairman, although he was acting in everyone's behalf.

### *Additional Studies*

There were two other conference simulations, each on a single occasion. The Television Center of Headquarters, United States Air Force, was the scene of one, in which two four-man teams carried on a negotiation, with a third group observing. One team represented management, the other scientists in a simulated research organization; they argued whether the scientists should have to travel tourist-class in commercial aircraft. (Although all the subjects were IDA professional staff and consultants, those who took management roles played them with impressive conviction.) There was no variation in experimental conditions. The TV presentations were split-screen, so each team could see both the chairman of the opposing team and its own chairman; the participants found this technique distracting and objectionable. The result, incidentally, was an agreement that the scientists travel tourist if they could enter and leave aircraft by the first-class passageway when they were under observation by family or friends.

In the second conference, twelve persons sitting at their own desks, in Washington, D.C., McLean, Virginia, and Cockeysville, Maryland, were linked together in a party-line telephone conference hook-up; there were no extra facilities. Their task was to agree on a date and place for a three-hour meeting; they were first notified about this task when they opened a sealed envelope as the telephone conference began at 9:15 A.M. The envelopes contained calendars of busy and free dates uniquely arranged so only one half-day period in the month would be available for all twelve participants. The subjects had been told to expect a conference call but nothing else. An examination of rating forms which the conferees filled in after they solved the meeting problem (in forty minutes) showed that none termed the telephone conference "very efficient" but nine called it "reasonably successful," and "all agreed that there was a chairman but disagreed as to who it was and the nature of the chairmanship."

In addition Stanford University consultants did some ancillary research. Bavelas (Sinaiko 1963) compared effects of telephone and teletype communication on the extent of interpersonal influence between two individuals, each of whom learned a concept about a set of photos of paintings when subjected to noncontingent reinforcement, then communicated with the other, then restated the concept. Three pairs were tested with a telephone link, three with teletypewriter. The amount of convergence seemed to be greater with the telephone link. Kite and Vitz (1966) compared face-to-face, telephone and simulated telephone, and written communications within approximately sixty-nine groups ranging in size from two to six subjects playing a game called "Crisis." This somewhat resembled SUMMIT–II. Resources consisted of poker chips, threat was established by a roll of dice by the experimenter, and each game had five trials or cycles for negotiation among the players. The experimenters reported almost no difference in negotiation behavior between face-to-face and telephone groups, but a large difference between telephone and written negotiations, the latter taking more time, tending to be more rigid, and being susceptible to developing intransigent positions. Several types of telephone networks were also investigated.

## SECOND PROGRAM

The second research program, labeled the "Secure Voice Conferencing Study," was undertaken for the Defense Communication Agency, as well as DDR&E (Sinaiko and Belden 1965). The context was reported as

(1) the use of a standard telephone conference network within our own offices at the Institute for Defense Analyses and (2) existing military telephone networks within and between various command posts. . . . The subjects vary widely in military rank and background. The experiments themselves are of very short duration, each test running less than 20 minutes including instruction time. Many of the experiments are unique in that in most cases a subject will have no prior notice that he will be participating in an experiment: he becomes a subject the moment he answers his telephone.

The participation of senior military and civilian officials and those less senior in various phases of this work led Sinaiko and Belden (1965) to the following observations:

One of the possible advantages of indelicate experiments is the ability to get the user of the results closely involved with the planning and conduct (possible even as a subject) of the experiment itself. If this can be accomplished several benefits (both to the user and the researchers) can be gained. First, there can be a clearer definition of the problem including limitations imposed by technology as well as policy. Such understanding at the beginning can do much to reduce the ghastly misunderstandings which too often occur at the end of research between the researcher and his customer. Second, it is often possible to break down myths, strong opinions, and even tradition, if the user participates in an experiment or can see the results as they take place in the laboratory. Third, the results become far more acceptable to a client who has been closely associated with the experiment. Finally, the customer in some cases becomes incidentally trained in the techniques of using the results.

*Unclassified Studies*

Although the formal report of this study was classified and unavailable for examination, some unclassified experiments have been reported in a number of Human Sciences Research Technical Notes. These had longer durations than those just noted and subjects other than very high-ranking military officers and government officials. In the five reportable experiments, a recurrent variable was the type of telephone network. This was either a common party-line conference hook-up or point-to-point channeling. In the former either all participants had continuous access to speaking on the party-line or they had successive access with only one microphone open at a time. The latter arrangement was known as "successive broadcast."

In an experiment by G. C. Bailey (1964), four three-person groups and four seven-person groups were netted in successive sessions in the party-line (continuous access) and point-to-point arrangements. Their task, like that of the twelve-person group in the 1963 program, was to schedule a joint meeting. Instead of sitting at their own desks many miles apart, however, the subjects (forty male college students) were isolated from each other in various IDA offices. For each group the scheduling task differed in the two sessions for the two communications arrangements, the order of which was counterbalanced. In this experiment the experimental design permitted analyses of variance. Not surprisingly, the larger groups took significantly longer to schedule their meeting. The party-line network became increasingly advantageous as group size increased, although the networks, considered by themselves, did not produce different effects.

In another Human Sciences Research experiment (Bailey and Jenny 1965), the functions of a chairman were examined in a successive broadcasting situation. Eight conferees included a chairman who controlled the access of the others to the party line by means of a console. In one condition he manned the console himself and in another he was remotely located and directed a console operator by telephone; in each he had an assistant to maintain a list of conferees waiting to speak. Each of four groups of college student subjects used each of the two arrangements in handling two problem-solving tasks, in a counterbalanced experimental design providing controls for learning. A conference lasted no more than 30 minutes. The tasks required coping with different crisis problems in a simulated public health organization. Measures included the duration and source of each interaction unit (a single speech or uninterrupted verbal expression by a conferee). Subjects filled out post-experiment questionnaires. Significance statistics could be applied to these results too. The console arrangement did not yield a clearcut difference in interaction rates. One interesting finding was: "Long speeches in the terminal phase of a conference were significantly related to dissatisfaction with the chairman."

An experiment by Kidd (1965*a*) with sixteen-person conferences tested some of the measures of conference behavior and evaluated some new measurement concepts; it also compared continuous access with successive broadcast arrangements. The results could be compared with prior data from eight-person conferences. In a counterbalanced design which permitted some significance statistics, four groups of subjects differed in each of four conferences, two per session. Two tasks consisted of the public health problem-solving task previously mentioned and one simulating a toy manufacturing company required to cope with a new doll developed by a competitor. (These tasks were evaluated in some earlier experimentation by Human Sciences Research.) Successive broadcast proved to be more advantageous for sixteen-man than for eight-person groups.

Another experiment by Kidd (1965*b*) again compared continuous and successive access; this time the latter had two variations. In one, any would-be speaker requested access when an active speaker had finished. In the other, requests could be made at any time and requesters were placed in waiting lines or queues. Each of the three party-line methods was followed in each of three sizes of conference: eight, fourteen, and twenty conferees. Further, each of the nine resulting conditions encountered four different tasks, known as "maps," "number trading," "stepping-stones," and "discussion." The subjects were male college students. Nine sessions were held at the IDA facility between 7 P.M. and midnight, one session per experimental condition. Among the many measures were ratings of (1) conferees' attitudes toward the chairman's control of the conferences; (2) the chairman's attitude toward the network; (3) conferees' attitudes toward the chairman; and (4) conferees' attitudes toward the conference. Other measures included average length of statement and distribution of talk-time among conferees. No significance statistics were reported.

The fifth experiment, performed by Teare (1965), investigated the differential effects of degrading continuous access and successive broadcast telephone communications by "peak clipping" and reducing loudness at the handset to a barely audible level. Twelve system analysts and computer programmers from the System Development Corporation served as subjects throughout eight runs, four with degraded and four with normal communications. Four tasks were imposed in each of these two main conditions: two public health problem-solving tasks, a "number trading" task, and a simulated military situation concerned with war escalation. The conferences for two of these could last no more than 10 minutes; those for the other lasted 40 minutes. Loss of intelligibility was slight in any condition.

When these experiments and the studies with high-level officials had been completed, the laboratory was dismantled, the telephone company removed the equipment, and the program ended. Approximately 680 subjects had been run in four months. A great deal of useful information had been accumulated, and it had considerable impact. The researchers felt certain that their indelicate experiments constituted the only available method for acquiring that information.

# 21

# Decision-making Research

The term "decision-making" is overused and underdefined. It can cover everything from a radar operator's discrimination between radar signals and noise, to an action by the President engaging the United States in war; from making a choice between two simple stimuli, to evaluating a threat in a complex operational environment. Decision-making has been the theme of much experimental research. This chapter will describe those experiments which simulated a complex operational environment, and especially those in which information was processed for the decision-maker by other members of a team.

Since Bayesian processing has inspired a great deal of this research, experiments directed at such processing will be outlined first. Programs with other objectives will be described in the second half of this chapter. Some of the programs covered in earlier chapters have also involved decision-making, particularly those in which commanders allocated resources, as in Chapters 16 and 18.

## EXPERIMENTS IN BAYESIAN PROCESSING

Three programs have been especially noteworthy, at Ohio State University, the University of Michigan, and the System Development Corporation. The first was outstanding. Although it will not be described here, some experimental research has also been done at the University of North Carolina, North American Aviation (Columbus, Ohio), and the Navy Electronics Laboratory.

### *Ohio State University*

In the period 1963–66, a dozen relatively complex experiments were conducted in a computer-based "Comcon" (command-control) simulation facility. Earlier, starting in 1960, seven studies, three of them experiments, were carried out in a pilot IPAC (information processing and control) facility, primarily to determine how to establish the subsequent facility and its program. Both of the programs and facilities were managed by Ohio State University's Laboratory of Aviation Psychology, subsequently called the Human Performance Center, with funding from the Aerospace Medical Research Laboratories (AMRL) of the Air Force Systems Command.

Although there has been no complete, single account of the two interconnected programs, the IPAC facility, IPAC studies, and Comcon facility have been described by Feallock and Briggs (1963). Howell (1967) summarized the results from the first nine experiments in the Comcon program, and Schum (1967) described the last three experiments. Reviews of the first four Comcon experiments and the Comcon facility have been published by Briggs and Schum (1965), while Schum, Goldstein, and Southard (1966) outlined the first five experiments. Four of the Comcon experiments were reported in technical journals and nine in AMRL reports, referenced below. In view of the availability of these reports, it seems unnecessary to go into excessive detail in the present account.

A large number of individual-subject experiments were conducted as part of the continuing program. From the outset technical support research of this nature was envisioned as complementing the more complex experimentation. These experiments on "simple and often abstract tasks" (Feallock and Briggs 1963) will not be reviewed here.

*IPAC Program, First Part.* The determination to construct a new laboratory for man-machine system experiments apparently stemmed from the Ohio State University air traffic control program described in Chapter 10. According to Feallock and Briggs (1963), the original aims were those "of establishing a human factors simulation facility and of formulating principles of human factors based upon experimentation with this facility." Thus, the program's objectives were very general. It was decided that experimentation would be directed at command and control systems, although appropriate parts of dynamic weapon system operations would be simulated and the principles emanating from the research should be of use to weapons system designers.

Before starting the IPAC experimentation, the Ohio State researchers visited the RAND Logistics Systems Laboratory (Chapter 13), the SDC SSRL (Chapter 17), and the Subsystem I facility (Appendix I), drew on experience in the Ohio State air traffic control studies, and reached the conclusion which was largely responsible for this book: "Where the design of system simulation facilities for research is concerned . . . the availability . . . of information is extremely limited. Systems research by the method of laboratory simulation is in its adolescence; consequently, there are few principles, precedents, or conventions such as established rules of thumb or tested formulas for simulation research that are available as bases for generating one's own developmental guidelines."

The researchers also concluded that "the development of the simulator should begin with exploratory work done on a pilot model simulator of simple design, construction, and operation," with successive short-term tests of equipment, environment, and experimental methodology; that the situational task should occupy two to five subjects, one of them to make decisions and the others to process information, all in mutual co-operation; that the simulated environment should be "dynamic, responsive and probabilistic"; and that the test model should provide work space "for as many as fourteen subject operators."

The first IPAC study involved two competing teams of two members each. A dozen control operators with whom the teams interacted manned a game board, recorded data, and established probabilities of events by drawing numbered pills from a bottle. Each team waged tactical air combat (air reconnaissance, air defense, and tactical bombing) against the other. The team outputs were flight plans. Sessions (trials) lasted four hours (one per day). The subjects, as in all IPAC and Comcon studies, were students. When it was found rather quickly that the task was neither interesting nor demanding, load and response alternatives were increased, but the main effect was to multiply the mistakes made by the overloaded control operators. The researchers concluded there should be more automation of control functions, including the determination and display of probabilistic events. The study lasted nine weeks.

In the second study, which continued for three months (eight hours a day, five days a week), the teams and their tasks were similar to those in the first, but the teams had access to more information and confronted more numerous and complex decisions. In place of the game board (on which manually moved magnetic markers represented aircraft), the control operators used the earlier simulation equipment (consoles and analog computer) developed for the Ohio State University air traffic control research. But now, in place of certain operator errors, the simulation suffered from undesired voltage changes in electronic components (electronic drift), and other operator errors continued. The need for the kind of control obtainable with a digital computer became obvious.

In addition, the researchers concluded that commitment of weapons should be eliminated from the task because students, lacking military command experience, were unable to allocate weapons realistically and effectively. Henceforth in the IPAC and Comcon programs, decision-making was limited to forms of threat evaluation; it did not include emphasis on action selection. The researchers also worried about the motivational uncertainties in using students for combat functions (Feallock and Briggs 1963):

> Another argument for deleting these responsibilities is that it is virtually impossible to attach real values to artificial consequences of weapons commitments. One may report to the decision maker the number of lives that would probably be lost as a result of his decision, the number of weapons that would probably be lost as a result of his decision, the number of weapons that would be destroyed, or the number of installations obliterated; but no matter how much stagecraft is used or how much the subjects are willing to be deluded, it is highly unlikely that these numbers for the average college-student would carry anywhere near the significance they would for a commander behaving in the face of real-world events and responsibilities. The problem of evaluating decision performance in these kinds of situations is equally unyielding, for to make the criteria realistic, one should have equations which relate values of material things to values of life and costs of death and injury.

In the third IPAC study, two teams again opposed each other, but they consisted now of three members each, and the tasks were limited to (1) a team's own reconnaissance and (2) frustrating the enemy's reconnaissance. Toward the latter aim, a team moved a probably detected installation and intercepted the enemy's reconnaissance aircraft. There was no more bombing; on the other hand, the study emphasized costs (fuel, aircraft-use time). Two small digital computers flew the flights automatically, in parallel with the air traffic control simulator. They also applied Monte Carlo random sampling procedures to the detection and destruction events, with printout feedback to the teams.

The researchers compared a technique of compressed time with continued reliance on real time. A real-time trial consisted of two three-hour sessions on successive nights; a compressed-time trial consisted of one three-hour session which included six hours of system time and events. Each team had several weeks of real-time experience during system shakedown, then four trials of compressed time, followed by four real-time trials. Compressed-time trials did not seem to alter system performance, but they increased the subjects' interest and attention, and they reduced void-filling conversation on extraneous topics. The researchers noted that some kinds of simulation can result in "the omission of minor operations that in the actual system absorb some time and attention of system personnel . . . . The technique of time compression can serve to reduce voids that occur" because of these omissions. It was decided to incorporate time compression into subsequent studies.

With the fourth IPAC study the researchers abandoned for the duration the concept of two competing teams in a responsive simulation. Now and henceforth teams played against a computer programmed to provide a hostile environment "which was neither occasioned by subjects' responses nor influenced by them." However, in this fourth study the environment was still dynamic to the extent of including predictable weather changes affecting reconnaissance. The researchers noted that responsiveness by means of competing teams had yielded "realistic, complex, and highly variable inputs . . . without the usual expenses of time, effort, and funds associated with producing them by hand or by a programmed computer." But the nonstochastic variability that resulted (as in number of aircraft available) was confounded with the specifically introduced independent variable and prevented the controlled evaluation of their effects on the performances of the teams (Feallock and Briggs 1963). Furthermore, the "accounting operations for handling interacting responses" required computer program space in the computers that was desirable for enriching the tasks otherwise and led to long updating intervals. In any case, the studies clearly demonstrated the need for a larger digital computer.

In the fourth IPAC study team-size rose to five and specialized functions were assigned to each processor of reconnaissance data, in order to investigate variables of functional differentiation within systems. Specialists had to exchange information to become maximally effective in advising the decision-making commander; however, it turned out that two of the specialists did not have much to do. Processing of reconnaissance data received still further attention. New dimensions of information were added. Airborne intercepts were eliminated, as were displays showing locations of simulated reconnaissance aircraft on a moment-to-moment basis. Thereby, the analog simulator could be foregone entirely and the control personnel reduced to eight, who operated the digital computer and provided an interface between it and the subjects. For example, one control person punched the subjects' handwritten outputs (flight plans) onto paper tape, and another read the computer's printout (reconnaissance report) over the intercom. Nevertheless, it was thought desirable to have displays that were more rapidly updated and flexible.

In summarizing the thirteen-month course of these four nonexperimental studies, the researchers noted that although they had some goals in mind, such as adopting digital computer simulation, "we were willing to follow fortuitous leads," especially in the first two studies. Subjects and control personnel provided comments and suggestions at formal debriefings after all sessions, and any resulting insights might be incorporated on a daily basis.

*IPAC Program, Second Part.* The end of the studies saw the start of design and construction of the Comcon facility. But the Ohio State researchers decided to do some more work with IPAC, this time within experimental frameworks. Each of three experiments involved the acquisition and processing of aerial reconnaissance data about hostile installations.

In the first of these (Feallock and Briggs 1963), each of two independent teams of six simulated officers experienced five trials of each combination of four load distributions and two team compositions. Each of the resultant eighty trials lasted three hours and three-quarters, including a half-hour for planning at the start and a quarter-hour as a rest break at midpoint. Team composition was varied by alternating the reconnaissance-operator function between two three-man subsets in each team; accordingly eighteen subjects were needed in all. Load distribution was varied by assigning different aircraft totals to members of the reconnaissance subset. The two crews were run at the same time to economize on time and cost in collecting data. One objective was to determine the feasibility of this technique. All subjects received ten practice trials. The independent variables and their order of presentation were organized in a Latin square design. A somewhat larger digital computer than the one used earlier simulated the reconnaissance environment and reconnaissance flights.

Both system and subsystem measures were obtained, system measures including number of installation detections, fuel used, number of flights flown, and aircraft-use time. (The experiment was supposed to be an investigation of measures to be employed later in Comcon. But, as will be seen shortly, such measures were actually not the ones on which the Comcon studies concentrated.) Results showed that system performance was not affected by load imbalance among the reconnaissance operators and they did not try to equalize load among themselves—presumably because it was not heavy enough.

A single, nine-man team functioned in the second experiment; it was composed of subjects from prior studies. Thirty four-hour trials followed four training trials. The independent variable, conditions of feedback, varied in five ways, each of which had six trials. Controlled randomizing of order placed each condition once in each block of five successive trials, with no condition occurring twice in succession. Conditions consisted of nonfeedback and four degrees of functional "remoteness" of the feedback data with regard to the reconnaissance operator to whom the feedback was directed. The data could describe his output or that of processors at two higher levels—sector and area commanders.

As in all three of these studies, results were subjected to analyses of variance. For most measures the conditions of feedback had no statistically significant effect, because, it was supposed, the feedback was noninstructional—it did not tell the subject how he might do better or the reasons for the particular output score. Actually, this study had as a major objective not this experimental inquiry but the development of a technology whereby the computer could fabricate the reconnaissance environment automatically. In the new technology, the reconnaissance maps which functioned as environment scripts were generated by the computer, through a program that specified most of the constraints used previously in manual map composition. The maps were somewhat formal compositions consisting of many cells; to vary them, cell characteristics were assigned randomly within specified limits. A library of such scripts would be accumulated to "provide the basic stimulus material for numerous system experiments. This approach was incorporated into the planning of Comcon and is considered to be one of the outstanding methodological features of Comcon research" (Feallock and Briggs 1963).

The third of the experiments also dealt with feedback, the nature of which differed in three ways; each indicated to certain operators the association between their output and actions more clearly than in the preceding study. This time feedback did have statistically significant effects, by guiding operators, the researchers said, "in the selection of response alternatives." Two seven-man teams were organized from ten subjects; four subjects were common to both teams. All but one had been subjects in the preceding experiment. Both teams were tested under each of the three feedback conditions, the order of which differed between teams. With four trials per team for each of two conditions and six for the other, the twenty-eight sessions followed four training sessions with a no-feedback condition. Sessions lasted three and one-half hours, with a 15-minute midpoint break.

*The Comcon Multiman Task Environment Simulator.* This laboratory occupied an area of approximately 5,500 square feet. Designated for subjects' experimental space was about one-third of it, one large room (28 × 26 feet) and six smaller rooms, two of which could be subdivided with a movable screen; generally the large-scale experiments used only the large room and one of the smaller ones. The largest single space in the laboratory housed an IBM 1401 computer and its peripheral equipment. The simulation was based primarily on an IBM 7090, subsequently a 7094, about 300 feet distant in the Numerical Computation Laboratory. The 1401 performed simpler data processing and served as a buffer, handling subjects' input requests and providing printouts. Display consoles (variously stated as four or five) with cathode ray tubes and selector buttons were linked directly to the large computer. Other spaces were devoted to a shop, parts and equipment storage, and a 28 × 28 foot area containing an experimental supervisor's office, a telephone exchange, tape recording, television and audio control, and a three-position TV-audio monitoring station. Visitors at this station could observe the experimental subjects. The entire experimental area was windowless and air-conditioned. In contrast to more luxuriously appointed laboratories, there were no false floors (cables lay on the floor), no one-way glass viewing panels, no movable walls. Some office space for the experimental staff was nearby.

The large computer generated the scenarios of a hostile environment (1,024 by 1,024 miles) containing military installations and various kinds of troop and vehicle movements detectable by simulated aerial reconnaissance. During an experimental run it also accepted inputs from the 1401 and the consoles and presented the console displays on a real-time basis. It was time-shared in the sense that part of its core memory for these purposes was locked out of other processing, and control was passed to the simulation program through interrupts activated by the 1401 or consoles. The simulation required seventeen man-years of analysis and programming (Briggs and Schum 1965), of which about fifty man-months were for programming (Feallock and Briggs 1963). The Ohio State University Computer Center devoted seven and one-half man-years to providing the OSU system monitor and operating system (Briggs and Schum 1965).

The subjects received tabular and map-like representations of the simulated, changing (but nonresponsive) environment on the console displays. They communicated among themselves by telephone, closed-circuit TV, and hand-carried messages, e.g., summary charts. Generally at least two of the consoles were used by the experimenters for monitoring and two were used by the subjects. Seven microphones could pick up face-to-face conversations. A television camera could scan the main experimental area; there were five monitoring locations. A 14-channel tape recorder could register face-to-face and telephone conversations, which could be monitored at five locations.

The simulated hostile (aggressor) area consisted of a latticework of supply areas and depots, forts, and airfields of various types, interconnected by roads and railroads. Air reconnaissance could detect up to sixty-seven types of ground vehicles and thirty types of vehicles; and on the basis of these detections subjects could make inferences concerning the status and movement of units down to battalion size. Subjects could obtain a description of what was happening within the area every ten minutes, in verbal and numerical form; it represented interpretations of photographic, radar, and infrared recordings from reconnaissance flights.

During an experimental session there could be as many as twenty-five different patterns of detectable objects and their movements; these patterns were called "developmental groupings." Each of these deployments culminated in some grouping at the area's border, remained there for a while, and disappeared; a few deployments would be in development at the same time. Each deployment could be described in terms of twenty-five attributes in four categories; main attack, combat support, logistics, and order of battle. An attribute had some fixed number of possible states. The state had to be inferred from the detected objects and their spatial and temporal characteristics, which evinced certain patterns in the course of the exercise. Then, on the basis of the inferred attribute states, it was necessary to evaluate the threat, that is, to judge which intent or strategy, out of some set of specified alternatives, was responsible for the deployment.

*Bayesian Processing*. In consequence of a paper by Edwards (1962), the Ohio State researchers decided to use the simulation facility they were building to investigate the application of Bayesian processing to command decision-making. In their research such processing was restricted to the evaluation of threat; as already noted, it did not extend to allocation of forces (action selection). In particular, it was applied to determining the hostile intent or strategy accounting for each of the deployments or developmental groupings at the aggressor boundaries during a session.

"Bayesian processing" means applying a theorem authored by the Reverend Thomas Bayes two hundred years ago to estimating the probability of some situation or event on the basis of fragmentary data or items of evidence. The theorem involves two steps. It first requires a probability estimate concerning each datum. This is not an estimate of how probable each of a number of alternative situations might be on the basis of the item of evidence, but is instead a statement of how probable it is that the item would have resulted from each of the alternative situations. Second, the theorem successively aggregates a number of these probability statements about data to produce an estimate of the probability of each alternative situation; in the theorem, each such estimate modifies a similar but prior estimate. All probability statements are phrased in percentage terms.

To those intrigued with applying this approach, a computer appears to be an efficient instrument to aggregate the datum probability statements; and the need for automatic processing of this sort seems to be a way to exploit computers in the context of threat evaluation and other forms of decision-making. Operationally, the computer output either could be the probability estimate which would actually be used by a commander, or it might merely constitute an advisory guide. In any case, experimenters, including the Ohio State researchers, have been interested in making comparisons between the probability estimations made about situations by human beings confronted with items of evidence and estimations from computer-implemented Bayesian processing.

Bayesian processing need not exclude all human estimation. It may, in fact, exploit it. For Bayesian processing to occur, there must be probability estimates about each datum, given each alternative situation. These can be derived from the frequency of a recurring item. They can originate from prior experience, or from expert opinion. The source can be objective, or it can be the subjective judgment of a human observer. In the latter case, some individual makes an estimate of the probabilities that an event would have occurred given each of a number of alternative situations, and these estimates are aggregated by the Bayes Theorem.

*Comcon Approach to Bayesian Processing.* In the Ohio State research, the situations whose probabilities had to be determined were always the intents or strategies of the aggressor. Each datum or item of evidence was one of the attribute states which had been inferred from simulated detections of vehicles and aircraft.

In the experiments two general human processes were at work, one to gather and present the data, the other to evaluate the threat on the basis of those data. Although the first process required substantial experimental activity and a number of subjects, the researchers indicated relatively little interest in measuring its outcome. In fact, in the latter experiments of the program this process was simulated within the computer rather than through subjects. Rather, the experimenters were concerned with the second process—the estimation of the probability of situations. They were particularly interested in comparing the usual way in which humans make such estimates (by inducing them from data) with computer estimation obtained by Bayesian processing and with human estimation when the decision-maker was given the Bayesian estimates as an aid. They were also interested in that part of Bayesian processing open to human participation—the subjective estimation of datum probabilities dependent on situations.

In all the experiments, then, the main focus was on the performance of a single individual—the threat evaluator; and although teams were involved in the eariler experiments, they functioned to supply complex information to the evaluator rather than as major objects of experimental inquiry in themselves.

The information-suppliers were called intelligence staff officers (ISOs). Each handled a particular, specialized set of information. The information the ISOs supplied was complex in that it covered all the states of many attributes for different deployments. It was provided intermittently as a deployment developed and at the end of that deployment. Each state of each attribute was given with a number which indicated the probability, in the ISO's judgment, that such a state was the actual one. This probability differed from those discussed above. It did not take into consideration the situation of which the state was a symptom; that is, it was not part of a Bayesian processing. However, it did aid the threat evaluator by suggesting how much to rely on the data.

The threat evaluator (TE) was also called the decision-maker (DM); he was the commander of the unit. In addition to the estimates from the ISOs he received other inputs. These might be predetermined, true conditional (situation-dependent) probabilities of the attribute states. They might be environmental rules which indicated the probabilities that various attribute states would arise in various situations. Sometimes, as an aid, he received from the computer a computed Bayesian estimate of the probabilities of the possible hostile intents or strategies. These might be derived from the true conditional probabilities or from conditional probabilities calculated and furnished by the DM himself; that is, the DM provided to the computer his own probability estimates that the states of attributes were due to various hostile intents. The DM evaluated the intent of each deployment intermittently as it developed and also when it had reached the borders of the hostile region prior to its disappearance. His major outputs were numbers, each a probability in percentage terms for each alternative intent or strategy for each deployment. The experimenters sought answers to two major questions. What process yielded the best probabilities—the DM's judgment by itself, that judgment aided by the computer's Bayesian estimates, or the computer processing itself? How did various inputs into the system or into the computer affect the outcome?

*First Comcon Experiment.* The first of the experiments in the Comcon laboratory has been described by Southard, Schum, and Briggs (1964*a*). From thirty male student applicants, thirteen were chosen according to academic grade minima, undergraduate class, long-term availability, and apparent maturity; they averaged twenty-three years old. They received 114 hours of lecture sessions, demonstrations, problem-solving sessions, and on-the-job training, required because the system and the experimental operations were considerably more complicated than it has been possible to describe here. Prior to fifty hours of on-the-job training, eight of the students were selected as system operators; the rest were designated as alternates, monitors, and helpers in preparing stimulus materials. Following training, a shakedown series of runs totaling about one hundred hours under various loads and procedural configurations aided the experimenters in establishing load limits and methods of collecting data.

The subjects played the roles of four intelligence staff officers, each handling a specialized area; an operations liaison officer, who among other things acted as a reconnaissance mission planner; a commanding officer—the decision-maker; and two aides to the DM. In all trials, the DM received from the ISOs their estimates, independent of situations, of actual states of attributes. He also had access to probability estimates of these states that were dependent (conditional) on the situations. The latter, which the experimenters derived in advance by applying a set of true contingency rules, were also given to the computer to calculate the Bayesian estimates of situation probabilities. In addition to comparing the computer's Bayesian solutions with the DM's estimates, the experiment contained two independent variables: (1) The DM received or did not receive the computer-derived Bayesian situation probability estimates as an aid. (2) The number of developmental groupings (deployments) completed in a session, which always lasted four hours, was one, two, or four. Since each of the six factorial conditions characterized four sessions, there were twenty-four sessions in all over a six-week period. This study posed twenty possible hostile intents or strategies, varying in penetration depth and in tactics (e.g., double pincer, multiple penetration, double or single envelopment) as well as representing either actual attacks or rehearsals.

In this experiment one of the measures was the accuracy of the attribute-state probability estimates provided by the ISOs. But the main measures concerned the DM's estimation of the probabilities of the hostile intents or strategies. The DM had to assign a percentage number to each of twenty alternative situations. In the Ohio State research, these estimations were measured generally in two ways. One measure was the actual probability he assigned to that situation which in fact characterized the hostile deployment. The other was a dichotomous scoring procedure by which the DM received full credit if he gave his highest probability to the true situation and none if he assigned it to some other situation.

Load had no consistent effect, although the Bayesian processing aid seemed to help the DM under heaviest loads. The Bayesian solutions were significantly superior to the DM's; however, as the researchers conceded (Southard, Schum, and Briggs 1964*a*), it was risky to generalize since there was only one DM. In considering the comparison it is important to bear in mind that the computer and the DM received the same true estimations of the probabilities of attribute states, given the various possible hostile strategies. These totaled 2,060, in a matrix of 103 attribute states and twenty situations. This large matrix of percentage figures was easily processed by the computer, but such a display was too unwieldy for the human subject; in fact, the experimenters said it was "a most excruciating task." Accordingly, they gave the DM in addition verbal statements of the contingency relationships (rules) which yielded the probability estimates.

*Second Comcon Experiment*. The second experiment, reported by Southard, Schum, and Briggs (1964*b*), was, like the first, described as "introductory"

(Briggs and Schum 1965). It ran for sixty four-hour sessions, in each of which four deployments (the index of load) terminated. The same subjects (apparently) played the same roles as in the first. An important change was the narrowing of alternative hostile strategies to four during any one session, although the experiment contained all twenty from the first experiment. Four experimental conditions occurred in an only partially counterbalanced order. In one condition the decision-maker received no aid from Bayesian processing, whereas in the other conditions he was aided. In the second he received such aid, as he had in the first experiment. In a third condition, he himself could revise the inputs which went into the computer's Bayesian processing, through a parameter change to make the most recent estimates of attribute-state probabilities more influential in the calculation of situation probabilities. Although in the fourth condition he could alter the inputs' values, he did so on an item-by-item basis, so that, in a sense, the Bayesian processing was now based on the DM's personal estimates. As summarized by the researchers, results indicated that the DM's performance improved during the experiment (independently of the aid condition) and became notably less "conservative"—the term given to a tendency to avoid extreme probability estimates; human and Bayesian estimations were strikingly similar, with equivalent accuracy.

*Third Comcon Experiment.* A new departure in this experiment (Schum, Goldstein, and Southard 1965*a*) was the production of situation-dependent estimates of the probability of attribute states by the decision-makers themselves. The load level was six deployments terminating during each of thirty four-hour sessions. The subjects were drawn from those in earlier experiments. In addition to the four ISOs and an operations officer (chief of staff), this study had four decision-makers (threat evaluators—TEs), acting independently, each receiving the same nondependent probability estimates of attribute states from the ISOs. Each TE had five four-hour practice sessions. The number of alternative hostile intents or strategies was fixed at eight, so the chance probability of each such situation was 0.125. The TE began with this probability and modified it as he acquired increasing estimates of attribute-state probabilities from the ISOs.

The experiment had two objectives. One was to see how a TE's estimates changed with increasing experience; at the start of each session he received feedback about the results of the prior session. The other was to compare his estimates with two kinds of Bayesian processing by the computer. In one of these the probability of an attribute state, given a situation, was estimated by the TE on a recurring basis for each attribute state; the computer used these estimates in its Bayesian processing, in contrast to the pre-established, true estimates in the two previous studies. In the other, the computer received the nondependent state probabilities estimated by the ISOs; its processing included a self-adapting feature concerned with state obsolescence and feedback proposed by Dodson (1961). All the Bayesian processing in this research followed a general modification formulated by that author to deal with more than two alternative situations and to allow for observational uncertainty about the occurrence of an attribute state.

The TE's evaluation of threat improved during the thirty sessions to a marked degree. So did the evaluations by both methods of Bayesian processing.

According to one measure, the human evaluations and the computer's evaluations were equally good, although Bayesian processing was superior in early sessions. Another measure indicated over-all superiority in Bayesian processing. The ISOs' estimations of attribute states became more accurate during the experiment and must have accounted for some of the improvement among the TEs.

*Fourth Comcon Experiment.* The fourth Comcon experiment had all the characteristics of the third and all its subjects except one, but it added three variations in input fidelity as an independent variable (Schum, Goldstein, and Southard 1965*b*). Verbal and numerical records were manipulated to simulate the effects of degraded photo, radar, and infrared sensor images. Three fidelity levels resulted from combining types of ISO access to such records with degrees of degradation. These levels led to differential uncertainties among the ISOs in their estimations of nondependent attribute-state probabilities. Human judgment turned out to be as good as Bayesian processing in the TE scores. Both suffered from the lowering of input fidelity, as did ISO estimations. Although Bayesian processing had been expected to show increasing superiority as input fidelity decreased, this did not happen; it was suggested by one measure in the case of the most degraded inputs, but statistical significance was lacking.

*Fifth Comcon Experiment.* This experiment (Schum, Goldstein, and Southard 1965*b*) had the simulation and methodological features of the prior two experiments and the same subjects as its predecessor. In its thirty-two sessions the fidelity of input data was again an independent variable, but took another form beyond that of degradation in sensor records. Surveillance halted before a deployment was completed, although the ISOs still had to make their nondependent probability estimations of attribute states. The amount of time between the end of surveillance and the completion of deployment was varied to help constitute four levels of input fidelity. Still another independent variable was introduced, "time stress," which had three values. The TEs had 1, 4, or 7 minutes to evaluate the ISO estimations before making their situation judgments. They learned, over closed-circuit television, how much time they would have only 30 seconds before they received the ISO estimations on which they had to base their judgments. As in other experiments, they received the ISO estimations on the same TV display.

The TEs could keep a record of the status of each developing deployment by processing interim estimations from the ISOs. This they did with such success that when they had to make a final threat evaluation they merely wrote down a percentage figure which, essentially, they had already derived from earlier ISO inputs. Consequently the time stress made no difference. Increasing loss of input fidelity resulted in marked, progressive decrements in both human and Bayesian threat judgments, as in the preceding experiment. This time, for the more sensitive measure of threat evaluation, the superiority of Bayesian processing not only held throughout conditions but increased as sensor records grew less reliable.

*Sixth Comcon Experiment.* In this study (Goldstein, Southard, and Schum 1967) the name of the decision-maker reverted to DM; there were eight of them, all subjects in prior Comcon experiments, divided into two groups for testing two new independent variables, amount and type of feedback. For all DMs the frequency of feedback varied between 0, 33, 67, and 100%. One group received knowledge of the accuracy of their own judgments only, while the other received, in addition, knowledge of the accuracy of the results of Bayesian processing. Within each group the DMs operated independently although concurrently (as in the preceding three experiments), receiving attribute-state estimates from the ISOs. A set of four ISOs and a chief of staff was common to both groups of DMs. For each percentage level of knowledge of results there were eighty-six deployments, six per session. A DM had to assign probabilities to eight hostile strategies (situations), which were labeled simply by letters.

The DMs also, as before, estimated for each of the eight strategies the probability that an ISO-reported attribute state would have occurred if a given strategy (situation) were in effect; and these percentage terms were the inputs to Bayesian processing. When the results of that processing were compared with DM judgment about the probability of the strategy, the Bayesian processing was superior (except at zero feedback) for both estimation measures. Percentage of feedback had relatively little effect except when it fell below 33%; such drastic reduction of feedback almost eliminated effective DM threat evaluation.

*Subsequent Experiments.* Additional variables in subsequent experiments included "prior uncertainty" and "amount of diagnostic evidence" (Schum 1966*b*); "several cost-payoff arrangements" (Schum, Goldstein, Howell, and Southard 1967); "non-independence" of ISO attribute-state estimations (Schum 1965, 1966*a*); "rate of accumulation of scenario evidence" (Schum, Southard, and Wombolt 1966); and "total amount of evidence" (size of the scenario) (Schum, Southard, and Wombolt 1966). The last experiment (Schum 1966*c*) "attempted to unscramble the confounding relationship between scenario size and scenario diagnosticity."

*General Comments.* The foregoing summary of the Comcon experiments hardly does justice to the magnitude of the program, the complexity of the simulation, the nature of Bayesian processing, and the results which were accumulated. As noted earlier, Howell (1967) summarized the results of most of the experiments. His summary adduced thirteen principles. One of these is that computer-processed aggregation of evaluated data according to the Bayes approach can "improve the quality of decisions by 10–15 percent," a figure based on nine experiments. In other words, the Comcon program taken as a whole produced substantial evidence of the potentials of Bayesian processing. Other principles dealt mostly with parameters influencing the extent of these potentials.

The Comcon research was concerned with probabilities and contingencies. In consequence, the reporting of the research was often phrased in symbolic shorthand. For example, the probability that a strategy, intent, or situation may exist in the light of certain attribute states or developing aspects of a hostile deployment was stated as P(H/D), meaning the probability of the hypothesis given the data. The probability that a certain attribute state or aspect of a deployment has occurred because of a particular strategy, intent, or situation has been stated as

P(D/H), meaning the probability of the datum given the hypothesis. Such shorthand is perhaps unavoidable for adequate treatment of this domain. Its omission in this account should be ascribed only to P(U/R), meaning the probability of understanding due to the nature of the readers, or at least many of the readers, of this book.

Nevertheless, if concepts of Bayesian processing are to gain acceptance and even understanding among the people to whose decision-making they might be applied, such as military commanders, it is important to present them simply. The same might be said about reporting the experiments which have investigated the concepts. A report is a display. Some display features can be optimized in spite of the complexity of the material. Orderly arrangement and simple phraseology are important. So are such formal aspects as line length and paragraph length, especially in single-spaced text if the material per se is difficult to digest. Perhaps if the reports of the Ohio State University experiments had been optimized, this ambitious program would have received more of the recognition it deserved.

*University of Michigan*

W. D. Edwards at the University of Michigan was the earliest and most articulate proponent of applying Bayesian processing to decision-making in operational situations, such as those in command and control systems. It will be recalled that his 1962 paper was the inspiration of the program just described at Ohio State University. In characterizing the differences between his own research in Bayesian processing and the Ohio State work, Edwards (1966) emphasized two points. First, the Ohio State program took place in "unquestionably the largest and most complex laboratory situation in which Bayesian ideas have ever been studied." Second, with its "large, frequentistic simulation" it was "primarily concerned with repeatable situations in which the set of possible observations is quite limited, so that subjects can reasonably expect to accumulate relevant relative frequencies linking data with hypotheses." Edwards described himself, on the other hand, as "primarily concerned with vague, verbal data, for which no hope of frequentistic information linking data with hypotheses exists."

Many of the experiments on Bayesian processing at the University of Michigan were structured around such tasks as taking red and blue poker chips from bags or urns and estimating the contents of those receptacles (e.g., Phillips and Edwards 1966); these studies were too far removed from the nature of man-machine system experiments to be reviewed here. However, Edwards and his associates conducted one set of three experiments which simulated, to a very limited extent, an operational situation; and Edwards (1963) formulated further experimental research with more complex simulation of a military situation while his work was still being sponsored by the Electronic Systems Division of Air Force Systems Command.

The set of three experiments, in a pseudomilitary setting, was conducted in 1961. The major one was first reported at a meeting the following year (Edwards and Phillips 1962) and published by Edwards and Phillips (1964) and Phillips,

Hays, and Edwards (1966). The other two experiments were reported in this last citation.

In all three, individual subjects performed at a console surmounted by a screen on which four large, subdivided circles were rear-projected. Each was said to represent a geographical area and was divided into twelve sectors. Dots appearing in one or more sectors represented predicted impact points of detected objects. One circle represented enemy attack, another friendly activity, the third a meteor shower, and the fourth enemy efforts to "spoof" the surveillance system. The dots were data or events (like the attribute states in the Ohio State research) on the basis of which the subject was supposed to estimate the probability that his surveillance system was detecting each of the four situations. Within each sector of each circle was displayed a number indicating the probability that an impact would fall there if the kind of activity represented by that circle was indeed occurring.

To carry out his task of estimating the probability that each of the four kinds of activities was being detected, the subject moved levers along 12-inch vertical scales calibrated from zero to 100, one below each of the circles. After the subject pressed a button to record his estimation, he reset his lever to zero and awaited the next display. Advance intelligence estimates of the probability of enemy attack were also displayed to the subject.

In the first experiment, each of five subjects (freshmen engineering students) saw sixty-four sequences of fifteen stimulus slides per sequence, in six to eight two-hour sessions per subject. The number of dots varied between one and fifteen and appeared in only three sectors. In one-half the sequences the number of dots increased in an ordered manner, in the other half the totals were scrambled. Prior probabilities of an enemy attack were 10, 25, or 67%. According to Phillips, Hays, and Edwards (1966), the experiment included three independent variables: number of dots (amount of information), order of their presentation, and prior probabilities. According to Edwards and Phillips (1962, 1964), it included five independent variables. One additional variable was the distribution of dots within three sectors, either five dots per sector or a 3–5–7 distribution. The second additional variable was embodied in four different sets of sequences according to the degree to which posterior probabilities converged on the hypothesis of enemy attack.

(In discussions of Bayesian processing, "posterior probability" is the output of the Bayes theorem. "Prior probability" is one of the inputs to the theorem, namely the probability of a situation or hypothesis before considering and incorporating the probability of the current event or datum. The Bayes theorem combines the prior probability with the aggregate of probabilities of events, given particular situations, to produce posterior probabilities.)

From this experiment the researchers concluded that four of the five subjects failed "to extract the certainty available in information" (Edwards and Phillips 1964); instead they displayed much "conservatism" in making only small changes in probability estimates from one stimulus presentation to the next. The subjects differed considerably in the extent of variability in their judgments.

The second experiment compared the distribution of dots among only three sectors during a presentation with distribution among more than three. Each of four subjects (students) viewed eight sequences of fifteen dots each. The third experiment introduced two new variables: (1) three levels of difficulty in the sequences, in terms of ambiguous and contradictory information; and (2) nonsequential vs. sequential modes of estimating posterior probabilities. A subject either reset the lever to zero after each estimation, as in the preceding experiments, or left it at the prior setting before making a new one. The modes were not counterbalanced to control for effects of order. Each of six subjects performed for somewhat less than two hours. Again the researchers (Phillips, Hays, and Edwards 1966) concluded that "conservatism is a very pervasive phenomenon, little affected by different stimulus displays or different response modes."

In these three experiments the decision-makers did not estimate the probabilities of the events, given the situations, as inputs into the Bayes theorem; instead, these probabilities were prepared by the experimenter and shown to the subjects. However, Edwards (1962) expressed the view that in operating systems these conditional probabilities should be estimated by people in a form of Bayesian processing called "probabilistic information processing" (PIP).

### *System Development Corporation*

Three experiments involving Bayesian processing were performed at the System Development Corporation in the Systems Simulation Research Laboratory and Command Research Laboratory (see Chapter 17); they were reported together by Kaplan and Newman (1964*b*, 1966) and individually by Kaplan and Newman (1963), Kaplan, Lichtenstein, and Newman (1963), and Kaplan and Newman (1964*a*). Interestingly, these researchers distinguished between "probability estimation," which they said was required in all three studies, and "actual decision-making behavior," which they said characterized only the second experiment, in which subjects had to make a binary choice. According to their criterion of what constituted decision-making, the experimental subjects in the Ohio State University and University of Michigan studies made diagnoses rather than decisions. As suggested at the outset of this chapter, the term "decision making" has been more fashionable than precise.

The three experiments had the following features in common. There were no teams, simply individual subjects, who were students (eighteen in the first study, twelve in the second, and thirty-two in the third). Their task was to assess enemy strategy from information about missile strikes on U.S. cities during a nuclear attack. One experimental condition was PIP, in which subjects estimated the probabilities that the data resulted from particular enemy strategies; these estimates then became inputs for automated Bayesian processing to determine the probability of each alternative strategy. Another experimental condition was non-PIP, in which subjects themselves directly estimated the probabilities of the strategies. These two conditions, PIP and non-PIP, constituted the principal independent variable in the research. (In the first experiment, all subjects encountered both conditions; in the other two, subjects were randomly assigned to a different group for each condition.)

The first experiment had three alternative strategies; that is, targets were military, civilian, or industrial. The second had two alternatives: two cities as targets. In the third experiment five alternatives were strategies called "Military,

Civilian, Industrial, Transportation, and Anti-recovery." In the second experiment the subjects had to pay out money (from an amount previously given them) to get more information and to determine when they had obtained enough data, and they could get money as pay-offs. Three levels of difficulty were introduced in two different ways into two of the experiments.

The experimental apparatus varied between experiments. In the first it consisted simply of a 40-page booklet of target maps, a nuclear detonation (NUDETS) report for each target, sheets on which each subject recorded responses, and a pencil. In the third experiment, all subjects recorded their responses by typing numbers between 01 and 99 on a computer-connected teletype keyboard; a computer-generated display on a cathode ray tube (CRT) showed aggregations of probabilities in the form of bars on a graph. The complex data in fifteen simulated reports of missile strikes against U.S. cities were presented to the subjects in a booklet.

In the second study, projected slides showed missile strike information to all subjects. The non-PIP subjects simply announced their confidence judgments aloud and an experimenter recorded them. The PIP subjects entered their judgments on a teletype keyboard so that a computer could quickly calculate the outcome of Bayesian processing after each input; the teletypewriter then typed out the outcome and the subject read it. The difference in methods of recording responses was confounded with the difference between the main experimental conditions, PIP and non-PIP. Further, there were numerous recording errors.

This research did not sufficiently resemble the kinds of complex, large-scale experimentation at which this volume is directed to warrant further detail. In any case, it has been fully and (especially by comparison) lucidly reported.

In general terms, Bayesian processing proved superior in the first and third experiments, but differences were not statistically significant in the second. PIP processing produced more false alarms than the non-PIP conditions, as well as more correct detections, and this greater proportion of false positives occurred with moderate totals of data items. The superiority in real detections was concentrated among the lower totals of data items, non-PIP subjects doing as well after about ten items of data. The researchers (Kaplan and Newman 1966) concluded that a PIP system "might have its greatest application to those situations in which diagnostic decisions had to be made quickly and on the basis of small amounts of information," the advantage in Bayesian processing diminishing as the amount of relevant data and time to process them increase. This notion runs counter to the proposition that automation of human evaluative processes is increasingly helpful as the number of data items to be aggregated grows larger.

Along the same line, one of the possible reasons why Bayesian processing displayed no superiority in the second experiment (and actually was inferior, although nonsignificantly), according to Kaplan, Lichtenstein, and Newman (1963), was that "the task for both groups was extremely difficult, in the sense that the implications of the data were very uncertain. Apparently, under such conditions PIP cannot be considered as a significant aid to a decision-maker." In other words, the more demanding the task, in terms of ambiguity of the inputs, the less effective is automation.

## OTHER TYPES OF DECISION-MAKING EXPERIMENTS

Experimental investigations of decision-making in contexts other than Bayesian processing have been legion; even those with a military setting have been numerous, the programs at the Operational Applications Laboratory and the Johns Hopkins Applied Physics Laboratory being extensive enough to deserve separate coverage (Chapters 16 and 18). To report the entire field would be beyond the scope of this book. Some examples will suffice.

### *Hayes' Experiments*

One approach is illustrated by the work of Hayes (1962), who commented:

> Currently, the name "decision making" is applied to a very large group of behaviors which ranges in complexity from predicting which of two lights is about to light to establishing plans for conducting a war. This report will be concerned with decisions similar to the decisions involved in choosing which one of a number of cars to buy or which of several apartments to rent. Most usually in such decisions, the alternatives will differ from one another in several characteristics, and these differences must be taken into account simultaneously in making the choice. For example, in choosing among alternative apartments, one may consider cost, size, appearance, convenience of location, quality of neighborhood, and possibly a number of other characteristics. The difficulty in making such decisions arises in trading the advantages of an alternative in some characteristics against its disadvantages in other characteristics. Such decisions might be described as multi-dimensional judgments. It is commonly assumed that the more relevant data one takes into account in making a decision, the better that decision will be. It is clear, however, that as one takes more relevant characteristics into account in comparing alternatives, the opportunities for confusion increase. If confusion were to increase rapidly enough as the number of characteristics increased, it is conceivable that decision makers would perform better if some of the relevant data were eliminated.

In four experiments by Hayes each of fifty-four subjects had to select which simulated aircraft would investigate a reported submarine sighting. The number of aircraft varied from two to eight and the number of pertinent characteristics of the aircraft also from two to eight, with each characteristic having one of eight values. One characteristic, for example, was speed. It was found that when the number of characteristics to be considered increased, so did the time needed to make the decision, but the quality of the decision did not improve—and when a limit was placed on decision time, the decisions became progressively poorer.

### *Electric Boat Program*

Another approach has been described by Sidorsky, Houseman, and Ferguson (1964) and Sidorsky (1966), who conducted an experimental program related to naval antiair warfare (AAW) and antisubmarine warfare (ASW). The frame of reference for decision-making is worth examining.

> The study of human decision making is complicated by the fact that in one way or another, every voluntary action reflects a decision. Thus, by definition, almost every observable human response can be construed to be within the subject matter of decision making. The inclusiveness of such a definition obviously

deprives it of any value in deriving a conceptual framework within which to examine the behavioral attributes related to optimum human performance in AAW and ASW tasks . . . .

To come to the heart of the matter, then, decisions are defined in this paper as those observable responses which reflect an intention to fulfill a vested or implied responsibility related directly to the needs of the Tactical Unit in resolving an interaction with an actual or possible enemy.

Further, these researchers established a taxonomy of decision tasks consisting of six classes: acceptance (learning about the enemy); change (increasing one's own relative advantage); anticipation (estimating the enemy's intention or future state relative to one's own unit); designation (maximizing one's own capabilities in the light of emerging requirements); implementation (resolving the tactical situation); and adaptation (preserving one's own unit in the face of unexpected circumstances). This taxonomy was named ACADIA, an acronym formed from the initial letters of the class names.

The following behavioral traits were set forth as pertinent to decision-making: stereotypy (persistently rigid and oversimplified behavior); perseveration (persisting with a particular response or interpretation when it's time for a change); timeliness (relation between time available and time used for making a decision); completeness (degree of exploitation of all relevant information); and series consistency (consistency of responses in a series of sequentially dependent or interrelated actions). It was proposed that decision-making responses can be scored in terms of their effects on spatial relationships between opposing tactical units, self-concealment, conservation of resources, information generation, and weapon utilization.

The site of the experimental program conducted within this framework was the SUBmarine Tactics Analysis and Gaming (SUBTAG) facility of the Electric Boat Division of General Dynamics. Five SUBTAG booths represented command stations. In each booth a moving pip of light on an x–y co-ordinate display simulated a hostile tactical unit, a ship, aircraft, helicopter, land vehicle, or missile. The moving light pips were generated by an analog computer. An experimenter sat at a problem director's console in a separate area which also contained a large situation display, control and monitoring equipment, data collection equipment, and a central communications panel. Subject booths had telephone connections between each other and with the experimenter.

Some of the most interesting experiments were those in which two opponents opposed each other, one in each booth. In one experiment (Sidorsky, Houseman, and Ferguson 1964) each simulated ship attempted to attack the other as they converged. The problem for each commander was when to fire his weapon. He was not told when the other fired. His probability of success increased the longer he waited, but so did the probability that the opponent would have fired and destroyed him first. Each commander had a display showing probability of success for himself and for his opponent. (The moving target display "was merely to heighten the realism of the situation.") By means of differences in the rates of probability increases, situations varied from those in which a commander possessed a 2 to 1 advantage over his enemy, to those in which the enemy had that advantage over him. Each of seven pairs composed of fourteen subjects (scientists and engineers) made 215 decisions, 43 in each of five one-hour sessions; 22 of these were one-minute data trials and 21 were masking trials. Pairings remained constant.

Extensive differences were found among the performances of individuals. These differences suggested the value of training for decision-making. Since the subjects' behavior tended to be stereotyped and predictable, it was suggested that a goal of training might be to "train out" such undesirable response characteristics as stereotypy and perseveration.

In another experiment (Sidorsky 1966), one opponent in each trial played the defender, the other the attacker. The attacker had to choose between two targets, and the defender had to decide how to apportion his resources to defend these. Each side made its decision during a 25-second period after a weapon had left its origin, but before it reached a point from which it could proceed to either target. The spots of light on the displays represented the weapon. Each player pressed buttons not only to register his decision but also to indicate his prediction of the opponent's choice. Targets had varying values and varying vulnerabilities, in twenty-six combinations. In one set of these, vulnerability of the two targets was the same but value differed; in another set, value was the same but vulnerability differed; and in a third set, both value and vulnerability differed. Each of twenty-four subjects remained a defender or attacker throughout, but pairings were changed. Each of five sessions lasted 80 minutes and included 108 data trials and three warm-up trials. One of the findings was that although subjects selected their own actions effectively, they predicted their opponents' decisions poorly.

### *Princeton Studies*

Several experiments that relied primarily on verbal simulation of a military situation were performed in the Group and Environment Design Laboratories at Princeton University, established by J. L. Kennedy. They emphasized how individual differences affected ability to integrate information required for making decisions. It was hypothesized that some individuals were able to employ many alternative organizations of the same units of information and to relate these organizations to each other in different ways by various rules, as a consequence of an individual's internal structure. Such persons were described as structurally complex or abstract. Other persons were said to use very simple and fixed rules which generate many fewer relationships among the information units. They were called structurally simple or concrete. The Princeton researchers believed it was possible to distinguish among these types by means of a relatively simple pencil-and-paper test for sentence or paragraph completion.

In the initial research, four-man teams were composed of either complex or simple information processors (Schroder, Streufert, and Weeden 1964; Streufert, Graber, and Schroder 1964; Streufert, Schroder, and Grenoble 1964). In a later study (Stager 1966), the four-man teams were composed of varying proportions of the two types of individuals to see what resulted if a team was not homogeneous.

Two teams, one of each type, were located in separate rooms and performed at the same time in seven 30-minute sessions. Each 30 minutes represented twelve hours of real time, shown on a clock in each room, which also contained a large topographical map of an island called Shamba. The four subjects in each team were told they were commanders of equal status in joint command of Army, Air Force, and Navy units approaching the island, which they were supposed to capture with their resources. These were known to them, but the resources of the enemy on the island became known only through feedback in reaction to their actions and through new information supplied during the session. Each team was told that a competing team of four commanders was playing the roles of the enemy commanders, but in actuality the experimenters played these roles. The team was not informed how long the sessions would last. A bonus of $3 was promised to the winning team. Information load varied between two and twenty-five units, differing among the seven consecutive 30-minute sessions.

Of some interest were the kinds of things measured. For example, in the first experiment the experimenters assessed the quality of decision-making according to the extent to which decisions were integrated into purposive strategies and the extent to which information feedback was integrated from diverse sources and across time. Such integration was an index of performance complexity. In addition, commanders and observers rated the team members, and the commanders rated themselves, for extent of contribution to the task. Experimenters also tallied the frequency with which the leadership role shifted.

A second experiment (Schroder, Driver, and Streufert 1967) varied the complexity of environmental inputs. Among other aspects it measured the number of integrations of self-generated information in decisions, amounts of information integrated, and time interval between the generation of information and the decision. There were ten teams.

Stager (1966) described a third experiment:

Verbal behavior of the groups was coded according to the predefined categories of perceiving and proposing the problem, requesting information, supplying information, suggesting alternatives, evaluating alternatives, autocratically deciding, and confirming decisions through consensus. Each category was considered as a functional role in the decision-making process; scoring, therefore, was concerned with the changes of a group member from one role to another. In order to provide additional analyses, the frequencies with which (a) new or novel information search was requested, (b) different alternatives were proposed, or (c) evaluations of different alternatives were given, while members were in the respective functional roles, were noted.

Individual members were rated on the extent to which their overt behavior was synergistic or facilitating, empathic with respect to the enemy, and conceptually integrative. Groups were rated with respect to their utilization or synthesizing of informational diversity, their generation of diverse and conflicting alternatives, the number of effective communication channels available in the group, and the type of group emergent structure.

### *System Development Corporation Studies*

Chapter 17 mentioned a few decision-making experiments at the System Development Corporation. One of these was called the "Force Allocation Experiment" (Wood and Friedman 1964). Since reports of this experimentation were restricted (although it was unclassified), it can be described here only sketchily. Three sets of four subjects each, all SDC employees, were matched on a pre-test, and each set performed in one of three experimental conditions. These consisted essentially of variations in multistage processing capability. This meant a computer-based evaluation of subsequent sequences of actions over many stages in a simulated war, in response to a tentative or test action. In one experimental condition, the subject could not look ahead more than a single stage to test the results of his action. In another he could keep repeating test actions to advance further and further into the future. The third condition was an intermediate arrangement. The subjects' task was to allocate weapon units representing one weapon type among targets having several dimensions of utility. The study was conducted in the Command Research Laboratory (described in Chapter 17). Subjects performed individually at a console on line with the laboratory's computer. A manually implemented experiment of a similar nature preceded it.

Two experiments described by Merrifield and Erickson (1964*a*, *b*, *c*) investigated how useful a computer-generated statistical summary might be in diagnosing enemy strategy. The summary was based on correlation, factor analysis groupings, and variance ratios. The experiments started with individual decision-makers examining a large data matrix containing values for fourteen resources at each of forty-nine U.S. cities. The enemy fired five missile salvos. After each attack the subject was told which five cities had been hit and was required to judge which resources constituted the enemy's objective (in terms of destruction or avoidance), that is, his strategy. The subjects also had to predict which cities would be hit in the next salvo. The subjects were thirty university students from advanced Army and Navy ROTC units. They performed in two four-hour sessions each at desks in the Command Research Laboratory with materials which had been computer-prepared; although two of the experimental conditions simulated on-line computer aid, such assistance was absent in the experiment.

For one group of subjects the computer-generated statistical summary was available in both sessions; for a second it was available only in the second session; and for the third it was available in neither session. This summary incorporated "the correlation of the salvo with each resource, the factor groupings most related to the salvos, and the ratios of variance of values for cities in the salvo to variance for all cities, for each resource." The subjects were given advance training in interpreting these calculations as displayed in a computer printout. In the first experiment this special training may have accounted in part for the substantial, statistically significant superiority of those subjects who were given the statistical summary. It was also suggested that the summary simply functioned as an efficient arrangement of data to compensate for an unwieldy arrangement in the displays of raw data given to all subjects. Accordingly, a second experiment replicating the conditions of the first was conducted with a redesigned display of the raw data as well as a similar redesign of the statistical summary. Again the subjects aided by the statistical summary produced better diagnoses of enemy strategy, but their superiority was not as marked.

*SAGE Battle Staff Study*. In 1960 an experiment in the Human Factors Laboratory (Chapter 17) illustrated some of the potential methods and problems which can characterize experimental research in decision-making (Meeker, Shure, and Rogers 1962). The battle staff at a SAGE Direction Center was played by a three-man team, considerably smaller than an actual SAGE battle staff. Each of four of these groups performed in five exercises, each exercise lasting about 90 minutes. Original plans had called for eight five-man groups; then the total was reduced to eight three-man groups in twice as many sessions, but laboratory occupancy schedules forced the further reduction and the elimination of one two-state independent variable—concerned with communication structure—which instead became a separate experiment in a different setting (see Chapter 17).

The subjects observed SAGE situation displays simulated by closed-circuit television CRTs. Being physically separated from each other, they communicated among themselves and with a team of simulators (experimenters) by telephone. About eighty-five hours of tape recordings provided the data which were analyzed; they included "thinking out loud" by the subjects, a practice encouraged in their training, which was intensive and continued for two months. Two two-state independent variables were planned: "overlap of information among group members" and "administration of group decision resolution." However, the published reports of the experiment did not present results for these variables.

In addition to narratives about two particular decision situations and outcomes (Shure, Rogers, and Meeker 1961), published results comprised a factor analysis exploring the role played by personality differences among the subjects (Rogers and Shure 1962) and a factor analysis of individual decision-making styles (Rogers, Shure, and Meeker 1962). The latter analysis was based on observer self-ratings and ratings of other members of the group, California Psychological Inventory scores, scores on scales measuring personal and material risk-taking, and measures of risk-taking in a gambling game involving real monetary risks. Four bi-polar decision-making styles were identified: "calculating-conservative versus intuitive-unsure," "orderly-active versus inhibited-autistic," "group-facilitator versus self-conscious-autocrat," and "responsible-enthusiast versus reluctant-participant." The subjects consisted of nine Air Force ROTC students and three SDC training specialists who had had military careers.

The researchers explained their selection of simulated operations thus (Shure, Rogers, and Meeker 1961):

> The choice of a SAGE Battle Staff for study was, in part, practical. We all had considerable experience in development and evaluation efforts with these groups, but there were even more important considerations which led to the choice of such a group for study. For one thing the military codifies and articulates the basis for decision behavior more completely than do most civilian organizations. This permitted us to make our analyses primarily in terms of documented statements of goals, procedures and responsibilities and even to some extent, of expectancies, without having to infer these from observation or interviewing. Secondly, a staff concerned with tactical decisions confronts, under conditions of battle activity at least, large numbers of situations in short periods of time. This feature was highly desirable for subsequent laboratory study. Third, the *raison d'etre* of such a staff is that they are necessary for making decisions which take into account the large picture of events.

They also explained why they wanted a realistic and complex setting for their experimental vehicles:

> We wanted it *realistic* so that (1) the decision processes as these are expressed in a real life setting would be minimally distorted and (2) so that an observationally

and inductively rich experimental context would be provided. We wanted a *complex* setting in order to (1) provide the opportunity for extended search and decision-resolving behaviors to occur and consequently to influence the final choice; and (2) to permit direct observation of those situations where rationalistic assumptions about the human decision maker would be most at variance with his actual capabilities. While simple decision situations are conducive to experimental control and precise mathematical formulations, we felt they failed to provide these opportunities or only provided them in an aborted fashion. Finally, we wanted to retain some degree of control over the character and occurrence of decision situations introduced under conditions of realistic complexity. As a result, a simulated rather than a real-life setting was selected.

The researchers followed three guidelines in designing their experiment. One was to distinguish between "choice in rule-following situations and choice in decision situations proper." To make certain the experiment concentrated on the latter, they performed a comprehensive procedural analysis of the simulated team's activities in actual SAGE operations to distinguish situations not covered by established rules or operating procedures from those so covered. Another guideline was to introduce decision situations that contained the classes of decisions categorized by March and Simon (1958) according to various combinations of good, poor, mixed, bland, and uncertain alternatives. The third guideline was to embed the decision situations in an on-going, realistic operational context instead of presenting them to the subjects as clearly defined, static options requiring choices among specified alternatives. This is where they got into trouble.

They designed thirty-four objective decision situations into the exercise for each team. They assumed that the teams would identify these and thereby make them into subjective decision situations. A detailed analysis of subjects' verbalizations showed that although the subjects did detect the objective decision situations in 90% of the cases, in only 29% of these did they consider the full set of relevant alternatives, and in only three situations were all of the available alternatives correctly evaluated. With so little control over the occurrence and identification of subjective decision situations, "no attempt was made to test predictions of decision resolution behavior." In other words, because they could not specify the stimuli, the experimenters could not describe results in terms that related them to the stimuli.

The researchers concluded that an emergent-dynamic decision context, in which different decision-makers extracted different problems from the same correctly processed information, resulted from six features of their simulation. These features were: (a) wide distribution (in time) of elements of the information indicating a decision problem; (b) difficulty in specifying when a problem began, and even more so, when it ended; (c) extended duration of problems and their overlap in time—instead of compactness and discrete order; (d) rapid changes in problem situations and in the utility and availability of alternatives; (e) pervasive time pressure on the decision-makers; and (f) unfolding of a problem "as a series of subproblems, each associated with subsets of the multi-goal structure of the problem." These six features suggest that decision-making in the context of a realistic military setting is more complex than many, perhaps most, experimenters have realized, or more complex than their experiments would imply.

# 22

# Other Areas: Vehicle Driving, Response-to-Disaster, ICBM Launch, Aircraft Navigation, and Space Flight

This chapter describes a number of man-machine system experiments in scattered fields which have had, individually, too few studies to warrant separate chapters. However, the smaller extent of research need not mean either that a field is insignificant or that it will not command more attention in the future. It is likely that man-machine system experimentation will extend into still other subject matter. For example, Van Cott and Kinkade (1968) conducted several feasibility studies of an information clearinghouse for biological scientists. In these studies, information requests were telephoned by actual scientists to a center at which humans performed the functions which simulated those essential in a manual or automatic information system.

## MOTOR VEHICLE DRIVING SIMULATION

Experimental research on motor vehicle driving has been increasing in recent years and can be expected to grow further. It may encompass large-scale, multi-vehicle experiments on highway traffic.

In one sense highway traffic should not be regarded as a man-machine system, since each driver is likely to have his own individual goal instead of a common system goal. However, to the people who design highways, intersections, and other features of motor vehicle transportation, their products are indeed systems, with the over-all objectives of safe and expeditious traffic flow. In addition, experimental studies may center around certain types of vehicles, their services, and their management, such as buses, taxicabs, and trucks.

Situations for doing experiments with actual drivers have been growing in variety and sophistication. By and large they fall into three categories, which omit an even more customary but nonexperimental situation—the observation and measurement of uncontrolled, on-going vehicle performance in the natural setting. This kind of research, as well as crash-injury research and all-computer simulations of traffic, will not be considered here.

The three experimental situations are:

1. Arranged driving of an actual vehicle in a real, uncontrolled environment. The driver follows experimental instructions and usually knows he is partici-

pating in an experiment. The environment is only approximately repeatable and often is rather complex, since it includes other, nonarranged vehicles and drivers on public roads. Generally only one driver-vehicle combination at a time is recorded and measured. There may also be one or more arranged vehicles and drivers to supply partially controlled inputs to the experimental situation.

2. Arranged driving of an actual vehicle in a real but controlled environment. The driver follows experimental instructions and knows he is taking part in an experiment. The environment is repeatable, but necessarily fairly simple; generally—but not always—there are no additional vehicles other than experimenter-operated ones which supply well-controlled inputs. In addition to little-traveled public roads, some automotive manufacturers, certain governmental agencies, and special military units have test sites at which they can do this kind of experimentation. By owning the site, they control the roadway and can restrict its use.

3. Driving a simulated vehicle in a simulated (and controlled) environment. Only one real driver is involved at a time. There are numerous variants within such simulation. The vehicle may be a complete automobile, or it may be just the driver station or a mock-up of one. The fidelity of simulation in instrumentation and motion varies in degree. The visual environment may be presented directly on the windshield or on a motion picture screen outside it. This environment is customarily dynamic, that is, it keeps changing; and in much recent simulation it is also responsive, that is, it changes in reaction to the behaviors of the driver and his vehicle.

If the depicted environment is some actual environment photographed at an earlier time, responsiveness is limited. If it is a scale model, other vehicles, pedestrians, etc. can be made to perform responsively and displayed to the driver by closed-circuit television. Still another technique is to seat the driver in an actual driver station, but connect his actions to a miniature vehicle in the scale model of the environment. He has to project himself, so to speak, into that small vehicle. His visual world is very different from what he would be seeing if he were in an actual automobile or viewing a picture screen through the windshield of a simulated automobile.

*Road Research Laboratory*

The principal research described here was British and employed the second arrangement, in multicar situations; these studies tend more than others to resemble the contexts of complex man-machine system experiments. The Road Research Laboratory of the Department of Scientific and Industrial Research, in England, initially used Northolt Airport, the Fighting Vehicles Research and Development Establishments at Chobham, and a number of airfields in disuse. In 1960 the laboratory's research track was opened at Crowthorne (Road Research Laboratory 1965). This facility had a central area 900 feet in diameter intersected by two large roadway loops, a straightaway, and a terminal area.

During four days in 1955 the laboratory undertook some experimental "weaving" tests at Northolt Airport in which 130 vehicles participated. They circulated continuously through a complex, pretzel-like configuration of roads. The laboratory's account said (Road Research Laboratory 1955):

The vehicles were divided into ten groups, each group containing about the same proportion of cars, taxis, buses, etc. The vehicles in each group performed a particular maneuver in the weaving section. Some entered from the left and emerged to the left, some entered from and merged to the right and others weaved from left to right or right to left. In almost all the tests the two weaving streams were equal in numbers. Each test lasted 10 to 12 minutes; during the first two minutes conditions settled down, and a queue of moving vehicles formed in each approach to the weaving section. In most cases the approaches were fully loaded throughout the remainder of the test and these conditions were regarded as giving "maximum" flow.

The flow of traffic in each of the four movements was recorded on teleprinter tape by observers stationed on a tower-wagon, using coloured labels for identification. The dimensions of the weaving section were varied over a considerable range of length and width, and the proportions of medium and heavy commercial vehicles (including buses) and of weaving traffic were also varied.

The results provided data on traffic-handling capacity for different design features and traffic compositions. Further tests were conducted in 1956 with about ninety participating vehicles. Results suggested that as more vehicles tried to get through the "weaving" section per unit time, fewer actually got through.

Tests were conducted on "car-following" at the Crowthorne track in 1962–63. A fixed number of vehicles drove on four circular single-lane roadways having radii of 50, 100, 200, and 415 feet, as well as on a straightaway. For example, twenty-eight vehicles drove on the 200-foot circle. Drivers could select their own speed or were given a speed. Results showed relationships between speeds, curvatures, and average concentration of traffic. The capacity of the circular roadway with the 50-feet radius, for example, was about 15% greater than that of the next larger one, "probably because drivers could see further ahead in the stream than on the curves of greater radius."

*Driving Simulators*

These provide the third situation that was outlined. Analyses of the requirements and advantages of driving simulators have been published by B. H. Fox (1960) and Schlesinger, Karmel, and Cohen (1964). Proceedings of a conference on "Mathematical Models and Simulation of Automobile Drivers" have been edited by Sheridan (1967). The most comprehensive review of the development and distribution of driving simulators has come from Hulbert and Wojcik (1964).

More recently a Driving Research Laboratory directed by R. K. McKelvey was established in Providence, Rhode Island, operated by the U.S. Public Health Service's National Center for Urban and Industrial Health. Laboratory plans initially called for the incorporation of three simulators (McKelvey 1967). Plans changed shortly afterward in a reorganization, and the laboratory was renamed. By means of a Radio Corporation of America device, moving belts were to carry fully simulated highways and miniature cars which the driver, thanks to an optical reduction system, could see through his windshield. In the second simulator, built by Goodyear Aerospace Corporation with additional equipment from Philco Corporation, the driving illusion was created by televising a highway terrain model; the image was received on a television monitor and the picture was projected on a large screen in front of the simulated automobile. The third device was an open-loop, shadow-graph (point light source) simulator.

The Goodyear simulator had been used elsewhere in studies of speed judgment and braking behavior. In one of these (Barrett, Kobayashi, and Fox 1968), a pedestrian (represented by a televised scale-model dummy) appeared unexpectedly in the automobile's path. Five subjects steered around the dummy, one struck it at a very low speed, and five hit the pedestrian with some speed. Eleven other subjects "became too ill to continue the study in the simulator." (A life-size rubber dummy of a pedestrian has figured in another study in which the vehicle, its driver, the roadway, and the environment were all real [Bidwell 1967].) The RCA apparatus was a modern version of the simulator first devised by Forbes (1938), who called it "The Miniature Highway Test."

Simulation equipment for driving research has also been developed by Ohio State University, George Washington University, Stevens Institute of Technology, University of California–Los Angeles, Massachusetts Institute of Technology (two simulators, in the electrical engineering and mechanical engineering departments), Liberty Mutual Insurance Company, General Motors Research Laboratories, and Cornell Aeronautical Laboratory, as well as in Japan. Additional simulators were being developed in 1967 at Cornell Aeronautical Laboratories and the University of Wisconsin. Sheridan (1967) listed university driver simulation facilities also at Harvard, Purdue, South Dakota, Southern Illinois, California (Berkeley), and Upsala (Sweden); others were said to be operated by the Grumman Aviation Corporation, the U.S. Army Tank Command, and the Renault Company. Simulation apparatus for training or testing drivers has included a device developed by the American Automobile Association, the Aetna Casualty and Surety Company's Aetna Drivotrainer, the All State Good Driver Trainer, the Miles Motor Driving Trainer (developed from R.A.F. World War II equipment), and the Sim-L-Car of General Precision, Ltd.

Two simulators at the Institute of Transportation and Traffic Engineering at U.C.L.A. were described by Hulbert and Wojcik (1964).

In the "fixed base driving simulator" a complete automobile was driven on steel rollers. Motion picture films projected on a screen presented responsive roadway environments. The driver saw them both through his windshield and on his rear view mirror. The other simulator was the "moving base driving simulator," described thus:

> This Driving Simulator consists of an automobile cab (includes controls, dashboard, seat, etc.) mounted on a steel structure that permits such movements as roll, pitch and yaw to simulate inertia forces as experienced during an automobile ride. The driver's view through the windshield is entirely filled by a motion picture projected on a cylindrically curved screen of four foot radius. This screen is also mounted on the steel structure so that the motion picture scene and the automobile cab with the driver may tilt as a unit. The cockpit thus formed is closed off by opaque curtains in order to limit the driver's visual environment to the cab and the motion picture scene as seen through the windshield area from which the glass pane has been removed.

Most simulators (and vehicles used in the other two research methods) have been elaborately outfitted with methods of recording such driver behavior as pedal movements and steering wheel movements. Hand signals and eye and head movements involved in visual behavior are more difficult to register, though they may be among the most important items of driver behavior. Cameras can be used for the purpose, but the data are difficult to reduce. Observations by a human monitor may be essential.

## DISASTER STUDIES

Although all experiments on military operations in simulated wartime, such as an air attack, could be regarded as disaster studies, here the scope is more restricted. It refers to the activities of civil agencies and personnel confronted by some catastrophe, which could include nuclear damage and destruction. This fertile area for experimental simulation-based research has been exploited in an experiment at Ohio State University and a two-experiment study at the System Development Corporation.

The Ohio State experiment dealt with the operations of a police force. Such operations, in nondisaster contexts, have also been simulated by a British non-profit research organization, Gordon Pask's System Research Ltd. In that organization's initial experiments, the simulated police force consisted of a team of detectives, a fingerprint bureau, specialist "scenes of crime" officers, and a records office. Centralized and decentralized structures have been compared for effectiveness in dealing with high crime rates.

### *Ohio State University Disaster Research Center*

An experiment examined the communication system of the Columbus, Ohio, police department by simulating routine "Friday night" information processing in the department's radio room, and then introducing a simulated crash of an airliner into an apartment house. This pioneering study, reported by Drabek (1965) and Drabek and Haas (1966), was funded through the Air Force Office of Scientific Research and carried out at the Ohio State University Disaster Research Center's Behavioral Sciences Laboratory.

Manning the simulated radio room were four subjects: a sergeant, a dispatcher, and two complaint clerks. Three four-man teams were formed from Columbus policemen who held these positions in real life and were assigned to the study. (The co-operation of the Columbus police department was exemplary.) The disaster occurred in the last of four two-hour sessions for each team. Although the subjects knew they were participating in a simulated situation, they did not know it was an experiment or that a disaster would occur. The name "Disaster Research Center" was concealed. All teams encountered the same session on the same day to minimize chances of exchanging information. The closest the subjects came to guessing what might happen was a notion several had that they might have to deal with a jail break. Some subjects said afterward they thought possibly they were being evaluated for promotion; others suspected a new radio room was being designed.

Eighteen simulator personnel, mostly students, played the roles of policemen in radio-equipped patrol cruisers, other law enforcement agencies, emergency organizations, and citizens who phoned in. The experimental data were the material collected on tape recordings and the latencies of replies to calls as logged by the callers. The simulator personnel were carefully and extensively trained in procedures, jargon, and geography. For example, they drove throughout the Columbus district where the simulated disaster would occur to learn the names of streets, landmarks, stores, and so forth. Cruiser simulators studied maps to become acquainted with their districts. According to the experimenters, the policemen subjects were impressed by the genuineness of the students' representation except that at first the pseudopolicemen stuck too close to specified communication procedures.

The simulated radio room was equipped with displays resembling those in the real room, which the experimenters studied assiduously. A set of 990 normal telephone calls was prepared, each call designed in detail and printed on a card. The calls were based on tape recordings of actual calls and on statistics of distribution. These differentiated between those that resulted in cruiser response and those that did not; they dealt with geographical location, type of event, sex of caller, time of day, and a real-life pattern of temporal cycling. An arbitrarily selected average rate of three calls per minute was lowered somewhat after the first session to conform to subjects' estimates of the rate on a busy Friday night.

In the last session, eighteen calls per minute were superimposed on the normal rate. The calls attributable to the aircraft crash were developed from a script and from detailed analysis of most of the police telephone communications following an explosion at the Indianapolis Coliseum in 1963; that actual disaster served as a partial model for the simulated one.

From the reports of this experiment it appears that these Ohio State University researchers had heard about the RAND air defense experiments and some gaming studies, but were unaware of the array of simulation-based experimental programs described in earlier chapters. The reports from the Ohio State University Disaster Research Center, for example, made no reference to the same university's decision-making experiments (Chapter 21) or air traffic control studies (Chapter 10). The Ohio State University Human Performance Center, which was conducting some of the decision-making experiments during the same time period, was situated very close to the Disaster Research Center. However, the researchers in the two centers represented different departments, the former psychology, the latter sociology.

Drabek (1965) formulated a number of methodological generalizations based on his experience in the disaster experiment:

Extensive field research on the system which is to be simulated must be completed.

Construction of an organizational simulate, or of a segment thereof, will require a clear division of labor among the research staff.

Researchers must remember that simulation is an art.

Researchers must continually view the simulate through the "eyes" of participants and seek to become aware of the values, criteria for decision-making, and so on, used by the participants.

Realistic simulation is a method of research and must not be allowed to become an end in itself, therefore, researchers must continually be mindful of their basic interest—theory development and testing.

If human subjects are to be used in the simulation, researchers must be mindful of their ethical responsibilities.

Realistic simulation, utilizing subjects from existing organizations, is a practical research tool which possesses much research potential.

The research was initially motivated by "a basic desire to explore the utility of simulation as a methodological technique for the study of organizational behavior under stress" (Drabek 1965). That author's earlier analysis of police communications following the Indianapolis Coliseum explosion apparently led to the selection of the police radio room as the focus of inquiry, despite some opposition from other researchers.

The results, presented in terms of stress, indicated that due to the aircraft crash the operations of the radio room were speeded up and the teams handled heavier loads than on a normal Friday night. The catastrophe calls received priority. These hardly sensational findings were embodied in a number of hypotheses that were stated as demonstrated according to significance statistics, although these hypotheses apparently arose after the data were gathered. Fastening the term stress on the catastrophe situation presumably lent special import to the findings.

It appears that much of the data was never analyzed (Drabek 1965), including television photography of the four-man teams in the simulated radio room. Television engineers were located at five different control points to operate two movable cameras, both of which had "pan and tilt" flexibility plus "zoom" lenses. Kinescope films were made from the video tapes so the researchers could easily view the visual record of a session.

*System Development Corporation Emergency Operations Research Center*

This laboratory, the last of its kind at the System Development Corporation (see Chapter 17), was the scene of two experiments in 1965–66 in a program supported by the Office of Civil Defense (Cusack et al. 1966). A total facility area of 1,924 square feet contained three 150-square-foot subjects' rooms, a simulation area subdivided into three parts, and an observation area. The principal equipment consisted of seven telephones for each subjects' room and associated simulators, a common simulated Emergency Broadcast System, and a different set of displays in each of the three subjects' rooms. The main purpose of the two experiments was to compare these displays for effectiveness.

Each set of displays served a four-man team of the city manager, his assistant, the civil defense director, and a display clerk (from the experimental staff), for a model city of one hundred thousand persons. These officials (in a protected office in "city hall") were isolated, except by a telephone, from their department heads, who were played by simulator personnel. It was assumed that a nuclear air burst had created blast and fire damage to one-fifth of the city, with medium fallout. Each experiment had one-hour sessions (in real time). The first experiment examined the warning-and-movement-to-shelter period, the second an in-shelter period.

One set of displays in one room consisted simply of a chalk-lined blackboard; it could be written on as seemed appropriate at the time. A second set of emergency action displays, in a second room, was comparable to those being proposed at the time for actual emergency operating centers. It was supposed to show all problems confronting the city departments and the actions being taken to handle them. In the third room a third set, based on the concept of "management by exception," was designed to show to the city's top management only deviations, good or bad, from a planned course or standard. The data on the displays resulted from incoming messages, including those over the Emergency Broadcast System.

During each session of each experiment three teams occupied the subjects' rooms which, with the different sets of displays, constituted the experimental conditions. Each of the teams, which remained distinct, rotated session by session through the rooms, that is, each used all the display sets. In short, all three conditions (display sets and rooms) and all three teams were active at the same time, and each team encountered each condition. This arrangement was a great timesaver but led to a major simulation problem. In the first experiment the same simulator acted his role—as police chief, for example—with all three teams. Because it proved so difficult for him to remember each team's actions and respond accordingly, in the second experiment a different set of simulators functioned for each subjects' room. Then the same simulator set remained with each display set throughout. Accordingly, the level of simulator ability may have been confounded with the type of display.

In three sessions in each experiment for each team the three three-state variables (display set, team, and order of presentation) were combined by a 3 X 3 Latin square. Team composition differed in the two experiments. Each team encountered each of three separate but equivalent disaster situations once; in addition, the effects of the attack were rotated around the city. As usual, much of the experimental effort went into creating the basic simulated environment—the model city. A matrix was constructed to organize certain physical and cultural features of Atlantic City, N.J., Binghamton, N.Y., Lancaster, Pa., Lowell, Mass., New Britain, Conn., and Stamford, Conn. An abstract of the matrix provided the features for the model city, which was laid out in map form by personnel with urban planning and cartographic experience.

The subjects were actual city officials from communities near the System Development Corporation's Santa Monica, California, headquarters. Their real-life duties approximated those they carried out in the experiments. Recourse to a model city eliminated special effects that might have come from basing the experiments on an actual city or cities with which some of the subjects were familiar. In addition, it was found that when the city in the experiment was an abstracted model, the subjects seemed much more willing than city officials had been in preceding investigations to accept procedures that differed from their own methods.

The subjects were not identified in the report of the experiments. One reason may have been their shortcomings in handling their jobs. They used "unbelievably poor" telephone procedures, failed to commit sufficient resources, overemphasized telephoned compared with displayed messages, and gave relative inattention to radiation effects, which they did not seem to understand well. It should be realized that these subjects "were selected as among the best City

Executives in the local area," with as much training as any would have had in the event of a nuclear attack.

The decisions which came from the subjects represented the primary experimental output. Since no scoring model existed for quantifying them, five SDC experts rated each decision on a seven-point scale for quality and importance. This procedure produced a composite measure that took into consideration the fact that some decisions were more important than others. Another innovation by the experimenters was to set the confidence level for statistical significance, before the experiment began, at .40 rather than at the customary .01 or .05 figures. B. L. Cusack and his associates wrote (Cusack et al. 1966):

> The rationale for 40 per cent is to prevent a finding of "no difference," even at the risk of declaring that a true difference exists between display systems, when in fact no difference is really there. In this research, it does not seem to be a serious decision error to suggest that one of three similar-cost systems is better if it really isn't, because one of them (or a comparable model) will be used in any event. The truly serious error, which this strategy is designed to minimize, is stating that there is no difference (the display systems are equal), when in fact one is better than the others. Even small performance advantages could mean the difference between life and death to many citizens.

The principal result was that none of the display sets proved to be very beneficial—that is, the performances were below average decision quality with all of them; and the differences were not very large. The experimenters recommended abandoning the proposed standard set for city executives and undertaking further study of the set based on management-by-exception, which showed some statistically significant superiority at the .40 level of confidence.

## ICBM LAUNCH SIMULATION

In 1958, Space Technology Laboratories (STL) proposed what became the Crew Performance Laboratory to conduct man-machine system research on the Atlas Intercontinental Ballistic Missile at Vandenberg Air Force Base (Space Technology Laboratories, no date). It was to be managed by the Air Force Ballistic Missile Division with STL in the role of technical monitor. When integrated weapon system training began, it was to be transferred to the Strategic Air Command (SAC) as a crew procedures trainer. It was part of a Crew Procedures Research and Trainer Development Program established by a SAC directive on January 10, 1958. This program had the following objectives:

Testing of alternative countdown and operations control procedures.

Testing communications channel and procedural requirements.

Testing organizational and personnel utilization requirements.

Determining and evaluating unit proficiency system standards and requirements (within the scope of the mockup).

Augmenting the operational equipment used during integrated weapon system training for procedural practice.

The proposal was put into effect. An extensive, integrated set of semidynamic mock-ups of all Atlas operating consoles was installed on the second floor of a two-story barracks-like building. The bottom floor housed simulators, central communications equipment, and offices. The mock-up area was partitioned into subareas and an observation area. Console mock-ups included a command control console, a guidance system console, a launch control officer's console, three launch operator/analyst consoles, and a complex facilities console. Equipment simulators included one for command/guidance, one for a computer, and one for launch. A station accommodating four operators could simulate personnel who were not manning consoles. Elaborate telephone communications were arranged.

The program at Vandenberg AFB continued for approximately three years. It combined research and training. Some exercises were regarded as serving both purposes; some were devoted more to one than to the other. The SAC crews that would man the actual ICBM launch sites were the subjects and trainees. Observations were made in a systematic fashion and much attention was given to the measurement of performance. Although there were constraints on the degree of control that could be imposed, the researchers, who represented a number of different contractors responsible for various portions of the Atlas missile and its ground-support equipment, regarded their work as experimental. They published a number of classified reports which have not been examined for this account. According to one of the researchers the program produced little of value (D. Meister, personal communication).

Exercises concentrated on missile countdowns at the launch location for several missiles. A substantial number of equipment malfunctions were preprogrammed and simulated. Signals indicating a malfunction appeared at a console, where a limited amount of trouble-shooting could be done to determine where the malfunction had occurred. However, since neither the missile itself nor most of its ground-support equipment was simulated, most of the maintenance operations -trouble-shooting on the hardware itself, replacement, and repair—were poorly represented in an exercise. Participants simply received allocations of time to correct malfunctions and sent messages indicating progress or completion. Preventive and reactive maintenance at the squadron level (including retargeting) and logistics activities (e.g., ordering spare parts) were also missing from the simulation.

All switch actions and telephone communications were recorded automatically. In addition to improving the human engineering design of the consoles, the aim was to optimize procedures and intercommunications and to devise the best ways to measure launch crew performance. Measures took the form of deviations from procedure, delays, and errors, which were assessed according to risk; an attempt was made to develop a composite error measure. The major criterion measures, of course, were whether a missile was launched at all and how much time was required.

This research program would have assumed greater importance if the activities it examined had been more complex and demanding. Starting and monitoring a countdown are perhaps the most dramatic aspects of human involvement in a ballistic missile system but certainly not the most fertile for significant man-machine system experimentation. For such research the big problem areas were off-console trouble-shooting and replacement, organization of an entire squadron's maintenance activities, and retargeting. To some extent these were the areas covered in one of the RAND Logistics Systems Laboratory studies (Chapter 13).

## AIRCRAFT NAVIGATION STUDIES

Although experimental research on pilot performance in aircraft has been extensive, much of it is already well documented, and its description would unduly lengthen this book. Further, it has placed relatively little emphasis on multioperator situations.

A large number of aircraft simulators of many types have existed for both research and training; many of these have been listed by the Society of Automotive Engineers SAE Committee AGE–3, Training (1964). Some of the most interesting simulations have been those of low-altitude, high-speed, terrain-following flight in which a pilot must follow a low-level path and avoid various kinds of natural obstacles. The simulated cockpit may be subjected to vertical accelerations as it would be by buffeting in actual flight. If the pilot navigates by some optical device or direct vision, the terrain may be simulated by a large scale model, such as the models at North American Aviation, Inc., in Columbus, Ohio. On the other hand, if he is using radar or other instruments, their signals may be generated by a computer-processed tape representing the characteristics of some actual terrain. Among study reports have been those by Ruby, Jocoy, and Pelton (1963), at Cornell Aeronautical Laboratory, and by Soliday and Schohan (1965) and Schohan, Rawson, and Soliday (1965), at North American Aviation.

The use of aircraft simulators for pilot training and for research on such training has been surveyed by Smode, Hall, and Meyer (1966). Krumm and Farina (1962) reported some interesting research on multiposition simulation. A B–52 flight simulator and a T24 radar trainer were electronically interconnected so the two pilots and two navigators could practice "a wide range of tasks requiring coordination among these four crew members," who also were "linked by a voice communication system similar to the actual aircraft intracrew communication system."

## MANNED SPACE FLIGHT

As is well known, a great many simulators have been developed for manned space flight, some for testing and experimentation, some for training, and presumably some for both of these purposes. Fraser (1966*a*) emphasized the distinction between "integrated mission simulators," "part-task simulation," and "environment simulation"; the names convey the general roles of the devices. Westbrook (1961) listed a large number of environment simulators for generating various accelerative forces, atmospheric pressures, temperatures, etc. Devices devoted to part-task simulation have been too numerous for anyone to have tried to itemize them. Fraser (1966*a*) mentioned eighteen integrated mission simulators, but his list quickly became incomplete. For example, it omitted the Apollo mission simulators and the LEM (lunar excursion module) simulators;

one of each was allocated to the Manned Spacecraft Center in Houston and the Kennedy Space Center. There were omissions also in the Society of Automotive Engineers list of simulators (SAE Committee AGE–3, Training 1964).

Rather than attempt to survey all studies using simulators, this account will concentrate on two simulation-based programs for the experimental investigation of system-oriented aspects of manned space flight.

*Grumman Program*

Seitz and Freeberg (1965) described the experimental work done by the Grumman Aircraft Engineering Corporation with LEM simulation up to the time of their report. They outlined three studies which occurred in the first of three phases of an over-all program. This first (preliminary) phase dealt only with certain elements of the mission. The second phase was to cover all elements in two separate simulators, while the third phase was to incorporate all of the mission in a single high fidelity simulation.

Concerning the first, three-part "Lunar Landing Study," Seitz and Freeberg noted that the fidelity of simulation of the cockpit and external visual field was limited. They added:

> Three studies were undertaken involving samples of three to four pilots per study. Factorial designs with as many as three factors per study and a maximum of four levels per factor were carried out. From eight to eleven dependent variables were used as measures of touchdown performance.
>
> The intent was to obtain initial data on overall crew member ability to exercise vehicle control—during the hover-to-landing phase—under various control system configurations; for various initial conditions (e.g., altitudes and ranges from the touchdown site).

The second study concerned orbital docking. North American Aviation, Inc., in Columbus, Ohio, provided simulator facilities through contract agreement. The objective was "to determine crew ability to achieve docking, visually, using various control system configurations and under nominal and degraded modes of operation." Ten performance measures were obtained under some twelve conditions and were "analyzed by standard univariate analysis of variance techniques."

The third investigation was the "LEM/CSM Rendezvous Study." The command service module (CSM) was represented simply as a flashing point source. The experimenters said:

> Simulated flights covered a range from 27 miles to 1800 feet separation between LEM and the Command Service Module. Again, a simple set of cockpit displays was used with only the basic information required for the manual rendezvous.
>
> Three studies utilizing four-to-seven pilots per study were undertaken. Multivariate analyses of twenty-five performance measures obtained during the rendezvous trajectories were carried out.
>
> Primary stress was upon procedures for achieving manual rendezvous. The study served to evaluate pilot control capability and various techniques for nulling vehicle rate as well as pilot ability to accomplish rendezvous by purely visual means. This capability was evaluated under various rendezvous trajectories, control system modes and degrees of control system degradation.

In this simulation, limitations in fidelity of the *external visual* display (i.e., noise in the gimbaled starfield) served to hamper an adequate evaluation of the purely visual techniques for rendezvous. The data obtained did, however, permit determination of the optimum rate nulling techniques using cockpit displays and an assessment of pilot ability to fly various rendezvous trajectories.

Seitz and Freeberg had the following to say about some problems and limitations in simulation-based research on manned space flight systems:

> Almost without exception the simulation configuration is not completely current with vehicle design. This means that the study being undertaken is not directly applicable to the current vehicle configuration. This is so because it requires time to build the simulator and during this time design changes are constantly made. This inability to stay completely current requires that, to the extent possible, studies be designed so that the results are generalizable. This is a desirable objective, in any case, but is often difficult to attain.
>
> A matter of no little concern is *subjects*. The astronaut population is very select and one wonders to what extent data collected on company test pilots and military reserve pilots are applicable! There is of course, the usual shortage of subjects and other pressures to keep the sample small. However, every effort has been made to run highly controlled studies with appropriate planning of study design.
>
> In spite of the engineering ingenuity and the money invested in simulation facilities, they are something less than perfect. Knowing that the study systems have no zero g or 1/6 g; that visual representation of the lunar surface is not adequately depicted with respect to surface character, the third dimension or light; that other aspects of the system are also imperfectly represented in the model, there will always be a little uncertainty in the conclusions reached on the basis of the studies. That doubt is, nonetheless, small when compared with the assurance simulation studies provide when design decisions must be made.

*Martin Program*

In 1964–65, the Martin Company in Baltimore conducted five seven-day simulations of the Apollo mission involving both the CSM and LEM vehicles. Each mission was performed by a different three-man crew of test pilots trained for five weeks at the Martin plant before undertaking the mission. During the mission they spent the entire time inside one or the other of the simulated space vehicles. This research has been reported by Grodsky et al. (1966), Grodsky (1966), and Grodsky, Moore, and Flaherty (1966). A somewhat similar but more modest program had been conducted at the Martin facility two years earlier (Grodsky and Bryant 1962). Two of the crews in the major study returned to the Martin facility after about seven and thirteen weeks, respectively, for an investigation of skill retention (Grodsky, Roberts, and Mandour 1966). All of the Martin work was sponsored by the National Aeronautics and Space Administration.

The simulation facility contained the following: (1) a mock-up of a command module; (2) a mock-up of a LEM which could be joined to the command module (CSM) and separated from it; (3) a four-station control room; (4) an analog computing and recording room; and (5) a physical conditioning laboratory.

The full-scale simulated command module consisted of a three-seat duty area with control panels; a sleeping area; a navigation station with visual simulations of starfields, earth, and moon outside the module; a sanitation area; and an off-duty area with special reading lights and piped-in music. Three closed-circuit TV cameras monitored the flight commander's, engineer's, and navigation stations, and live microphones picked up conversations. Headsets and telephone lines simulated radio communication with ground control. A separate loudspeaker played simulated engine noise.

Like the command module, the simulated LEM and its out-of-the-window displays were housed in a large sound-and-light-isolated mission room. Unlike the command module, the two-position LEM was supported by a three-axis gimbal system to permit travels of plus-or-minus 40 degrees in pitch or roll and plus-or-minus 180 degrees in yaw. The yaw travel allowed the LEM to rotate from its normal position so that its forward hatch connected with the command module's forward hatch, for transfer of personnel. Pitch and roll motions were used only in the docking phase. A complex combination of a movable spherical screen and a projector displayed starfields for out-of-the-window viewing by the astronauts.

The four stations in the mission control room were a capsule communication console, a flight director console, systems operation consoles, and data recording consoles. Three large analog computers solved the trajectory and control equations and provided the signals for the flight instruments and out-of-the-window displays. The physical conditioning laboratory–where the experimenters participated during the subjects' training–included bar bells and a sauna bath.

The seven-day mission consisted of the phases which would occur in an actual lunar mission: earth ascent, translunar coast (somewhat shortened), lunar orbit, separation, LEM de-orbit and coast descent, LEM braking and hover, LEM powered and coast ascents, rendezvous, docking, transearth coast (shortened), and earth entry. All the missions involved the same tasks and parameters, including simulated malfunctions and work-rest cycle; in other words, no independent variables were introduced to make them differ. All the training programs were also alike except for one change following the second mission: the crews thereafter received feedback about their errors of the day before and then discussed these errors.

"The conduct of the mission was based upon operational procedures as of December 1963," the researchers wrote (Grodsky et al. 1966). However, the crew tasks were not entirely the same as those initially envisioned. Manual control by the crew was increased. Instead of the closed-loop automatic primary flight control being planned for Apollo, the pilot was given complete manual control of vehicle attitude and translation during all phases except earth ascent. The purpose was to furnish "a sufficient variety of tasks to allow for generalization of the data to other types of missions and systems." In addition, the operational mission cycle was changed to "provide each pilot during the mission with an opportunity to perform each dynamic mission phase."

### *Experimental Results*

A great deal of data was recorded from training trials and mission trials. As a matter of fact, "a total of 170,000 verifiable data points were collected" for elements of four general tasks: flight control, guidance and navigation, malfunction detection, and switching (operation of two-state switches). Physiological data were also collected. Training trials consisted of various portions of the mission. Some of the training trials, when performance acquired a steady state, constituted "baseline" trials. In the published results the principal comparisons tested for statistical significance were between baseline performance and mission performance. These comparisons indicated the degree of reliability, which suffered only "minimal losses," except during the LEM period in switching and flight control (under a stringent flight control criterion). The high levels of performance were attributed to the skill and training of the subjects.

In another document, in contrast, Grodsky (1966) reported that the seven-day mission showed the value of simulating a long-duration mission in its entirety. Switch activations, he said, "seemed to reveal a very definite mission time effect." He explained:

> From lift-off to 67½ hours performance appears to be equal and in most of the sampled time periods better than baseline performance. From 73½ hours to touchdown at 166½ hours, there certainly appears to be less stable performance with significantly poorer performance than baseline at 88½, 106½, 132½, and 155½ hours. The causal factors which gave rise to these data are not easily determined. However, with a simple psychomotor task such as switching which is conceptually similar throughout the mission a number of hypotheses are possible:
>
> a. Fatigue due to the long duty periods from 75 to 102 hours. (An 18 to 22 hour awake period for each pilot).
> b. A systematic chronic effect from initiation of the mission which lowered the performance reserve of the crews and made them susceptible to errors after a long duty period.
> c. A demonstration of chronic stress which might have become progressively worse and further affected performance had the mission duration been extended.
>
> These data are further substantiated by looking at other tasks such as flight control which indicates a similar drop in performance when compared to baseline at 90 hours and which continue to the end of the mission at 166½ hours. It, therefore, is clear that the techniques and measures used are sensitive and are indicative of a general degradation in performance due to the application of a stress. Certainly it would be possible to pinpoint the causal factors involved in this degradation if required.

In addition, a human factors analysis sought to trace differences in performance to mission time, mission phase, subsystem, and workspace. However, the analysis "yielded no consistent trend of the effects of the variables studied on pilot reliability" (Grodsky, Glazer, and Hopkins 1966). Further, the results of a correlation study "on switching and flight control tasks and their relationship to the human engineering ratings, work element count, pilot ratings, error scores, and task load showed no consistent trend and no total subjects significance." The analysis brought out two points of human engineering interest. Subjects' performance, e.g., errors, failed to correlate with ratings of equipment items by specialists and pilots for human engineering quality. Checklist errors occurred due to "an inconsistent format of data presentation," which led the pilot reading the list sometimes to miss items.

In the skill retention study, the two crews that participated went through part task trials on selected mission phases and also an integrated fast-time mission during four days of testing for each crew. The fast-time mission simply eliminated some of the mission phases, such as the coast phases. The researchers reported (Grodsky, Roberts, and Mandour 1966): "In general the results indicate that each pilot maintained a level of performance comparable to that obtained during the 7-day lunar mission, and was lower than the baseline level in relatively few instances."

In the study conducted two years earlier (Grodsky and Bryant 1962) four experienced test pilots took part in three simulated lunar flights. Two of these lasted three and one-half days and one seven days. Some variation in the duty-rest cycle was introduced, as well as a maintenance task and a battery of behavioral response tasks.

Subsequently, Grodsky (1966) cast doubt on synthetic task batteries as indicators of human performance during space flights, for reasons set forth in Chapter 24.

# 23

# Related Research: System Testing, Small-Group Studies, Gaming, and All-Computer Simulation

Chapter I explained that man-machine system experiments could be identified through a clustering of characteristics rather than any single attribute. Various other kinds of research share some of the characteristics of such experiments, but this book has excluded them from membership. However, at least four approaches seem closely enough related to man-machine system experiments to deserve some brief description. Let the reader be thereby reassured that the author knows they exist and are important.

## SYSTEM TESTING

One is system testing, as part of system development. Military departments and their contractors perform a great variety of tests to assure the acceptability of a procurement or to diagnose design features which should be changed. A number of questions may be asked to reduce the confusion resulting from this wide range of testing. Does a test deal with equipment components, with subsystems, or with the entire system? Does it occur while the system is being developed or after (or as) it becomes operational? Are physical characteristics or design tested exclusively, or are there also humans in the test, operating or maintaining the equipment? Is the purpose of the test to see whether the item or system meets some criterion or specification, or to predict its capability, or to ascertain problems and their causes so they can be resolved?

As the next chapter will note, the distinction between test and experiment has not been widely examined, and the two terms have often been used interchangeably. As for system tests and man-machine system experiments, there is an overlap where either term seems proper. In this overlap area fall system tests dealing with a subsystem or the entire system, more often as the system approaches operational status and thereafter, where people play significant roles as operators. Further, such tests are more likely to be the kind investigating system capability as a function of input, or system difficulties and their origins. This kind of test requires variables to be manipulated and controlled to a much greater extent than do tests matching performance against specifications.

The overlap of man-machine system experiments with certain system tests suggests that people familiar with conducting the former should be also involved in the latter. For one thing, they may know how to organize complex tests or experiments. For example, they should be helpful in developing sophisticated experimental designs. For another thing, they should realize what the injection of human elements requires in regard to their selection and training, experimental design, and test operations. These kinds of knowledge may otherwise be in short supply.

The people with such knowledge are often found among engineering psychologists or human factors specialists. These may be called on for another type of support in system testing. They can provide ways of examining how well the system and equipment design has conformed to human engineering requirements, how effective the techniques are for training system operators, how useful the operating and maintenance handbooks are, and whether the right number of individuals with the requisite skills are being assembled to man the system. These are important tasks in the testing process. They may be carried out through analyses, inspections, checklists, observations of performance, examination of records and other documents, interviews, questionnaires and rating scales, and even experimental evaluations of components. But they are not the same thing as the over-all planning and conduct of tests which are equivalent to man-machine system experiments.

*Categories of Tests*

What are some of the system tests required by the military departments? They can be distinguished according to test objectives. Sackman (1967) listed twelve of these: capacity testing, degradation, demonstration, design verification, normative, procedural, quality control, "realtime optimization," reliability, rehearsal, retrofit, and shakedown. According to Sackman, these are overlapping categories and not exhaustive, but cover the main types of test goals in common practice.

Meister and Rabideau (1965) listed ten factors to differentiate three test categories of exploratory, resolution, and verification proposed by Shapero and Erickson (1961). These factors were the stage in the system's design and development cycle when the test category occurs, the extent to which independent variables are manipulated or controlled, the number of measures recorded, the repeatability of test conditions, the number of conditions compared, the extent of control over the test environment, the number of dependent variables, the reasons for testing, how much of the system is tested, and how closely test conditions resemble operational conditions.

There have been so many system tests it would make no sense to try to describe them here beyond the frameworks in which they fit. For example, the Air Force has had a number of large centers for testing both airborne and ground-based systems and equipment. Its testing falls into Categories I, II, and III. The basic arrangements are set forth in Air Force Regulation AFR 80–14 and are summarized thus in Air Force Systems Command Manual 375–4:

The Category I test phase is concerned with the testing and evaluation of the individual components and subsystems of a system. In general, the Category test phase ends when the evaluation assures that the component/subsystem of a system, including the operational, utility and support computer programs, meet the minimum performance requirements of the system specification. Category I test effort is predominantly a contractor effort and in many cases a large part of the test planning is done by the contractor.

Category II testing examines both system and subsystem performance, including "the ability of the system to provide at a given time a proper output with the inputs as stated in the system specification." It is a joint contractor–Air Force effort under Air Force control, with the Air Force effort becoming predominant during the testing and with military personnel performing the system operations and maintenance. A Category III test is performed by the operating command that acquires the system. The entire system is tested in an operational context.

Clearly, Category III tests resemble man-machine system experiments, and so do some Category II tests. But Air Force documentation has not pointed this out or suggested calling on those experienced in such experimentation to plan and run such tests. Rather, in emphasizing the personnel subsystem (i.e., human factors) aspects of system development, it has limited itself to calling for human factors specialists in category testing to check on human engineering, training, and manning. Although it must be granted that in this respect the Air Force was ahead of the Army and Navy, all three military departments have been unaware of the help that might come from human factors people in over-all test planning and conduct.

A systematic attempt to relate human factors requirements to Army testing (and other system development aspects) was undertaken by McGuire et al. (1966). This report spelled out the various kinds of tests the Army had been conducting; it was implied that operator considerations had been neglected, notably in engineer design tests and R&D acceptance tests, because those conducting them presumed that questions of design for operation by people were inconsequential in these early tests.

Army R&D acceptance tests have included preliminary and formal qualification tests, reliability tests, and engineering critical component qualification. They have extended from components to subsystems and even to entire systems, as set forth in Army Regulation AR–10 and Army Material Command Regulation AMCR 70–7. In addition, the Army (like the Air Force) has required installation and checkout tests, essentially the same as engineer/service tests, as well as special tests: check tests, confirmatory tests (Type I and Type II), and troop tests. The last "is a test conducted in the field for the purpose of evaluating operational or organizational concepts, doctrines, techniques, and procedures, or to gain further information on material" (McGuire et al. 1966).

Another report on Army testing (Army Research Office Professional Summer Study Group 1964) listed tests at the Army Electronic Proving Ground (USAEPG) under performance tests and quality control tests. The former included eight engineering performance tests (research, engineering design, engineering, electromagnetic environment, susceptibility, vulnerability, compatibility, and component development) and five operational performance tests (feasibility (1), feasibility (2), service, integrated engineering service, and troop).

The quality control category included five production tests (initial production, production, acceptance, comparison, and renovation), and four compliance and other tests (military potential, R&D acceptance, confirmatory, and product improvement). The total was twenty-two.

*Methodological Problems*

The same draft report distinguished between tests which compared two or more items of equipment, those which compared equipment against a standard, and those which ascertained performance characteristics. It described experimental and measurement methods and suggested guidelines for Army testing. These resulted from examining seventeen tests at the Army Electronic Proving Ground in the preceding twenty-three months. The report included the following observations about some or all of the seventeen tests:

Essentially, we are talking about the lack of control of the test situation in such a way that the intent and the integrity of the design for the test are diminished or in some cases, lost. We are talking on some occasions about the failure of test execution personnel to carry out the design of the test . . . some of these things happen . . . at USAEPG . . . much, much too often, seemingly in every test in some measure or another. . . .

It would appear that part of the appropriate operational environment of equipment would be its use in actual combat operations. However, most of the evaluations attempted to simulate only to the extent of "normal simulated routine" of the Army. It would appear that the equipment and the men operating it are being evaluated for use in a peace time Army. Yet the equipment the Army possesses at any given moment is the equipment with which perhaps the decisive battles will be fought. . . .

The word "hypothesis" was not seen in any of the 17 evaluation reports received. Rather, the approach at USAEPG is to state the purposes or objectives of the test. . . . The usual statements of objectives and purposes as seen in USAEPG documents are simply not amenable to proof one way or another. Objectives are usually quite general, in themselves, and are only more or less defined by stating criteria and methods in following paragraphs. . . .

When the equipment is to be evaluated against certain accepted standards such as manufacturers' specifications or standard electromechanical criteria there appears to be no difficulty at USAEPG in accomplishing reasonably good tests. However, when such things as ease of installation, suitability of location in vehicle, ease of operation, human factors, and overall system performance are the questions under evaluation then there is often either nothing but uncontrolled individual opinion used as a criterion or, in the case of the human factors aspects of the evaluations, the use of a few items in a checklist. . . .

In one of the 17 evaluations reviewed there was an excellent application of the counterbalanced method of control. Therefore, the technique is at least partially known and valued locally. However, there are many, many other evaluations in which the method could have been used but was not. . . .

Not enough attention is being paid to the selection of test personnel who will approximate those who will eventually operate or maintain this equipment in the Field Army. It is possible that these Field Army personnel are either more or, as is more probable, less capable of operating the given equipment than the evaluation subjects at USAEPG. . . .

It was considered curious that not one single test of the significance of difference between two mean performances, nor one correlation attempting to ascertain the relationship between two sets of measurements was found in the entire sample of 17 tests . . . this means that USAEPG is conducting evaluations which are governed by the limited background of its evaluation personnel and as such is not using the most modern, efficient and, in most cases, necessary techniques in the accomplishment of its mission. . . .

In many of the evaluations the dictates of the test plan and test execution plan are simply not carried out by those who do data collection or analysis . . . where only a token response is made to a requirement of the test plan, it was noticed in many evaluations that only one value of an independent variable was used in determining the effects of the variable on system performance. It is true that this represents an effect of the variable on performance but only under a single condition of the variable. However, unless the range of the variable as it is expected to occur in the operational situation is explored in a systematic manner, its effects on performance in the operational situation are definitely not yielded by the evaluation.

Although the report included many other critiques of testing at the United States Army Electronic Proving Ground, the foregoing should suffice to illustrate methodological problems. Seldom have these been stated so explicitly and comprehensively, yet those familiar with testing in all the military departments could cite similar shortcomings at other test centers. Although the seventeen tests covered equipment configurations ranging from the simple to the complex, the problems apparently centered on tests which were analogous to man-machine system experiments.

## SMALL-GROUP RESEARCH

It might be claimed that man-machine system experiments actually constitute a subset of small-group experiments, of which there have been many indeed (Cartwright and Zander 1953; Hare, Borgatta, and Bales 1955; Thibaut and Kelley 1959). What seems to distinguish most of the small-group research from man-machine system experiments is the latter's system and operations context. However, in one rather heterogeneous grouping of small-group research which has *not* placed its experiments in a setting of system operations, these researchers either asserted or implied that their findings could or should be applicable to man-machine systems. Among indicative characteristics of such studies, the group of subjects in an experiment might be called a "team" or "crew." The terms could imply a similarity to "system," since the components of a team or crew, like those of a system, have a common objective.

A somewhat arbitrary selection of small-group research related to man-machine system experiments will be described under the headings of communication and information processing, feedback and reinforcement, confinement, work-rest cycle, and bargaining and negotiation.

### *Communication and Information Processing*

Chapter 20 has already referred to some of the well-known network studies which examined the effects of group structure on communication, or of communication structure on information processing within a group. Associated with this research have been such names as Bavelas, Leavitt, Smith, Heise, Miller, Guetzkow, Simon, Dill, Goldberg, Trow, Shaw, Gilchrist, Walker, Rothschild,

Strickland, Christie, Luce, and Macy. As observed earlier, there is no need here to duplicate the comprehensive review by Glanzer and Glaser (1961).

Another topic has been the distribution of functions in a simulated team, also covered in the Glanzer and Glaser review. Lanzetta and Roby (e.g., 1956, 1957; Roby and Lanzetta 1957) described experiments in which a team task was abstracted from a bomber crew situation. Three subjects sat in separate booths in each of which were two instrument dials and two control switches. The subjects adjusted the switches in response to changes in dial readings. Since the source dial might be in a different booth, subjects had to relay information over an intercom. The research inquired how performance was influenced by such factors as the degree to which a team member depended on team-mates for information; the degree of predictability of the input; operating procedures or rules for disseminating information; and input load.

In a number of studies multiman teams have performed a monitoring task (e.g., Bergum and Lehr 1962). Wiener (1964), who has reviewed these, himself did an experiment in which he varied the size of the team between one, two, and three individuals. He also compared joint and individually isolated monitoring in three-man teams. Moore (1961) investigated the performance of forty-eight two-man teams operating a simulated taxi-dispatching station under two conditions of load and two conditions of information access. Smith and Duggar (1964) studied twelve four-man groups working on a series of problems that involved searching and counting visually displayed items.

### *Feedback and Reinforcement*

When an experimental subject gets knowledge of the results of some performance (feedback), this information can exert either reinforcing or discriminative effects, or both. That is, depending on its nature, it can increase (or decrease) the likelihood that the particular performance will occur again, or it can indicate in various ways what the performance should have been, or it can do both. In team situations, the results about which knowledge becomes available may concern only the team output, rather than individual output. In this case, knowledge of results (KR or KOR) is said to be "confounded." On the other hand, information may be provided about individual output, either alone or in combination with KR for team output.

A three-experiment program reported by Klaus and Glaser (1960), Glaser, Klaus, and Egerman (1962), Egerman, Klaus, and Glaser (1962), and Egerman, Glaser, and Klaus (1963), was carried out at the American Institutes for Research Team Training Laboratory. It concentrated on the feedback's reinforcing rather than its discriminative effects. It applied the concepts of operant conditioning. Individuals in three-man or two-man teams had to make a lever-pressing response. They were reinforced when the duration of pressing approximately matched a duration established by the experimenters (two or four seconds); the latter was not communicated to the subjects. The reinforcing stimulus, for team output, was the advancement of a visible counter.

In one situation, if either of the first two members made the correct response and the third member also responded correctly, the team was reinforced.

Thereby, one of the first two members could be reinforced by the confounded feedback after he had responded incorrectly. Presumably this reinforcement would strengthen incorrect performance on his part, making it more likely to occur in subsequent trials. This in time would lead to incorrect team performance. The experiment demonstrated this outcome.

Hall (1957), Rosenberg and Hall (1958), and Rosenberg (1958, 1959, 1960) explored the relative effects of confounded and nonconfounded KR. Members of two-man teams turned a concealed knob a required number of turns during a timed interval. Zajonc (1961) investigated the effects of direct and confounded feedback under conditions which did not permit the mutual compensation of errors by team members.

*Confinement Studies*

Fraser (1966*b*) summarized fifty-three studies of confinement (some studies involving more than one experiment) conducted up to the time he wrote his report; these did not include the Mercury, Gemini, Apollo, and Vostok manned space flights.

Among these fifty-three studies were investigations of the work-rest cycle and manned space flight simulations (see also Chapter 22). The number of persons confined varied from a single individual to thirty in a fall-out shelter and one hundred in a submarine. Fraser did not include a large number of experimental investigations of sensory deprivation which have appeared in the psychological and physiological literature, although it would seem that these could qualify as extreme confinement studies.

Among more recent projects have been investigations of diver performance and responses to tests in the Navy's Sealab studies of long-duration, deep submergence (Bowen, Andersen, and Promisel 1966 and Radloff and Helmreich 1968) and an experimental program (initially called "Argus") directed by W. W. Haythorn at the Naval Medical Research Institute (Altman and Haythorn 1967; Haythorn and Altman 1967; Haythorn 1967). The latter program was based on an elaborately instrumented laboratory which included a number of isolation spaces as living quarters for individuals or pairs of subjects. Some of the research examined human "territoriality"—the development in an individual of claims on particular locations within the isolation space. The research also investigated how personality attributes, degree of homogeneity within pairs of subjects, and extent of stimulus variety affected tolerance to confinement. In one study, nineteen of the forty most stimulus-deprived subjects were unable to complete the seven-day period of confinement.

*Work-Rest Cycle Studies*

Seven studies with the Lockheed-Georgia Company's crew compartment mock-up in 1960-64 investigated the effects of different arrangements of the work-rest cycle on the performance of groups of subjects. The subjects engaged in various types of tests and team co-ordination tasks. Most of the studies involved six or ten subjects. This HOPE program, reported by Adams and Chiles (1961), Alluisi, Chiles, Hall, and Hawkes (1963) and Alluisi, Chiles, and Hall

(1964), was sponsored by the Aerospace Medical Research Laboratories. Overviews have been published by Alluisi, Chiles, and Smith (1964) and Fraser (1966*b*). Eberhard (1966) has reviewed all studies which have dealt with sleep requirements and work-rest cycles for long-term space missions.

*Bargaining and Negotiation Studies*

Bargaining (or negotiation) is a situation in which the participants have mixed motives toward one another. It is neither purely co-ordinative (or cooperative), nor purely competitive. It can be argued that many conflict situations are basically bargaining situations. In these the ability of one participant to gain his ends depends greatly on (1) the effects of threat on the other participant and on himself, and (2) the other participant's decisions. Bargaining may be explicit or tacit.

Experiments investigating this interesting two-person or dyadic situation began with those of Deutsch and Krauss (1960) and Borah (1963). They developed into an extensive research program at the System Development Corporation, initiated in its Systems Simulation Research Laboratory (see Chapter 17). The SDC program had many characteristics of interest.

Questionnaire queries and rating scales probed a subject's intentions, his reasons for the actions he took, his beliefs about the other participant's motives, and his expectations of what the other participant would do. Thus, the researchers investigated the subjective states of each subject in the dyad by requiring the subject to verbalize these states intermittently during the experiment.

Innovation in methodology resulted from on-line computer support. In an experiment the computer first paired up as many as two dozen subjects. Then a subject sent messages (moves, bids, threats, offers) to his paired opponent by moving switches (e.g., pressing buttons). These produced computer displays on the other player's TV console. Computer programs assisted in umpiring legal moves, displaying relevant information, recording all moves, and making the probes about subjective states.

Because massive amounts of experimental data had to be processed, a comprehensive computer program, TRACE, was developed to help classify, group, and summarize data. TRACE made it possible to explore relationships among complex sets of data rapidly and easily, as well as check hypotheses about patterns in particular subsets of the data.

The experiments used various games: "Communications," "Territories," "Pacifist," "Prisoner's Dilemma," and "Bartering." The research, funded by the Advanced Research Projects Agency of the Department of Defense, has been reported by Shure, Meeker, and Moore (1963), Shure and Meeker (1963), Shure, Meeker, and Hansford (1965), and Shure and Meeker (1967).

## GAMING

"Gaming" sometimes refers to all-computer simulations but here its use is limited to situations in which people participate. A game is generally a contest between two opposing individuals, or sets of individuals, who play the roles of problem-solvers, policy-formulators, or decision-makers. The situations with which they deal are usually represented by verbal descriptions, which may be written on paper, spoken by an opponent or experimenter, or sometimes displayed at a computer-linked console. In some games there may also be pictorial representations, such as a map or some kind of abstractly patterned game board. The objective is to win, or to achieve some specified goal, perhaps even a mutual one. Umpires may determine the victor. A game may be undertaken in the context of some theoretical model from game theory, but gaming is by no means limited to such contexts; thus, "gaming" and "game theory" are not the same thing. The course of a game can be preplanned only to a limited extent, through rules and other stated requirements. What one player does depends on what the other player does. It is useful to conceive of a game as a tree pattern, with each player making choices at alternate branchings. It is most unlikely that a subsequent game would follow the same path.

Because of this lack of reproducibility and the consequent loss of experimenter control, games are seldom played as experiments. There is another difficulty in embedding a game in an experimental context. Since the policies and strategies players employ are difficult to assess quantitatively, games frequently fail to measure what would be regarded as dependent variables. This does not mean that games lack quantitative processing as they proceed. As a matter of fact, computer support may be helpful and even essential to prepare rapidly a summary of the effects of one side's move so the other side knows what it confronts when its turn comes to make a move.

When a game does possess both experimenter control and the measurement of outcomes, it is really the same thing as an experiment characterized by responsive or reactive simulation. For example, Willis and Long (1967) described a three-sided game (a "truel") in which three two-state independent variables were knowledge of the source of attack, knowledge of the number of trials, and the sex of subjects. Since these were unchanged throughout the game, they remained under experimental control. When games have been played as experiments, subjects' characteristics have been more than likely to be the independent variables, because these characteristics are, or are assumed to be, constants unaffected by the different courses the game may take in successive plays.

As a research technique, gaming has been acclaimed for the insights which it provides and for the subjective judgments it generates in players concerning alternative courses of action. A number of experts may compare procedures, tactics, policies, or strategies, or may develop new ones. They may be perfectly satisfied with these insights and judgments, finding no requirement to subject them to further investigation of an experimental nature. The players are, or at least think they are, enlightened about potential actions which they or their opponents may undertake, and about the possible results of such actions. In this sense they are being educated. Gaming is more widely viewed as a pedagogical device than as a technique for research.

The effectiveness of gaming for teaching is thought to rest in part in the involvement of the players while they enact their roles. So popular has gaming become for teaching in schools and colleges that the Western Behavioral Sciences

Institute, a pioneer in the introduction of games to high schools, has published an "Occasional Newsletter" to acquaint game developers and users throughout the United States with the great number of new developments in games for education.

Most of what has appeared about gaming has concentrated on particular areas of application rather than on over-all analyses of the technique. One exception has been M. Shubik (unpublished paper 1964), who surveyed and collected cost figures concerning a large number of facilities where gaming programs have been conducted.

Most of the situations within which games are played have fallen into three categories: war–strategy or tactics; business and management; and political and international relations. But educational games have recently been extended to such themes as caribou hunting, labor unions, life careers, judicial processes, and consumers, to cite only a few examples.

### *War Games*

An extensive literature exists about war gaming, and a great many different war games have been developed–for the Air Force, Navy and Army–by a large number of organizations. The U.S. Army Strategy and Tactics Analysis Group–STAG (1962) has published a 130-page directory of such organizations and their games. More recently, the Joint War Games Agency of the Joint Chiefs of Staff (1966) issued a directory of about one hundred different war games, both all-computer and manual.

Among the most useful general descriptions of war gaming have been those by Weiner (1959*a*, *b*, 1960, 1961), Hausrath (in publication), McHugh (1961), Kahn and Mann (1957), and Abt (1964). The earliest history of war gaming was compiled by Young (1957). Harrison (1964) published an extensive annotated bibliography, together with a discussion of computer involvement (all-computer games and computer support). Individual papers may be found in the proceedings of the War Gaming Symposia of the Washington Operations Research Council (Greyson 1964; Overholt 1961) and in the numerous proceedings of the Military Operations Research Symposia (MORS).

### *Business and Management Games*

Business and management games have rivaled war games in number, if not in complexity. In their comprehensive (but relatively uncritical) review and analysis of this gaming, Kibbee, Craft, and Nanus (1961) included a directory of ninety-two games. They summarized the content of the games thus: "Most games concentrate on general management principles, such as organization theory, long-range planning, decision-making, communications, and the effective utilization of time, men and materials. Other games aim at teaching specific skills and techniques, particularly those built around the production planning and control functions."

Another survey of management games has come from Cohen and Rhenman (1961). In contrast to war games, which have had a considerable history, development of the first widely known business game was initiated only in 1956, by the American Management Association. It explicitly attributed its innovation to war gaming. The AMA game was quickly followed by an International Business Machines game, a RAND game (Bellman et al. 1957), and a game developed by J. R. Jackson (1958, 1959) at the University of California, Los Angeles. The educational value of management games should be approached somewhat cautiously. Cohen and Rhenman (1961) observed:

> Although we have reviewed a great many kinds of things which we think can be learned by participation in business games, we must again caution the reader that no objective empirical evidence has been amassed which proves either that these concepts can be actually taught by the use of management games or that they can be taught more effectively by games than in some other ways. It is always well to remember that the use of business games is not free; in fact, it can be quite expensive.

One of the questions, of course, is whether the value of gaming as a pedagogical technique varies with the level of sophistication of the players in relation to the problems and processes simulated in the game. It is possible that games are educationally most effective when they constitute the initial assaults on inexperience. On the other hand, their value as origins of insights and hypotheses may be greater when the opposing players are experts.

An interesting variant of the business game was developed at Princeton by J. L. Kennedy (1962*a*, *b*) in Project SOBIG. Forty-two three-man teams of undergraduate and graduate students attempted to make as much money as possible in the stock market. They played the roles of investment committees of fictitious banks over a three-month period that represented ten or more years, with interteam competition and intrateam co-operation.

### *Political and Inter-Nation Games*

Political and inter-nation simulation in manual games appears to have originated in the United States at the RAND Corporation about 1954, according to Goldhamer and Speier (1959). These authors pointed out that such gaming was known even earlier in Germany, Japan, and the Soviet Union in connection with military factors. Discussions of the use of political and inter-nation games and descriptions of particular games have been published by Guetzkow (1959, 1962), Guetzkow et al. (1963), and Kaplan, Burns, and Quandt (1960). Among many accounts of individual projects, Davis (1963) and Boguslaw, Davis, and Glick (1964) described an elaborate simulation for manual play for studying arms control, which they developed at the System Development Corporation.

## ALL-COMPUTER SIMULATION

By "all-computer simulation" is meant simulation by means of a digital computer rather than by analog or hybrids of analog and digital machines. Analog simulation has figured in some of the earlier programs of man-machine system experimentation, and it continues to be used for representing certain dynamic man-machine interactions, such as vehicle control in tracking. But it does not play a significant role in all-computer simulation of the kinds of systems to which man-machine system experiments have been addressed.

A detailed review of digital computer simulation would be both inappropriate to this volume and presumptuous. Not only is it too voluminous a domain, but its complexity and continuing development require treatment by someone with an extensive specialized background.

"All-computer simulation" in the present context means the simulation of a man-machine system entirely within a computer for the purpose of experimentation. Detailed structure and processes of the system, including both its machine and human components, are modeled in a computer program along with their inter-relationships; certain inputs are introduced; and the computer generates the outputs of the components, subsystems, and system.

The principal distinction between all-computer simulation and man-machine system experiments is obvious. In the former are no real human beings as subjects. People are represented solely within the computer. This means that considerable information must be known, or assumed, about many aspects of human beings, their relationships, and their processes, both in general and in the system being investigated. This information is needed by the computer. It must be quantified. The quantification must include both means and distributions. It must consider variations both within and between individuals. This can be a rather tall order, as the next chapter will indicate. More often than not, all-computer simulations of systems have handled human components in an aggregated or restricted fashion, or they have minimized their relative importance, thereby avoiding the problem of comprehensively simulating the human being through something other than himself.

General sources of information about all-computer simulation of systems include Flagle (1960), Morgenthaler (1961), Naylor, Balintfy, Burdick, and Chu (1966), Evans, Wallace, and Sutherland (1967), Hollingdale (1967), and Martin (1968). One of the first to report on the potentials of such simulation for system design was Goode (1951).

Experimentation with all-computer simulation has become a complex art. There are a number of contrasts between it and the kinds of experiments described in preceding chapters (and most experimentation with physical objects, like people or plots of land).

1. A great many variables and values of variables (which, of course, can include inputs) can be introduced, more than are practicable in other experiments. The total is limited largely by computer capacity and cost of processing time. Experimental design is multivariate and factorial, although many versions do sacrifice some interactions to reduce the total amount of calculation. Good experimental design and tests for statistical significance are being made more widely known to certain populations (e.g., engineers and computer programmers) through all-computer simulations.

2. A great many replications can be run in fast-time processing in the computer at relatively small cost in time and money. Many, many more are feasible than in experiments that must operate with human subjects in real time, whether or not those experiments are computer-supported by on-line methods. Hence, the certainty that results are nonchance outcomes can be much enhanced when

such certainty depends on replication. The number of replications can either be established beforehand or the required number can be determined in the course of the experiment by applying a sequential statistical test.

3. By means of Monte Carlo techniques, random distributions of variables can be easily generated for situations where stochastic variation of some variable can be safely presumed. As a result, in the many successive replications of experimental conditions, randomly determined values of such variables come into play. Then, the measurements of dependent variables acquire stochastic distributions with parameters that otherwise could not be predicted. In one Monte Carlo procedure, the repeated selection of random numbers enables the researcher to designate one of two possible outcomes at critical decision points—or one of several possible outcomes if he knows the probability of occurrence of each (Martin 1968). In another procedure, he can randomly select values from a distribution if he knows its parameters, such as mean and variance. Various probability distributions figure in Monte Carlo applications, including rectangular (uniform), triangular, binomial, Poisson, and Gaussian (normal). Monte Carlo techniques are applicable not only to problems which have a probabilistic base, such as consumer demand, but also to completely deterministic mathematical problems which cannot be solved easily, if at all, by deterministic—analytical—methods. In addition, nonprobabilistic problems may be investigated by means of deterministic computer models, but the latter "are illusory for psychology," according to Adams and Webber (1963), because "intra- and inter-subject variability make psychological data inherently statistical."

Discussions of experimental design in all-computer simulation have been published by Jacoby and Harrison (1960), Burdick and Naylor (1966), and Naylor et al. (1966). Geisler (1962) analyzed an important design problem—the size of simulation samples required to achieve confidence and precision; this problem has been receiving increasing attention recently, along with stopping rules and the determination of steady states. Special simulation languages have been developed to write computer programs for all-computer simulations. It is possible to write such programs also in general-purpose assembly and compiler languages, but the special languages tend to embody structures which can describe common aspects of the systems to be simulated. For example, in the vocabulary of the special language, "system" may consist of sets of entities, each of which can have a number of numerically described attributes; the entities may be differentiated into those which are permanent throughout the simulation and those which are evanescent. Teichroew and Lubin (1966) have compared six "discrete flow" simulation languages: GPSS, CORC, CLP, GASP, SIMSCRIPT, and CSL. They have also discussed the implications of different languages for users, language implementers, and language designers.

Very little in the way of human institutions and organizations has escaped all-computer simulation. Programs of military operations and conflict are innumerable.* Pool (1964) has outlined some of the simulations of social and

*See *Computer Simulation of Human Behavior*, eds. J. M. Dutton and W. H. Starbuck (New York: Wiley, 1970) for a comprehensive review.

political situations, ranging from predictions of presidential elections and vehicular traffic on highways and city streets to analyses of communication in community controversies and the impact of advertising. Although most all-computer simulations have handled human beings in sets or aggregates which functioned by themselves or interacted, in a substantial number isolated or interacting individuals have been modeled. There have been computer simulations of personality (Tomkins and Messick 1963), of elementary social behavior (Gullahorn and Gullahorn 1962), of role conflict resolution (Gullahorn and Gullahorn 1965), of international relations (Benson 1962), of individual and group economic behavior (Clarkson and Simon 1960), of business (Sprowls 1962), of tracking (Adams and Webber 1963), and of thinking (Hovland 1960), to cite just a few earlier instances. The number has been expanding rapidly.

A few of the numerous simulations of the kind of man-machine systems with which this book is specifically concerned are of particular interest. Some have already been mentioned in earlier chapters (e.g., Chapter 13). In addition, Siegel and Wolf (1961, 1962, 1963) described all-computer simulations of two tasks: the in-flight refueling of an F8U receiver aircraft by an A4D tanker aircraft, and an intercept by an advanced supersonic aircraft. Siegel (1967) and Siegel and Wolf (1969) also described simulations of carrier landings and of the activities of a large shipboard crew to predict the number of required crew positions.

All-computer simulation must be distinguished from situations in which computer simulation and human subjects (and equipment) function in the same experiment. The computer may contain models with which the subjects interact by providing inputs and receiving outputs. The models in the computer represent subsystems or other systems, processes, environments, or organizations which are not physically represented otherwise in the experiment. This kind of computer simulation, exploited in the work reported in Chapters 11, 13, and 17, for example, can complement other types of simulation with great effectiveness.

Another method of combining computer simulation with human and equipment simulation is to run one or more experiments using all-computer simulation in the same program with one or more experiments based on human and equipment simulation. (These latter experiments may include computer simulation in the manner noted above.) An all-computer simulation may come first, with the man-machine experiment following to validate it. Or the man-machine experiment may come first, thereby furnishing data about human and machine performance for the all-computer simulation. Both types of complemental combination, within experiment and within program, deserve greater study of the techniques and trade-offs involved—as Chapter 25 will point out in discussing experimental strategy. The other, associated question—which kind of simulation to use when—will be discussed to a limited extent in the next chapter.

# 24

# Man-Machine System Experiments in Systems Research: Commentary

At this point the reader may well ask how man-machine system experiments, defined in Chapter 1 as a particular constellation of features, fit into the over-all domain of systems research. The last chapter made it clear that certain features of the constellation are shared by other methods of systems research, such as all-computer simulation, gaming, system tests, and small-group research. Hopefully, the twenty chapters that preceded it, by describing sets of experiments each of which possessed the man-machine system constellation of features substantially if not entirely, provided enough illustrative examples to give a partial answer to the reader's query.

What have man-machine system experimenters themselves said about the relationships between their particular technique and other methods? Although they have said relatively little in their reports of experiments, they and others have discussed these relationships in a number of papers. A review of this published commentary may help the reader better understand this book's central theme.

In most of this literature the principal dimension for relating various research techniques to each other has been simulation. There has been relatively little effort to explore other relationships. Although much of the general commentary has aimed at classifying simulation, it has also discussed simulation's purposes, its advantages, its objects, and its methods.

## TERMINOLOGY

In the literature of systems research there are many terms whose meanings tend to overlap, or they are sometimes used in an arbitrary and inconsistent fashion. These terms include simulation, game, game-simulation, simulation-game, model, Monte Carlo, experiment, test, field experiment, field test, and exercise. This chapter will start with commentary about the usages of these terms, not in order to formulate clear-out definitions, but rather to demonstrate that if the reader is perplexed about terminology, confusion stems from the literature itself.

Obermayer (1964) has observed, with illustrative examples, that "a variety of definitions of games, models and simulation exist in the man-machine systems literature. . . . While no small amount of difficulty would be met in clarifying and extending these definitions, the problem is compounded since there is little consistency in the literature. . . . Without proceeding further it is clear that various disciplines using similar techniques apply somewhat different terminology." Obermayer's somewhat cursory review of a rich domain also touched on levels of abstraction, fidelity of simulation, and measurement validity. He pointed to the need for validation of analogies, whether they be called simulation, models, or games.

In his penetrating and engaging essay on models, Chapanis (1961) suggested that in its current usage in research, "model" means analogy or representation and so can be regarded as a synonym for simulation in experiments. He pointed out an important distinction between a model in this sense and a model as employed by mathematicians. In the latter case conclusions are deduced from what is already in the model, whereas in experiments "We can ask an open-ended question and the answers we can get are almost unlimited in variety." Chapanis also warned that "At best a model represents only a part—and usually only a small part—of the thing being modelled."

*Examples of Terminological Usage*

In attempting to create "A Systematic Framework for Comparison of System Research Methods," McGrath, Nordlie, and Vaughan (1960) set forth three research stages: model development, information collection, and information synthesis. They distinguished between research models and design models, characterizing the former as describing variables and relationships relevant to system performance rather than describing the system itself. (They conceded that "model" lacked "a single, clear and unequivocal meaning" in the literature.)

Mathematical models, they said, can be either deterministic or stochastic. Direct empirical methods were classified as "studies in operational settings," "field experimentation," and "laboratory experimentation"; and the last was broken down into "simulator studies" and "laboratory experiments." Simulator studies were said to involve close experimental control.

In field experimentation, these analysts wrote, one would find "major experimental interference with on-going operations—sometimes by restrictions of the scope and range of environmental conditions, sometimes by restriction of the range of system tasks which are included. In laboratory experimentation only limited portions of the system yield data, while "major portions of that system are represented under close experimental control." In contradistinction to modeling, information synthesis was illustrated in mathematical techniques of closed analytic solutions, iterative approximation techniques, and stochastic estimates.

The literature of operations research abounds in instances of inconsistencies and parochialisms. For example, in *Operations Research and Systems Engineering*, edited by Flagle, Huggins, and Roy (1960), the topic "simulation" in a chapter on "A Survey of Operations Research Tools and Techniques" (Page 1960) referred exclusively to player-enacted decision-making in war games or management games; this was called "operational gaming or simulation." A few pages later, in a chapter on "A Survey of Systems Engineering Tools and Techniques" (Kershner 1960), the topic "simulation" dealt with "general purpose analogue or digital machines"; and the author stated that during hardware development "the resulting compromise between a computation and a test is called a simulation, and a special purpose computer designed to incorporate a certain number of actual 'hardware' system components is called a simulator," usually analog. Then in a chapter on "Simulation Techniques" (Flagle 1960), one finds exclusive reference to the use of a digital computer and the statement that "simulation may be regarded as one of several forms of Monte Carlo techniques."

*Terminological Distinctions*

Distinctions between experiments or tests in the field and those in a laboratory have apparently seemed self-evident to most writers, presumably because of the physical connotations of "field" as "outside the laboratory." But this distinction weakens when some piece of terrain is designated as a "laboratory" and particularly when "field" and "laboratory" are considered along some other dimension, such as simulation. To some authors, such as Meister and Rabideau (1965), the presence of simulation seemed to characterize "the laboratory" and its absence "the field," although they conceded that "the field test and laboratory situations are merely points on a continuum of test situations." On the other hand, Sackman (1967) noted that field tests (or experiments) could incorporate simulation; and a number of these have been summarized in earlier chapters of the book (see Chapter 11, for example). Distinctions among various kinds of field tests have occasionally been attempted (see M. I. Kurke's exposition of Army terminology in Chapter 9); and some writers have preferred the term "operational experimentation" to "field experimentation" (e.g., Garner 1950).

The two terms in systems research whose differentiation has probably aroused the least interest have been "experiment" and "test." Morgan (1950) discussed "system tests" by means of which "new equipment and new systems are set up experimentally and evaluated." Garner (1950) defined an "experiment" in human engineering and systems research as "the systematic measurement of performance of men, machines, or these elements in combination under controlled conditions of operation" and "the systematic variation of certain operating conditions so that performance can be measured as a function of these conditions." He said that observations under known and controlled conditions and systematic manipulation of important conditions distinguished an experiment from casual observation; but he did not suggest what distinguished an experiment from a test.

Chapanis (1959), who also refrained from distinguishing a test from an experiment, had the following to say about the latter: "Despite the importance and universality of the experimental method in science, it is difficult to find two scientists who will agree on how to define it. As a point of departure, however, we can say that in human research an experiment seems to have these important features: It is a series of *controlled observations* undertaken in an *artificial situation* with the *deliberate manipulation of some variables* in order *to answer one or more specific hypotheses*. To Meister and Rabideau (1965), the differentiating aspect appeared to be the isolation and manipulation of variables, since: "The field test is most often designed not to manipulate variables but rather to verify their predicted effect on system performance."

The preceding chapter looked at developmental tests of component equipment, subsystems, and systems as an area related to man-machine system experiments and overlapping with them. It appears that "experiment" and "test" in general overlap and are used interchangeably much of the time. Yet many scientists and engineers, though unable to specify the differences, would probably assert that these terms often enough involve different operations. Perhaps the principal distinction is that whereas experiments generally attempt to link one or more outcomes to causes, only the diagnostic type of test serves this objective. Yet even here the distinction is sometimes tenuous. Some experiments, such as those which through psychophysical techniques establish limits of human discrimination, could be regarded as simply descriptive. Others look for correlations. So cause-and-effect relationships are not the exclusive concern of either experiments or tests.

Another distinction may lie in the degree of assurance which can be associated with any outcome. The assurance level is raised by the kinds of counteractions described in Chapter 2, to prevent or limit "confounding" and "contamination." Those who conduct tests may sometimes be less concerned about these assurance methods than experimenters, because the methods seem less important to nondiagnostic testing. Frequently an engineer's primary concern is to make something "work" or "run" or "fly." He tends to look for a yes-no answer, little realizing that whether to ascertain capacity (if yes) or to troubleshoot the difficulty (if no), he must also diagnose. It is also possible that those who test are equally as concerned about assurance as experimenters, but less knowledgeable about assurance methods.

The absence of terminological clarity in systems research should not be too surprising. Such research is still so young that clarifying conventions have not been widely adopted.

As often occurs in the absence of such conventions, terms which share some of the same attributes are employed interchangeably. If this is true for "experiment" and "test," it is also the case with "simulation" and "model." Each contains the notion of representation, frequently through analogy. Yet "model" in its usage in the literature, as Chapanis (1961) indicated, would seem to contain still other concepts. These are abstraction or simplification and the interrelating of two or more variables or system features. Thus, three concepts—representation, abstraction, and relatedness—seem to intersect in the term "model." That they can be so conjoined is due no doubt to the symbolic structure of those models that are composed of words or numbers.

## CLASSIFICATION

Instead of trying to provide simple definitions, some commentators have turned to various kinds of classification to relate different types of simulation to each other and to other research techniques. Some have created dichotomous or trichotomous schemata, others continua or spectra. Still another classification approach has been the matrix of research methods. Although simulation can also be categorized according to purposes and objects of simulation, these will be discussed separately.

### *Simulation Schemata*

Harman (1961) took note of a number of dichotomous schemata. These, he said, included deterministic-stochastic, deductive-inductive, analytical-physical, and computerized-manual. More recently, Crawford (1966) has suggested that one can distinguish between open-loop and closed-loop simulation. The former means the simulation of "gross characteristics of the external environment," which man cannot alter. Within closed-loop simulations, human beings play a role in one type and are merely represented in another. Further, in those simulations in which human beings do take part there may be two categories. One mostly emphasizes machines, the other the people.

According to Zelditch and Evans (1962), simulation as employed in games incorporates some real properties of the natural setting and transforms or makes substitutions for other properties. Transformations in scale are called iconic; substitutions where functional similiarity is maintained are called analogue; and substitutions where only a surface or structural similarity remains are called homologue.

Morgan et al. (1963) distinguished between computer simulation and mock-up simulation. In the former, "a computer stores various transfer functions or anticipated response characteristics of various components and their interactions with each other. . . . The computer is then given data to determine what the expected output of the system would be with various inputs." In mock-up simulation, "operators are given the equipment (or facsimiles thereof) that they would have in operational use. Then, with the aid of some artificial inputs, they are asked to operate the system as they would under operational conditions."

Grodsky (1966) distinguished among "soft mockup–a cardboard display layout," "hard mockup–an actual three dimensional layout," and "functional mockup–a hard mockup with a design layout and actual or prototype equipment and with some function associated with the man-machine interface equipment."

### *Simulation Continua or Spectra*

Harman (1961) concluded that the best proposed classification was an ordering of simulation by I. J. Good (1954) according to degree of abstraction, although Harman thought it still did not discriminate sufficiently. The real world (providing identity simulation) stood at one end of a continuum. Next to it came "an operational model of the system in its normal environment" (replication simulation, or "replica simulation" in Good's terminology). Following this was a laboratory model consisting of the inclusion of some elements and symbolic representation of others (laboratory simulation). Still further along the continuum came a computer's programmed operations and decision rules of a model of mathematical functions (deterministic) or including probability distributions (stochastic); this was computer simulation. The continuum ended with a set of equations constituting a mathematical model and a solution (analytical simulation). Similarly, Davis and Behan (1962) said that experiments could range along a continuum of abstraction, depending on how much of the normal environment was simulated.

Haythorn (1963*b*) presented a "spectrum of methodological tools in information systems research" to provide a framework for relating complex man-machine system experiments to other research methods with similar purposes. The methods were arrayed along a continuum of increasing abstraction, symbolization, and generality, and decreasing validity and detail. The first method considered consisted of observation and measurement of the real world, although in some cases this real world might lie in the future. The next method was the field study. Although here again the world must exist, access to it "is frequently impaired through inadequate reporting, deliberate distortion of fact, and inevitable lags in information gathering." Next Haythorn placed "laboratory experiments," followed by "game-simulation." The four methods considered so far were concerned with information collection. Although the author did not fully explain the distinction between laboratory experiment and game-simulation, there is an implication that the latter differs because it includes some aspects of computer simulation, such as Monte Carlo modeling. Monte Carlo models, in fact, appear as the next step or level in the spectrum, followed by analytic models and, finally, mathematical models. The three types of models were construed as information synthesis.

"It has seemed desirable," Haythorn commented, "in the development of a program of research on information systems to range back and forth along this continuum, using data collection procedures to provide the inputs to data synthesis techniques and to use the data synthesis techniques to direct additional data collection."

Haythorn pointed out that symbolization increases when one proceeds through the types of models; and the increasing generality of the statements one can make is matched by a decreasing amount of detail, a loss of specificity which carries the risk that "conclusions will be invalid because of simplifying assumptions that have been made and important variables that have been ignored."

In earlier papers, Haythorn (1959*b*) and Geisler (1960) presented a similar continuum in which field studies were omitted, detailed simulation substituted for laboratory experiments, and abstract simulations (in Geisler's case, analytic simulation) stood for what later became game simulation. In addition, analytic models were absent; and Geisler listed mathematical solution instead of mathematical models. However, in a subsequent spectrum Geisler and Ginsberg (1965) reverted to mathematical models; and all-computer simulation replaced Monte Carlo models, game-type simulation replaced game-simulation, and something called one-to-one simulation took over from what had been detailed simulation or laboratory experiment. The evolution in nomenclature over the years provides further evidence that terms in systems research have not been handled uniformly. Researchers become absorbed in other interests than in trying to clarify labels.

*Matrices of Research Methods*

Another RAND researcher (McGlothlin 1958) developed a matrix for interrelating various types of research methods and for outlining the dimensions along which the researcher must make decisions to develop a man-machine system experiment. The rows consisted of "characteristics." These included the extent of using a computer (for preparation or play); extent to which reality was simulated; degree of manual participation; extent to which decision rules were specified; degree to which an outside arbitrator participated; degree to which embedding organizations were needed; extent of time compression; and amount of flexibility. The columns were types of methods classified according to objectives. One was heuristic games (simple games to provide insight), another the estimation of quantitative solutions as in Monte Carlo. Two were called "developmental." One of these, identified as "participation" and illustrated by war games, aimed to produce ideas rather than solutions. In the other (observation), the aim was to draw conclusions about measured and observed behavior that could hopefully be generalized to similar situations. Still another method was said to be that which provided a prototype or feasibility demonstration. This matrix approach indicates that the classification problem is not easily resolved.

Geisler and Ginsberg of RAND (1965) also tried a matrix approach. Their rows were roles of the players, level of detail, amount of analysis, extent of time compression, and cost. Each of their three columns combined label and purpose. The first specified game, for training and education; the second, game-simulation, to investigate tactical policies and feasibility; and the third, all-computer simulation, to provide quantitative answers to specific questions. This matrix reduced the alternatives for research to game-simulation and all-computer simulation. The authors have discussed the advantages and difficulties of the former at some length (see Chapter 13).

## PURPOSES AND OBJECTS OF SIMULATION

Both of the foregoing matrix approaches included some classification according to objectives or purposes. Other commentary concerning the purposes of simulation follows, along with views about the aspects of systems which may be simulated, i.e., the objects of simulation. Both purposes and objects may be regarded as additional approaches to classification.

*Purposes*

Harman (1961) said that simulation objectives included evaluation, training, and demonstration. Crawford (1966) also specified three purposes, but selected design, training, and testing of occupational proficiency.

In an unpublished paper, G. C. Bailey has outlined the purposes of simulation in man-machine systems at somewhat greater length. One class of objectives is to make organizational decisions, either in operational contexts such as job shop production, or in planning. Another class, training, includes executive training, skill training, and operational unit training. A third class concerns system research, which Bailey has subdivided into design, development, evaluation, and theory.

Davis and Behan (1962) suggested that the experimental evaluation of a system could have the following purposes: predicting how well the system could achieve its aims, comparing it with one or more alternative systems, or improving it. There might be combinations of these purposes.

According to Holmen (1963), one may distinguish between special-purpose and general-purpose simulation according to the number of uses to which the inputs can be put within the system. He commented that "if good simulation is provided for program testing, subsystem testing, system testing, and evaluation, then the simulation capability required for training is essentially paid for since it has been subsumed in the skillful design of the overall simulation." (Presumably Holmen was discussing only simulation capability and did not mean to imply that the considerable cost of simulation materials for training exercises was included in the cost of simulation for testing.)

Haythorn (1957) included among the purposes of RAND simulation-based experiments the acquisition of knowledge to permit allocating tasks between man and machine and to estimate manpower requirements by skill types and levels. Haythorn felt that simulation was necessary because task analysis was not an adequate methodology. Siegel (1967), on the other hand, believed task analysis is a feasible method for generating a computer simulation which can produce estimates of manpower requirements.

The most ambitious type of simulation, Enke (1958) declared, is to compare different sets of policies and practices; it requires parallel simulations. A simpler approach is to have people play out a set and thus test the ability of people to implement them. Another is to stipulate only broad policies and let the people evolve the details and daily operating procedures; this requires much intralaboratory or floor observation. "Generally," Enke wrote, "we have found that the economists prefer to compare policies (stressing numerical outputs) while the psychologists prefer to test and evolve (emphasizing human behavior); hence the economists try to transfer as much as possible to the enactment of a computer, while the psychologists resist this on the grounds that organizational conflicts are no longer being simulated."

Zelditch and Evans (1962) said that simulation aimed to reduce the number of variables and their permissible values, and to isolate the investigated system from some effects of a varying environment. They said that such control could be necessary because in natural environments there could be canceling, additive, irrelevant, and confounding factors. The first of these counters the effect of an observed variable; the second obscures its relative importance; the third may be highly correlated with the observed variable but unrelated to any other; and the fourth may generate a spurious correlation between two variables.

### *Objects of Simulation*

Chapman (1965) observed the range of system aspects which may be simulated: (1) various combinations of a system's machines, communications, people, their operating environment, or their procedures; or (2) inputs to the system; or

(3) its environment; or (4) combinations of elements from all of these. Furthermore, two or more systems may be involved in co-operation or competition with each other, in a common environment. Chapman also noted that real-life details might be represented in real time or aggregated, and that a simulation could be performed by people, by a computer, or by a combination of these. He expressed doubt that the availability of facilities or personal interest principally determined the choice of techniques; rather, such choices would seem to depend on the character of the particular problems and desired answers.

A taxonomy which is rudimentary but seems to be among the best available appeared in a paper by Fitzpatrick (1962). He suggested that the aspects of systems which might be represented through simulation could be categorized into equipment components, personnel, organization, input data, output data, system procedures and processes, and environment. After discussing such simulation characteristics as representativeness, validity, precision, cost, stage of system development, mathematical vs. physical, and whole vs. part, Fitzpatrick cogently commented: "It is commonplace to suggest research *using* simulation. It is far less usual to suggest research *about* simulation. But, knowledge about such an important tool of research and development may act as a lever to increase the value of research to a degree which would more than repay the cost of gaining that knowledge."

Davis and Behan (1962) said that simulation was used in system evaluation either to provide the inputs to the system or to represent its internal functioning. The latter type of simulation is usually all-computer simulation. They asked what system inputs should be included in the simulation and concluded that they should be those that were relevant and that satisfied the demands for outputs. For example, the system boundaries must be extended to include the surrounding community if an output of interest was one that might affect that community. Can the inputs omit noise? Weather on a scope is noise to the controller, information to the weatherman. (But noise can also provide a type of information to the controller as well as block information; signal and information are not necessarily the same thing.)

Forms of operational simulation in command and control systems have been enumerated by Holmen (1963), who meant by that term the combination of simulated inputs with actual men and machines in the operational setting of, for example, an air defense system. Such input simulation may be static or dynamic; in the latter case "the inputs are modified as a result of system response." Inputs may be created within the system computer itself, or by equipment (e.g., consoles) directly associated with that computer, or by external equipment linked only indirectly (e.g., radar signal generators).

## METHODS OF SIMULATION

Closely related to objects of simulation are methods of simulation. Much commentary concerning methods seems to have concentrated on some of the disadvantages of relying exclusively on all-computer simulation. Along a complementary line, observers have outlined the advantages of simulation with human participants. Turning the coin, some of the same observers have described the drawbacks of such simulation and the superiorities of the computer.

*Drawbacks of Computer Simulation*

In comparing simulation of controlled inputs to operational systems with all-computer simulation of the entire system, Davis and Behan (1962) said that the latter runs into trouble with psychological and sociopsychological variables, and instead of establishing true values for these, mathematicians tend to "circumvent the problem . . . by sets of simplifying assumptions about human behavior."

One of the earliest researchers to voice alarm about the lack of empirical validity in simulations of human behavior in mathematical models was Kennedy (1952), whose views substantially influenced subsequent man-machine system research at the RAND Corporation:

> If we require the mathematician to provide solutions to systems questions without giving him the necessary data, he will make some assumptions on his own in order to wrap up his package. Then we find that we cannot stomach the assumptions, that the mathematical model is too simple, that our wisdom tells us that life really isn't that way, and we become suspicious of the whole mathematical model concept. I would argue that the fault, if there be one, in this controversy, lies mainly with the provider of data for mathematical treatment. We stick to our small components when the mathematician needs quantitative *system* information for his special brand of magic. . . .
>
> The great leveler of theory construction is the laboratory, where concepts can be put to operational test. If a laboratory can operate with systems rather than component criterion measures, I believe that we will achieve the optimum climate for solution to our pressing problems of complexity.

Along the same line, Bray (1962) commented as follows:

> Operational research and system analysis studies have been developed and led by mathematicians, physical scientists, and engineers. As a result, mathematical models of man-machine systems sometimes have been oversimplified with respect to men (e.g., when it has been assumed that certain human functions are linear), an oversimplification which has been almost inevitable in any case because of the limitations on the knowledge of men's properties. Available mathematical models have particular difficulty, furthermore, in dealing with the reprogramming in which the human components of systems indulge. . . .
>
> The way to avoid mathematical oversimplification and to check the accuracy of assumed constraints in system analysis, it is commonly recognized, is through physical simulation of systems (or subsystems) as wholes, supplemented by actual tests of critical system operations. In a general sense, this is to say that the mathematical, solution-oriented approach needs to be supplemented by the laboratory, variance-oriented approach.

Criticisms of mathematical modeling have not been focused exclusively on human factors. Sackman (1967), summarizing some opinions expressed at a symposium on operations research approaches to testing, at the University of Michigan in 1959, said that some of the limitations which were expressed concerning mathematical models included esoteric contexts that are not understood by users, overselling by originators, and the misleading glamor of computerized models. One symposium panel reached the conclusion that "little realism could be obtained in an analytic model." This panel was concerned with noise, jamming, and countermeasures, that is, "dirty inputs," which apparently created difficult requirements for realistic modeling.

(It can be presumed that certain kinds of noise are difficult to simulate in mathematical models because they are difficult to describe and predict. In this sense, certain kinds of human behavior in systems might be regarded as noise, notably input-associated behavior which degrades the desired system output.)

### *Advantages of Simulation with People*

Inability to represent the dynamic and adaptive behavior of humans in a complex system by means of more analytical abstractions has been conceded in some instances by nonpsychologists as the reason for including human beings in simulations. Geisler and Steger (1962), mathematician and economist, have noted further that the problem is compounded when the system is a future one so there is no current behavior to observe: "adaptive behavior of the system's ultimate human operators on the basis of conventional heuristic extrapolations from previous experience is *not* considered sufficiently reliable." But these writers did not rest their case with just a general assertion about adaptive behavior. They examined such behavior further, classifying the learning processes evoked by new organization or hardware into four groups. One was learning a set of system-pertinent tasks; a second was defining a system's goals and subgoals requiring co-operation and co-ordination; a third was learning how to deal with a system's uncertainty through search, innovation, and decision-making; and the fourth was inductive learning through heuristic reasoning.

Enke (1958), another nonpsychologist, acknowledged the need to obtain human performance data empirically. Decisions, Enke said, can be unpredictable in the real world. Also there may be "simply no action at all by some unit that is unwilling to comply and too cautious positively to disobey; it is this sort of all-too-human behavior that prevents a modeler writing a complete and realistic computer program."

"Putting people into the simulation," Geisler (1960) asserted, "helps to insure the completeness, compatibility and workability of the model being constructed. People thus provide quality control, feedback, and learning."

Ernst (1959) commented that in simulating a system, "mechanical as well as human elements may be used in place of their mathematical descriptions when the latter are impracticable." One consequence is to proceed in real time, without significant delays in the computational portions. Further, "the use of human subjects, and other real elements of systems, introduce requirements for coupling the 'physical' to the 'mathematical' portions of the system representation. The necessary coupling devices (or transducers) tend toward the special-purpose as contrasted to the more flexible general-purpose equipment suited for treating the mathematics."

Human subjects are used because of human variability and nonlinearity. This nonpsychologist explained: "Both skill and capacity vary from operator to operator and from time to time. The human equation or transfer function is probably not only nonlinear but also modified by learning and experience." Variability imposes various demands for replications, large amounts of data, and design tactics such as using a subject as his own control.

Information flow in operating systems includes feedback loops "in which actions taken by human operators modifies the information fed back for future action." Thus humans influence system performance whether they are executing simple control actions or making command decisions. Performance in a simulation must be comparable to that in the real system. An operator may time-share various tasks, within one or many feedback loops, and "the condition of time-sharing may significantly affect the subject's capacity for performing any one task" by the division of attention or due to operator nonlinearity. "Hence," Ernst commented, "care must be taken in partitioning the system so that the features of the human environment that are not a direct part of the control loop under study, but possibly affect that loop significantly, are included in the simulation." Ernst summarized his views thus: "Simulation dictates a functional description of a system with particular attention to the functional structure of the jobs, tasks, and operations to be performed by the human elements of the system. The more difficult problems involve these human elements, whose transfer functions cannot be prejudged and whose presence is therefore essential to realistic experimentation."

Geisler, Haythorn, and Steger (1962) called attention to a number of other advantages of having human subjects in man-machine system experiments, or "game-simulations" as they called them. For example, the subjects (participants) in an experiment exploring a future system may work in the organization which will operate that system. Then the experimenters can draw from them ideas for improved system design and can train them in system operations. These authors observed:

> In addition to using people to impart greater realism to the model, game-simulation also puts people inside the model who can perform useful functions. First, they can use their creative abilities to design decision rules or strategies that help to advance the defined goals of the model builder regarding the realism of the simulation, the feasibility of the policies and procedures assigned to management, and the way in which other personnel who participate in the simulation should be indoctrinated and trained. . . . .
>
> As we mentioned earlier in this section, the game-simulation representation of a system produced an environment very conducive to training and learning. This has been a consistent part of of the evaluation reports of participants who have been in the previous Laboratory experiments. There are several reasons for this evaluation. First, the participants spend all their time working on problems and material relevant to their assignment in the simulation. Second, because of time compression, the participants obtain a much more intensive experience than they would normally receive in the real world. Third, also because of time compression, the managers receive the effects of their decision-making faster. Fourth, it is more feasible to assign causes to effects in simulation than in the real world so the manager can more directly determine the impact of his decisions.
>
> Management games provide advantages similar to game-simulation for training or exposure of managers to problem situations. However, game-simulation attempts to be a more accurate representation of the real world so that the environmental conditions, the organizational relationships, the data system, and

the participants' roles have substantive significance. The results and experience thereby derived from game simulation have interpretive meaning for real world policy and system design.

The same authors pointed to a different kind of advantage:

It is also much more feasible to re-program the functions, activities, or even decision processes of a person used in a game-simulation than those of a computer model. Re-programming can even be done when the instructions for guidance to the person are broad or general. In fact, this characteristic is an important requirement of complex systems since we often do not have decision rules developed for such complicated interactions, and so we can only lay down broad guide lines to the management. We cannot usually give such broad guidance to a computer and have it apply the guidance to each situation. Rather, we are compelled to produce much more specific decision rules which probably means that under certain circumstances our computer rules will not do as well as a person who can take more explicit and complete account of the relevant variables and criteria. We have already seen this happen in LP–II where the participants achieved substantially higher levels of alert than the all-computer simulation of a comparable missile system predicted.

### *Drawbacks of People*

Turning the coin, Geisler, Haythorn, and Steger (1962) analyzed "undesirable features of game-simulation":

Finally, there are some definite disadvantages in game-simulation, particularly when compared to all-computer simulation. For one thing, game-simulation is very costly because of the requirement for precise representation of the simulated system. A realistic model is important to obtain valid responses from humans. Hence, with people in the simulation, the demands for realism, and therefore detail, are even greater than in computer simulation.

A careful representation of the system involves a number of costly operations. First, it necessitates the collection and processing of large amounts of data. Second, it involves large programming costs because of the need for detailed representation of the several organizations used in the simulation. Third, it requires much computing time to provide the data used during the operation of the simulation. Fourth, it uses a large clerical staff to perform the variety of clerical operations such as doing hand calculations, drawing charts, filing reports, and following detailed instructions as part of the simulation activities that cannot readily or feasibly be machine programmed. The professional staff in the Laboratory also tends to be large to perform the variety of studies required and to provide the mix of professional training and experience used in designing, operating, and analyzing the experiment. Finally, the number of participants and their time required also adds substantially to the cost of an experiment.

Because of these high costs it seems appropriate to use the game simulation method for studies of very significant systems which tend to have very high system costs, and for which a high level of effectiveness is very important. The purpose of the simulation then is to suggest policies, management systems, resource allocations, or operating procedures which will have a significant impact on the cost or effectiveness of the system under study. . . .

Because of the high cost of running game-simulations, it becomes infeasible to do sensitivity testing of the results. This limitation is a serious defect in game-simulation because we know that any such simulation involves a large number of uncertain features, and it would be most important in assessing the stability and conclusiveness of the findings to be able to determine their behavior over the range of foreseeable conditions. Such validation of the findings applies not only to the computer element of the simulation and the parameters used in that part of the model, but also to the human element. We would like to know if certain decision patterns or results tend to be reproduced if the participants in a simulation are changed. Thus, there is a need to consider the possibility of human replication.

Since there are many ways of classifying the participants that can affect results (e.g., with respect to personality, experience, amount of pre-training and instruction), adequate allowance for both sensitivity testing and replication of the experiment with different human participants leads to the need for more runs and consequently to longer operation of the experiment.

*Advantages of Computers*

Haythorn (1963*b*), a psychologist, has set forth the advantages of computer modeling as a technique complementary or supplementary to game simulation (which actually has included within the same experiment both man and computer simulation):

One of the shortcomings of the "game simulation" is the excessive demands on time and personnel. One technique that reduces these demands has been the Monte Carlo modelling. In essence, a Monte Carlo model identifies the essential elements of a complex system, ties them together symbolically in the way one understands them to interrelate, and includes the random variance that one believes to exist in the system. Such models have been computerized and can, of course, be run many times under a wide variety of conditions. The Monte Carlo model avoids many of the costs of the laboratory experiment and game simulation, but runs graver risks regarding the validity and generalizability of results. Nevertheless, it has proved helpful as an information synthesis device to put together what one believes he knows about a complex information system and to examine the implications of this alleged knowledge. The process of modelling has itself proven to be useful in that it identifies additional information collection requirements.

In discussing game simulation's greater realism, Geisler, Haythorn, and Steger (1962) commented that it "can include those important characteristics which are human, as well as those more readily simulable on a computer, including the development of insights, heuristic problem solving, negotiation, competition, ambition, communication between organizations, and the capacity of individual decision makers." They were referring to combinations of man and computer simulations within an experiment. Later, in discussing combinations within a program, in which one or more experiments were game simulations and one or more used only a computer, they observed that due to the high cost of game simulations, "intensive repetition or variation of runs becomes infeasible." They went on:

This raises the need for exploring ways of dealing with this limitation in game-simulation. The most feasible alternative that has been used in past experiments is to try to model the structure of the game-simulation as an all-computer simulation, and then to use the latter model for sensitivity testing, since it has those characteristics which make sensitivity testing a much more economical and practical procedure. However, this approach does not help in the replication of human participants because we do not know how to model human characteristics very well. The effects of such replication must be inferred from the floor runs even though these runs may not be designed or arranged specifically to test human replication. Thus post hoc analysis still seems to be the only feasible procedure with this method at the present state of research knowledge.

In addition, Geisler and Steger (1962) wrote that game simulations, "by their very nature, produce at best highly preferred–but not optimal–policies. Significant betterment in management control design is, of course, nothing to belittle." They suggested that "all-computer simulation and analytic models based on the manned simulation as a breadboard model should help considerably to produce even better results."

*Graphical Simulation*

It is clear that human beings can represent human beings, hardware can represent hardware, and computer digits can represent everything. However, there is still another method of simulation called graphical, schematic, pictorial, or symbolic (in a nonmathematical sense). Fortunately this has produced some commentary, although it has probably deserved more.

Kinkade et al. (1963) described types of graphical simulation, some of it introduced by Franklin Institute to CAA research in 1949 (see Chapter 15). This was altitude/time plotting, which showed, minute by minute, the location of each approaching aircraft in an air traffic control situation. Then came space/time plotting, such as flight distance versus time. The distance was related to some fixed point, such as an approach gate or runway threshold. Altitude data could be entered on the curves at various points. In 1958 a sequencing chart was developed to show queuing phenomena. Map simulations of air traffic control processes were added more recently. People moved markers representing aircraft according to rules.

Another, more ingenious type of graphical simulation, mentioned in Chapter 17, is schematic simulation, described by Alexander and Cooperband (1964*a*). This might be characterized as a dynamic but non-real-time mock-up to trace through operators' actions in designing a projected computer-based system. Approximations of console control panels and display surfaces drawn on placards were posted in operational sequence along a wall. Personnel representing the system operations sat facing their consoles. A sequence of actions would begin at one console and progress along the line. An operator took an action by pencil-marking a control which would activate, in the actual system, a computer. But in this simulation the computer was represented by the programmers who had been designing the computer programs. A programmer would process in his head the action taken and indicate its effect, such as the appearance of something on a display surface. When an action at a console generated actions required of another operator, the item would be manually indicated at this operator's console.

This simulation could operate only in expanded rather than real time and thus could not test reactions to load conditions. Nevertheless, it helped revamp some of the computer programs by demonstrating sequences of desired actions which had not been included in the program design; and it showed action sequences needing redesign. Of particular significance for potential application in other contexts, this schematic simulation as a physical and behavioral referent made it easier for members of the design team with differing technical and professional backgrounds to communicate with each other. Words alone sometimes failed as a medium.

As noted in Chapter 17 this simulation method also was compared as a training device with the operation of the actual hardware. The experimental comparison of graphical and dynamic simulation described in Chapter 15 indicated that the former could function as an economical predictive and design method. Other methods of graphical or schematic simulation are described in connection with Subsystem I (Appendix I).

In simulation-based experiments graphical simulation has been introduced only infrequently to represent something which would not appear directly to the users of the real system. But in nonexperimental exercises such as war games, some kind of pictorial counterfeit of the "world out there" has often been the key simulation device. It has taken some abstracted form such as a checkerboard, or some realistic form such as a three-dimensional model of terrain, or some form of intermediate realism, such as a map. Although more of this type of simulation of the invisible world beyond the system might well be incorporated into man-machine system experiments, it has been rarely mentioned as a type of simulation in general commentary.

On the other hand, as noted in Chapter 2, symbolic representation which is already part of an actual system has frequently been copied or otherwise represented in simulation-based experiments. Within the actual system symbolic or graphical representation occurs through some kind of transformation. For example, radar echoes on a PPI could be regarded as simulations of aircraft, a map display as a simulation of a ground environment, and discrepancy reports as simulations of actual hardware malfunctions. In this sense, there already exists in many real systems considerable simulation. If this is copied or otherwise represented in simulation-based experiments, it would be proper to term this process "simulation of simulation."

What is meant above by "otherwise represented"? As we have seen in Chapter 2, the simulation in experiments may be generated synthetically rather than by copying what has been transformed in the actual system. The radar echoes, map display, or discrepancy reports are simulated through some method of synthesizing them. The degree of realism in the synthetic products depends on the extent to which these products of synthesis match the products of transformation in the real system.

When the transformations in first-order simulation in the actual system take a symbolic form, it is not too difficult to match them through synthesis in the second-order simulation. The nature of its first order simulation helps determine how readily it can be investigated, as Haythorn (1959*b*) has astutely pointed out. "Symbols," he wrote, "are easier to simulate than things. . . . Organizations which deal with symbolic data and transformations of such data are fair game for systems simulation."

### *Word Simulation*

In all the discussion about simulation in experiments it seems strange that almost no one appears to have pointed out the significance and potentials of a type of simulation which is with us continually. This is word simulation. When we describe something in words, we are simulating it. In this sense a novel, a briefing, a task analysis, a name are all simulations. Words may simulate other words or other symbolic material, in the manner Haythorn meant in the preceding paragraph. Or words may simulate physical objects—people, equipment, environment—and events. How well can words do this? How can they be made to do it better?

Word simulation seems to be explicitly acknowledged as such only (1) in games where environment and events are represented by words, and (2) in mathematical models where the words are coded by symbols standing for quantitative properties. As a matter of fact, however, in many man-machine system experiments much of the world and its events is represented—simulated—by descriptive words, whether these occur in scripts to be followed or in instructions and other preliminary material which sets the scene. Surely words deserve a place in the classification of simulation. We should look harder at what composes this type of simulation and how it may be improved. Hopefully, incorporating word simulation into the recognized realm of simulation may make such study more likely.

## CRITERIA FOR SIMULATION SELECTION

Although various researchers have described relationships among simulation techniques and have expressed particular advocacies, general guidance is still needed as to when and where to use what, at least according to Devoe (1963), who commented:

There is a need for an integrated treatment of models per se, from simple flow-diagrams through deterministic and stochastic mathematical models, to mock-ups and simulated systems. The strengths and weaknesses of the various types of models for various applications need to be summarized systematically. For example, the dependence of stochastic models on the validity of their sampling distributions needs emphasis; the inherent differences in simulation for design, simulation for evaluation, and simulation for training require explanation. . . . A guidebook should . . . establish criteria for deciding on when and how to use models.

The production of such a guidebook would be a major undertaking. This book contains a data base and a number of guidance elements which might support one.

In an integrative approach in the direction Devoe suggested, Chapman (1965) tied the selection of simulation method to the nature of an experiment's objective—and, by implication, at least, its design. He distinguished between simulation studies which searched for hypotheses or variables, those which aimed to test hypotheses, and those which tried to show consequences or implications. He also distinguished between investigators who courageously confronted uncertainty in order to map a domain, and those who preferred to select problems where they could apply rigor. He suggested that the strategy of selecting among simulation methods to apply to the development of information systems should consist of choosing the appropriate method for the particular purpose. This would change with stages of system development and with information gained from preceding simulation studies in a step-wise process.

In an earlier paper Chapman (1961*b*) said the use of all-man simulation might be advocated for exploratory studies when the researcher is looking for critical variables in the operations of a particular group. But all-computer simulation can help locate "man's degrees of freedom in a technologically-enriched group." Chapman advised handling each problem individually, because "No simple rules stand up under scrutiny."

*Simulation vs. Real World*

In addition to the choice between methods of simulation, the researcher may have to decide whether to simulate at all. The enthusiasm for simulation among those who have developed its techniques is exceeded only by that among those who have manufactured simulation devices. Although these enthusiasms may very well prevail, it is better to weigh the advantages and disadvantages of simulation against those of sticking to the real system or its actual environment.

Geisler (1960) gave the following reasons for man-machine simulation. Although such simulation is very expensive and cannot be iterated very rapidly, "it is the only systematic way available for studying organizational, communications, complex decision-making, and information problems." Such simulation "stresses the study of organization interactions, conflicts, communication, data flows, created by decision-making."

Bogdonoff et al. (1960) listed the following reasons for *not* using the real world but simulating it instead:

Cost
Uncontrollable variables
Unavailability of the environment
Complexity of the environment
Danger to the individual and equipment
Hypothetical nature of the system
Infrequency of events
Difficulty of administering the experiment

Among advantages of simulation, Davis and Behan (1962) mentioned controllability of inputs, precise replication of stimulus conditions, relative cost, and availability of inputs on demand. Geisler, Haythorn, and Steger (1962) had this to say:

> The game-simulation model is more flexible than real world service tests and all-computer simulations. The use of a simulation technique makes it more flexible than the real world because it is possible to change more readily either the computer part or the man part of the simulation. It is also possible to study situations or environments that have not existed in real world systems, or to try out hypotheses or proposals that would be unsafe or hazardous in real world arrangements.

Meister and Rabideau (1965), however, urged caution in comparing simulation with field testing for evaluating system performance. They pointed out that when the simulation is being done by engineering groups for other purposes than human factors investigation, problems arise in control, schedule, and cost similar to those which occur in a field test. Since "management pressures may strongly resist special simulations," these authors recommended making use of field testing to the extent feasible. They warned that special simulations for human factors evaluation must produce results in time to introduce them into the development of the system, but it may be difficult to do this when it is necessary to create an elaborate simulation laboratory and its components. Further, if the system being investigated keeps changing, it becomes necessary, but difficult, to keep its laboratory representation up to date.

Meister and Rabideau stated that field test and laboratory situations can be compared according to a number of trade-off factors. One is the extent of control through knowledge of system inputs. These can be restricted in simulation to provide more certain knowledge about those included, but then there is the danger of eliminating some inputs which critically affect system performance. The researcher faces a dilemma. To make the proper selection for testing, some kind of testing is needed to establish the proper selection. Another factor is control of variables through their isolation and manipulation: these may not play a significant role in field testing. Other factors include cost, time, and the stage of system development.

Sackman (1967) pointed out that formal testing, by providing impartial and shared information, curbs the tendency to manufacture subjective evaluations to fill gaps and attain closure. But "experimental system tests are difficult to establish, controls are only partial, performance data are hard to get at, test costs may be prohibitive, and the final results are often ambiguous. This has tended to discourage extensive system testing. . . . Field tests often become feasibility demonstrations of gross effectiveness levels of system behaviors rather than controlled experimental measurements under standardized conditions."

Nevertheless, Sackman found great advantages in the real world over the laboratory due to the former's "experimental analysis of natural events" instead of "artificial isolation of hypothetical events," "conditional real time" instead of "temporal invariance," "open systems" instead of "closed systems," "heuristics" instead of "algorithms," "operational simulation" instead of "symbolic models," "psychological" reality instead of "logical," "open experimental exploration" instead of "tight experimental control," "probabilistic" situations instead of "deterministic."

In contrast, many researchers (e.g., Garner 1950) have pointed out that experiments could be better controlled in the laboratory than under field conditions. Garner weighed the relative advantages of operational experimentation and laboratory experimentation. By the former term he meant experiments conducted in the operational field. He preferred laboratory experimentation because it had greater generality of prediction, which he said was always related inversely to precision of prediction. Garner felt that "the one kind of prediction which can be done only with operational experiments is the determination of *absolute average* performance characteristics. But even with this type of prediction, the importance of the prediction is questioned because of the high probable error which must be attached to all figures of average operational performance. . . . In terms of relative cost and time, certainly, it is difficult to justify operational experimentation as being superior to laboratory experimentation," whose predictions have generality and accuracy.

The difference in viewpoints concerning experimental control is worth developing. Garner favored the laboratory and control. Sackman preferred the field and exploration. (It might be mentioned also that Garner thought of system experiments as manipulating variables, whereas Sackman sought and found correlations.) The divergence indicates that different objectives may lead to different locations and methods. The laboratory provides the kind of control needed to give assurance about the linkages between outcomes (dependent variables) and causes (independent variables). The field furnishes a diversity of variables to explore and discover.

There can be simulation in field experiments as well as in the laboratory. Then the real world is not entirely "real." By being simplified and repeatable, preplanned and synthetic simulation in each situation contributes to experimental control. The more the simulation as in the laboratory, the greater the control. The less that is simulated, as in the field, the greater the diversity. Thus the extent of simulation plays a significant role in the relationship between objectives and methods.

*Simulation plus Real World*

In Sackman's field studies and those of other System Development Corporation researchers in air defense (Chapter 11), there was in fact considerable simulation. Synthetic inputs represented the air environment of an actual operating system under field conditions. This was one way to combine simulation and the real world. Conceivably, Sackman would have been less enthusiastic about real-world experimentation if his "world" had been less contrived and thereby less controlled. (Additional control came from restricting the system under investigation to the data processing portion of a computer-based system.) Further, by means of regenerative recording, described in Chapter 11, Sackman (1967) found it possible in such a situation to "capture events in real time" and to "replicate or vary such events under controlled conditions."

Another method of combining simulation and the real world has been explored by Siegel (1967) and Swain (1967). The former created computer simulations of manned system performance by first analyzing the human actions required and then inserting the performance parameters derived from these task analyses into the computer simulation. Siegel validated the results of the computer simulation by conducting a field test (or a laboratory study) to determine the congruence of results when actual people perform in the real world. Swain specified a technique of field-calibrated simulation. The data on parameters of human performance were derived empirically from either a field test or a real-time, simulation-based exercise, then put in mathematical models for computer simulation. The results of the computer simulation were compared for congruence with those of the field test or simulator exercise. Both Siegel's and Swain's approaches attempted to get better data about human performance for computer analyses.

A viewpoint somewhat similar to Swain's has been expressed by Geisler and Ginsberg (1965) and other RAND researchers (Chapter 13). They exploited human behavior data acquired from laboratory simulation experiments by using it for computer models. However, the RAND researchers also tended to emphasize additional complementary aspects of the two kinds of simulation, such as the capability of the computer simulation to provide a large number of replications at low cost.

*Other Relationships between Methods*

Man-machine system experiments based on simulation have been related to other investigative techniques by Grodsky (1966), who has discussed on-the-scene (inflight) evaluation, analytical studies, engineering or expert opinion, mock-up techniques and laboratory tasks, and synthetic task batteries. Grodsky expressed strong views concerning "laboratory tasks and synthetic task batteries," considered together as the same sort of thing. Although this technique, he said, has "demonstrated internal validity and sensitivity to stress situations, the general difficulty in generalization makes the tasks somewhat useless in an operational situation. The lack of an agreed on taxonomy of operational task elements certainly contributes to the difficulty in generalization." He ventured several additional criticisms of this approach, which has been used by a number of other researchers.

Instead, Grodsky favored the technique of integrated mission simulation as "applicable to system design problems," because it required an astronaut's actual task behavior. He cited among its advantages "performance in real time, hardware and system dynamic fidelity, appropriate sequencing tasks... use of the actual flight personnel, measurement of operator capability relative to system operation and system tasks." He noted that "part-task simulation" was a companion type valuable in the detail design of specific systems but unable to produce "the effects of task sequencing particularly during the long duration missions." (In other general commentary the real-world time boundaries selected for experimental sessions have received little attention although occasionally descriptions of specific experiments have considered this question.)

In further comparing the task-battery and mission-performance methods, Grodsky observed:

> It would appear that the acceptable approach is not at all clear because of the existence of two separate goals: system design and general investigation of human behavior in unique environments. Further, neither of the approaches have demonstrated inflight validity. Two general sources of criticism may be directed toward the synthetic task approach: its lack of face validity and the difficulty of applying the obtained data to system and mission tasks. Criticism may be directed toward the system measurement approach (mission simulation in ground based evaluations) primarily in that the stimulus which initiates a response is never precisely known. Whether this is an important consideration is conjectural and neither is it known what tolerances are allowable in our knowl-

edge of the stimulus. Another difficulty involved in the use of system or mission measures concerns the sensitivity of measures required and the wide tolerable system boundaries which make difficult extrapolations to general performance capabilities in unique environments.

Morgan (1950) also related man-machine system experiments to other methods of investigating man-machine problems. He distinguished between operator problems and interoperator problems. To the former one could apply such techniques as basic research, design consulting, design appraisals, and field evaluation. Morgan felt that systems tests, i.e., man-machine system experiments, could investigate inter-operator problems, in addition to basic research and nonexperimental field studies. Despite their difficulty and cost, he recommended to the Navy "tests in which new equipment and new systems are set up experimentally and evaluated. Only through such tests can we hope to discover experimentally any new principles of systems design and only through such tests can we be fairly certain of the performance of a system before it is formally installed in a ship."

## CONCLUSION

In summary, the views of a number of writers may in one way or another be helpful in orienting this book's man-machine system experiments to research on manned systems in general. It should be understood that manned system research does not encompass system aspects where people have no impact through operation, maintenance, or decision-making. Many types of engineering tests and mathematical analyses where human behavior is properly excluded do not fall into the category of man-machine system research. Yet in many tests conducted by engineers and in much operations research performed by mathematicians the object of inquiry is a manned system or subsystem. The techniques of man-machine system experimentation should interest these professions as well as psychologists.

It seems a pity that such experimentation generally has failed to make an appearance in texts on systems research by or for these other professions (although some human engineering research methodology occasionally has been included). It is also unfortunate that the literature relating such experimentation to other research techniques has been fragmentary, as the material in this chapter has shown. The lack of any comprehensive account indicates that empirical systems research is still in its youth.

# 25

# Objectives, Strategies, and Accomplishments

What have man-machine system experiments tried to do? How? What have they accomplished? The data in Chapters 3–22 have already answered these questions, but the heterogeneity of these data calls for some structuring and summarizing in this concluding chapter.

The domain will be structured by classifying research objectives in several ways. Some strategies for reaching these objectives will be noted in passing, and other strategies will be discussed in connection with research decisions and improving the cost-benefit ratio. Accomplishments will be related to objectives, followed by a brief look at the future of this kind of research.

## OBJECTIVES

One way to classify the objectives of man-machine system experiments is according to fields of knowledge. Another is to distinguish between the specificity and generality of the knowledge sought. Earlier chapters have referred to ad hoc experiments seeking knowledge pertinent to a particular system and experiments seeking more generalizable knowledge. A third classification is the distinction between the objectives of discovery and certainty, or between the processes of exploration and verification for achieving these. This distinction has also been mentioned in earlier chapters. It will become apparent that the differences between ad hoc and general knowledge objectives, and between the goals of discovery and certainty, are not as clearcut as the terms may sound. In a sense each pair marks the ends of a dimension or continuum. Nevertheless, they are convenient. It is hoped that these three schemes of classification will reduce the confusion that occurs when man-machine system experiments are discussed as though they all had the same objective.

Before going further it might be wise to take a look at the reasons why various man-machine system experiments came about in the first place. How did any objectives at all come to be adopted?

### *Origins*

In most instances the first experiment in a program of interrelated experiments, or one which stood alone, grew out of some real-world problem or problems. This was the case whenever the experiment sought an ad hoc objective, as would seem logical, but it was also true when the over-all objective was general knowledge. In the ad hoc cases the need for the particular experiment might be expressed by some external agency involved operationally or developmentally in the real-world problem. Alternatively, it might be deduced either by the experimenters themselves or by other individuals in the same research organization—or jointly. It was the researchers alone who determined the need for a general knowledge experiment. Such an experiment was related to the real-world situation through its context.

Subsequent experiments within a program might have the same kinds of origins, but many were inspired by a prior experiment which indicated the need for further experimentation. Programs of interrelated experiments seemingly have developed much in this fashion, rather than in accordance with a master plan. Some amount of planning may have been in the heads of the researchers, if only to the extent of one or two experiments beyond the current one, but such projections did not appear in the reports of the experiments.

(This last statement is subject to several qualifications. (1) Some documents with limited circulation may not have come to this author's attention. (2) The accounts of some very large experiments stated that further phases which had been planned had to be abandoned. (3) Other accounts indicated that certain lone experiments were meant to herald an extensive program. In fact, reports which projected a series of interrelated experiments that were never undertaken seem to have been the only long-range planning documents published!)

Why have research efforts ceased in spite of the self-perpetuating nature of programs of interrelated experiments? Although one experiment led to another, or additional variables came to be investigated in further studies, sooner or later the program terminated. Perhaps the lode it mined petered out, or so it seemed to the sponsor. Whatever the reason, to stay in business the research group had to stake out another claim with equal appeal, find financial sponsorship, and get management backing. In general knowledge experimentation this meant developing a new area. In ad hoc experimentation it meant involvement in a new system. This did not always happen.

In contrast, there have been long-lived programs of ad hoc experiments which were not interrelated but dealt with the same general technical area, such as air traffic control, forces on a battlefield, or logistics. New needs for experiments continued to emanate both from operational agencies and from the research organizations themselves. The three programs of man-machine system experiments characterized by the highest expenditures and greatest longevity have been programs of this nature.

### *Fields of Knowledge*

Davis and Behan (1962) were quoted in the last chapter as observing that a system experiment had one of three purposes: comparison, evaluation, or diagnosis. These terms can be interpreted variously. According to one view, if the experiment compared systems so a choice could be made, the principal independent variable consisted of the systems, each being one of its states. Evaluation revealed the capacities of a single system in relation to goals under various levels and types of load; input load was the major variable. In diagnostic experiments the independent variable might consist of system configuration, policies governing system functioning, or information flow through the system.

In discussing "Man-Machine System Evaluation," Taylor (1959) used the term "evaluation" to cover both comparative and diagnostic experimentation. He also included a limited set of equipment or a single equipment item under "system." He described a man-machine system experiment thus:

> It is comprised of comparative tests of experimental, prototype, or operational man-machine systems, where the emphasis is on the evaluation of the system relative to some other system or to itself modified in some way. Here, experimental methods are employed to determine how system variables affect performance. Often the results are specific to the devices employed, and frequently they do not permit scientific generalization.
>
> Because of this, and also because many of the evaluative studies are carried out under a security classification, little of this work is available in published form. Yet it is an important part of human engineering and it is often the most costly research, both in terms of money and manpower, carried out in this field.
>
> Evaluative studies have been performed on headphones, range-finders, gunsights, fire control and missile control systems, aircraft instruments, radar sets, information plotting systems, CIC lighting systems, target designation equipment, combat information centers, and airplane control rooms, to name but a few. In some instances, the tests are performed in the laboratory with system inputs being simulated; in other cases they are carried out in the field. But in both situations the attendant complexities and difficulties of control make this necessary variety of research as trying as any in which psychologists are likely to participate.

The distinction between "comparison" and "diagnostic" may have some advantages but it is easily blurred. A substantial number of experiments have made comparisons between systems. In pursuit of this goal, some of these tried simply to establish the capability of the current system as a baseline for subsequent comparison. That capability had not been adequately ascertained in the field by the organization operating it, at least not with the variety and range of inputs required to give a complete picture. When the current system was operated in the laboratory it was possible at times to improve it. Then the experiment could also be called "diagnostic."

Comparison experiments have frequently contrasted a current manual system and a proposed replacement or a number of replacements characterized by greater automation. In some instances this involved mechanization or analog computation but the principal contrast concerned digital computers. (However, not all new system proposals for introducing more automation have been put to such experimental test, an outstanding example being SAGE.) The difference between the systems being compared lay in their total design. Diagnostic experiments also made comparisons–between procedures, for example, or training techniques–but the over-all physical design of the system remained the same.

Rather than pursue these semantic interpretations of "comparison" and "diagnostic" further, experiments will be considered according to various fields of knowledge represented in their objectives. These coincide with some of the categories of independent variables listed in Chapter 2. They also help relate man-machine system experiments to the human factors fields of human engineering, procedure development, decision-making methods, training techniques, personnel requirements, and organization.

*Design.* As noted, the degree of system automaticity has been the theme of a number of system comparison experiments. New machines were introduced into man-machine interactions, including computers. When human engineering is viewed broadly as including the allocation of functions between men and machines, such experiments are seen as embracing this human factors field. These experiments contrasted operator stations consisting of consoles and displays representing the differing systems. In relatively few experiments was the optimization of a particular station or its elements a primary goal. Where it was, a more restricted kind of human engineering in the design of equipment focused on display content, coding, alternative displays, console configurations and capacities, and feedback for switch actions. Design improvements were more often subsidiary aims or by-products of the experiment, and the data were suggestive rather than conclusive. Similarly, as another human engineering venture, experimenters have collected some information about the capacities and limitations of individuals in performing various system tasks. This occurred more often when the system was a small one or the data could be collected by the system's own computer. But the determination of operator capacities and design of individual items of equipment have not been a main objective of most experiments described in this book. The reasons are clear. Individual actions are not easily controlled and manipulated in system experiments, and much within-equipment design is not easily varied. Such fields of knowledge are better investigated through other kinds of experiments.

*Procedures.* These cover a broad area. They are the rules to follow in performance. They are followed by computers faithfully, by human beings with varying rigor. Team tasks in systems possess them in great variety. They indicate who does what, how, and when, to whom and to what. Policies are composed of procedures and are guidelines for their formulation. (Akin to procedures are tactics, and similar to policies are doctrines.) Procedures often depend on equipment and its design. A procedure cannot be used which equipment renders impossible, and the equipment gives some procedures greater effectiveness than others. Many or indeed most procedures relate to the operation of equipment. Yet the same design, perhaps with only minor changes, permits procedural alternatives. For example, information flow in a system can be varied by applying different procedures. An experiment on information flow could be viewed as one with a procedural goal, although some redesign or rearrangement of equipment might be required as well. A system reconfiguration may consist of changing procedures and rearranging equipment; since such a rearrangement involves human engineering, procedure development and human engineering are closely linked. Experiments on system policies are experiments on aggregates of procedures. In man-machine system experiments procedures–and policies–have been varied as states of an independent variable to see which were the more effective. Other experiments have inquired how closely individuals would adhere to established procedures, including those in a computer program. Procedural flexibility has also been investigated. Some experiments have studied a system's generation of its own procedures as a consequence of its operations.

*Decision-making.* Procedural behavior and decision-making are closely related. Decision rules are procedures to follow where the situation is unambiguous; no decision-making is needed. But otherwise an individual at some apex in a man-machine system must decide whether the rule applies, or which procedure to follow, or what to do if none is available. Experiments which have studied acceptance or rejection of procedures are perhaps better regarded as decision-making experiments than as procedural behavior experiments. As in law, precedents can be viewed as prescribing procedures. Some man-machine system experiments have investigated whether one commander would accept the decision made by another, or even by himself on a previous occasion. Command decision-making has been somewhat arbitrarily divided into threat evaluation and resource allocation. Experiments have concentrated on one or the other, although they clearly interact. Including the interactions in an experiment—through feedback from resource allocation, for example—introduces the problem of uncontrolled reactivity. However, some experiments have required the subject to forecast the adversary's reaction to his own tactic. The effectiveness of Bayesian processing for making threat evaluations has been a prominent theme. Relatively few experiments have considered personality variables, staff organization and staff procedures, or the broader processes of which decision-making is a part: planning, negotiation, and problem-solving in a system context.

*Training Techniques.* This human factors field has received some attention in man-machine system experiments, though not as much as might be desired. In such experiments, the emphasis has properly been placed on training teams rather than individuals. Evaluation of techniques has occurred in several programs. Although one technique led to joint training exercises of many interacting units of an entire system, the technique's experimental evaluation was far more limited in scope. Evaluation did not progress to finding out which in a complex of methods was primarily responsible for system improvements, or how profitably the technique could be applied in systems other than air defense. Other experimental programs investigated the distribution of input load over time during training and the value of specific training in interactive performance of a task which involves interactions between operators. Further research on training interactive performance in team situations seems needed. As a base, there will have to be better definition of such performance.

*Personnel Requirements.* Any man-machine system requires that some number of personnel be allocated to each task, with some level of skill. In other words, the system has certain manning requirements. A few man-machine system experiments investigated team size directly, varying the number of operators assigned to a joint task. The team size was small even at its maximum, and team efforts were concentrated on a single function. In the discussion of system design and human engineering, it was pointed out that system experiments do not characteristically get data about individual operator performance. Yet when this has been possible, the information about capacities has helped determine the number of operators needed in the team and their skill levels. Few studies have investigated personality factors, whether in decision-making or other tasks. Yet more information about individual differences—in capacities and personalities—is needed for manning a system. A by-product of numerous man-machine system experiments has been the demonstration of substantial performance differences between teams and sometimes between individuals. Although such differences have usually been disregarded by researchers, they have important implications for improving systems through the way they are manned. Some aspects of performance capacity and personality may have to be studied in the kind of context available only in a man-machine system experiment.

*Organization.* Organization is a field overlapping with personnel requirements and procedural development. Team composition has been the target of a few of the experiments in this book. Team composition depends on the tasks assigned to the team and the number of individuals assigned to each task. Another aspect of organization is how organizations evolve and adapt when given demanding situations to handle. Such demands may be called "stress." Experiments have shown that organizations develop new procedures or change old ones to cope with difficult situations. Finding the best way to give an organization the opportunity to do this can be an interesting research objective. Other experiments have dealt with organizational structure. For example, a logistics organization was either centralized or decentralized. Procedures differed as well. Some of the experiments on communication netting and relaying might be viewed as studies of organizational arrangement.

### *Specificity-Generality*

Most man-machine system experiments should be placed in the ad hoc category. They diagnosed a particular system to improve it, evaluated a system to see how well that system performed, or compared two or more systems to determine which was better. The other experiments, the general knowledge variety, were not concerned with a particular system as such. Clearly the results would pertain to the system in which the experiment was emplaced, but the researchers were really looking for knowledge which would extend beyond it—although how far beyond it was not always clear. General knowledge experiments concentrated on decision-making, training techniques, and organizational adaptation. Ad hoc experiments embraced the knowledge fields of design, procedures and policies, personnel requirements, and organizational structure. This differentiation is only approximate, but it does suggest that researchers were a little uncertain whether general knowledge experiments could be organized for some of the fields.

Because generality is a relative matter, as stated earlier ad hoc and general knowledge can be thought of as occupying two ends of a dimension. Some ad hoc studies have had a wider application than others. For example, a study of a control procedure common to the entire air traffic control system yielded greater generality than an experiment about the airport arrangements for a particular city. Results of an experiment on a particular system might be generalized to the class of systems to which it belonged—if an acceptable taxonomy of systems were available. In any case, one approach to generality is to experiment on many systems. Generalizable statements can then be induced from the data yielded by many experiments.

Certainly it does not seem necessary to regard the objectives of specificity and generality as antagonistic. If generality might be derived from a number of ad hoc experiments, the results of a study with a general knowledge goal might be applied to a particular system. Such was the case, for instance, with the initial experimentation in the RAND Corporation's Systems Research Laboratory. The aim had been to find out how organizations adapt to demanding situations. But the research results led to a new training technique for the particular system which provided the context in which the research was conducted.

*Bounding the System.* Regardless of where the experiment lies along a specificity-generality dimension, one of the strategy questions which researchers have faced is bounding the system. How much of it should be represented in the laboratory by the experimental subjects and their equipment? How much by computer models or quasi subjects as other parts of the system or as other systems? How much by the simulation inputs? Finally, how much should be disregarded altogether?

Changes which have occurred during experimental programs testify to the significance of this problem. The system being manipulated in the laboratory has been enlarged or reduced before or even during an experiment. More often the projected scope has been changed as the program developed. Plans to increase the scope and complexity of the system were abandoned. In any case, the system in the laboratory usually was only part of the total system in the real world. For example, if the total system consisted of a number of similar units, only one or two of these would be operated in the laboratory. Since every system is part of a larger system, no total system can ever be fully represented. Bounding the system in the laboratory becomes a matter of research judgment. In considering those elements whose interactions could influence the experiment's outcome in some important fashion, the researcher must weigh the cost of allocating them among the laboratory's subjects and equipment against the cost and adequacy of simulating them through simulation inputs, quasi subjects, or computer models. Further, different approaches may be taken depending on where the experiment is found among the fields of knowledge. If it concerns training techniques, for example, the researcher may want to concentrate on nodal operator positions or on critical linkages between positions. But if he is conducting an experiment comparing system designs, the system boundaries will have to be wider, especially when positions or linkages do not correspond between systems.

As indicated in Chapter 2, in bounding the system the researcher may be tempted to restrict himself to those portions (such as computer data processing) which present the fewest obstacles to representation, either in laboratory operations or in simulation. In some cases, he may have little choice. The "effector" parts of a system, such as aircraft or other vehicles, cannot be directly represented in a closed-space laboratory. To include them, either the experiment must be conducted in the field, or they must be simulated symbolically. The same is true of sensor parts such as radars, the "front end" whose functioning—as Wolin

(1959) has observed—is so critical. Again as Chapter 2 pointed out, realistic symbolic simulation of either effectors or sensors may be difficult to achieve. Special-purpose equipment may be needed to provide the required inputs and feedback to the data processing portion of the system and to handle its outputs. Although this requirement is by-passed if the system in the laboratory consists only of the data processing portion, the experiment may then disregard some critical problems related to the sensor and effector elements.

If the system is one which is designed to deal with an adversary, the researchers must decide whether to bound the system in a way which includes that adversary. Most military command and control systems, for example, must cope with hostile forces. These forces constitute what has been called the "anti-system" (Parsons and Perry 1966). A case can be made for considering the anti-system as part of the command and control system, to compel concern about its effects on the command and control system in all aspects of the latter's development, test, training, and operations. A man-machine system experiment about a command and control system might be required to represent the anti-system not simply through simulation inputs. Experimental subjects, role-playing the adversary, would react to the command and control system's operations, which in turn would have to respond to their reactions. The difficulties created by reciprocal reactivity in experiments have been discussed in Chapter 2. Because of the obstacles to full experimental control, experiments incorporating the anti-system might have to be limited to those of the exploration type.

Chapter 2 also indicated that in bounding the system the researcher may want to represent operationally only one of its subsystems. In addition to the smaller investment required, a more complete picture of its functioning may be obtained if the subsystem does not have to depend for its inputs on outputs from another operating subsystem. On the other hand, in some situations it may be extremely difficult to simulate these with sufficient verisimilitude. It may be advisable to bound the system differently in successive experiments. The first, emphasizing discovery, would include two or more interacting subsystems. A following certainty-oriented experiment would be confined to a single subsystem. The "partitioning problem" is not a novel one. Ernst (1959) discussed it thus:

> The effective representation of even a relatively simple system may require substantial simulation capabilities. It becomes quite impractical, if not impossible, to represent a complex both *effectively* and *completely*. Since the representation *must be* effective, it becomes necessary to study the system piece-by-piece. However, by definition, a system contains no independent problems. Discerning acceptable lines of cleavage and making proper allowance for the effects of such cleavage constitute the Partitioning problem.

*System Phases*. During its life cycle a man-machine system goes through conceptual, developmental, and operational phases. Each of these has figured in ad hoc experiments.

The potential systems which have been tried out in the laboratory in their conceptual phase were those which for the most part could be simulated symbolically. These included a number of logistics and air traffic control systems.

The term "new system" must be interpreted here to consist essentially of a new policy or new set of procedures rather than new equipment. Obviously it would be impossible to place in an experiment any equipment which had not yet been designed. Except for computers, representing future equipment in a meaningful manner incurs difficulties. Although it may be assumed that consoles and displays in an operator station are an adequate forecast of what they would be, this is only an assumption. But a current computer may simulate a future one to a sufficient degree, and programs may be written to simulate future programs.

Experimenting during the development phase has had varying success. It has been possible to experiment with a prototype set of equipment either to compare this with the equipment it was designed to replace or to establish design parameters and try out new design features. Both a tight time schedule and firm organizational control over the equipment and experimentation have been necessary to do this effectively. When researchers have tried to bring systems of greater scope into the laboratory during their development, various problems blocked experimentation. For example, while the design of the system was still undergoing changes, it was difficult to keep the simulation and data collection current; as the system equipment in the laboratory became outdated, the laboratory's metasystem lost its congruence with the system. There could even be competition between experimenters and system developers (and testers) for simulation facilities and funding.

Another problem has been to get experimental results fast enough to influence the design process with respect to some of its increased automation or human engineering aspects. In addition, in fairly complex systems experimenters faced increasing difficulties in coping with the interdependencies among changing equipment design, changing computer programs, changing system procedures, and changing skill levels of the subjects. Since each new design of equipment or programs could call for new procedures, there seemed no point in testing the procedures until the new design was available in the laboratory. But that design had to be available to develop the procedures in the first place—and the design could not be tested without the appropriate procedures. Crews of subjects had to be trained to acquire necessary skill levels, yet it might prove impossible to retain them long enough.

System tests which resembled man-machine system experiments have usually been deferred until the system reached its operational phase. Then trained crews of actual user personnel could operate the system, and it would interface, as intended, with other systems. By this time, however, there would be great resistance to making any equipment changes diagnosed as advisable by the test or experiment; such changes would be exceptionally expensive and might delay deployment. Yet such diagnosis and improvement were really the only purpose of experimental testing. Evaluation to determine whether the system met required standards seemed to hold little value, since the system had already been bought and produced.

Once a system had become fully operational, its ongoing operations could provide much of the information needed for developing new procedures, modifying equipment, or changing computer programs. Yet considerable experimentation might still be useful. An experiment might be required, for example, to reveal a flaw which simply did not make itself apparent to the operational users, such as the lack of feedback about computer response to an operator's action. The operational system might have to be reproduced in the laboratory to establish a baseline of performance for comparison with a proposed replacement system. Experiments on a system in the operational phase might also evaluate new training methods for that system. Finally, an operational system might become the vehicle for a general knowledge experiment.

From this overview of experimentation and testing during the system life cycle, it should be apparent that the value and feasibility of conducting a man-machine system experiment during a particular phase depend on many factors, including the experiment's objectives. Caution is advised concerning simplistic concepts, such as viewing a system laboratory as a test bed for investigating a wide variety of systems throughout their life cycles.

*Generalizability.* There have been understandable aspirations among some researchers to conduct experiments which would yield generalizable knowledge about systems. A fair number of experiments have resulted, but fewer, as noted earlier, than ad hoc experiments.

Aspirations have at times taken the somewhat wistful form of a search for principles of systems in general in the sense of underlying relationships, or human factors principles, or principles of system design and development. Such expressions have characterized proposals for new laboratory facilities. Analyses of feasibility and strategy have been strikingly absent. But researchers may have expressed their ideas about feasibility in what they chose to investigate. It has already been observed that aside from communication channeling, those who initiated general knowledge experiments steered clear of system design and procedures as fields of knowledge. Rather, they showed interest in training techniques, organizational adaptation, and decision-making. Even in these fields, it has not been entirely clear how widely researchers expected their findings to generalize.

What strategies can be adopted for achieving generality? These would seem to concern the domain represented in the experiment. This is composed of subjects, inputs, and the system. In each case, one strategy might be to select an example which the researcher felt typified some range of instances in the real world. An alternative strategy would be to vary the examples over a range, within an experiment or over a series of experiments. To a degree, the problem is the same as achieving external validity, discussed in Chapter 2; it is larger, however, because the researcher has to achieve generality as well as validity.

Consider, for illustration, how to get generality in an experiment on the decision-making involved in evaluation of threat. First of all, the researcher must select subjects as decision-makers. If a single subject serves as decision-maker, how can the researcher be assured he represents decision-makers even in the kind of system being simulated, much less in other kinds of systems? Perhaps if the researcher could persuade a commander from an analogous, operating system to take part in the experiment, he would implement the first strategy—typical representation. But even this commander might not be typical of most. More likely the researcher would use a college student. How far can the results of the experiment be generalized? The second strategy would call for a number of college students, and later a number of actual commanders. Although stratified sampling or proportionate stratified sampling would be too much to expect, the diversity would increase the likelihood of representativeness. In the absence of this second strategy, the researcher would have to claim such universality for the effects of his independent variable that the representativeness of subjects was not a factor.

The second strategy has actually been pursued in some experiments. There have been teams of civilian experts, or a diversity of military teams. In one program initiated with a general knowledge objective, the subjects were college students at first, then military personnel in similar experiments. Some generality could be claimed for the results, which were similar from one experiment to the next.

The decision-maker who has to evaluate threat must base his decision on the simulation inputs he received. Should these be a single set that mirror a best estimate of actual threat? That would reflect the first strategy. But how good would that estimate be? Alternatively, according to the second strategy, the simulation inputs ought to reflect the variety of demands that might be imposed on the system. The inputs might include a systematic variation in load, some of the noise features which were discussed in Chapter 2, and a range of adversary reactions. This sampling could provide a better indication of how decision-making would occur in a variety of real-world situations. Again, generality would be served, as well as external validity. In decision-making experiments, the second strategy has been favored over the first, but some man-machine system experiments have been based on the first.

In what systems should the decision-maker make his decision? According to the first strategy, it could be a single system. But should this be an operating one, a discarded one, a system synthesized from parts of actual systems, or an entirely hypothetical system? Would a hypothetical system provide more generality? The researchers would escape particularity, and they might feel they could capture the essence of some set of systems in this fashion. But some configuration of sensors, communications, information processing functions, and displays would still have to be simulated. These would have to have particular features, somehow typical of actual systems. Hypothetical systems have been concocted for the purpose, but for some man-machine system experiments researchers have instead resorted to using an actual system whose particular features were already in existence. They seemed satisfied as long as they could get the maximum of the behavior they wanted out of the system. The current one seemed as typical as any, if not more so. Another program exploited a prototype system which had been discarded. Still another approach has been to modify a current system with some hypothetical features, such as displays.

A system has also been created from parts of similar systems so the subjects, who were familiar with some of the particular systems, would be confronted by one unfamiliar to them. The same rationale has been advanced for hypothetical systems, which have been advocated also on the methodological grounds that they can be designed to evoke more of the kinds of performance the researchers want to examine and to minimize other behavior. Appropriate design can make them easier to learn to operate or more meaningful to the subjects. It has been argued further that situations can be better standardized for repetition, and performance can be measured more accurately and completely.

If our second strategy were followed, general knowledge experiments would be replicated through a variety of systems. This would be done first among those that seemed similar to the initial system, and then among systems with greater divergence. In determining similarity the researchers would have to pay more attention to common characteristics than common labels. An example is command and control systems. Comprehensive surveys (e.g., Parsons and Perry 1966) have shown that those with this label are a fairly heterogeneous lot. It might be possible to vary the characteristics of a hypothetical system to create a number of similar systems. To a limited extent this has been done. How far researchers would want to proceed in experimenting on a diversity of systems concerning the same system phenomenon or independent variable would depend on how much generality they wanted to attach to system principles. In the past the strategy of replication across systems has not been noteworthy among general knowledge experiments.

A third strategy has approached system generality through abstraction. The system has been simplified by a reduction in detail. Some variables have been disregarded. Symbolic elements stand for aggregates of what would be encountered in the real world. When detail is lost, an abstract system sometimes tends to become a hypothetical one. Geometric patterns, such as checkerboards, represent a battle area, and markers represent parts of the system. Computer models aggregate elements into larger units. Complex communication nets are reduced to simple patterns. Interorganization dealings are epitomized in two-person dyads. The researcher hopes that the relationships between the aggregates will reflect corresponding relationships in the world to which he wishes to generalize. But he runs the risk that in the process of abstraction so much is lost that the relationships are not representative, or when expressed they are difficult to relate to reality. Abstract approaches to generality have to be validated. Would the same results occur if the system were presented in all its complex detail? How well does the abstracted system predict the performance of the real one?

The same sort of question must be asked when batteries of test tasks, such as those discussed in the previous chapter, are substituted for mission performance in experiments on confinement. The rationale for such batteries is that they represent certain essentials of human performance in system operations. But how good is that representation? How does it compare with simulating the requirements of an actual mission?

Closely related to seeking generality through abstraction is the use of analogies. Here the same kinds of human performance are called for as in the real world, but the inputs may be somewhat simplified and the task characterized in more general terms. For example, instead of radar signals, there are simply dots on a display which has only some of the characteristics of a plan position indicator. Four of five operators linked together by telephone can engage in viewing, reporting, receiving, plotting, comparing, and converting data much as they would in a radar site. Can this arrangement provide more generality about information processing by operator teams than a crew in an actual radar site, or is it merely a convenient substitution? If it is only such a substitution, it may be the basis for capabilities in experimentation that would otherwise be difficult to create.

The use of analogy is related to abstraction through the concept of common properties. If two or more systems have certain characteristics in common, and if these are the key characteristics, then it may be possible both to generalize from one to the other and to strip one of them of other characteristics as extraneous detail. This reduced system can be the locus of experiments to generalize to the other, which may be a current or future operating system. But the burden of proof lies on the researchers to justify (in writing) both their selection of joint key characteristics and the elimination of detail. This may not be easy. The composite man-machine nature of system and subsystem performance, described in Chapter 2 in the discussion of measurement, compounds the problem, as it must be considered in selecting key common properties.

*General Purpose Laboratories.* The urge to achieve generality has led to proposing general purpose laboratories for man-machine system experiments, and some have been established. Perhaps the most charitable comment that can be made about these is that they have not fulfilled their promise. After a single experiment, which may not have been a particularly impressive one, the laboratory as originally conceived ceased operations. The general purpose facility was converted through subdivision to the support of other needs, such as experiments on a more modest scale, all-computer simulations, development of computer programs, and office space. Why has this happened?

One reason has been the inversion, mentioned in Chapter 2, in the reasons for creating the laboratory. The facility was established first, and then the management attempted to develop the research which would make use of it. But programs of large-scale man-machine system experiments have not appeared simply because a facility was there waiting for them. According to the record, significant experiments have sprung from problems and ideas. The facility has been created to support the projected research. Apparently some challenging theme or abiding set of problems must exist first for a man-machine system research laboratory to prosper.

It seems likely that when much effort is first devoted to building the facility, too little is directed at acquiring the aggregates of brains and imagination needed to envision, plan, and conduct experiments. A few gifted individuals have struggled to live up to the fanfare marking the laboratory's dedication, but scientific creativity cannot thrive on bricks and mortar alone.

A related reason for the short life spans of general purpose laboratories has been confusion and vacillation on the part of the research and development organization's management in specifying the facility's aims. Was it in fact created to demonstrate the organization's capability and thereby get new business, or was its purpose to find answers to important questions through research? Although the researchers tried to use it for research, was the management most influenced by the facility's potential impact as a symbol of competence and prestige? Did the management understand the distinction between ad hoc and general knowledge objectives and the nature of the various fields of knowledge related to systems? Were the laboratories proposed for amorphous investigations of systems, rather than for studying various knowledge fields? Finally, was more interest concentrated on the techniques of simulation than on the experimentation which might incorporate them? In these questions can be found the problems of the general purpose laboratories.

Further questions must be asked about their feasibility. Versatility, flexibility, and expandability are desirable for a laboratory, along with reliability and reasonable economy. But how far can they extend? What are such a laboratory's limitations? How well-equipped can a general purpose laboratory be to investigate a real system, or a variety of these, especially in ad hoc experiments?

The principal requirement for a general purpose laboratory is a general purpose computer, although this may be used also for other purposes and may not even be regarded as a component of the laboratory. Computers differ in their utility for on-line operations in an experiment according to such features as direct storage capacity and processing speed or multiprocessing capability. They need buffer equipment—another, smaller computer or special hardware built for the purpose; its capacities vary according to design. Some limitations on the laboratory's generality of purpose can come from the computer and its buffering equipment. But the major limitations are found in peripheral apparatus for introducing inputs into the computer and for receiving outputs.

Consider the system being represented. If in that system the computer receives data from elsewhere in the system or from another system and if it transmits data elsewhere, in the laboratory the peripheral apparatus for transducing inputs and outputs must be specific to that system, as pointed out by Ernst (1959). It must either be that system's equipment or simulate it. The general purpose laboratory is then no longer so general in purpose.

Less constraint is imposed if the input and output equipment is limited to what is used by operators in the real system's data processing portion. This consists of control panels and displays. How "general" can these be? To the extent that the functions of panel switches and the contents of displays can be determined by the programming of the computer, considerable flexibility exists. Programming or reprogramming does require time and money but may call for less than redesigning and constructing hardware. However, some aspects of control panels and displays are built into the hardware. The laboratory's general purpose nature is diminished according to the extent to which these influence system operations by constraining human performance. The control panels and displays should be analyzed to find out how widely they can be generalized.

How well can a general purpose laboratory support experiments with general knowledge objectives? This may depend on the inventiveness of researchers. If the strategies of creating hypothetical or analogical systems are effective, they can be put to work in such laboratories. But it should be realized that a general purpose laboratory based on a general purpose computer would normally represent only the data processing portion of a computer-based system. If other portions of the system or other kinds of systems must be represented, this may be possible only by analogical simulation. If researchers wish to simulate the inputs from these, they may encounter the difficulties of noise and transformations described in Chapter 2.

A general purpose laboratory is expensive to keep operating. It requires staffing, programming, and preparation of simulation inputs. Means of limiting costs and increasing benefits are discussed later in this chapter. Although a high occupancy rate is desirable to reduce the prorated cost of the facility per experiment, this can greatly increase the total costs. Further, it is not clear how high the occupancy rate can be, but the limit is probably well below past aspirations. Whatever is projected may be colored by wishful thinking. When the concept of a general purpose laboratory is voiced, any proposal should carry with it the evidence that such a facility should be built. Such evidence should include estimates of upkeep and productivity, and assurance of long-term support.

*Discovery-Certainty*

The objectives of discovery and certainty are accomplished through the processes of exploration and verification, respectively. In either pair the terms can be thought to signify the ends of a dimension, like ad hoc and general knowledge. Increasing emphasis is placed on verification processes as one moves closer to complete certainty—a point never reached. Certainty is relative. Some must be achieved during exploration to warrant the announcement of a discovery. Conversely, discovery may occur in an experiment oriented toward certainty.

This classification of objectives, like the preceding one, gives structure to man-machine system experiments. Most of these have been directed at certainty, the goal most commonly attributed to experimentation. But in some the researchers were explorers, a role equally praiseworthy in research, although its methods have been less developed and its advocacy less eloquent.

The exploration-verification classification relates objectives to methodology. Chapter 2 gave considerable attention to methods of achieving both internal and external validity. These are methods of verification to achieve certainty. They concern certainty about outcomes within the experiment's framework and certainty about their representation of the domain investigated. The methods are ways of handling both independent and dependent variables. The degree of certainty rests on both experimental design and measurement.

The concept of this classification is not novel. Rauner and Steger (1961*b*) urged the joint pursuit of discovery and certainty when they discussed the experiments of the RAND Corporation's Logistics Systems Laboratory. They said that these studies

> are designed to permit and encourage the generation of ideas and problem-solving heuristics while, at the same time, keeping the variables under sufficient identification and control that quantitative analysis is still possible. . . . It is evident that obtaining the right balance of the two characteristics of experimental heuristics and quantitative analysis is a delicate and difficult matter. . . . As with the question of the level of detail and scope of the overall model, however, we have no final standard to which we can refer for guidance as to the ideal amount of these two opposing elements to build into any given study.

The term "heuristic" has been defined by dictionaries generally as "serving to discover" and "stimulating investigation." Its growing use in the data-processing community has arisen, it would seem, from the need to contrast processes which create some degree of understanding with those which embody specific problem-solving procedures. The latter are called "algorithms." Precise operations in a computer program are algorithms. As we saw in the preceding chapter, Sackman (1967) also generalized from the computer usage, suggesting that heuristics could be found in the real world, algorithms in the laboratory. It is true that exploration tends to occur more often outside the laboratory, verification in it. If the reader prefers more old-fashioned terminology, he may be happy with "suggestive" and "conclusive."

As the preceding chapter also brought out, Chapman (1965) distinguished between studies which searched for hypotheses or variables and those which tested them. He had touched on the same theme earlier (Chapman 1961*b*). His distinction between mapping a domain and applying rigor is really the same as that between exploration and verification. His apparent preference for mapping a domain was suggested in his contrast (1961*b*) between "the impatient search for an immediate answer" and "the unhurried exploration of the domain of truth." Chapman also made it clear that some man-machine system experiments neither sought nor tested hypotheses, a point made in Chapter 2 in the discussion of reasons for conducting an experiment. Alternative states of a variable, such as alternative systems, may be compared without a hypothesis that one or the other is superior, or even that a difference exists between their performances. The requirement for certainty is just as strong, however, as it would be if the comparison were stated in the form of a hypothesis. When an experiment is purely descriptive, the need for certainty still exists, but the emphasis is placed on the process of measurement rather than on both measurement and experimental design.

In discussing inter-nation games as a mode of research in the social sciences, Snyder (1963) wrote that "quasi-experimental exploration" permitted a "decrease in rigor" in "semi-controlled exercises in contrived situations." Such studies, he said, "belong in the discovery phase of science-building, not in the verification phase . . . concentration on the discovery potential or heuristic values of experimental devices like simulation or gaming frees us from the strictures of Mill's Canons, which properly concern verification. . . . One is led to different implications if one regards simulation as a flexible mode of discovery and clarification rather than as a mode of rigorous test or validation."

An interesting analogue is the administration of criminal justice in the English tradition. Following a crime there is first an investigation phase, in which tips, hunches, presumptions based on prior experience, and evidence (data) lead to an arrest or indictment. Arrest is supposed to be based on adequate grounds, indictment on a prima facie case. The trial phase follows. It seeks certainty about the guilt of the indicted or arrested individual. Now the analogy weakens. A trial's methodology differs from that in experiments. It is based on adversary proceedings rather than the counteractions of good experimental design and the methods of collecting and measuring data outlined in Chapter 2. Nonetheless, an effective attorney may try to show that some of these counteractions were necessary but not invoked, or that the data are unreliable. Outcomes may be only relatively certain; a verdict may hang on reasonable doubt.

*Man-Machine System Experiments.* Although the discovery-certainty classification applies to research in general, it has special implications for man-machine system experiments. One has already been suggested in the quotation from Snyder. Like games, such experiments employ simulation. Simulation permits a great deal of exploration. In most experimentation on human behavior the experimental situation excludes the environmental richness and variety that exist in the real world. Such experimentation has seemed a poor method of exploration to ecological psychologists interested in complex environmental effects on behavior. But input simulation, even though it has to leave some gaps, can provide much of the real world's complexity. In fact, it can often do better. It can furnish a full range of situations. It can present rare and future events. Thanks to input simulation man-machine system experiments are well-equipped to serve as vehicles for exploration and discovery.

At the same time, input simulation can be arranged to preclude unforeseen and unscheduled stimulation from the environment that would jeopardize the experiment's internal validity. It also makes possible the systematic repetition of the same situations. Thus, by aiding the counteractions of preclusion and replication it serves the objective of certainty. It supports verification, even during exploration.

Much experimentation is based on data from preceding experiments or on theories induced from such data. Man-machine system experiments of the general knowledge variety constitute too new an area of research to be able to profit in the same way from prior research. They need more exploration than other kinds. An ad hoc experiment may have even less background, since it is often investigating a new system or a new feature which has just been designed. An exploration experiment tries to find out where trouble exists. A verification experiment seeks the best or a better way to remedy it.

Man-machine system experiments may encounter more obstacles than most experimental research in achieving satisfactory verification, for reasons indicated in Chapter 2. Systems have a multiplicity of independent variables, states of variables, and interactions between them. Due to the newness of the area, it is difficult to identify all the pertinent variables and to know the relative importance of those identified. Often it is uncertain, as in other kinds of experiments, which states of a variable to introduce. How, then, does the experimenter know what he should seek certainty about? The field also lacks enough prior research to give the experimenter assurance as to which data to collect and which measures to use.

The cost or urgency of experimentation may make it difficult or impossible to include all the assurance tactics the researcher might want to bring to bear during the verification process. Which can he curtail? How much certainty is enough? Cost or urgency may also make it advisable to incorporate both exploration and verification in the same experiment.

When exploration is conducted as a separate experiment, the usual case, the very nature of systems as large and complex contexts tends to make the experiment an extensive one. The pilot study of other research is dwarfed by comparison. Because of the cost of experimentation, the researcher will want to get as much discovery out of the exploration experiment as possible.

*Nature of Exploration.* What characterizes the exploration process? It is necessary first of all to ask what kinds of things the researcher hopes to discover.

It is not enough to say he hopes some new phenomenon will emerge or wishes to ask questions of the data. What kind of phenomenon is he looking for? What kinds of questions should he ask?

The answer may lie at the descriptive level. Some aspect of system functioning may become apparent which hitherto attracted little attention; its interest may be due simply to its resemblance to other aspects—the relationship of classification. The data may suggest some new measure which is associated with another measure. At the analytic level the researcher is trying to discover some new independent variable whose variations can differentially affect system performance, or some new dependent variable which will reflect the influence of independent variables, old or new. Since in either case he is interested in uncovering something pertinent to causal relationships, it is safe to say he is hoping to find new causal connections. Inherent in the exploration for independent variables is the determination of which states of a variable merit further research. After all, if no differential effects come from any variation in these, the independent variable is hardly worth worrying about and may not even be worthy of its name. Much exploratory experimentation looks for the states of an independent variable to incorporate subsequently in a verification experiment. Similarly, inherent in the exploration for dependent variables is the determination of which measures should be used later in a verification experiment embodying a particular dependent variable.

Why is it possible to discover new independent variables and states? In man-machine systems these involve new units of the performing entity. Since the performing entity is the aggregate of machines and men, a state of an independent variable has the same bounds as that aggregate, be it system or subsystem. The units are no longer individual machines and individual persons. System variables describe the entire system, or at least substantial parts of it. Different methods of information transfer, for example, can extend throughout the system. The degree of automation can be a crucial system aspect. It is the system which interacts with other systems, receiving information and feedback. Policies and the procedures of which they are composed may be systemwide. The entire organization of people in a system may adapt to new demands placed upon it. A new method of training may call for the participation of all the system, or at least an entire subsystem, in exercises. The new boundaries for units of manipulation open the doors for the discovery of new independent variables and the exploration of their states. In view of the diversity of man-machine systems, the discovery potential may be great indeed.

Along much the same line the dependent variables to be explored are the performances of systems or subsystems, not of individual machines or human beings. Measures indicate the production of the entire aggregate, the system, or the relative achievement of the system's objective. Other dependent variables which go beyond individual performance include the interactive performances between subsystems or between people. Thus there are new measures to be discovered in the data because there are new units of performance.

In short, the development of man-machine systems has made possible the discovery of more molar variables than have characterized experiments on individuals. To some extent the shift from molecular to molar units has occurred also in the presentation and manipulation of the environment. In studies of individual behavior, stimuli have customarily been few and discrete. The influence of large environmental aggregates or configurations (other than groups of people) has been examined by ecological psychologists and engineering psychologists interested in unusual environments, but by few others. In man-machine system experiments, however, the environment has often been broadly represented. True, its representation in inputs has usually concentrated on collections of particular elements, such as aircraft, which have been simulated. But some experiments have been based on the real air environment, containing actual aircraft, and others have used actual terrain. In any case, environmental variables have been introduced and manipulated as large complex sets of stimuli, in contrast to a single, simple stimulus serving to signal some response from an individual subject.

The identification of sets of influential stimuli and the selection of important variations in them can be regarded as among the aims of exploration in man-machine system experimentation. In other words, the researcher tries to discover what inputs to give the system he is investigating. Since these inputs are what drive the system, their discovery for variation and manipulation may be desired for descriptive as well as analytic experiments.

One way to look at exploration in man-machine system experiments is to compare it with other kinds of research. In controlled observation the researcher is passive. He systematically records the activities he can observe in the real world, possibly only in samples or with emphasis on certain aspects, such as linkages. He usually quantifies the data he gathers. But he does not arrange or control what occurs.

As Chapter 23 indicated, another vehicle of research—gaming—generally follows a scenario and uses simulation, but this may lack great detail. Referees rather than rules can prescribe courses of action in contingency situations. Since two or more sides react to each other in unpredicted ways, the game cannot be repeated with each side getting the same inputs. Thus, the researcher has only limited control over what occurs. Outcomes are likely to be stated as subjective judgments rather than as quantified data. Independent variables and their states are usually neither manipulated nor identified, and in any case assurance tactics receive little emphasis. As a result, any statements about relationships between states of variables are phrased as insights rather than conclusions. Since those who achieve the insights are the participants, a game is a teaching as well as a research device. In fact, it may be mostly pedagogical. Exercises resemble games in numerous ways, except that real instead of simulated forces operate in a real rather than a simulated environment.

The process of exploration in experiments bears a close resemblance to gaming and exercises. This is especially the case in experiments which do not manipulate system variables but rather expect them to emerge as a result of the inputs—which may be manipulated for the purpose. Games (and exercises) differ methodologically from man-machine system experiments mostly in the extent to which the contents are controlled and repeatable and the degree to which results are objectively obtained and quantified. If an experiment and a game agree in these aspects, the labels could be exchanged.

There seem to be two strategies that can be followed in experiments oriented primarily toward exploration. In one, the researcher abstains from structuring in the hope that something will emerge. In the other, he tries things out in the spirit of "What if I did this?" With the emergence strategy he hopes to detect a new independent variable of importance, and he collects reams of data which may suggest to him a new measure. With the try-out strategy, he introduces various states of an independent variable to determine if differences between them lead to differences in system performance. He also adapts data collection to meet the needs of possible measures he has in mind before the experiment starts. The two strategies can be combined by starting with the first and then moving to the second to try out what emerged in the first.

More planning is required by the second strategy than by the first, but less planning in general is needed for exploration than for verification. Exploration is marked by flexibility in directing the course of the experiment while it is in progress.

Greater flexibility is achievable because exploration does not call as much as verification for the counteractions which help assure certainty. This point has already been made in defining the discovery-certainty distinction. Yet exploration, particularly of the try-out type, does not mean the complete sacrifice of certainty, another point made earlier. Experiments still incorporate counteractions, even if they are invoked with less vigor and rigor. It is also important to understand that diminution of these assurance tactics in an experiment does not in itself justify calling the experiment exploratory. This leads one to ask what helps make an exploration experiment effective.

*Aids to Emergence Exploration.* There has been far less development in the methodology of exploration than in the methodology of verification, as the literature on experimental method attests. The methodology of verification received exclusive attention in Chapter 2. Here the best that can be done is to indicate some of the strategies which may aid the exploration process.

To support "emergence" exploration, the experimenter apparently must refrain from intervention within specified domains. For example, in presenting inputs he might simply simulate situations rather than designate clearly defined alternatives or options for decision-making. This would mean that the subjects themselves would have to identify the decision points. Although in many cases they might fail to do so, the experiment could become a probe of system problem-solving as a process rather than a highly structured investigation of the decision-making portion of that process. As another example, the subjects might be given only limited instructions about the procedures they should use in handling the demands inherent in the inputs. If the behavior of the subjects was thus left relatively unstructured, the procedures that emerged might be unpredictably appropriate to the particular system being investigated. If the experiment was oriented to the general knowledge objective, it might yield knowledge about the processes of procedurization and adaptation in systems.

It should be evident, however, that if the inputs are not stratified or organized to indicate requirements, they still should be bounded and replicable. Similarly, although interactions between subjects as system operators can take many forms, all these forms would ostensibly be serving the known aims of the system; they, too, would thereby be bounded. Although the same procedures might not emerge in experimental repetitions, divergences would be limited because of the co-operative demands of the system tasks.

A phenomenon does not emerge into view unless a researcher detects it. What factors facilitate that detection? Although these are difficult to specify, they must include the experience of the researcher in conducting system research, his familiarity with the particular system and its operations, and the availability of data from similar experiments. A number of qualified researchers should be involved together in the detection process, not only to increase the number of possible detection sources but also to stimulate each other through discussion. The researchers should observe all the experimental sessions carefully to try to get impressions and insights which might be transformed into statements about a new variable.

The likelihood that something worth detection will occur may be increased by varying the demands on the system. Inputs can include very heavy or very light loads, rare circumstances and events, and situations which call for non-routine performance. One result can be such variations in performance that the data will suggest new measures. Observations of critical incidents may also suggest these. The researchers must keep reviewing the data and applying measures which might have relevance.

A more systematic technique favored in other fields of research is the use of correlation statistics to determine whether a significant association exists between two or more variables. In man-machine system research this technique can show the degree of association between two or more measures of system performance. Thereby the value of some component or intermediate measure can be tested for predicting total or final performance. But an even more interesting application is to ascertain the associations between input or system variables and performance measures. In correlating performance measures with each other, there is no presumption that one causes the other. In correlating them with input or system variables there may exist some uncertainty whether the particular variable being correlated was the one responsible for the differences in performance. In the absence of an experimental design some other co-varying but unmeasured input variable or system variable might be responsible. The same uncertainty arises in the case of ex post facto experiments, in which the independent variable and its states are defined subsequent to the set of events—such as an exercise—in which they were situated. It would be extraordinary if all other variables which could have influenced the results had been precluded or held constant; and it would not be feasible to achieve equivalence among these through some procedure of matching.

When correlation statistics indicate a substantial and statistically significant association, the researcher has discovered something of apparent importance. Correlation statistics give quantitative expression to subjective impressions that relationships exist, or even reveal a relationship which had been completely obscure. The "something" may be either an independent variable or the measure of a dependent variable. Either can be subsequently introduced into another experiment aiming at certainty.

One of the advantages of correlation statistics is that they can be applied as an afterthought. It is also possible to set up an investigation in which the use of such statistics is planned in advance as the analysis tool. The counteraction of preclusion can be designed into the investigation. Whether this should be called an experiment is largely a question of semantic bias. It does not present different discrete states of different variables. But a number of variables can be defined, including input variables. Their values are not categorized according to a small number of states. Rather, the values of each variable extend over a range, as do the values of any measure of a dependent variable. The extent of association can be expressed in a correlation coefficient. Of more interest in the present discussion, by the technique of regression analysis the co-variation of an independent variable and a dependent variable can be expressed in an equation. Multiple regression analysis relates a number of variables to some dependent variable and by means of factor analysis groupings among these independent variables can be identified as having similar associations with the dependent variable. Each assembly of similar variables is called a factor, which receives an identifying label from the researcher on the basis of his understanding of what they have in common.

The advantages of multiple regression analysis and factor analysis in man-machine system research have been set forth by Sackman (1967) and Sackman and Munson (1964); their use of these techniques has been described in Chapter 11. The data for the techniques are easily obtained in computer-based systems from the system's own computer. It can record what it receives, its own processing operations, and its output. If the dependent variable in an investigation is some aspect of the computer's operations, such as processing time, it is possible to relate this to various kinds of inputs and operations. All values of variables are known, and hopefully no unidentified variables intrude. Processing time can be predicted in terms of the inputs and operations in the investigation. The regenerative recording of these permits replays. In principle, at least, new values of input and operation variables can be programmed as mutations, although the programming effort for some of these may be exorbitant.

When correlation statistics are carried to this point, the researcher may feel there is little to be gained by proceeding further to a verification experiment. But the question may be debatable. Factor analysts and experimenters do not always agree on the relative merits of their approaches. This is not the place to settle their argument.

*Aids for Try-out Exploration.* Other aids may be required for the try-out strategy in exploration experiments. This strategy implies introducing many independent variables, each with a number of states, or many states for one or more independent variables. Each variable or each state should get some exposure. How should this be done? Within some fixed constraints of cost and time, the more variables or states explored, the more the counteractions against confounding must be curtailed. This means that experimenter judgment must be exercised in selecting variables and their states. The researcher may have organized his domain by setting down all the variables he could think of, and the full range of states for each. But he must review the variables to exclude those which are manifestly trivial though discriminable. This, of course, is a matter of judgment. In selecting the states of continuously varying variables, he will probably want to include those which represent the ends of the range and enough points in between to discover whether the states generate a curvilinear function and what its characteristics are. Later, in a certainty experiment, he can reduce the number of states to those sufficient to describe that function.

In the trade-off between variables or states and counteractions, which of the counteractions can be curtailed? There seem to be no certain answers, but some strategy considerations can be noted.

If the exploration is aimed at trying out states, the counteraction of contrast will be necessarily included in the experiment. If it is trying out variables, contrast may be omitted in the case of two-state variables for which the absence of the variable is the zero state. In other words, the familiar control condition or group required to test what would happen if the variable were not introduced at all may be deferred to a subsequent certainty experiment. But this omission can be justified, if it can be at all, only if (1) a certainty experiment is assured before any favorable results are accepted, or (2) there would be some value in simply finding out how the system would perform with the new variable.

Refinement is a prime candidate for curtailment. A pair or set of independent variables can be varied as such in the exploration experiment—for example, a procedure which consists of a number of subprocedures, or a policy made up of a number of procedures—and the components isolated in the certainty experiment later on. Curtailed refinement means less manipulation of different variables.

Randomizing of crews is so difficult to implement anyway in man-machine system experiments that its curtailment is hardly at issue. If it were, it could probably be a counteraction to sacrifice in exploration. Orthogonality can be curtailed to an acceptable extent by means of Latin and Graeco-Latin squares and fractional factorial designs, which have been discussed earlier, thus reducing the experiment's duration.

Replication, constancy, equivalence, preclusion, and counterbalancing have several aspects in common. They can be designed without much difficulty into the input simulation, so they may not have to be curtailed for input variables. They can also vary in extent. The number of repetitions of unique conditions can be as many as the experimenter thinks desirable. He may sacrifice most repetition in exploration, although some replication remains desirable. Inputs aside, constancy and equivalence largely concern equipment and subjects, one important variable being level of skill through training or prior experience. Although some curtailment may be necessary, experimenters should be most cautious regarding the diminution of these counteractions. Preclusion should also be curtailed only when essential. Although it is deliberately foregone with regard to some aspects of emergence exploration, the same rationale by no means applies to the try-out variety. Careless diminution of preclusion is a well-established curse of try-out exploration experiments. Circumstances may necessitate modifications of counterbalancing, but these are permissible only within a modular framework for organizing the experiment. Counterbalancing should be maintained within a module and sacrificed only between modules, as in a serially designed experiment which seeks steady states between modules.

A facet shared by constancy, equivalence, preclusion, and counterbalancing is that they do not in themselves increase the scope and thus the cost and time of the experiment as much as some of the other counteractions. They do require a certain amount of helpful advance planning. In modular experiments this is at least the planning of a module. Sequential planning within the experiment should occur only from module to module, not within one. Among other benefits, advance planning helps organize computer usage and subjects' employment.

Again it must be emphasized that if counteractions are relaxed too much the researcher can have little confidence that something has been discovered when results do suggest effects of variables or of differences between nonzero states of variables. Nor can he have much confidence there was nothing to discover when the data showed no apparent effects. The researcher must achieve enough assurance to decide whether or not a certainty experiment is called for. This is what distinguishes legitimate exploration from the spurious claims of cranks, kooks, and quacks.

The legitimacy of an exploration experiment is heightened if it is labeled as such. This is probably more painful to overpretentious investigators than to more conscientious ones. The latter may have to resist considerable pressure to inflate the certainty of their findings. The report of an experiment, however, gives a clue to its legitimacy.

Because in exploration try-out experiments some of the emphasis on counteractions does have to be curtailed, there is always the danger that the experiment will degenerate into disorder and just "fooling around." This can happen also, the record shows, as a result of experimenters' naivete or circumstances beyond their control. There was so little to say about the outcome that nothing was published; or if something was published, its distribution was limited and it received little critical scrutiny. The degree of success of a discovery experiment of any type is reflected in what is written about its results by the experimenters. They must have discovered something which can be reduced to words and numbers for communication to a peer audience. The findings must be such that they can be aggregated and summarized in some fashion to make them understandable to professionals who lack the experimenters' detailed and comprehensive knowledge about the system and experiment. The test is succinct communication of objective data, not just greater understanding on the part of the experimenters as expressed in "gut feelings" or sharing in some mystique.

It is certainly permissible, in fact advisable, to include any personal insights gained from the experiment in the report describing it. But the experimenter should also indicate what gave rise to the insights. Thereby the reader can evaluate an insight better, and the researcher may become more self-critical and less eager to offer an insight as a conclusion.

Another requirement in reporting an exploration experiment is to make explicit the limitations in its design. This will assist the reader to determine how much assurance to place in the results. It will also indicate whether poor design arose from ignorance, carelessness, or unavoidable circumstances.

Technical managers of man-machine research programs have an obligation to seek help from people knowledgeable about experimental method and behavioral science. By "help" is meant more than casual advice. There have been notable instances where programs have benefited from such assistance and probably could have benefited more if help had come earlier and been more substantial. Managements have the responsibility to make sure that experiments do not turn into demonstrations for visitors but instead yield reports of findings, even if these are tentative. They must realize that safeguards should accompany try-out experiments.

Further aid to such experiments can come from de-emphasizing their external validity; its consideration may be left to a subsequent certainty experiment. For example, in the exploration experiment there may be only a single crew, or a single decision-maker, with no assurance that either is representative of the population of crews or decision-makers for the system. This approach reduces the number of repetitions of other experimental conditions that would be required to combine them with different subjects; within an experiment of a fixed length more time can be devoted to introducing variables and their states.

A related economy strategy, in exploration experiments of the ad hoc type, is to limit subjects or inputs to a certain type. From one viewpoint this should be the most favorable type. That is, the subjects should be experts, and the inputs should lack difficult features such as noise. If the results for a proposed improvement are negative, it should abandoned; if positive, a certainty experiment should examine the innovation further with subjects more like the operators of the real system, and with difficult inputs that the real system will encounter. This strategy has some appeal within the larger one of first exploring, then verifying, but it can be disastrous otherwise. If no subsequent certainty experiment takes place—and there might well be pressure from the hopeful to forego one—an unfortunate innovation may be accepted because of positive results in the exploration experiment.

From the opposing viewpoint, the subjects should be as inexpert as the potential operators, and inputs should include noise and other unwelcome aspects if these are to be expected in the real world. If the results are negative, the innovation should be shelved (or improved); if positive, the follow-up certainty experiment, while desirable, is not crucial. Presumably the choice between strategies should depend on the amount of wishful thinking among the developers and the likelihood of sticking to the two-experiment approach.

The last aid to try-out exploration to be mentioned here is the use of alternative methods of simulation. One, of course, is all-computer simulation, which can explore a larger number of variables and variable states than man-machine simulation. Its value, it has been noted, resides particularly in the ability to demonstrate how a great many variables, each with a variety of states, affect each other. Its disadvantage lies in the inadequacy or uncertainty with which human performance—and much machine performance—can be represented. Since this is a question of relative certainty, there is a persuasive logic in using all-computer simulation in exploration and man-machine simulation in verification.

Although programming for all-computer simulation entails cost and time, it can be cheaper and quicker than preparing for a man-machine system experiment in exploration; and the data-taking is much more economical. Similarly more economical is graphical or schematic simulation, in which the system is represented diagrammatically in ways which permit humans to simulate its operation. This simulation, discussed in the preceding chapter, admits human subjects into an experiment, although it cannot register the time their activities would take in operating real equipment or the frequency of certain kinds of errors. All-computer simulation and graphical simulation might be combined in some fashion for exploratory work.

*Aids to Verification.* It does not seem necessary to comment at this point on the process of verification to the extent that exploration has been described. It is essentially no different in man-machine system experiments than in other kinds, simply more difficult to carry out; it has been amply described elsewhere. Verification means making as certain as possible that a difference between two results really stems from the difference between two states of a variable in the experiment. It also means that the same difference would occur in the real world.

Why verification is more difficult in man-machine system experiments has already been discussed. How can all the states of all the variables pertinent to complex system functioning be included in an experiment? How can confounding and contamination be prevented?

Varying viewpoints have been held about the extent to which variables can be specifically and systematically addressed. For example, Chapman (1960*b*) commented thus concerning the plethora of variables: "At least in the design phase, there is the possibility of modifying the number, kinds, and skills of personnel, training methods, machine characteristics and numbers, communication patterns, procedures and programs. To attain the comfort of the systematic variation of all these variables in a factorial design is utterly preposterous. At what point, then, can one say that the prediction is valid?"

Haythorn (1963*b*) felt it was both desirable and possible to design a very large number of independent variables into an experiment. He observed:

> One of the problems facing investigators in systems research is the fact that any real-world information system contains a very large number of important variables. In the early days of our research it seemed that all one could do in the face of this complexity was to represent, as nearly as he could, an existing or contemplated system and assess its performance under anticipated environmental conditions. While this approach allowed ready applicability of results to specific systems under specified conditions, it did not allow generalization to conditions that were not represented in the study. It was considered desirable to attempt to parameterize important aspects of this system in order to exercise better experimental control, to provide a better ability to extrapolate to conditions not represented in the experiment, and to provide the beginning of a science of information systems. . . .
>
> The point I wish to make is that even in systems as complex as Air Force logistics systems, it is possible to construct experimental designs that control stimulus variables, and that such designs considerably increase the predictability of one's results. Conversely, they significantly decrease the amount of random variance in the system. More importantly, they provide parametric information which permits one to extrapolate his results to situations other than those he has included in his experiment. It is this latter parametric information that has been so sadly lacking in much of empirical systems research.

The question of incorporating variables systematically is really twofold. First, how far can a researcher go in a single experiment? Although everything cannot be included, it has been possible to introduce a large number of variables and their states in multivariate designs. The problem for the researcher is partitioning the variables. He may concentrate on one field of knowledge at a time, yet these fields overlap, and the relationships and trade-offs between them must also be investigated. Second, should some variables be left unmanipulated even when they can be? This course might be favored to permit and demonstrate system adaptation, a process discussed in connection with exploration experiments.

As for strategies for aiding verification, the obvious one consists of putting into effect the various counteractions against confounding and contamination described in Chapter 2. The cautions voiced about their curtailment in exploration experiments must be emphasized even more vehemently for verification experiments. Yet some diminution is inevitable. The researcher can be caught between those who even advise diminution and those who are surprised by it. The latter may have a low tolerance for relativism. If they dislike an experiment's results, they may exploit an experiment's imperfections to reject them.

Two ways suggest themselves for coping with the level-of-certainty problem. One is to face it before the experiment is undertaken. The researcher should explain to those providing the funds and facilities the safeguards that are required and the consequences of any compromises. He should also develop his own position as to the line that must be drawn, and he should stick to it. He should be prepared to forego the experiment if safeguards are lowered below that line and to terminate it if this happens after the experiment starts. Although this can be a discouraging development, it is not as depressing as producing experimental results which cannot be justified.

The other action is to express in the report of the experiment precisely what was done to assure certainty in the results, and what was not done but might have been. Academic researchers conventionally describe how they conducted an experiment in a university laboratory, and such accounts of precedures have often been included in reports of man-machine system experiments. But they also have been omitted, or have lacked adequate content. In any case, even in university laboratories researchers are not likely to point out in their reports any aspects of design or measurement which could raise questions about internal or external validity. It is suggested that reports of man-machine system experiments do this explicitly.

Such reports have at least two audiences, as Chapter 2 noted. One consists of the system users or developers, the other of professionals familiar with experimental and statistical methods. If the latter point out an otherwise unmentioned flaw in the experiment to the former, these can become most concerned. They themselves cannot evaluate it as a major flaw or a minor one. It is better if the experimenters describe and evaluate it in the first place. In addition, a reporting requirement of this nature might influence researchers to insist on sufficient safeguarding of the experiment.

Reports might also clarify the role of statistical significance. The outcomes of significance tests, when favorable, have often been assumed by report readers to indicate that the experiment has incorporated all the necessary safeguards against confounding and contamination. It was explained in Chapter 2 why this is not necessarily so. At the same time, it was suggested that results of significance tests be stated in terms of the confidence levels achieved instead of indications whether they fell within the conventional limits. Such a practice would strengthen the concept that certainty is relative.

*Exploration and Verification.* If it is natural for man-machine system experimentation to progress from exploration to verification, this can happen in a number of ways. (1) Between programs: within one program one or more large experiments may be oriented to discovery, while experiments in another program build on them to get certainty. (2) Within a program: in a similar succession within the same program, the exploration experiments may be large and complex, or they may consist of small pilot or exploratory studies when only a modest try-out search is needed. (3) Within an experiment: the experiment proceeds by sets or phases; its design is modular. In the first one or two sets, for example, a number of states of a key variable are tried out to determine which should be put into the following set or sets. Similarly, a number of measures may be tried out to learn which should be used later. These try-out phases should be distinguished from such preliminary parts of the experiment as sessions in which the experimental staff rehearses laboratory operations, simulation, and data collection, and the sessions in which the subjects are trained in operating the simulated system.

If exploration takes place in the first part of an experiment and verification in a subsequent part, the researchers must avoid experience bias. They must make certain that the subjects who participate in both parts become equally experienced with all the states introduced into the later part. If different subjects perform in the different parts, the researcher could still say the parts together constituted a single experiment, or he could call them two experiments.

What is an experiment? There appear to be no hard and fast rules for bounding one, either in man-machine system research or elsewhere. Experiments have varied from miniature ones to large studies consisting of many phases, some combinations of which could be viewed as individual experiments. An experiment might be regarded as all the data-taking sessions producing results which are compared with each other. Yet the results in some phases of a multiphase experiment might not be compared with those in other phases, and results from different experiments might be compared with each other, all the more legitimately when they have had inputs, subjects, and other features in common.

A kind of progression from verification to verification can occur in a multiphase experiment without advance planning. During a verification phase the researchers, as a result of something they have observed, decide they want to introduce a new variable. They add on a phase in which they do this, combining the new variable with others already in the experiment. They have to realize, of course, that performance of the same subjects with the new variable may be affected by the experience they gained in prior phases. But they may have reason to believe this prior experience will affect all states of the new variable equally except for the zero state (absence of the variable entirely).

If the researchers introduce a new state of a variable already in the experiment, they are on much thinner ice. Prior experience will almost certainly not exist to benefit performance with the new state as it does the old. The researchers may continue the added phase until performance on the new state reaches a steady level. Then they may assume—perhaps with risk—that the subjects have learned or adapted to the new state to the same extent as the other states. This method of continuing until a steady level is reached may be adopted also when a new variable is introduced.

One way in which new variables may be introduced during an experiment is through refinement. What had been a composite variable is split into its components, each becoming a new variable with some number of states. But experience with the composite variable in preceding phases can exert considerable influence on performance with the new variables, and some of the influence is likely to be differential between states.

It must be made clear that sequentially developing a verification experiment in this fashion during its course is risky business. The process resembles the intermodule alteration discussed in connection with exploration experiments. A phase can be regarded as a module. The effects of experience are not counterbalanced between phases or modules, although the researcher may resort to the technique of continuing the new phase until performance with the new variable or state levels off. In any case he may find it advisable to view the new phase as exploration. As in exploration experiments, the new phase must include all experimental conditions under appropriate control, as in a complete experiment. There may be a single run-through of these in a single phase, several in a phase, or one per phase for several phases.

Alternatively, new subjects can be introduced along with the new variable or state; the researcher may want to describe the sessions in which they perform as a new collateral experiment rather than a new phase. Although this might be a more customary approach in some research, in man-machine system experimentation it can be difficult to bring about because of a scarcity of subjects as teams and the time required to train them to operate the system.

## STRATEGIES

Many of the strategies which researchers may follow have been brought into the preceding discussion of objectives. This section will review other strategy decisions which must be made and methods of improving the cost-benefit ratio. But first some attention should be given to the constraints under which man-machine system experiments have been conducted.

### *Constraints*

Notable among constraints, as evident elsewhere in this book, have been costs, durations, and scarcity of teams of subjects. These are inter-related. The duration of an experiment increases cost. So does a large number of paid subjects. Other kinds of subjects may be scarce for different reasons.

It has been possible to put together an experiment with relatively little outlay. The scope was limited, facilities available, equipment simple, duration brief. With similar ingenuity many experiments of a like nature could be conducted in the future, depending on objectives, system settings, and experimental methods. But other experiments have entailed substantial expense. A few have cost as much as a quarter of a million dollars or more. The facility has not been the only major cost item. Simulation and data collection, computer time and programming, subjects and experimental staff, all have contributed to costs. As the record has shown, preparation for the running of the experiment and the analysis of the data have each taken considerably longer than the occupancy of the laboratory; the cycle time for some of the largest experiments has been several years.

Need costs continue to be as much of a constraint? Various methods of effecting greater economies will be discussed in connection with the cost-benefit ratio. At this point it should be observed that many of the difficulties encountered by man-machine system experimentation lay in the analog simulation equipment which was the only available resource in earlier days. Other difficulties arose from the growing pains of developing simulation with digital computers. The high cost of simulation can be attributed in part to such difficulties. In addition, the facility or the instrumentation has sometimes been unnecessarily elaborate.

All system testing tends to be expensive, and man-machine system experiments are either closely related to such testing or can be viewed as part of it. The cost of testing unmanned equipment is seldom questioned. Perhaps what is required is wider understanding of the need to learn more about complex systems under the circumstances where they are operated by people.

The absence of such understanding became clear when new systems were proposed as improvements on older ones, or new features were advocated for the same reason. Many proposals assumed superior system performance because there would be less need of the human element. To evaluate the proposed change required knowledge about the current system as it was operated by people, but it was discovered that such knowledge had not been assembled. One of the constraints on experimenters in investigating a new system was that they had to put the current one in the laboratory also, to see how it worked. The very fact that no one really knew this suggests the esteem in which such knowledge was held by system designers and builders.

Experimenting on men and machines working together in a system is a relatively recent concept, as this book has shown. The concept has not penetrated deeply into the behavioral science, engineering, and data processing communities. As a field of research, man-machine system experimentation has remained relatively unknown. This is partly understandable. Since most of the experiments have had an ad hoc objective, attacking problems in some particular system, only the people involved in that system knew about the experiment and profited from it. To describe the experiment to a wider audience, if security limitations permitted, called for describing also the complex and unfamiliar system with which it dealt. This was difficult. If results were generalizable, either little effort was made to broadcast them, or the potential audience lacked interest. Security constraints limited the dissemination in many cases. Prior to the study which led to this book, research agencies had not assembled the history of this research.

The youthfulness of the field has been a constraint otherwise. It has been professionally undermanned. Those who could do it well have been relatively few, though the demands of such research are heavy. It is not taught anywhere, and there are no textbooks. The technologies of simulation and measurement for the research have been poorly documented. The literature of engineering, psychology, and operations research has not filled the gaps. Lack of experience in methodology has at times been an obstacle to good research. Even the interchange of information among experimenters has been limited, as noted earlier. Practitioners have not known about other experiments, even those concerned with similar or related systems. It is hardly surprising, then, that nonpractitioners who might need such research were even less familiar with it.

More widespread familiarity cannot guarantee interest. Other barriers between disciplines must be overcome. The experimental method for obtaining knowledge is not universally appreciated or understood among military personnel, engineers, programmers, operations research specialists, and others. Indifference to criterion selection and measurement has resulted in lack of support for attempts to make methodological improvements. If a particular discipline dominates an organization, its interests may compete successfully with experimental evaluation. Engineers may prefer to design new equipment, data processors to build new computer programs.

But there is another side to communication and persuasion. On occasion (some say, typically), experimenters have compounded their problem. They have failed to convey their approaches and rationales adequately to military customers, engineers, programmers, and managers who were really interested and would have lent more support if they had encountered greater lucidity and candor. Many written reports have left much to be desired. In addition, the grandiose nature of some proposals, and divergences between aspiration and accomplishment, may have led to a certain amount of skepticism.

If the technical people have not always displayed the most advantageous qualities, much can also be said about managements. Such research certainly gives managements of several varieties unusual responsibilities. One of these is to co-ordinate a great many people: members of the sponsor or user organization that will be affected by the experiment's outcome; laboratory supervisors and technical personnel who gather information, develop and maintain equipment, prepare simulation inputs, produce computer programs, and collect and analyze data; and the subjects and quasi subjects. For all these to function effectively and in harmony at all times in all experiments would be most unlikely. Organizational frictions have occurred. The goals of different groups may differ. Management's problems have been exacerbated when it has lacked the talent for reviewing the planning and design of experiments. In some, technical comprehension failed to match interest. In others, interest was narrowly concentrated on managing.

Not all managements have been sympathetic to man-machine system experimentation. For example, when one general knowledge experiment led to other experiments and then to a large-scale application of the results, further general knowledge experimentation ceased (although ad hoc studies were taken up later). Man-machine system experiments investigated many of the Air Force and

Navy projects to automate air defense. But during its development the principal system adopted by each service was not given such scrutiny. A commitment to automate could be found in connection with those experiments which did take place. A conviction about the miraculous nature of computers was matched by indifference to demonstrations of the relative effectiveness of manual operations—and of ways in which these could be improved.

Managements, of course, have had to deal with other considerations. These are found not only where development is especially urgent or the system is "one of a kind" but also to some extent in all system development. An experiment might compete for personnel, money, and prototype equipment with other needs. The management might understand these better or value them more highly. An experiment might delay development unduly from the sponsor's point of view. It might even raise doubts in the sponsor about the merits of the innovation.

System development has been a competitive enterprise, with competition sometimes not only among development and production contractors but also among the sponsoring and using agencies. Funding responsibility has shifted, or funding was reduced. Required information was not exchanged. Some projects suffered because of the tangle of conflicting interests. A multiplicity of contractors has reduced the integrated administrative control necessary for effective experimentation. Professional support was lost due to discontinuities in organizational associations. System design and goals have been changed unpredictably. Slippages occurred in schedules. Managements have sometimes found it difficult to provide clear and consistent policies and objectives for man-machine system experimentation.

It should be realized that a large-scale, ad hoc experiment is not only an operation of considerable magnitude, it also interfaces with many different groups, all with their own concerns. Field commanders might think they were being evaluated. The conclusions from an experiment could influence the acceptance of a proposed system or its design. Substantial funds might be thereby rechanneled. Professional reputations could be affected. The management of the research and development organization or of the sponsoring organization might discourage or postpone the publication of experimental results or limit the distribution of the report. It might simply disregard the published data or fail to draw attention to them. Enough instances exist in the folklore of man-machine system experiments to justify a suspicion that when an experiment's results are unfavorable to some established interest, they may be resisted, perhaps successfully; since experimental methodology cannot be perfect in system research, a rationale for resistance would not be difficult to develop. In a sense, this applied science gets involved in big business. To those researchers trained in university psychology laboratories, this may come as a surprise.

### *Strategy Decisions*

Since many of the strategy choices which an experimenter is called on to make have already been noted, this summary necessarily covers some old ground. Strategy decisions are required both for the arrangement of each experiment and for the organization of the research program in which each is situated. What are some of the decision points?

*Scope of Experiment.* The researcher must decide how many and which states to give each variable, and how many times to repeat each combination of conditions (replications). The exploratory researcher must also decide what to abstain from controlling, to encourage emergence. As indicated in Chapter 2, these decisions determine the size of the experiment, which obviously affects its cost. Testifying to the difficulty of making a durable decision, the scope of a number of experiments has been reduced subsequent to initial planning. Considerations governing these choices have been discussed earlier in this chapter and in Chapter 2.

*Extent of Counteractions.* The number of replications is one of the counteractions the researcher has to review in deciding how ambitious to be about certainty. His biggest dilemma can become whether or not to proceed with an experiment when he knows he cannot invoke counteractions to the extent he would prefer. This dilemma characterizes both exploration and verification experiments, the difference being that the former can have a lower threshold of assurance. Where should this be set? The experimenter must resist the temptation to set it too low just to conduct the experiment.

*Boundaries of System.* How much of the system under investigation should be reproduced in the laboratory? This is the partitioning problem referred to earlier. Since every system can be regarded as a subsystem of another, there is no question of putting the *entire* system on the laboratory floor. But what is not there still has to be represented to the extent of simulating its inputs to what is there and the outputs it receives from what is there. It may be better to put more on the laboratory floor instead of relying on quasi subjects to represent embedding organizations or other subsystems, or on simulation inputs to provide the information from these. On the other hand, there are factors of feasibility, ease of engineering, capital investment and other cost factors, timeliness, and experimental control. If the research reproduces only a critical subsystem or even a nodal operator position, the teams of subjects can be smaller, the cost lower, and control tighter. Another decision about boundaries concerns experiments which focus on competition or conflict and co-operation or co-ordination. How and to what extent should the competing or co-operating system be represented?

*Simulation.* Intertwined with decisions about bounding the system are choices of simulation agent, degree of verisimilitude, level of detail, and extent of time and organizational compression. All of these factors can raise or lower costs. They have been sufficiently discussed elsewhere in this book.

*Measures.* Strategy decisions are also required concerning the number, objectivity, and precision of the measures for assessing the data. The selection of particular measures is one of the problems of methodology reviewed in Chapter 2.

Other strategy decisions are related to an entire program of experiments, or are equally applicable to such a program and to individual experiments.

*Locus of Experimentation.* Should existing facilities be used or a new laboratory be built? Should the program be part of a larger research effort or be independent? Should it be tied into an actual system—current or future—or exist autonomously? These decisions can determine whether the program of experiments will be assured of long-term support, adequate funding, and a steady supply of matters to investigate.

*Planning and Authorization.* How orderly and systematic should planning be? How flexible? What is feasible? Experiments that have been planned have been dropped. So have phases from an experiment. An experiment has often come about because of a preceding one. Certainly the planning of a program cannot be as rigorous as, for example, the development of a new system. Then what kinds of projections can be made to get long-term support? If the program is tied into a larger research program or an actual system, who authorizes individual experiments, with their considerable outlays? What authorization procedures should exist?

*Relationships between Man-Machine System Experiments.* One strategy has been to try to incorporate a great deal into a single experiment, another to resort to supplementary, ancillary, or side experiments smaller in scale and subsequent to the main study. Occurring in ad hoc investigations, these have attacked a different area, such as training, or examined another variable, such as some unusual condition, rare event, local requirement, or extreme situation (e.g., saturation or equipment outage).

A program may be arranged, as noted earlier in this chapter, to start with an exploration experiment and proceed to a verification experiment. The exploration experiment may be large or small. The program may also include experiments—other than preliminary check-out phases of a study—to investigate methodology.

When verification experiments succeed each other, the researcher can adopt the strategy of progression in independent and dependent variables. Experiments become successively larger as the researchers gain experience in preparing and conducting them. Progression in verification experiments may also be marked by refinement—the reduction of a composite independent variable into its constituents.

When a number of experiments deal with the same objectives in a program, the researcher may wish to do what he can to make their results comparable. (He faces a similar problem in trying to make an experiment in one program comparable with one in another.) A subsequent experiment may be either an attempt to replicate an earlier one, or a treatment of another state of one of the variables. In either case it is possible in a man-machine system experiment to achieve considerable equivalence by using the same simulation inputs, experimental operations, laboratory equipment, and measures. In making any comparisons, of course, the researcher is obligated to point out divergences. Because of the complexity of the experiments there are sure to be a number of these.

Instead of a comparison between laboratory experiments, a program may contain both a laboratory experiment and a field experiment with the same theme. The latter, whose purpose is to validate the laboratory study, should come second so researchers will have gained methodological familiarity during the work in the laboratory. There will be enough additional problems in the field, including co-ordination, administration, weather, sufficiency of resources, and reliability of equipment. Although the field experiment may also use simulation inputs and these may be the same as those for the laboratory, equivalence in other respects is likely to be approximate at best, even when congruence is sought. When the inputs come from real sources, comparisons should be made even more cautiously. However, if certain salient inputs, goals, subjects, and equipment are the same or equivalent, subjective comparisons may be warranted. It may be impossible to reproduce in field situations the experimental design used in the laboratory, and the field study may not justify the term "experiment." Nevertheless, it can at least imply the extent of the external validity in the data from the laboratory experiment.

*Relationships to Other Research Approaches.* The other research methods to which man-machine system experiments are related include observational or questionnaire field surveys, individual operator experiments, technical support experiments, and all-computer simulation. The strategy questions are various. How can these support man-machine system experiments, and vice versa? What can one of these do better than a man-machine system experiment, and vice versa? What combinations are desirable, and in what order is it best to place the components?

The function of the field survey is to indicate the variables and their states that should be investigated in the man-machine system experiment (or through one of the other methods), rather than to provide conclusive information. This limitation is not always well understood.

Experiments which investigate individual operators performing component tasks which are important in the system can provide information for skill training, operator selection, and the design of component equipment; such information may not be easily derived in a large-scale experiment. The inadequacy of a single component in a system can degrade its performance to an intolerable degree. The component-task experiment may be essential to show where the trouble is. If it precedes the large-scale experiment, it can furnish a certain amount of familiarity to the researchers about simulation and measurement. If it precedes an all-computer simulation, it can provide data about individual or interactive performance to put in the computer model. Individual operator studies have comprised some of the technical support research that has accompanied some experimental programs. This research may depart from specific tasks in the system to examine variables of general interest, such as properties of displays. Such studies may need a separate laboratory.

Much has already been said in Chapters 23 and 24 concerning all-computer simulation. Undoubtedly its importance will continue to grow as more is learned about human performance, so it becomes increasingly feasible to model this in a computer. Differing viewpoints exist concerning this trend, one holding that virtually all system experimentation will become all-computer simulation (with some cross-checking against data from system tests), the other believing that a strong complementary relationship will prevail.

It has been pointed out that not only can human participation and computer simulation occur profitably together in the same man-machine system experiment but also that one or more experiments and one or more all-computer simulations can strengthen each other within a program. An all-computer simulation is an experiment in which all the simulation is within the computer. The point has been made that it has a higher capacity for independent variables and states of variables and it can repeat experimental conditions a large number of times. Human behavior can be represented deterministically or probabilistically. In either case, however, the computer must be told what the human responses will be, either precisely or in terms of a particular distribution. As more detail—or less aggregation—is required in this information, it becomes increasingly difficult to furnish the information to the computer with sufficient accuracy.

How does the researcher choose between experiment and all-computer simulation? As Chapman (1961*b*) observed, it is difficult to state simple rules. A number of suggestions were quoted in the last chapter. It does appear that the researcher should consider the research objectives of exploration and verification, the stage of a system in development, the level of detail in the simulation (macro vs. micro), its validity and generality, its utility, the kinds of human variables and performance in the study, and the differences between static-analytic and dynamic representation.

For example, from what has been said earlier it would appear that all-computer simulation would be helpful for try-out exploration but not for emergence; it could contribute to certainty in verification studies; it should concentrate on early stages of system development which lack detailed design; and it should be used only with caution where neither particular individual performance nor distributions of responses can be specified.

Interactions between individuals and between teams of operators would seem especially difficult to represent satisfactorily in a computer model. No adequate taxonomy of such interactions presently exists for man-machine systems, nor is knowledge about the variations within particular interactions available for computer modeling. Individual differences between operators, between machines, and between operator-machine combinations are also insufficiently known and documented, as are variations within individual operators, machines, and combinations thereof. Data must be obtained on the scene from current systems, but such data may not be valid for a future system.

This is not very much in the way of guidance. More analysis is needed in the new field of computer modeling of complex, interactive human performance. Validity is an especially important aspect but not a simple one. Witness the manifold approaches to it: contact validity—as in the comprehensiveness of the simulation; construct validity—of a hypothetical construct about behavior, for example; predictive vs. concurrent (related to a present system) validity; and empirical (data-based) vs. face (apparent) validity.

As previous discussion has shown, all-computer simulation can either precede or follow an experiment. When it precedes one, it can indicate what the experiment should incorporate. When it follows one, it can incorporate what the experiment has indicated about the effects of variables on human behavior. Through all-computer simulation it is also possible to project the future, based on the past. Past performance of actual subjects is put into the computer model, which then projects trends. These projections can be compared with subsequent performance of the subjects, without or with the intervention of trend-changing events. All-computer simulation can also demonstrate what ideal performance might be—without errors and with minimum time lags.

All-computer simulation can benefit man-machine system experimentation in a particular way. Because of its potential use, the researcher can be required to state what he expects to learn from an experiment that he could not from an all-computer simulation. He must try to set forth the kinds of human and equipment performance and changes in performance which he thinks cannot be predicted for modeling in the computer. Then he must design his experiment to examine this performance. Such a requirement can sharpen the experimentation. There would seem to be no point in conducting a man-machine system experiment which one was sure in advance would simply duplicate an all-computer simulation. The requirement to indicate as explicitly as possible what the experiment may accomplish that the all-computer simulation cannot should provide help in designing both.

*Cost-Benefit Ratio*

How can the cost-benefit ratio of man-machine system experiments be improved? (The cost-benefit ratio is similar to the cost-effectiveness ratio which has also been applied to the evaluation of systems and techniques.) One way is to reduce cost, the other to increase benefits.

*Costs.* As the discussion of strategy decisions brought out, costs can be reduced by limiting the scope of the experiment, but this could also lower the benefits. The same is true, but perhaps to a smaller extent, for constricting the boundaries of the system. A plea for "more efficient and economical experimental designs" to reduce the number of required runs and the length of each run has been entered by Geisler, Haythorn, and Steger (1962). They also urged "the development of criteria for selecting the level of detail" in computer models and "the development of more rapid and less costly ways of programming simulation models." The latter aspiration may have been met by the creation of a number of simulation-oriented languages for computer programming in more recent years. The extent of time compression and organization compression would seem to be another cost-saving simulation technique worth greater study.

Costs can be reduced in a variety of other ways. One strategy is to hold concurrent sessions. In a number of experiments two systems or alternative methods of operation have been run at the same time with the same inputs, in neighboring laboratory spaces. Since the stimulus situations and temporal factors are the same in the two locations, experimental control is heightened while laboratory occupancy time is cut in half. This technique of concurrency can be exploited either when the inputs are simulated or when they are signals (e.g., radar) from actual objects, such as aircraft, made available to two operating systems. It should be realized that heavy demands are placed on data collection and laboratory management. Limitations on the total of demand-inducing inputs can help prevent the systems from getting out of phase with each other due to procedural errors by subjects.

Two kinds of ventures have been responsible for large expenditures unaccompanied by much return. One was the creation of general purpose laboratories, discussed earlier. These were intended to investigate a broad range of systems or problems with considerable generality, especially in command and control. It seems reasonable to predict that new ones will have as little pay-off if the impediments outlined earlier in this chapter continue. Better forecasts for use are required. The other venture has been the attempt to assist the development of a system by a program of man-machine system experimentation during that development. The obstacles in such a path were too little appreciated when these programs were envisioned. Manifestly, managements must do a better job of integrating experimentation with other types of testing, but also they may find it advisable to simulate more modestly and invest only in items not subject to drastic change or only in ones over which they exert control. Perhaps such experimentation should be restricted to a few operator stations and nodal positions.

It has been suggested that experimental facilities should have a high degree of occupancy, such as (1) two or three shifts per day or (2) continuous use for data-taking sessions during the year. This strategy would lower that part of the cost of any one experiment attributable to the cost of the facility. However, personnel could not be similarly shared. In fact, any scheme for multiple shifts or continuous usage requires very large outlays for personnel. Continuous usage seems more feasible than multiple-shift usage and has been approximated in some locations where the needs for experimentation have been heavy; it follows that the generation of needs is a way to improve the cost-benefit ratio.

Ingenuity has yielded some relatively simple and inexpensive simulation methods. Judicious use of schematic or graphical simulation, as well as all-computer simulation, could result in some economies. Other economies have been introduced through modest design of the facility, even though this made it less of a showpiece.

Money has been saved by using facilities primarily devoted to other purposes. These varied from offices to operational sites. Another stratagem was to conduct experiments with equipment from a discarded prototype system, obtained at no cost. It became a research tool for general knowledge studies. Along a similar line, simulation equipment developed for one program of experimentation was subsequently used in two different programs. But probably the greatest economy of this nature can come from using the system's own computer and ancillary equipment to produce simulation inputs, present these to the subjects in an experiment, collect performance data, and analyze these data. This capability should have a marked effect on the cost-benefit ratio of man-machine system experimentation directed at computer-based systems. Perhaps such research can be integrated with other investigation in what Sackman (1967) has called omnibus testing.

The concept of multiple use can be carried further than exploiting the system computer for experimentation and testing as well as system operations. Often the data about system operations expensively gathered for an experiment can assist other kinds of research, including all-computer simulation. It helps, for this and other reasons, if the laboratory is part of a multidisciplinary organization. The same costly production facility for generating simulation inputs for an experiment can generate inputs for a training program also based on simulation. On occasion, even the same inputs can be used for the two purposes, although usually the needs will differ. If they are designed for such dual use, simulation transducers and other devices can support both training and research; it is less likely—though conceivable—that those developed just for training will be appropriate for experimentation on system effectiveness, and vice versa. Actual sensor recordings acquired for evaluation may contribute to training or proficiency testing; radar, sonar, and optical image recordings probably can be exploited more widely than they have been. The research facility itself can be used for training and proficiency testing during otherwise idle periods. And of course if a research program uses a nonsystem computer this can and almost certainly will serve a multiplicity of other enterprises.

*Benefits.* In the instances just cited, a prorating of costs lowers that for man-machine system experiments. Multiple use also effects benefits. Multiple objectives in an experiment raise the experiment's returns. Human factors aims should be tied into the experiment. For example, information acquired during the experiment can be applied to improving the human engineering design of some of the equipment in the new system or making some of the procedures more effective. In an experiment comparing a proposed system with a current one, human engineering analysis should be applied to the current one to optimize it for the experiment. Important improvements may result. The same may be done for procedures. These by-products can come about either directly or through the task descriptions and task analyses to which experience in the experiment contribute. Improvements in equipment design of a non-human engineering variety may also result. So may changes in computer programs. The data may incidentally yield information about skill levels and operator capacities that can be exploited for determining manning requirements, although studies of individual operation are the preferred source. The training methods employed to indoctrinate and train the subjects before the experiment can help create a training program for the system, and other experience from the experiment can be tapped for that program.

A multiple objective of particular importance is one that seems to have been seldom realized. Even during its operational phase a system undergoes continuing evaluation; for example, operating subdivisions are tested for proficiency. An ad hoc man-machine system experiment—or a general-knowledge type experiment based on a particular system—develops measures of system performance to express results. The analysis responsible for the data-collection techniques and measures chosen, and the assessment made of their usefulness, can be put to use for selecting criteria, measures, and data-collection methods for later evaluations of the system. As a matter of fact, the development of such measures should accompany the creation of the system along with the development of training programs. Both developments should be closely associated, because much of both system evaluation and system training will depend on exercises involving configurations of men and machines responding to either simulation or live inputs.

The support of human engineering, training, procedurization, manning, and evaluation need not consist of by-products of experiments aimed at some other goal. As this chapter pointed out earlier, they can be objectives in their own right. Then the entire program of experiments includes some studies which have one of these objectives, some which have another. At least one major program (see Chapter 10) did just this, thereby improving the cost-benefit ratio of the program.

Benefits can be increased in various other ways. Conducting a thorough field survey *before* an ad hoc experiment makes the experiment more likely to attack critical questions and possess external validity. Experienced system operators can also serve as advisers. Combining an experiment with other types of experiments and with all-computer simulation in the manner discussed a few pages back can augment its effectiveness. Those who act as subjects learn a great deal about the system as well as about the experiment. It makes sense to involve them in subsequent development and operations as well as in communicating experimental results. At times these can be senior individuals who share in deciding about the application of these results. Through participation the outcomes become more understandable to them, the problems become more clearly defined, and unfounded opinions and myths about system procedures and design are dissolved.

The advantages gained from an experiment depend in part on its reporting. Ad hoc experiments should be reported in two ways to get maximum circulation and critical inspection, as suggested earlier. One version goes to the sponsor and system user, emphasizing results. The other, giving technical explanations of methodology, is circulated among professional disciplines. In addition, it is helpful to issue a nonclassified supplement if most of the report must be classified, or vice versa. This arrangement assures more readers for important nonclassified material. Any information from an ad hoc experiment that seems to have some degree of generality—about the design of displays, for example—should be so designated and perhaps published separately. As a precedent, the sponsor of the Ohio State University programs in air traffic control and Bayesian processing in decision-making wisely required that the researchers put down on paper what they believed could be generalized from the experimental results. Researchers should also exploit other methods of information distribution besides reports.

One of the most important ways to make man-machine system experiments more effective is to optimize their methodology. This means several things. It has been emphasized that pre-experiment sessions should rehearse the staff in laboratory operations and check on simulation and data-gathering techniques. Separate methodological experiments may have to be encouraged to investigate effects of certain aspects of experimental design and relationships between measures. Staffing is critical. When a verification experiment badly fails of certainty due to confounding, it produces no benefits. If an experiment lacks external validity, it too yields little. Experiments require professional competence in experimental methodology. Its absence has marred research programs and delayed their inception. Such competence must be accompanied by an understanding of how to apply that methodology to system experiments, knowledge in depth about systems in general, and intimate familiarity with the particular system being examined.

The required capabilities cannot come solely from past practitioners. One resource is to search out up-to-date, improved methods of simulation, data collection, and data analysis. But the best source of expertise will be on-the-job experience. This can be gained in quantity only if a program is a continuing one. Conversely, the program must continue to exploit it. In this sense whatever assures a program's longevity improves the cost-benefit ratio.

What has favored longevity? This question can be answered only with respect to ad hoc experiments. Continuity of funding is one obvious support. A multidisciplinary organization seems to have been a hospitable location. Above all, the past record suggests the value of institutionally close associations between the research facility and developmental agencies, so the facility will have a continuing series of problems to investigate and an assured consumer of its output.

## ACCOMPLISHMENTS

Undoubtedly, it would be easier to improve the cost-benefit ratio if it were possible to establish completely and conclusively what man-machine system experiments have accomplished. That has not been possible, although accomplishments will be assessed to a limited degree shortly. If a Project Hindsight (Sherwin and Isenson 1967) were instituted for the purpose, it might be feasible to trace the effects of all the ad hoc experiments, or the lack of effects; in investigating the impact of government-sponsored research Project Hindsight had a staff and authority unavailable to the author of this book. As for the pay-off from experiments with a general knowledge objective, this would be difficult to track down, even through an official inquiry.

Even more elusive would be the serendipitous effects of experiments. For example, in one program it was claimed that an experiment pulled together the research in the same field in the department in which the laboratory was situated. Along with this integration, the requirements for simulation compelled a specification in detail of the system investigated, and this was valuable for other applied research. It has been observed concerning a number of programs that an experiment educated the experimenters about the system as well as about such experimentation. Many have also learned about using computers for research. (The education of the subjects was noted in the preceding section.) From the repetition of some of the names in this book it should be apparent that a by-product of early programs was to train experimenters to conduct subsequent ones.

Another by-product has been the technical support research which probably would not have been funded if it had not accompanied a program of system experiments or followed one. A substantial amount of generalizable knowledge about human engineering design and human performance has come from component experiments thus supported.

In a research domain where so much money has been spent, one would think those responsible for its expenditure would want to know the pay-off. However, tracing and recording pay-offs seem to be neglected processes in research generally. No explicit requirements exist, the skills needed are scarce, and there is always the hazard of publicizing negative consequences. It would be desirable to assess not only man-machine system experiments but also all-computer simulation, gaming, and the mathematical analyses used in operations research. Not only might these other techniques be helped by a scrutiny of their accomplishments and their cost-benefit ratios but man-machine system experimentation could gain respect by comparison. This kind of comparative approach would enhance fair competition between research techniques and suggest where the research dollar should be invested.

Recently increasing emphasis has been placed on making certain that the results of human factors research do get implemented in system development (Mackie 1968). Why not have an on-going examination of consequences? Associated with every research and development organization might be a requirement to keep track of the benefits attributable to the work performed. This might have to be done by an agency independent of the research organization, and that agency would have to keep an eye on more than hardware. That is, it would have the tantalizing task of showing linkages between the procedures, personnel requirements, and training techniques developed experimentally and those eventually put into effect in systems.

### *Criteria*

What are the criteria of accomplishment? One is indeed the fact that experimental findings were implemented in the system. Another is the matching of what was done in the experiment or experimental program to what was projected.

A third criterion was suggested by Rauner and Steger (1961*b*). The system savings and benefits which an experiment demonstrated should be matched against the cost of the experiment itself; the experiment would be justified if the former exceeded the latter. (With this approach it need not be asked whether the experimental findings were actually adopted.) One of the difficulties in using this criterion was the need to include a "benchmark" study to represent the system or policies to be superseded. Unfortunately, this study would increase the cost of the research, and the benchmark's validity might remain uncertain. This criterion applies to comparison experiments. Analogously, in a diagnostic experiment the researcher might try to show how much degradation in performance would result if the experiment's findings were *not* implemented.

A fourth criterion of accomplishment could be a program's longevity. Presumably this would reflect success. (The converse is not implied. A management might terminate a program out of short-sightedness, poor planning, or the need for money elsewhere.)

### *Ad Hoc Objective*

If we look back at the programs of ad hoc experiments described in this book, we can divide them into a number of categories: those whose accomplishment is known; those whose lack of accomplishment is known; and those whose accomplishment or lack of it is unknown, at least to this author. General knowledge objectives and findings will be discussed later. This categorization of ad hoc experiments may help remove some misapprehensions the author has encountered. One is a failure to realize that the bulk of man-machine system experiments have had ad hoc objectives or effects. Another misapprehension arises from some spectacular failures; these have produced an aura of futility.

The fact is that according to the criteria listed above, considerable accomplishment can be credited to man-machine system experimentation, certainly more accomplishment than otherwise. Perhaps those who have read Chapters 3-20 have come to the same conclusion for the same reason—the record of what was done. Although no attempt will be made here to go through a systematic assessment of each program and experiment in those chapters, a quick review of some should be illuminating.

Much accomplishment can be attributed to the four largest programs, the RAND air defense experiments (Chapter 8), the RAND logistics studies (Chapter 13), the air traffic control investigations of the CAA and FAA (Chapter 15), and the program of the Army's Combat Development Experimentation Center (Chapter 14). The first of these led directly, albeit serendipitously, to the creation of a very large training project, probably the largest ever established to train teams and systems of men and machines through simulation; some of the experiments in the program were run to support that innovation. The researchers' own accounts of the RAND work on logistics systems have included assertions about the acceptance of experimental results and consequent substantial savings in logistics organizations. Other improvements and economies were indicated by the studies and may have been instituted. The continuation of the program after the first couple of studies suggests there was a consensus regarding benefits. Even greater longevity has characterized the CAA-FAA studies, which have dealt with both particular geographical areas and systemwide problems. It can be assumed that researchers continued to receive requests for experimentation year after year because previous output was helpful to the development and operational arms of the agencies. The same may be said about the CDEC program; an attempt to ferret out implementations from that extensive effort, however, would be particularly difficult due to security restrictions.

Beyond these, a number of SDC field studies were beneficial (Chapter 11). Investigations of computer processing time and feedback demonstrated that processing time did not constitute the problem many feared and they brought about a major programming change to eliminate the feedback difficulty. The AZRAN study also brought a major improvement through a change in programming, as well as some procedural alterations. Two large experiments on the System Training Program, one in the manual system, the other in SAGE, provided evidence that this training technique was helpful to the nation's air defense. (Since the installation of the training program was well under way when these experiments were conducted, it is interesting to conjecture what would have happened if they had yielded contrary results.) Two other SDC field studies concerning subsystem training and evaluation methods led to innovations in SAGE system training operations.

The research by Psychological Research Associates (Chapter 9) produced tests and training techniques which the Army adopted, according to the reports of the investigators. An important equipment design parameter was established through the experimentation at the Electronics Research Laboratories (Chapter 7). This also indicated that the proposed new system would be as effective as the old system in some respects and more so in others—a somewhat lukewarm finding which failed to inhibit the new system's adoption.

In contrast, some projects stand out for their failure of accomplishment. Among these have been three major laboratories established by the System Development Corporation (Chapter 17) and one by the MITRE Corporation (Chapter 19). By the criterion of what they contributed to the development of particular systems it is difficult to discern much benefit from them. To be sure, quantification of benefit in these instances is as infeasible as it is in the positive cases previously mentioned. By the criterion of comparing the man-machine system experiments conducted with the research projected, these laboratories were most disappointing. Another perhaps more spectacular instance of non-accomplishment was the TRW laboratory in Colorado (Appendix I) that never got off the ground. Some of the SDC field experiments in the manual system failed to produce useful data. One of the projects at the Willow Run Laboratories (Chapter 9) apparently was virtually nonproductive. The first major program of man-machine system experiments, the Cadillac Project (Chapter 4), became productive after a slow start, but there seems to be little indication that its experimental products were ever used.

What about the programs concerning which this author finds himself unable to provide indications of benefit or lack of it? These include one category composed of studies on systems which were never built and operated. Obviously in these cases it would be impossible to trace effects of an experiment to a new system, and it cannot be said with assurance that the experiment was one reason why the system was rejected. This category includes early studies by the Lincoln Laboratory (Chapter 6), the Naval Research Laboratory (Chapter 5), the Willow Run Research Center (Chapter 6), and the Operational Applications Laboratory (Chapter 6).

Other programs with unknown effects—on systems adopted or not adopted—have included those at Ohio State University in air traffic control (Chapter 10), the Naval Research Laboratory work on CIC information processing (Chapter 5), the Willow Run Laboratories research for the Army (Chapter 9), the SDC research on manual system electronic countermeasures and on civil defense (Chapters 17, 22), the MITRE experiments on air traffic control (Chapter 19), the Martin and Grumman simulations of space flight and moon landing (Chapter 22), and the IDA studies on communications (Chapter 20). It does seem likely that some degree of benefit has come from at least some of these.

### *General Knowledge Objective*

Finally, a brief look at experiments and programs which had general knowledge goals or results is also in order. The OSU air traffic control experiments had the dual objectives of ad hoc information about a radar-based system and generalizable information about team performance. The OSU decision-making studies (Chapter 21) were concerned with Bayesian processing in a fairly wide context, rather than a particular system. The OAL decision-making studies (Chapter 16) also sought a fair amount of generality, as did those at APL (Chapter 18). Some of the SDC laboratory experiments, in the Human Factors Laboratory and SSRL (Chapter 17), sought and produced generalizable data. The IDA studies (Chapter 20) tried to get general information about communication linkages and processes as well as data about a particular system. For the RAND air defense studies (Chapter 8), the actual purpose initially was to investigate how information-processing organizations functioned and changed. The outcomes which inspired an air defense training program were regarded as extendable to other kinds of organizations. Some experiments directed at methodology—which might be applicable on a wide basis—were conducted for the FAA (Chapter 15) and CDEC (Chapter 14).

There might be two ways to assess the accomplishments of these experiments. One would be to weigh the importance of what was learned from them. The other would be to say what was done with that which was learned. Either is beyond the purview of this book, and neither may be achievable by mortal man. However, it has seemed advisable to place in an appendix some of the generalities about systems which can be derived from this research. One reason for doing this is the relatively limited distribution that reports of this research have been given, notable exceptions being the two OSU programs. Appendix III also attempts to record some of the generalities that might be inferred from experiments of a strictly ad hoc nature. Since it is debatable how general and how certain these "lessons learned" are about systems, they will not be called "principles."

## FUTURE RESEARCH

Should more man-machine system experiments be encouraged, conducted, funded? If so, under what circumstances? What kinds of problems should they encompass? In short, what are the needs? The applicability?

### *Ad hoc Experiments*

The need for ad hoc experiments will depend largely on the man-machine systems that are developed. Increasing computer automation seems sure to be one of the design decisions for many new systems. Its extent and various kinds of symbiosis between man and computer should furnish the content of many experiments on proposed systems. Alternatives to automation may also be investigated. These include better human engineering in the current system, improved procedures, more selective matching of personnel to system tasks, and training with more effective techniques.

When new systems are developed, with or without more automation, experiments may examine trade-offs among particular design features, training, procedures, and manning to find out through which of these the system's performance can be enhanced at the most favorable cost-effectiveness ratio. This approach to system development would be a novel one. Although the existence of trade-offs between these human factors approaches has been acknowledged, they have not been forthrightly addressed in system experimentation. But they seem to be a natural target for man-machine system experiments. Too, researchers must deal expertly with these approaches right in the experiment. For example, they have to select subjects and determine their proficiency, train them before the experiment starts and check on their learning during it, and specify the procedures they should use; if the experiment concerns some equipment design variable, they must also be experts in human engineering. Such versatility is not necessarily encountered otherwise among human factors specialists.

Whether or not the trade-offs are examined, in each new system there will be some need to investigate experimentally the areas noted. Hardware elements must be distributed in some fashion. How? Arrangements can be tested in an experiment. They will be closely related to the team's procedures. Not only do the interactive, co-ordinative, and collective procedures and tactics of teams have to be designed initially, but ways have to be established for procedures to grow and change after the system becomes operational. Experiments can examine both the procedures and the methods of procedurization. All systems have rules, and people try to beat the rules. It has been suggested that this propensity might be exploited in experiments to see which rules or procedures were durable. Each new system will also have its training requirements. Ways of adapting team training to the particular system have to be investigated by experiment. Each system must be manned. The optimum numbers of individuals have to be worked out for each system task according to expected input loads, the skill requirements must be determined, and the organization of the individuals must be designed. Here again man-machine system experiments may be needed.

But why experiments? Why cannot these problems be solved satisfactorily either through initial analysis or by trial-and-error after the system has been designed and built? The answer is that on occasion they can, but often they cannot, due to various circumstances. Analysis may not work if knowledge is lacking or dynamic situations are too complex. The option of trial-and-error is excluded when it is too difficult, costly, or late to change a system or innovation after it has been installed. For example, rather than alter the layout of airport runways and other fixed or semi-fixed features, it is preferable to vary them in a simulation-based experiment and select the patterns found desirable—before installation. The optimal tactics and composition of an infantry unit might be ascertained in actual combat, but again this might be too late. Alternative methods of training can be evaluated on the scene over a long period of time, and often are. But for an air defense system, it would be highly desirable to do this as quickly as possible, and before attack—through experimentation.

If simulation-based experimentation can provide answers sooner and less expensively than trial and error, it can also incorporate rare events like crises and catastrophes, or infrequent situations like very heavy traffic loads, which might occur too seldom for their impact to become known through the normal course of events. Further, because it can compare innovations or new systems with current ones under the same circumstances, a man-machine system experiment offers greater opportunity for developmental decision-making than tryout in the real world. These are some of the considerations which make experimentation preferable.

When it comes to choosing between a laboratory study as the vehicle on the one hand and a field test or all-computer simulation on the other, the researcher must take into account the relative advantages and disadvantages of each method discussed earlier in this book.

It does appear that some new systems as well as improvements in human factors areas will continue to call for ad hoc experiments, Other developments besides automation will be responsible for the new systems. In general, what does history suggest will characterize the innovations which will be examined through man-machine system experiments? For one thing, the physical features to be varied in an experiment must be objects that can be easily manipulated. Consoles and vehicles can be moved around but otherwise not easily altered. Because they are relatively immutable, physical environments can be varied only through selection of those readily available. Some experiments will continue to investigate physical objects directly, in a physical environment, within the foregoing limits. But most will, as in the past, rely on the symbolic and pictorial representation of objects and environments. These are easier to manipulate than objects and environments themselves. In turn, experiments will favor those systems and parts of systems in which objects and environments are transformed into symbols and pictorial representation, their simulation being relatively straightforward. Within the limitations which have been discussed, computer-based systems are obviously eligible for ad hoc experiments. Not only are they built-in laboratories, but their role is to handle symbolic and pictorial representations as inputs, outputs, and stored data. Communication systems of all kinds are also candidates, for the same reason; these include information collection and distribution systems. They often have tie-ins with computer-based systems. Systems based on signals such as radar echoes will continue to be experiment-prone. Also needing experimentation will be those systems that use teams of human operators who interact significantly both as individuals and as groups in making the system perform.

### *General-Knowledge Experiments*

In pleading for generality as one of the criteria for doing man-machine system experiments, Chapman (1960*a*, 1961*a*) also specified relevance. He urged that experimenters investigate causes, not symptoms such as morale. Perhaps the best way to get at causes is to look for new variables. Thus, exploration experiments should consist of the emergence type, and the results of these would then be further investigated in verification experiments. The major emphases might be placed on processes which occur within groups of operators in system settings. Both independent and dependent variables would be stated in system or group terms rather than individual terms, although the processes themselves might bear the same names as those which occur within individuals.

One of the processes is decision-making, which has been a leading theme of general knowledge experiments so far. Although these experiments have been cast in system settings which required that other subjects or quasi subjects provide information to the decision-maker or respond to his decisions, the critical performance has been that of one individual. Additional experimentation needs to be done in that kind of context, particularly further investigation of interactions with a computer. But experiments should also give more attention to the interactions between individuals—to the procedures, training, and personality variables (such as motivation) which affect the way a decision-maker and his staff work together. Another decision-making area that needs further research is that of interpersonal differences as these are related to various categories of decisions, information quality, risk, and action requirements. To achieve generality, research on decision-making in systems must also be investigated across systems.

Risk, incentive, penalty, and pay-off influence choice—in detection, evaluation, and action. To simulate these at all realistically, however, is a challenge which calls for great ingenuity in man-machine system experiments. Research in decision-related human motivation would be likely to attract funding. No one is prepared to deny its importance, even though—or possibly because—it is so ill-defined. The wide interest exhibited in stress is a case in point. Grant and Hostetter (1961) took note of the need to consider motivation variables in manned system research thus:

> While much of the work being produced by the decision and game theorists contributes significantly toward the basic aspects of man's decision-making behavior, the laboratory situations generally used in decision and game studies are in many cases too artificial to be generalized to surveillance systems with any degree of dependability. For example, in many of these studies, the motivation is produced by relatively small monetary gains, and the risk is provided by potential loss of money. There is also the problem of having only a very few alternatives from which to choose. It can be seen that behaviors under these conditions could not be used to predict behavior in a surveillance situation where the motivation, risks, and over-all situation are quite different.

The theme of decision-making might be broadened from making choices between selected alternatives to problem-solving, in which choice is one step. "What shall we do now?" replaces "Shall we do this or not?" In fact, it may first be necessary to find the problem, that is, to discover that one exists and then define it. In a major criminal trial, to pick an analogy, the verdict is the decision, and jury decision processes are dramatic. But the over-all system of law enforcement solves the problem of a crime by first discovering it, then going through investigation, indictment, arrest, presentation of evidence, and attorneys' and judge's activities before the decision; and after a finding of guilty come further investigative actions, sentencing, appeals, punitive actions, and rehabilitative actions. The antecedent and consequent proceedings not only influence the decision and are influenced by it, but are interesting in their own right. The same is true in man-machine systems.

More experiments might be addressed to the consequences of error, accident, mishap, and disaster (and recovery from it). Except for systems devoted to dealing with disaster, such circumstances seem to have attracted little experimental study. This may be because they are rare events or unpalatable occurrences. The proper strategy may be, through simulation, to force the accident (or error, or malperformance, or misjudgment) in the experiment and then let the system try to cope with it. Problem-solving would result.

Akin to problem-solving is the process of planning. This can also demand a co-ordinated team effort. It involves examining files and displays, filling information gaps, asking and answering questions, distributing tasks, conferring and exchanging information, dealing with probabilities, and weighing risks and payoffs. Computer-based management information systems have been created to help executives and commanders plan better, but more research is needed to show how these systems can mesh with the total planning process. Of particular value might be experimental investigation of making inquiries, for example. When should they be addressed to the computer, when to managers? What should be stored in the computer's data file, what in "paper" files?

More research is needed to get generalizable knowledge about team training methods, team composition, and team procedures (and adaptation). The generality problem here in part is that these all vary, at least on the surface, with the task the team performs and the system in which the task is embedded. Better classification of tasks and systems is needed. At the same time, however, a body of knowledge must be built up from experiments—which have no choice except to incorporate particular tasks in specific systems—from which it will be possible to generalize to the categories to which the particular tasks and specific systems belong.

General knowledge experiments may also be directed at multisided situations of competition and conflict. A side may include a number of "nodes." The difficulties of experimenting in a sufficiently controlled manner on such situations have been mentioned earlier in this book, but ingenuity in design may resolve some of them; assistance can come from the branching capabilities of computers and their programs.

Can the computer-based laboratory be a source of increasing understanding about how men and machines—especially computers—should work together in systems? The computer's versatility, including on-line recording and evaluation of performance, can be a great boon to general knowledge experiments; as Shure (1967) observed, new techniques can detect order and pattern in complex events.

But more than computer technology is required, and generality, as observed before, can come also from wider resort to ad hoc experimentation. The need for greater understanding about man-machine systems does exist. There are ways to acquire it through experimentation, for direct application or for the expansion of knowledge. The future of such research depends on scientific imagination and perspicacity. It is hoped that this book has provided a fund of information to help these prosper.

# APPENDIX I

## Experimental Facility Proposals

In the main body of this book the laboratories in which man-machine experiments were conducted have generally been described along with the experimentation. Where a new facility was proposed (but not built) in connection with an on-going program, the proposal has been outlined in the review of the program, as in Chapter 9.

There have also been proposals for facilities which were never built because the experimental programs with which they were associated were never initiated; and there has been one case where the facility was built but no man-machine system experiments were conducted in it because the program was dropped. These various facilities, proposed or never exploited, will be described here.

### DEPARTMENT OF DEFENSE–SMITHSONIAN INSTITUTION PROPOSAL

As Bray (1962) has chronicled it, "a series of planning studies of the research on human behavior required to meet long-range needs of the Department of Defense" was initiated in 1957 by the Advisory Panel on Psychology and the Social Sciences of the Director of Defense Research and Engineering. The studies were contracted in 1959 to the Smithsonian Institution, which established a Research Group in Psychology and the Social Sciences. Recommendations of subjects matter emphasis and mathods of support came from six task groups, functioning under the following labels:

- Design and Use of Man-Machine Systems
- Human Performance Capabilities and Limitations
- Decision Processes in the Individual
- Team Functions
- Adaptation of Complex Organizations to Changing Demands
- Persuasion and Motivation

In the project's final report, the programs advocated by the first four "were consolidated into a single program on Man-Machine Systems, Intellectual Skills, and Team Functions" (Bray 1962). For each of these three areas, the report proposed a large, well-staffed, and well-financed laboratory. In addition, a nonlaboratory institute of organization research was proposed for the organizations area. The report did not clarify the dividing lines between the three recommended laboratories. One of these was to be a man-machine system laboratory and the description implied that it would conduct complex, multioperator man-machine system experiments as well as individual-subject experiments. It did not make this aim as explicit as it might have, although research in simulation techniques was urged specifically. The proposal for a team performance laboratory did not mention simulation and confined its attention to face-to-face interactions in subsystems.

The separate laboratory proposals were probably due to the backgrounds of the task groups that made them. The members of the group proposing the man-machine system laboratory were mostly human-engineering oriented and some had been associated with man-machine system experiments. Small-group studies had greater interest for those advocating the team performance laboratory.

In its original report (Miller et al. 1959), the task group on design and use of man-machine systems proposed a multipurpose simulation facility housed in a man-machine system research institute. The facility should have a computer, the report said, able to generate "the complex situational programs representative of complex system." It should attack the following areas, with supplementation from outside research:

a. Theoretical and empirical models of systems, some of which may be developed by intra-system research. Mathematical theory applicable to large numbers of interacting variables and parameters.
b. Methodologies for evaluative prediction of system performance from multiple criteria.
c. Methodologies and principles for design and development of major system types.
d. Methodologies for partitioning systems into subsystems and independent study and evaluation.
e. Simulation methodology applied to large systems.
f. Task taxonomy and performance theory.
g. Displays for decision making.

Because the final report left the proportion of complex system experiments within the total experimental program ambiguous, it is not certain to what extent the manning, size, and funding requirements it set forth for the man-machine system laboratory were meant to be those for the kinds of experiments reviewed in this book. Nevertheless, the estimates in the final report (Research Group in Psychology and the Social Sciences 1960) deserve mention:

Staff size. The work proposed requires a staff whose size is as follows:

| | |
|---|---|
| Key scientists and engineers | |
| Simulation technique | 7 |
| Systems theory | 5 |
| Inventive research | 16 |
| Computer (not including lower level programmers) | 12 |
| | 40 |

| | |
|---|---|
| Technical support | |
| Laboratory technicians and programmers | 50 |
| Clerical and stenographic | 20 |
| Shop | 50 |
| | 120 |
| General, administrative and custodial | 40 |
| | 200 |

Roughly half the key scientific and engineering personnel should be psychologists from the related fields of engineering psychology, the experimental psychology of sensation, perception, learning, and measurement, and physiological and social psychology. The remainder should include engineers, mathematical statisticians, sociologists, physiologists, and operations research specialists. Included in the positions listed above are approximately five for key scientists and engineers who would work temporarily in the Laboratory while on "sabbatical" leave from government, industrial, and academic positions. Similar provision is made for approximately five technical support positions for internship training of junior scientists and engineers.

*Laboratory Characteristics.* The Laboratory should provide for the observation and measurement of the performance of large numbers of human subjects in the operation of simulated information-processing systems. For economy of computer use and to meet the needs for replication of observations, it should be designed for two-shift operation, up to 100 subjects per eight-hour shift, giving an average of 150 subjects per day, 200 days per year. A high speed, scientific computer with large storage capacity, highly flexible input-output equipment, and provision for multiprogramming is required. Extensive model shops must be available. The Laboratory should include several large, open, "playing" spaces for system simulation, as well as extensive facilities for related studies of individual human subjects. Interior walls should be readily rearranged; a heavy initial investment in partitions which can easily be moved will speedily be repaid.

A facility is proposed of 76,000 square feet total space, 48,700 square feet in laboratory and 27,300 square feet in office space.

*Location.* The Laboratory should be located in a metropolitan area, convenient to public transportation. It should also be convenient to a large military installation, which might occasionally furnish special subject groups of particular backgrounds for periods of 30–60 days in length.

*Cost.* The cost of the Man-Machine System Laboratory with its capital equipment is estimated at $4,500,000. Annual operations costs are estimated at $5,400,000. An initial budget of $2,000,000 is proposed to cover the facility and an initial five years of operation during which the staff will be built up slowly.

The estimates of the task group on team functions for a team performance laboratory were given in the final report as follows:

*Staff size.* The work proposed requires a staff whose size is estimated at:

| | |
|---|---|
| Key scientists and engineers | |
| Team effectiveness and task analysts | 4 |
| Team composition and organization | 2 |
| Team training | 6 |
| Field research on teams | 4 |
| Computer (not including programmers) | 6 |
| Total | 22 |

| | |
|---|---|
| Technical support personnel | |
| Laboratory technicians and programmers | 30 |
| Clerical and stenographic | 8 |
| Shop | 20 |
| General, administrative and custodial | 20 |
| Total | 100 |

Of the key staff, about one-half should come from the fields of social and experimental psychology, including evaluation and measurement. The remaining half should be engineers, sociologists, mathematical statisticans, and operations research specialists. Included in the staff estimates are a few positions for key scientists on sabbatical leave. A number of technical support positions should be filled by graduate students in order to improve the flow of scientists into this multidisciplinary field. A larger staff would be proposed for this Laboratory if there were not a severe shortage of qualified scientists for this type of work.

*Laboratory Characteristics*. The Laboratory should provide observation and measurement of the performance of teams of men. The work will concern small groups of men simulating the operation of those parts of systems in which men work in a face-to-face relation, helping one another. Several fairly large, open, "playing" spaces are needed, as well as smaller, closed laboratory observation rooms for associated studies of individuals or very small groups. The subject flow should average 100 subjects per day. A computer of "intermediate" size is required for data reduction and control of input and feedback to the team members. As with the laboratories described above, interior partitions should be readily arranged.

A facility is proposed of 47,000 square feet total floor space. Of this total, 32,300 square feet are laboratory space proper, and 14,700 square feet are office space.

*Location*. The Laboratory should be an integral part of a university in order to stimulate the production of technically qualified scientists for team research. As with the other laboratories proposed above, it must be located in a metropolitan area, convenient to public transportation, in order to insure the flow of subjects in the numbers needed.

*Cost*. The cost of the Team Performance Laboratory is estimated at $2,000,000. Normal annual operations costs, when the Laboratory is completed and staffed, are estimated at $2,700,000. A budget of $10,400,000 is proposed for the facility and an initial five years operations.

It was proposed that the intellectual skills laboratory, the third facility proposed in the final report (not including the institute of organization research), would have thirty key scientists and engineers, eighty-five technical support personnel, and thirty general, administrative, and custodial personnel. Subject flow requirements were identical to those for the man-machine system laboratory. A total area of 50,000 square feet was envisioned, slightly more than half in laboratory space. Cost with capital equipment was set at $2,100,000 and annual operations costs at $3,600,000; an initial budget of $13,200,000 was proposed for the first five years.

## NATIONAL BUREAU OF STANDARDS FEASIBILITY STUDY

In 1956, the Aero Medical Laboratory at Wright Air Development Center asked the National Bureau of Standards (NBS) to study the requirements and feasibility of a laboratory facility employing dynamic simulation for the experimental investigation of man-machine systems. Most of the work of NBS was devoted to building what Ernst (1959) called a scale model of such a facility. It consisted of a manned cockpit simulator associated with an analog computer and a one-position or two-position display console for directing airborne intercepts in an air defense mission. The console served as the manned input-output station for NBS's SEAC digital computer, which calculated interception commands and data. The digital computer and cockpit displays were linked by simulated data link; the pilot and the intercept director were linked by simulated radio communications.

A few modest exercises were run with this setup, three Air Force pilots playing the pilot role and NBS personnel acting as intercept directors. In some, the commands were sent directly from the SEAC computer to cockpit displays; in others, the intercept director observed target and interceptor tracks, estimated the proper course-to-steer commands, and transmitted these to the pilot. No data were analyzed. The purpose of the exercises—which Ernst (1959) called "experiments"—was "merely to demonstrate the capability of the NBS facility." They led the NBS investigators to conclude "indisputably" that it was feasible to study man-machine systems in a laboratory facility using dynamic simulation.

During the same time period, as various chapters of this book have indicated, simulation-based programs of man-machine system experiments were investigating or had investigated the ground control of interceptor aircraft, although they omitted the linkage to a cockpit simulator. The simulation and laboratory facilities in these were more elaborate, more sophisticated, or more pertinent to real systems than in the NBS study, in some cases much more so. Yet the NBS study report (Ernst 1959) made no mention of any other simulation facility or air defense study.

The NBS study report acknowledged that its so-called "scale model" did "not incorporate all of the functions which preliminary studies indicated would be desirable" or contain the "capacity desired for some of the functions." As a matter of fact, from the point of view of the simulation of an air defense system its design was rudimentary. The study report based its assertion of feasibility on two factors: "The first is the successful combination of analog and digital computers for dealing with the mathematical models of systems. The second is the identification of nonessential requirements based upon an appropriate qualification of the problem." This second factor was not clarified.

The kinds of systems at which the study was directed "include those for air traffic control, ground control of interceptors, missile launch and control, and command systems in general." Design objectives of a laboratory facility for research on such man-machine systems were stated as flexibility (in changing experimental conditions and varying system design), versatility of application, reliability, and expandability (for growth potential and to minimize "both the initial investment and the time required before operations can begin"). Subsequently, the "objective of minimizing cost" was added.

Components of a laboratory facility were listed as a digital computer, an analog computer, operators' work spaces, central control, monitoring and communication, data recording, and interconnecting devices (e.g., cables). Both kinds of computers might be regarded as general purpose. But, "Work spaces tend to be specific to a particular system, and to that extent they limit the generality of the research facility." Concerning these locations where the operators interface with the system, the report noted:

> The effort required for preparing work spaces is comparable to that required for analyzing the system and preparing the mathematics. In order to maintain reasonable cost and time schedules, it is always desirable to make work spaces no more elaborate than absolutely necessary. The degree of necessity is determined by the complexity of the subjects' tasks, the desired variability of these tasks, the realism required, and the generality to be achieved for adaptability to other experimental applications. . . .
>
> One shortcut is to obtain elements of the work spaces from an existing system and provide the appropriate activating information synthetically. Some electronic displays have intrinsic generality in that the kind and nature of information presented are largely a matter of computer programming. In any event, the preparation of work spaces can become a substantial effort with respect to requirements upon staff and operating funds.

At the time of the NBS study, digital computers were by no means so widely used for experimentation and other purposes as they later became, but the NBS investigators were alert to their potential. They were alert also to the fact, not always given sufficient emphasis, that man-machine system experiments cannot depend on computers alone:

> The general-purpose digital computer is without doubt the most important individual equipment of the facility from the standpoint of cost, general utility, and its effect upon the detailed design of the balance of the facility. However, even the most powerful digital computer would have little capacity for simulating systems in real time unless provided with input and output capabilities which would qualify it as a "special purpose" computer according to present standards. . . . For the present application, such a machine would have to be equipped with specialized input and output equipments which have the necessary characteristics and which are designed to be comparable with the computer itself.

The report failed to discuss at any length some of the critical requirements of an experimental facility, such as staffing, preparation of simulation materials, and provisions for large numbers of interacting system operators and their accompanying equipment. Above all, it did not ask which should come first, the research questions or the facility. Perhaps the project's engineering orientation was responsible for the tacit assumption that the laboratory should be built first and then someone should figure out what to do with it.

## PRC'S TEAS TSRF FOR AFCRL

As mentioned in Chapter 16, the Planning Research Corporation (PRC) in 1961 developed, for Air Force Cambridge Research Laboratories (AFCRL), concepts of a threat evaluation action selection (TEAS) simulation research facility (TSRF). Such a facility was aimed at primary and secondary research objectives (Dodson et al. 1961). The primary objectives were to "serve as a test facility for TEAS prototype system concepts," to evaluate "subsequent suggestions for the overall TEAS concept originating both within and without the research establishment," and to conduct "degradation studies." These would investigate system degradation resulting either from equipment malfunction or enemy action, overt or covert. The secondary objectives were to support basic research and investigations of subsystem operations. The basic research area would be decision making.

Facility characteristics, it was said, should include flexibility (to deal with an environment many years in the future and to permit experimentation on factors not ultimately incorporated into a TEAS system); "multiple read-in-read-out points" (for subsystem research); input degradation capacity; and design factors pertinent to TEAS itself, including a gamut of TEAS models and the generation of the TEAS environment.

Space requirements were specified "for the experimental or playing areas, for the control and recording equipment center, experimenter control or management areas, experimenter and visitor observation, and storage" (Blanchard 1961). The playing areas should be high enough to accommodate large vertical displays, easily partitionable by movable walls, possessing control of temperature, noise and lighting, and large enough to provide 50 square feet to each subject (or 300–350 square feet where there was large equipment). Elsewhere in the same report (Blanchard 1961), an area of 150 square feet per player was advised, and a similar area for each of the experimental staff. A construction cost of $25 per square foot was assumed to cover power, light, air conditioning, and similar needs.

Other equipment requirements and estimated costs (for experiments involving fifteen interacting subjects) were: special construction (e.g., temporary walls and observation booths), $50,000; a purchased telephone system, $5,000; two closed television circuits with two extra monitors, $10,000; five microphone circuits, $2,000; fifteen disc recorders, $5,225; one twelve-channel tape recorder, $12,000; six audio monitor stations plus patch boards, $2,500; a rented time-signal generator, $65 per month; an intercom system, $300; and miscellaneous equipment including terminal and equipment racks and patch boards, $10,000. Concerning a digital computer, Blanchard (1961) said:

> While an essential part of any TEAS study, the computer is not considered part of the TSRF laboratory. Preferably, the computer facility should be adjacent to the TSRL, but not an integral part of it, unless it is installed expressly for the TEAS studies. Usually, a research program cannot maintain the continuous load necessary to warrant exclusive use of a computer. Remote input-output equipment should be used to communicate with the computer installation. Staff programmers and operators would provide the necessary human link between the laboratory and the computer facility.

A ratio of one individual on the laboratory staff to each player was assumed to be required during an experimental run. For each senior staff professional, there should be two to five technical and clerical assistants. For each experiment, the staff would build up during twelve months preceding the prerun training, then phase down during the six months after the runs, which might last from a week to several months. A laboratory could be steadily employed if there were "three to six teams of experimenters, scheduled into and out of a laboratory as rapidly as possible"; and such an arrangement would insure a "steadily employed staff of trained people." Blanchard wrote, "Economic use of an expensive facility dictates a high occupancy rate." He also warned:

> Two kinds of time pressures, (1) to get something going in the laboratory, and (2) the programmed arrival of subjects and visitors, exert a great deal of stress upon the staff. Careful planning and attention to detail do not suffice, unless *all* conceptual issues are discussed and resolved in the early stages of game planning. Tremendous interaction exists between decisions concerning an environmental input model, the physical environment, a research strategy, and implementation. Many a researcher has discovered to his dismay that leaving an important decision to be dealt with last often necessitates drastic coverup action.

Although AFCRL never built the TSRF proposed by PRC for TEAS, the preliminary documentation contains useful messages for any future planners of such a facility.

## R. L. CHAPMAN'S ESTIMATES OF REQUIREMENTS AND COSTS

Confronting the reality that large-scale man-machine system experiments can be a demanding and expensive type of research, R. L. Chapman has produced two sets of estimates for a large laboratory facility. Chapman drew on his experience in the RAND air defense experiments (Chapter 8), in the Cadillac Project (Chapter 4), and in the aborted work on Subsystem I (reviewed in this appendix). One set of estimates concerned a laboratory capacity for twenty-five interacting subjects and their equipment (Chapman 1960*a*, 1961*a*):

> Before we can consider how to manage simulation studies of human behavior, we must have some idea of the resources needed. The requirements I am about to outline come not only from the practical experience of building a number of labs but also from the discomfort of committing both kinds of mistakes—being too plush, on the one hand, and miserly, on the other.
>
> To conduct studies of human behavior with simulation, laboratory space, instrumentation, and staff are needed. Space is needed for a playing area (where the subjects work); a management area for observation, control, and embedding organizations; storage areas; equipment area; and office space for the staff. If care is taken to get the best possible relations among these areas and if some of them can be double-used for different purposes, a laboratory adequate to handle up to 25 subjects simultaneously can be obtained with 8,000 to 10,000 square feet of floor space. Depending on whether good space is available, whether extensive modifications are required, or whether construction must begin from scratch, this requires a capital investment between \$50,000 and \$200,000.
>
> Several kinds of instrumentation are needed. First and foremost, a computer—small as one of the desk-types or as big as a Ramo-Wooldridge RW-400 polymorphic computer.
>
> Communications might be handled by an intercom or extensions off the existing dial system—or a custom telephone net might be necessary.
>
> Input-output equipment for the computer is another requirement. For simpler simulation exercises, the printed output from the computer and standard

methods for putting information into it might suffice (as it does here in the AMA Business Game). On the other hand, more ambitious simulations, such as of the air defense system, involve the use of special equipment.

Also needed is observation and data collection equipment. Direct visual observations might do for smaller installations, while closed circuit TV might be needed for larger ones. For recording verbal behavior—a prominent human activity—office dictating equipment or single- or multiple-channel tape recorders might be used. But to do this involves tapping telephone lines (which takes additional equipment) or using microphones with their associated gear. Instrumentation to permit the observers to monitor this verbal behavior while it is taking place is also needed.

In more extensive installations, automatic time-control equipment to pace the inputs or mark the recordings is needed. Automatic equipment for tabulating the use of the communications net might be necessary. Special data processing equipment might be needed to speed the analysis of results.

Stagecraft is sometimes employed to attain the illusion of the reality desired. In that case, standard stage equipment such as daises, flats, lights, and lighting control equipment is used.

As you can tell from the possibilities I've mentioned, the costs of instrumenting a laboratory varies widely—perhaps from $20,000 to $200,000 for a twenty-five man lab, exclusive of the computer.

A rough rule of thumb is that the number of subjects participating in the study must be matched one for one by staff members. This would mean a staff of 25 to support the size lab we've been talking about. The ratio of technical and support personnel to professional people on the staff should be about 7 or 8 to 1. So three qualified research people are needed for the twenty-five man lab. The staff cost might therefore run to half a million a year, overhead and the like included.

In addition, there is the expense of the subjects' time, a significant item in itself.

What we're talking about is a capital investiment in the range of a quarter of a million to more than a million dollars and a yearly operating cost of a half a million. This could be scaled up or down by building a laboratory of larger or smaller capacity.

Let's consider alternative approaches to cutting these costs. One approach, and a usual one, is to try to cut down on the space required per "player" from 200 to 50 square feet, eliminate storage areas and the like, use cheaper and fewer recorders, microphones, data-collection and data-processing equipment. Sizeable savings can obviously be made in this way.

Another approach, and one that I believe leads to more significant savings, is to amortize these capital costs more rapidly against many studies. To do so, full laboratory occupancy is needed, up to 24 hours a day, seven days a week. We need to put experimentation on a production line basis—and this, of course, violates scientific mores. Another rule of thumb is that a staff will spend approximately one-third of its time preparing a run, one-third of its time in the laboratory, and one-third of its time analyzing and reporting results. So that full lab occupancy would require three times the staff I mentioned for each shift, or ten or twelve times for twenty-four-hour operation, seven days a week. And, of course, many people working on many projects must have their efforts scheduled, coordinated, and supported by an effective management.

To cut costs, we also need to keep the staff on the job for some five to ten years so that they can profit by their experience and accumulate the wisdom that permits them to do more effective studies with greater efficiency. Over a period of four years at the RAND Corporation, for example, we cut our data analysis expense to one-fourth of what it cost initially.

Also, we must find or train the kind of personnel that merits our confidence so that we don't have to disrupt their progress by heckling them for elaborate explanations of what they are going to do and justifications of what they have done. This, as you well know, can take from 20 to 50 percent of the working scientist's time.

In the other set of estimates, which reflected the costs and salaries of earlier times. Chapman (unpublished report) developed the following estimates for a laboratory capable of running experiments with 50 subjects:

1. *Floor Space Required*

| | |
|---|---|
| a. Playing area–based on 50 to 350 sq. ft. per man or an average of 200 sq. ft. | 10,000 sq. ft. |
| b. Management area (double use of part of playing area is possible) | 7,500 sq. ft. |
| c. Storage areas<br>Live storage–2,000 sq. ft.<br>Dead storage–6,000 sq. ft. | 8,000 sq. ft. |
| d. Equipment center | 1,000 sq. ft. |
| e. Computer area | 5,000 sq. ft. |
| f. Office space–210 sq. ft. per man for 200 men (based on the assumption that half of 200 technicians will work in other areas and will not require office space) | 42,000 sq. ft. |
| g. Shops<br>Woodworking & metalworking–500<br>Photo and electronic–500<br>Simulation preparation–1,000<br>Miscellaneous–500 | 2,500 sq. ft. |
| h. Air conditioning | 4,200 sq. ft. |
| i. Miscellaneous | 4,800 sq. ft. |
| Total floor area required | 85,000 sq. ft. |

Estimating the cost of the playing area at $35 a square foot (based on the requirements for special floor, ceilings, lighting, and utility distributions), and the other areas at $25 per square foot, building costs will be $350,000 + $1,875,000 or $2,225,000. Assuming that the land will cost $200,000, total building costs will be $2,425,000.

2. *Installed Property and Equipment*

| | |
|---|---|
| a. General-purpose | |
| Communications<br>Monitoring–$30,000<br>Mikes, amplifiers–$15,000<br>Patch panels–$30,000 | $ 75,000 |
| Data Collection<br>Recording (50 channels, disk or tape including time signal generator, mixers, racks–$30,000<br>Data recording keyboards–$25,000<br>TV & monitoring facilities–$50,000 | 105,000 |
| Staging equipment | 25,000 |
| Total General-Purpose | $ 205,000 |
| b. Experimental equipment<br>Analysis consoles–$500,000<br>Displays–$100,000<br>Telephones–$30,000<br>Miscellaneous–$400,000 | $1,030,000 |

| | |
|---|---|
| c. Shops | |
| Woodworking–$3,000 | |
| Metalworking–$10,000 | |
| Electronic–$10,000 | |
| Photo–$25,000 | |
| Simulation Preparation–$100,000 | 148,000 |
| d. Office equipment–assuming a cost of $2.50 per sq. ft. for 42,000 sq. ft. (200 men) | 105,000 |
| e. Air-conditioning equipment (200 tons @ $750) | 150,000 |
| Total Installed Property & Equipment | $1,638,000 |

The total cost of the building plus installed property and equipment is thus $4,063,000.

3. *Staff*

Assuming an occupancy rate of 50 percent and three shift operation and three staff members per subject (one getting ready for an experimental run, one conducting the run, one analyzing the results of the run), the size of the technical and professional staff is thus 1½ X 3 X 50 or 225. Assuming a ratio of eight technical and support people to one professional, the staff breaks down into 25 professionals and 200 technical and support. We can assume five consultants, five interns for short term indoctrination, and five professional people on sabbatical leave as relatively short-term (one-year) observers with a fresh point of view, which brings the total of professionals to 40. Figuring administrative personnel at 25 percent of the total professional, technical, and support (240), this adds another 60 people to handle travel, purchasing, security, finance, personnel, and plant engineering. The total staff is thus 300. The annual direct labor charts are thus:

| | |
|---|---|
| 40 professional @12,000 | $ 480,000 |
| 200 technical @7,000 | 1,400,000 |
| 60 administrative @7,000 | 420,000 |
| Total direct labor cost | $2,300,000 |

4. *Subjects*

Assuming again an occupancy rate of 50 percent and three-shift operation, the costs of the subjects for the experiments can be estimated by assuming that they will work eight hours a day at the rate of $2.00 an hour and that there are 200 working days in a year. The cost will then be, for the assumed 50 subjects, 50 X 1½ X 200 X 8 X $2.00 or $240,000.

5. *Cost Summaries*

| | |
|---|---|
| a. Initial cost | |
| Building | $2,225,000 |
| Land | 200,000 |
| Installed property and equipment | 1,638,000 |
| Total initial cost | $4,163,000 |
| b. Annual operating costs | |
| Computer rental | $1,260,000 |
| Special equipment for experiments | 500,000 |
| Supplies for experiments | 100,000 |
| Direct labor costs | 2,300,000 |
| Subjects | 240,000 |
| Laboratory modifications | 100,000 |
| Miscellaneous | 50,000 |
| Total direct operating costs | $4,550,000 |

| | |
|---|---|
| Overhead @ 50 percent | $2,275,000 |
| Direct cost plus overhead | 6,825,000 |
| Fee @ 10 percent | 682,500 |
| Total annual cost | $7,507,500 |

6. *Schedule of Expenditures*

| | |
|---|---|
| Facilities cost (first year and a half) | $4,163,000 |
| First year's operation (with only a key staff, half professional, half technical, averaging 50 people @ $10,000) | 500,000 |
| Second year (one-third normal operating budget) | 2,500,000 |
| Third-year (four-fifths normal operating budget) | 6,000,000 |
| Fourth year (full operating budget) | 7,500,000 |
| Fifth year (full operating budget) | 7,500,000 |
| Five-year total | $28,163,000 |

## THE THOMPSON RAMO WOOLDRIDGE DATA SYSTEMS LABORATORY

In 1959–60 an ambitious program in man-machine system experimentation for an intelligence data processing system led to the construction of a large laboratory by Thompson Ramo Wooldridge Inc., with Air Force funding, in Littleton, Colorado, near Denver. But the program and the laboratory were discontinued before any of the projected large-scale experiments were conducted.

The object of the laboratory research was Subsystem I, a code term for a portion of a highly classified system to obtain various kinds of information about certain parts of the world, using a number of sensors whose ability to obtain data was based on recently developed technology. Since very little knowledge about either Subsystem I or the larger system of which it was a part has reached the unclassified literature, the present account will be somewhat sketchy. Some information has been gleaned from unclassified papers by several of the principal professional personnel in the aborted project (Blanchard 1961; Chapman 1962; Davis 1960).

According to Blanchard's unclassified Air Force report,

> a computer–centered information processing retrieval and reporting system was to be tested . . . using simulated photography and electromagnetic sensor data as inputs. A large real-time operational data processing system would have been tested under controlled conditions, in a fully instrumented laboratory prior to becoming operational. A computerized model of a synthetic country was to be used to create instructions for changes in a large photo mosaic and associated electronic emitter model that would illustrate various strategic and tactical postures. Simulated data sensors following hypothetical collection routines were used to create realistic system inputs.

A two-story laboratory with more than 20,000 square feet of floor space was created by modifying the end of a large computer manufacturing plant in 1959. It was replete with sound-recording equipment, closed-circuit television, an intercommunication system, a balcony area where the experimenters could manage the experiments and view the subjects operating the Subsystem I equipment, and various prototype components of that equipment, eventually including processing and display units. A staff of about two hundred consisted in large part of an engineering section to check out these prototype components in Category II tests and support the human factors-oriented section which was to conduct Category III testing. (These kinds of tests are discussed in Chapter 23.) Several dozen Air Force officers and enlisted personnel were assembled as experimental subjects and advisers.

According to Davis (1965), Subsystem I was the "first computer system with the capability of a true man-computer interaction (in terms of permitting a real dialogue to ensue between user and computer)"; and it incorporated the first of the "programming language for display, or display languages as they are occasionally called."

A number of illuminating problems arose. For one thing, the on-going engineering tests constrained the would-be experimenters' access to use of the equipment. Second, slippages occurred in the delivery of prototype (operational) hardware. Third, changes occurred in the design and objectives of some of the system equipment, requiring changes in the design and planning of experimental approaches based on earlier versions of that equipment.

These problems moved Davis (1960) to warn against experimentation strategies based on "the concurrent development of all facets of a metasystem so that these fit neatly together at some point in the future . . . particularly with respect to the simulation vehicle." By metasystem, he meant primarily the simulation and recording complex to be used by experimenters, who, he said, should not make "commitments which may be irreversible and disastrous if the mission or nature of the prime system changes dramatically."

Accordingly, wrote Davis (1960), one should "invest most heavily in those equipment aspects of the metasystem which are not subject to drastic change as a consequence of progress elsewhere in the system." Further, one should "begin with a relatively modest simulation vehicle which is sufficiently flexible to grow with the changing system and which will have fast payoff in terms of its impact on the design of the prime system."

The researchers resorted to innovation. They mocked up units of equipment, including operator consoles. The processing to be performed by a computer was simulated by a human being.

As a by-product, a paper and pencil exercise was developed to examine the "flow of material through the Formatting and Duplicating Subsystem" (Chapman 1962). Squares, circles, and triangles represented eleven different machines of three types that physically processed material which was loaded, unloaded, and monitored by an operator. In this subsystem, input formats were transformed into output formats. A process control group, represented by a hexagon, supervised the quality and timeliness of the products by means of information to and from each component. The processing chain contained two other functions, indicated by parallelograms.

The material which the subsystem was processing was represented by "material-in-process" cards. A master process card listed, for each batch of material, the address sequence shown in a flow chart; and a "Master Program Card stated the operation sequence for each material at each machine along with each time:

for loading the machine, for running a unit quantity through the machine, for unloading the machine, and for reporting job status." Rules (instructions) also had to be formulated. Chapman added:

> It took perhaps two weeks of technical time to specify the master cards and instructions, utilizing documented information and expert opinion when data were not available. Even though the process was to be simulated manually, addresses, operations, and the like were coded as if the process were to be done by machine. Attention was given to specifying the structure so that particular flow sequence and operation times could be amended without redoing the entire structure of the simulation. After several false starts due to incompleteness of rules, it was possible to simulate 24 hours' worth of system operation in about three work days, or something close to real time.

This simulation of an entire process was able to expose gaps in the system design. Alternative procedures could be envisioned. The information from this simulation had particular importance for the actions of the processing control group. In this sense it examined the management requirements.

The main thrust of the Subsystem I laboratory initially made progress with the gathering of vast amounts of simulation materials for large-scale experiments and many other preparations. Some component studies were performed, but, before any of the projected large-scale experiments could be conducted, shifts in funding and policy resulted in decisions concerning Subsystem I that led to the disbanding of the laboratory and its staff.

Before this termination, the would-be experimenters had learned many lessons of value to future practitioners of this kind of research in addition to those already mentioned:

1. It may be difficult to determine the required outputs of the system being simulated if the agencies which are developing or will use the system either will not describe them (due to security and other obstacles) or cannot.

2. To simulate the inputs into the system may be equally difficult if information about the inputs must come from a hostile area, or if they depend partly on unpredictable technology, or if some of their characteristics are being established by other groups, e.g., other contractors.

3. A system simulation facility can expand the role of human factors personnel not only in the design of operator procedures and in human engineering design of equipment but also in other respects. These include other system, facility, and equipment design matters; the selection of system operators; training (plans, material, equipment and actual training); preparation of operating and equipment manuals; logistics support and reliability; and evaluation and testing. Such expansion of role reflects an integrated approach.

4. The operations of the research facility may be influenced by situations over which its personnel have no control. These are obstacles between the organization developing the system and the one which will use it; conflicts between the user and the research and development (R&D) agency, among users, among R&D agencies, among associated contractors, and between prime contractors; and the strategy toward the customer adopted by the organization in which the facility is embedded, a strategy which may depend on the organization's financial health.

5. On the other hand, the operations of the research facility are also subject to the conflicts within the facility and to competition between design and evaluation interests.

6. When goals and objectives change and schedules slip, replanning and explaining demand time needed for technical work. Integration may not be effectively organized or enforced. The level of aspiration may be set too high.

# APPENDIX II

## Guides to Methodology

Some of the practitioners of the art of man-machine system experimentation have drawn on their experience to record their recommendations about the methodology and strategy of this kind of research, that is, how to do it. Some of these recommendations have appeared in their reports of experiments; and most of these recommendations have been summarized or quoted in earlier chapters. Other general advice on methodology and strategy (e.g., Chapman 1960*a*, 1961*a*; Davis and Behan 1962; Geisler, Haythorn, and Steger 1962; Kidd 1962; and Rauner and Steger 1961*a*,*b*, 1962) has appeared in Chapters 2, 24, and 25. This appendix contains other material from three fairly detailed, checklist-type guides which may be useful to practitioners-to-be.

### KIDD AND MICHELS (1959)

In 1958–59 a staff development course in "Research in the Analysis, Design and Evaluation of Man-Machine Systems" was presented at the Technical Development Center of the (then) Civil Aeronautics Administration by J. S. Kidd. A companion course in "Statistical Techniques" given by K. M. Michels was also under the aegis of Courtney and Company. This company's contractual work with CAA included efforts to make the methods of applied science more widely used among air traffic controllers, engineers, and related research personnel in CAA's simulation-based testing of air traffic control features (see Chapter 15). Although much of Kidd's material covered research methodology applicable to all kinds of experimentation, the following excerpt may be considered especially pertinent to large-scale man-machine experiments, such as those in which Kidd was participating at that time (Chapter 10). To the sophisticated researcher some of his observations may seem self-evident, but those who undertake this kind of research have not always been that sophisticated.

*Research Management*

A. Planning

1. *Apparatus.* In the planning stage, we must exert planning control over apparatus and the critical apparatus for experimentation falls into three

categories: (1) that which is involved in task generation—the core of the simulation for our purposes; (2) data recording apparatus; (3) data processing and reduction equipment. These equipments must be compatible and the linkages established between them prior to the experimentation itself.

2. *Personnel.* The second main category in planning is personnel. There are certain rules of personnel management which are almost common knowledge these days, but it helps to establish a clear organizational layout and a clear notation of assignment of role and function to each individual involved in a research effort, again, prior to the actual accumulation of data.

a) *Line Functions.* In research operations, line functions consist of supervision of the data collection process. These functions should be in the hands of people with a very strong feeling about control, consistency and reliability because if these factors break down in the data collection operation, the results of the experiments lose their usefulness.

b) *Staff Functions.* Staff and line functions in research operations should be as clearly differentiated as they are in any other kind of work activity. Your staff people, statisticians, electronics technicians, etc., should not be given attitudes of inferiority by emphasizing the ancillary role that they plan. Rather, they should be involved in planning activities at the beginning of an experimental program, because the experiment itself can be designed to facilitate or to inhibit the application of their skills to the main operation.

c) *Subjects.* A third category here regards the subjects or experimental participants. We have many problems here which are often-times overlooked in the set-up of an experimental program. We have already talked about the problem of selection to some extent, but we have not emphasized the problem of whether we should use professionally skilled operators for systems investigations, or whether we should make efforts to train people to participate in the research work without what might be called operational biases.

Operational personnel as subjects in a research setting bring with them certain advantages in that they carry a certain amount of realism with them. They have incorporated in their own thinking the established values that are required for effective operations. Professional ATC controllers, for example, are deeply concerned about midair collision prevention, whereas the naive subject trained only in the laboratory will not have the emotional commitment to this value. On the other hand, operationally trained personnel may bring into the situation a certain inertia that is not present with laboratory trained subjects. For example, if a man has been trained in a routine or in the use of certain types of equipment, and the experimental program is centered on the evaluation of new equipments or new procedures, there may be considerable emotional reluctance on his part to learn these new procedures or to learn to use the new equipment. Thus, the productivity of the combination of an old operator and new equipment, or old operator and new procedures may be severely degraded. There are, therefore, pros and cons in the matter of subject selection. Ideally, the criteria for subject selection will be a consequence of the intent of the program or the specific experiment. We have already touched on training and indoctrination of subjects. These matters come up regardless of whether the subjects are skilled in the general requirements of an operation or whether they are totally naive. One of the main aspects of training is a familiarization with the intentions and purposes of the research program itself—this is, over and above the skills that are required for their participation. One must enlist, in other words, the sympathetic cooperation of subjects if any level of generalization is to be attained. Motivation is still a highly pertinent variable, and the participants in the simulated environment must be comparably motivated to actual operators in the real-life task.

Another consideration with regard to subjects is the volume—the number of subjects that you use. This relates to the nature of the subject participants. Usually, if we use operational personnel, we must be satisfied with limited numbers. A fundamental rule of statistical analysis specifies that the larger the number of subject participants employed, the greater the confidence one can place in the outcome, or, in effect, the greater the reliability. The trade-offs here involve factors such as the control of non-participating activities, that is, what the subject does when he is not actively being observed. The larger the numbers the more difficult it is to control this portion of the participant's life, which is a potential source of bias. All these factors must be weighed and considered before the quality, numbers, extent of control, and the training and indoctrination program can be mapped out.

3. *Procedures*. We have gone over many of the procedures that must be planned for in setting up an experimental operation—such things as problem definition (which usually means a translation from field terms into laboratory terms) and a considerable emphasis on the measurement activity so that the variables involved can be specified.

a) *Statistical Design (Latin Square)*. We have talked a little bit about statistical design and this is the opportunity to make sure that you are aware of a somewhat abstruse but very useful statistical technique which may not come up in the statistics portion of the course. This technique, known as the Latin square, is a modification of standard factorial design, which we touched on previously, but one which has certain advantages especially where a limited subject population is available and one is not particularly concerned with the learning aspects or with skill acquisition as such. The Latin square allows you to utilize the same subjects throughout the experimental program under different experimental conditions. Therefore, it not only gives you the most extensive utilization of your subjects, but also adds to the precision of control in the sense that you are able to factor out—or balance out—individual differences effects in this way. Such characteristics commend its use in systems research, where typically we have a limited number of subjects available and where the number of test trials that we can allocate to a given problem may also be severely limited, thus requiring the maximum experimental power. We have found in our work at Ohio State University that we can reliably detect differences as little as five to ten percent between conditions—a practical difference from an engineering standpoint—with the use of a very severely limited number of subjects, when we use such techniques as Latin square design . . .

b) *Scheduling*. Another matter regards scheduling of test conditions and the main recommendation here is that a balance of conditions across time is necessary. Even though we may be dealing with very highly skilled personnel, there is usually some residual practice effect that appears in the data, and unless conditions are replicated along the time sequence so that each condition has an equal opportunity at various levels of practice on the part of the subjects, we stand a chance to substantially bias the outcome of the experiment.

c) *Parameter Definition*. With regard to parameter definition, the decision criteria must be made explicit in the planning stage; that is, if we choose a certain type of aircraft or certain types of aircraft to employ in the simulated operations, the reasoning that we went through to choose this aircraft type must be laid out. If the parameter is supposed to be representative, we must indicate that such is the case. If we are attempting to utilize parametric values that represent future conditions, then this must be made explicit also.

d) *Data Handling and Storage*. The last point to be discussed here is data handling and storage. Very specific recommendations along this line are that for any one criterion, multiple data recording devices be employed whenever possible. This helps to insure the reliability of the data and also its objectivity. For instances, if we have a manual recording of flight time, it is sometimes possible to obtain the same information from automatic equipment. The point is that just having the automatic equipment does not necessarily mean that one should drop the manual recording process, since the duplexing of this critical data recording operation adds to the soundness and the confidence that we can attribute to the information we have acquired. The storage of data is a rather minor matter, but should be managed in the sense that people have had a sad experience where they have not stored data in a way which made retrieval simple. Storage in an unorganized fashion can effectively destroy the usefulness of an experiment, especially where analysis or report writing is delayed over a period of time after the experiment is completed.

B. Controls

Let us now consider experimental control in a little more detail and consider what happens when there are breakdowns at various times. For example, we may be right in the middle of a data collection activity or an exercise or a problem and have a piece of apparatus break down, or a participant may become ill or fail to show up at his scheduled reporting time. We must establish some criterion in advance for whether or not we will continue to collect data; whether we can substitute equipments or subject participants; whether we must abort that particular sequence of tests; or, in the most unfortunate instances, whether we must abort the entire experiment. Many of these criteria are strictly arbitrary, but they must be explicit, again, in order for a rational interpretation and evaluation of experimental results to be made.

By and large, the rule is against substitution, especially substitution of subject participants. This is because the state-of-the-art, as far as measuring subject participant characteristics is concerned, is not developed well enough so that we can actually make a very precise evaluation of what effects such substitution would have.

Another factor that should be specified carefully in advance is the role of the study supervisor. Ordinarily, it is your supervisory personnel who make or break an experimental program. Their role as decision makers, their reporting requirements to the research director or manager, etc., should be laid out well in advance. Preferably, report forms should be established so that the information will be consistent and readily interpretable.

Another consideration is the use of shake-down runs, since they are a potential source of bias. While on the one hand it is very often necessary to test out equipment in a preliminary fashion or test out experimental ideas and controls prior to a formal test series, one can lose some experimental precision by this process. For example, if a shake-down run is to be accomplished with the same subject participants who are going to be active in the formal test run proper, a certain amount of selected indoctrination can occur and attitudes and biases on the part of your participants can be established during the shake-down process. The selection of conditions under which shake-down runs are accomplished requires, therefore, some very judicious planning and control.

We have already discussed briefly the requirement that the manager or research supervisor keep on top of the situation as it progresses with regard to such things as subject participant inadequacies, equipment breakdowns and failures, and other types of contingencies. It is also possible for data processing to take place concomitant with data collection such that preliminary evaluations can be made of the quality of the experimental test program itself or the adequency of the original hypotheses before a study is entirely completed. There are dangers involved in this, of course, if a balanced schedule is employed because the preliminary evaluation of the data that takes place before the total sequence has been completed may not be valid because of the unbalance. However, if experimental procedures and methods are inadequate, leading to a severe amount of error variability, this fact can often be detected very early in the process and means can be employed which will correct the condition. The study may then be reinitiated. It is helpful, therefore, to have regular evaluations of the data being produced while the study is going on and not wait until the study is complete before any summarization is undertaken.

## KINKADE, KIDD, URBACK, ICHNIOWSKY, AND WIDHELM (1963)

In a sense following up the earlier efforts of Courtney and Company, a handbook "to describe the scientific methods that are applicable to air traffic control (ATC) system research" was compiled by Aircraft Armaments, Inc., for National Aviation Facilities Experimental Center (NAFEC) personnel of the Federal Aviation Agency. This handbook is summarized here to indicate its contents.

It first dealt with basic concepts, such as variability and its measures, additivity of variance, variance ratio, system error, and linear and other relationships. Then in outlining the study planning phase it discussed defining the level of effort and research objectives, defining test conditions, steps involved in evaluation studies (control and experimental conditions, effects on system factors, logical relationship between test conditions and objectives), steps involved in exploratory studies (levels of test conditions, interactions), specifying measures and system performance, and defining the precision of the experiment (significance level, erroneous conclusions, some steps in increasing precision).

Next, the handbook considered test strategy selection under the following headings:

- Defining Potential Sources of Bias
  - Traffic sources
  - Component sources
  - Environment sources
  - Organizational sources
- Types of Biases
  - Discrete biases
  - Systematically changing biases
  - Randomly fluctuating biases
- Equalizing Bias Effects
  - Procedure for controlling discrete biases
  - Procedure for controlling systematically changing biases
  - Processes for controlling randomly fluctuating biases
- Selecting and Defining Constant Factors
- Sampling Procedures
  - Random sampling procedure
  - Upper limit sampling procedure
  - Critical-position procedure
  - Dependent-team procedure
- Outlining an Experimental Design
  - Single variable designs
  - Confounding
  - Multi-variate designs

Of particular value, if allusions to air traffic control research are generalized, is a "Dynamic Simulation Checklist." It is reprinted here in case interested readers are unable to obtain the handbook itself.

## STUDY PLANNING PHASE

### FORMULATING THE PROBLEM

*Defining the Level of Effort and Research Objectives*

1. Is a simulation experiment the most suitable test approach?

*Comment*—The status of an issue may be such that simulation is not the right tool. Some problems are too vague for rigorous experimental treatment and require preliminary definement by means of field research or other research methods. Some problems are so highly specific, on the other hand, that a simulator approach is too elaborate. Other methods such as small scale experimentation, graphic analysis, fast-time simulation, etc. should be considered.

2. What is the long range objective of the experiment?

*Comment*—The experimenter will do a better job if he understands how the results of the experiment are to be used. He should be aware of the relationships between the research and the policy goals of the branch, division, and service, as well as the FAA as a whole. If policy goals are incompatible with good research technique, this potential source of trouble should be made explicit before the experiment is started.

3. If multiple purposes are involved, is a single experiment enough?

*Comment*—Killing two birds with one stone is frequently attempted in ATC simulation research. However, there are times when either policy or technical objectives are so complex that they generate a large number of testable issues; too many for a single experiment. Moreover, some issues do not fit well in the same experiment.

4. To what ATC facility or class of facilities are the results of the experiment intended to apply?
5. Will the schedule of test trials yield results while the need for answers is still salient?—Or will the decisions be forced before the experiment can be concluded? If so, should the experiment be initiated?
6. Have prior research findings related to the objectives been studied and evaluated?
7. What are the consequences of the decision which will be based on the information supplied by the research?
8. What is the immediate technical purpose of the experiment?

*Comment*—Some issues are not easily translated into testable alternatives. The logic of experimentation requires that some cause and effect statement be prepared, which becomes the "character" of the experiment. Another way of stating the matter is: What are you trying to prove?

9. Can the final statement of the problem, expressed in the language of experimental test, be recognized as the same problem used to initiate the planning process?

*Comment*—Problem definition for experimentation involves abstraction and simplification. The experimentally oriented interpretation of the problem should be reviewed by the originators of the issues—thereby preventing unpleasant consequences.

### STATEMENT OF THE PROBLEM

*Defining Test Conditions*

1. Does the problem require an evaluation or an exploratory study?
2. How many test conditions must be specified?

3. Is the problem concerned with the direct effects of the experimental change or with both the direct and indirect effects?

4. Will the experiment result in clear recommendations regardless of the statistical outcome?

*Comment*—It is the responsibility of the researcher to make sure that the experiment will make a positive contribution to the management decision regardless of the content of the results. Provisions must be made at the start for the interpretation of negative statistical outcomes.

5. What is the control or reference condition of the test?
6. What factor or factors are to be varied in the experiment?
7. Can the changes be described by a single dimension?
8. Can the change from condition to condition be described numerically, or are qualitative changes involved?
9. Is there likely to be an interaction between the factor being investigated and other factors?—If so, can different levels of these factors be investigated?

*Specifying Measures of Performance*

1. What kinds of changes are expected?—Are the experimental manipulations intended to influence the work pattern of the controller, immediate system effectiveness, safety?—Some aspects, or all aspects of system performance?
2. What measures will reflect these changes in performance?
3. Are the contemplated measures directly relevant to the purposes of the experiment?
4. Are the measurement operations susceptible to recording error?
5. Does the list of contemplated measures include any "hedge" or insurance against unexpected effects?

*Comment*—If the research provides unexpected outcomes, the experimenter frequently is at a loss to explain why such effects were obtained.

6. Do the measures under consideration lend themselves to clear interpretation and presentation in a research report?

*Comment*—Some performance measures, derived from a combination of other measures, involve complex transformations. These may be so extensive that the experimenter is hard-pressed to explain why they were used and how they should be interpreted.

7. Have the primary system performance measures been specified on the basis of the experimental objectives?
8. Has an interpretation plan been formulated in case the results obtained from different measures are not compatible?

*Defining the Precision of the Experiment*

1. Are gross effects of major interest or should the experiment be very precise?
2. Will it be necessary to conclude that the experimental change will not affect performance?—If so, will the experiment be sufficiently precise to permit this conclusion?
3. How large an error variance and an experimental effect is expected? Will these expected figures provide statistically significant results?
4. Which of the two types of erroneous conclusions is the most acceptable?
5. Has the significance level been specified?—If so, is it compatible with the experimental objectives?

SELECTING A TEST STRATEGY

*Practical Considerations*

1. Is the schedule of test trials compatible with the availability of equipment and test personnel?

2. Is there sufficient time to complete the study, including some time for unexpected delays?

3. Are funds which are available for the study sufficient?

4. How many people can be used on the project to help run and implement the study?

5. How much equipment is currently available, will be available within a short time, or should be purchased?

6. Do these practical considerations so constrain the study that it cannot be completed—and, therefore, should not be started?

*Sources of Bias*

1. Have all the potential sources of bias which could obviate the experiment been defined?

2. Have the types of biases been anticipated?

3. Have bias control procedures been firmly established?

*Selecting and Defining Constant Factors*

1. To what area, areas, situation, or situations will the results be applied?

2. Is the sample of constant factors representative of these areas and situations? In particular:

a. Are wind and weather to be held constant?—If so, are the levels representative?

b. Are traffic load and composition to be held constant?—If so, will they contribute to decreasing the error variance?

c. Are route geometry and area geography factors to be held constant?—If so, are they representative, but not overly complex?

d. Are ground-rules and procedures to be held constant?—If so, are they representative, as well as easily understood and followed?

*Sampling Procedures*

1. Are the sampling procedures related to the research objectives and practical considerations?

2. If the random sampling procedure has been adopted, will there be a sufficient number of subjects available?

3. Is the effect of the experimental change under "fair weather" conditions an important consideration?—If not, has an upper-limit sampling procedure been adopted?

4. If an upper-limit sampling procedure has been adopted, is the load likely to be so great that it will lead to a system breakdown?

*Comment*—The "test to breakdown" is a valuable research technique and provides important information. However, if the intent of the experiment is to measure system performance under adverse conditions, the load should not be so great that the system will cease to function. This would result in a loss of data and might destroy the logic of the experimental design.

5. If the critical position sampling procedure has been adopted, which positions are "critical"?

6. How many critical crews are to be used in the test?—Is this compatible with the required precision of the experiment?

7. If the dependent-team sampling procedure has been adopted, what kind of crew composition program is applicable?

8. Have steps been taken to insure that effects like learning and fatigue will not invalidate the assumption of statistical independence?

*Outlining an Experimental Design*

1. What bias control procedures, constant factors, and sampling procedures must be incorporated into the experimental design?

2. Does the test design conform to the logic of the experimental problem?

3. Will interruptions or forced changes in the schedule ruin the logic of the experiment?

4. Is the design expandable or modifiable in the event of emergencies like equipment failure or loss of controller crew members?

5. How many test trials will be conducted under each experimental condition? Is this compatible with the required precision of the experiment?

6. How many trials will be repeats of the same controller crew under the same conditions?

*Comment*—The reason that some experimental designs use "repeated trials" is to obtain a better estimate of crew performance under the test conditions. This technique is applicable where crew performance is relatively stable in the test conditions. When it is not, little can be gained from using this technique. The research time would be better spent in testing different controller crews under the test conditions. This use of a larger sample would insure greater generalizability of results, and increased precision.

7. What is the length of each test trial, and what is the total number of trials for the whole experiment? Is this compatible with practical considerations?

8. Is the design compatible with the statistical treatment plans?

*Comment*—A frequent mistake is to design an experiment in such a way that the assumptions of standard statistical tests cannot be met. The experimenter then pays the penalty of lost precision by having to use a less powerful test, or, worse, in not being able to assess the reliability of the obtained differences at all.

9. Has a restricted random arrangement of test conditions been selected?—If so, will this allow the objectives of the experiment to be met?

10. Have different factors been so confounded in the design that their individual effects cannot be assessed?

11. Is the design so complex that there will be problems in data reduction, statistical analysis, and interpretation of the results?

## TEST OPERATIONS

### PREPARATION OF TEST PERSONNEL

1. Are all target generator operators (TGOs) and TGO supervisors thoroughly briefed on the specific ground rules of the present experiment?

2. Are all TGOs competent in flight and voice procedures?

3. Has agreement been established between project staff and TGO supervisors with respect to coordination procedures?

*Comment*—Two matters are of particular importance: moment-to-moment trip assignment coordination with special emphasis on terminal departures; and adjustment in pilot response to compensate for position misalignments between pilot and controller displays. A third factor is the recovery and re-setup after an error in starting position or starting time.

4. Are sufficient numbers of qualified controllers available for scheduling the planned number of crews?

*Comment*—Limited numbers of controllers may force resort to controller rotation procedures. Several examples of feasible rotation procedures are included in Appendix A of this handbook.

5. Have all participating controllers been briefed on the general objectives of the study, the ground rules in effect, and the schedule of runs?

*Comment*—For motivational purposes, the controllers should be told as much as possible about the experiment short of providing them with information that could bias the results.

6. If peculiar procedures, new equipment, or unusual traffic factors are to be included in some test conditions, have all controllers been given specific famil-

iarization practice so that learning on the job and experimental bias is minimized?

7. Are emergency standby personnel available to fill in for TGOs, TGO supervisors, or non-critical controller positions in case of illness or annual leave?

8. Does everyone concerned with the actual runs know what he is supposed to do and when he is supposed to do it?

*Comment*—A small amount of "dress rehearsal" is usually a good way of insuring that all the bugs are out of the operation. One procedure which has worked well is for the test director to inform everyone that the formal data runs are starting on a particular day. However, he has scheduled this start-day just prior to the actual scheduled start-day. If everything goes smoothly during the dress rehearsal the data can be retained, but if any forgotten details show up, there is time to remedy them.

## PREPARATION OF TEST ENVIRONMENT

### *Equipment*

1. Have all equipment components been tested (as a system) and found to be in operating condition?

2. Have maintenance personnel been alerted to the test schedule?

3. Have calibration procedures been established with the experimental design in mind?

4. Have provisions been made to insure that all situation factors will be held constant?

*Comment*—The experimenter may become concerned when he finds that there are certain amounts of voltage drifts and other equipment malfunctions occurring during the test trials. He may even decide to "peak up" the equipment in the middle of the experiment to obtain more constant test conditions. This procedure could defeat his purpose. Since certain minor equipment malfunctions occur randomly during every trial in a system as complex as the ATC simulator, they may be considered as a constant factor. By changing the condition of the equipment in the middle of an experiment, the experimenter would be increasing, rather than decreasing, the variability in performance measures.

5. Are standby equipments necessary or feasible in the event of catastrophic equipment failure?

### *Traffic Factors*

1. Will there be a sufficient number of simulated aircraft to perform the experiment?

2. Will the traffic composition and the input schedule contribute to increased variability?

*Comment*—This topic is specifically discussed in Method Development for ATC System Study.

### *Geometry and Geography*

1. Is the geometrical configuration so confusing that it will increase the variability without contributing to close up generalizability?

*Comment*—In an effort to be realistic, complex area geometries and geographies are frequently employed. It should be realized that the experimenter pays a high price for this surface realism. Extensive familiarization training is usually required and learning bias effects have to be strictly controlled in the experiment. Some supervisors say that, in the operational situation, it usually takes a controller from one to two years before he is considered to be completely familiar with the area. The experimental disadvantages of using a complex, realistic geometry probably outweigh any purported advantages in generalizability.

*Others*

1. Are the other constant factors overly complex, contributing to increased variability without contributing to the objectives of the experiment?

## PREPARATIONS FOR DATA COLLECTION

1. Have clear, concise data recording sheets been designed?

*Comment*—There seems to be a tendency to have the people who record data obtain as much information as possible. Frequently, this interferes with the persons' other duties and critical mistakes are made—both in recording the data and performing the other duties.

2. Have provisions been made to insure that all stations, including data recording positions, are manned at the start of each run?

3. Have provisions been made to assemble, label, package, and store all data record forms immediately following each run?

*Comment*—Proper labeling of data record forms is essential. While the data record forms are being handled, the people involved are very familiar with abbreviations, codes, symbols, etc. However, as these people work on other projects and a certain period of time passes, the meanings of these codes are lost and this frequently means that the data are also lost.

4. Are immediate spot checks of critical items of data required?

5. Have provisions been made to meet all important contingencies including the discovery that the experiment is not working as expected?

6. Has a clear-cut assignment been made with respect to the data reduction function?

*Comment*—Data reduction, in this case, refers to the summarization of each test trial. The tabulation of indices like average delay, number of flights processed, number of conflicts, etc. is the end product. The task may be assigned to a computer, a clerical group, or both. Both statistical clerks and computers must be "programmed" and activated. Valuable time and continuity in data reduction are frequently lost when this step is neglected.

7. Have provisions been made for detailed review of the data as they are being gathered, as a check on the completeness and as a search for unexpected outcomes?

*Comment*—Researchers often fail to extract all the potential information from an experiment by relying solely on routine procedures for data processing. At the end of an experiment, they may find that a critical performance measure should have been recorded, but was not. This problem might be avoided if the data are reviewed after completion of portions of the experiment.

## PREPARATION FOR TEST RUN MANAGEMENT

1. Is the number of shakedown runs sufficient to establish routine procedures?

2. Have provisions been made for telling the project manager about on-the-spot decisions made by other supervisors?

*Comment*—The project leader cannot make all of the decisions concerning minor details. However, he is the one person who has the "big picture" in mind and some decisions regarding some details could ruin the experiment. Therefore, a log or some other reporting device should be employed so that he is aware of the decisions that have been made and can assess their consequences in relation to the experimental objectives.

3. Have provisions been made for annual leave, sickness, equipment failures, or unexpected outcomes in the data?

*Comment*—It is usually a good idea to have a back-up plan available in case things go wrong. Valuable simulation time will not be lost and something could be salvaged from the experiment.

4. Has a list of the specific procedures which should be followed during the run been prepared?

The handbook did not neglect "interpreting and reporting," an aspect of man-machine system experimentation to which researchers might well pay more attention. (If they always had, this book would have been much easier to write.) It was suggested that experimenters should include insights or circumstances which were not reflected in the reported data and should state recommendations and how they were reached. It cautioned against the predictive or operational use of absolute numbers derived in simulation-based studies, since these can reliably indicate relative effects but lack precision. The handbook also cautioned against the inclusion of too much detail, since the purpose of an FAA report is simply to inform and not to enable the reader to duplicate the experiment. (Left unsaid was how much detail is required to provide confidence in the research results to the discriminating reader.) The handbook did suggest a general format, as follows:

Introduction
- Statement of the problem
- Review of previous work
- How the problem is going to be studied

Methods
- Task description
- Apparatus
- Subjects
- Test conditions
- Statistical design
- Performance measures

Results
- Main effects
- Interactions
- Subordinate effects

Discussion
- Restatement of the problem
- Conclusions
- Implications
- Explanations
- Concrete recommendations

## HAYTHORN (1963*b*)

Considerable reflection about the methodology of man-machine system experimentation has come from W. W. Haythorn and the other researchers in the RAND Corporation Logistics Systems Laboratory. The twenty-five research functions which Chapter 13 noted as identified by Haythorn (1963*b*) are presented here in their original detail. They form a useful framework for anyone undertaking this kind of research, although some aspects are oriented to a particular type of system and experimental objective.

1. Field research: The observation and measurement of relevant real-world systems preparatory to representing them in the laboratory.
2. Written descriptions of field functions: The organized presentation of results of the field research, documenting the observations made.
3. Modeling: The construction of computer or other models to represent real-world functions in the laboratory. This is a topic worthy of extensive discussion in its own right. The availability of improved programming languages, general purpose simulation languages, standardized flow-charting procedures, etc., has greatly increased the researcher's ability to construct models quickly. Some of the more significant decisions that must be made in the modelling process are the degree of time aggregation to be attempted; whether to use event- or time-interval pacing of the model; whether to computerize particular decision points or leave them to human decision-makers; whether to assume a static, pre-canned input or a dynamic relationship between inputs and system performance; and what records of model performance should be retained, and how.
4. Information system representation: The modelling and/or simulation of information system design considerations for laboratory representation. This includes the determination of what information is to be given to decision-makers in the system, the form it is to be given, specification of assumptions regarding the sources of the information, the processing performed on it before its presentation to the decision-maker, information storage assumptions, the degree of currency and accuracy assumed to exist in the information, quality control procedures to be built into the information system, etc.
5. Computer programming: The actual programming and coding for computer representation of those features of the modelling and information system representations requiring it.
6. Demand generation: The preparation of experimental control procedures for exercising functions of the system under investigation—stimulus generation in a sense.
7. Policy justification: The written presentation of a justification, including background research, for including particular policy alternatives in the laboratory study. This was felt necessary since the number of possible policy alternatives is very great, and costs prohibit laboratory exploration of all of them.
8. Laboratory implementation of policy: The identification of requirements and development of techniques for including policy alternatives in the laboratory system.
9. Real-world policy implementations: The identification of requirements and development of techniques for including policy alternatives in the laboratory system.
10. Input data requirements: The identification of data required for the preparation of stimulus inputs to the system.
11. Input data collection: The acquisition of data required for input preparation.
12. Laboratory data production: The data processing and manipulation necessary to prepare laboratory inputs from the background data obtained through the preceding two functions.
13. Experimental design: The development of a set of experimental conditions to obtain answers to the questions generating the research.
14. Participant or subject orientation: The instructions or training provided to participants before the beginning of the laboratory exercise. This is a frequently neglected but crucially important part of any such study.

15. Participant manual preparation: The writing and publication of manuals required to provide participants with information concerning operating procedures, system policies, and so on.
16. Performance evaluation: The determination of criterion measures required to assess the effectiveness of system performance.
17. Cost evaluation: The determination of procedures for assessing the cost of system design and operation.
18. Analysis: The preparation of plans for, and the conduct of, analysis of the performance and cost data collected during the experiment.
19. Documentation: The write-ups of all functions undertaken with regard to the system. We have felt it to be highly desirable to document our activity as thoroughly as possible. This is important not only for presentation to other researchers and to the customers, but for one's own review to help recall why things developed as they did.
20. Spin-offs: The identification and development of by-products useful to the customer, the researcher, or the scientific community.
21. Laboratory operations: The conduct and control of experimental runs, including the operation of imbedding organizations, environmental inputs models, monitoring of human decision-makers, etc.
22. Implementation aid to the customer: The post-experimental assistance usually required in communicating laboratory notions or results to customers in such a way as to facilitate their implementation in the real world.
23. Briefings to the customer: The usual round of formal and informal talks required to communicate broadly to one's customer the fact that research has been completed and that results of possible value have been obtained.
24. Interface specifications: The identification and specification of input-output relationships between and among functions served in the system under investigation.
25. Final reports: The terminal process of writing up for public display the objectives, conduct, and result of one's research.

# APPENDIX III

## Generalizations

Although most man-machine system experiments have attempted to produce knowledge pertinent only to a particular system, both these and experiments with general knowledge objectives have yielded results which can be generalized. Some of these generalizations are summarized in this appendix. How far the generality extends is not readily determinable, and in any case the fact that there is any at all is largely the opinion of this book's author. The generalizations will be organized according to the fields of knowledge described in Chapter 25. At the end are noted several system phenomena concerning which the author believes there is too little understanding.

### DESIGN

Greater automation does not necessarily increase system effectiveness. Although this may appear obvious to the reader, it has not seemed so to some system designers.

Design decisions as to what should be automated and what should not may be based on invalid assumptions about human capabilities relative to those of machines.

Alternatives to automation include improved design of the current system, better procedures, more training of personnel and better training techniques, and improved personnel selection.

Relative effectiveness of automatic and manual modes can depend on the kinds of inputs (e.g., noise), the level of load, the measures used, and the system function.

An automatic mode is not necessarily superior with heavier loads or complex inputs. Human operators and decision-makers can apply capabilities in pattern perception that are lacking in the automatic mode.

Automatic modes are free of the human "maybe" reaction—to do something with a possible but improbable pay-off—and the human survival reaction—to be oversensitive to the possibility of one's own destruction.

When a human mode is back-up for an automatic mode, operators need a display of information accumulated before the back-up operation takes over. Training is also necessary.

One version of man-computer symbiosis is to give operators options to override or reject automatic mode (computer) actions or recommendations. Such interventions can be expected to vary in extent according to the personality of the individual and the load; heavier load may actually lead to more human interventions in decision situations.

Automatic mode actions or recommendations can be set by the operator to occur only on demand, or automatically at some maximum rate (such as at six second intervals). The design can give the operator one of four options: doing nothing; accept by doing nothing and reject by switch action; accept by switch action and reject by doing nothing; and accept and reject by switch action. The interval during which the operator must make his choice can be fixed or varied by the operator; it may be preferable for the computer to pace the operator.

Although the computer might be required to repeat a rejected recommendation, this seems inadvisable. The design can enable the operator to ask the computer for its reasons; these are probably best stated in the form of the data that led to them. The computer may be required to display the reliability of the data.

Another form of man-computer symbiosis is to give the operator or decision-maker the ability to choose the computer program in the first place, or to change it. The operator's choice would depend on the characteristics of the problem situation to be resolved.

When a large set or matrix of numbers must be processed so a person can make a decision, this is best done automatically. This is true also for statistical treatment of data.

In handling noise, the automatic mode should dispose of inputs about which no doubt exists, both noise and signals, according to cut-offs in the equipment or program which may be alterable by the operator. The operator can be assigned the marginal cases to process with his pattern-recognition and judgmental capabilities.

When significant unquantifiable or unforeseen factors enter into a choice, human processing should supplement or displace the automatic mode. In the supplementary role the human operator can weight these factors.

Design should give special attention to the potentials of human filtering of sensor data before they are converted into digital inputs, and quality control of human-generated data.

Human operators often fail to detect the absence of a signal when the absence is due to (1) the signal's elimination during processing in an automatic mode or (2) nonoccurrence or delay of the event that would produce it. Operators can be helped by receiving contextual indicators that the signal might have been expected.

Humans make mistakes in converting data terms from one frame of reference to another, e.g., from azimuth-and-range co-ordinates to grid co-ordinates. Such conversions are much better done automatically.

Humans are liable to high error rates in making switch-actions to communicate with a computer, whether on a function-switch panel or typewriter keyboard. Error frequency rises with the load and complexity of messages. There may be useless repetition of function-switch actuations.

Human operators can make better predictions based on geometric relationships (spatial, angular) than is generally realized—for example, predictions of courses, collision points, and impact points. Man-machine symbiosis calls for operators to make initial estimations when these need to be only approximate, and for computers to make them when they must become precise; "approximate" varies according to the particular system and task.

As input load increases, human operators do less well in making predictions which must consider the passage of time—such as speed—than in making spatial predictions. This suggests it is a good idea to automate estimations of rate and duration.

Input load may be expressed as the ratio between some unit of time (including time available) and the number of events to which an operator must respond. Increasing load has three effects: (1) Up to a point, there is a pacing of operator performance to increase its rate. (2) Some cues are disregarded or performance with some task elements is omitted or becomes poorer, while response to major cues and performance with seemingly more important task elements are maintained. (3) Errors increase.

Although it is difficult to generalize about task capacity because of differences between complex tasks, well-selected, well-trained system operators have higher capacities than may be realized; for example, they can track or control a half-dozen aircraft during the same period of time or make more than that many successive, complex judgments in assigning resources during one minute.

*Displays*

When an operator has to keep shifting between different tasks or task elements, his performance is aided by displays which indicate what he has done and should do in each instance. Otherwise his short-term memory becomes overburdened.

Data should be categorized by a computer program for display to an operator according to the usefulness of the categories to the operator—not according to the ease with which they can be programmed.

Display design can profit from grouping together the data for a task, demarcating groupings of data, displaying at the same time (instead of serially) items among which a choice must be made, and other formatting features. Formats and coding should be consistent between related displays. These can include checklists, handbooks, and computer printouts, as well as wall displays, console displays, and cockpit displays.

When numeral coding is used within a display, those items from different subsystems and functions that are regularly or frequently associated with each other should share the same numerals, although the total designators may differ; such items include paths that lead into each other, and locations and the units stationed at them.

When two system elements remote from each other interact, co-ordination may be enhanced if related or similar displays are available in each.

A major design decision is whether to display all planned operations and the steps being taken or only deviations from plan—a display for management by exception. Attention is drawn to this here because often only the former is considered.

For scheduling and co-ordination tasks which depend on each other all planned operations should be displayed, along with (1) feedback about starts, stops, and delays; and (2) time requirements for essential and nonessential tasks (so designated) with emphasis on the time requirement (and deadline) for the longest task.

Designers should keep in mind the trade-offs that can be made between displays and intercommunication; the more of one, the less of the other.

Feedback should be displayed by a computer to a console operator to indicate (1) that the computer has accepted his input, (2) that the computer regards his input as inadmissible, and (3) that the computer is busy and there will be a processing delay. If there are processing delays but no feedback, the operator will make many switch-action errors.

Input devices should incorporate feedback displays to show an operator what he is entering into the computer so he can detect and correct his own errors.

In communications between individuals representing organizations, a communication method which produces a "hard copy" display and record (e.g., teletype) may be preferred by users in taking or maintaining a firm position to methods which do not (e.g., face-to-face discussion, or telephone).

To transfer pictorial data manually from one display to another, it seems better to give the operator at the second display direct visual access to the first display (by rearranging equipment and positions or through a closed-circuit television relay) than to have an operator at the first display convert the data to words and telephone these to the operator at the second display, who must then convert the words back to pictorial data. When one operator transmits data in words to another operator to place on a display (perhaps after conversion to pictorial data), the first operator often receives no feedback about his own or the second operator's actions to help him detect errors committed during the transmission process. Such "invisible" functions contribute to system error.

(Note: Other display design considerations have been suggested in Chapters 10 and 11 and are described in the reports referenced there.)

## PROCEDURES

When a team is given latitude to do so, it will develop its own procedures and change those under which it had been operating to adapt to new situations. Such situations include increases in load, or the team task when the team is first formed.

The course of a team's self-procedurization will be shaped by its goals. One of these may be to achieve success (or avoid failure) or to escape the stress of the task. More specific goals are the criteria of success given to the crew as instructions.

Self-procedurization consists of evolving a new procedure or procedural change, in contrast to the acquisition of skill in putting it—or any procedure—into practice. Different factors facilitate these different processes.

Among the factors which facilitate self-procedurization is an opportunity for the team to receive feedback information about system performance and to discuss possible procedural changes to solve problems which occurred during that performance.

One way in which teams procedurize to handle heavy loads is to assign priorities to inputs and drop tasks which have low priorities or are viewed as nonessential.

A manageable input rate can also be achieved by interposing a buffer procedure and device. Inputs reaching it at a high rate can be removed at a lower rate. Another procedure for coping with heavy loads is to sequence inputs or tasks in advance so inputs occur at a manageable rate and tasks are scheduled in relation to each other and their required durations.

A third way of handling heavy loads is to schedule in an opportunistic fashion, combining planned activities with unplanned activities that occur without warning—as in doing periodic preventive maintenance when it becomes necessary to do emergency maintenance.

Procedural flexibility can help dealing with heavy loads in carrying out recurrent system functions, such as the routing of aircraft; but it is probably better if emergency procedures and communications between operators are standardized.

Intercommunication of data between two operators can take many different forms other than a transmission or relay. Either the sender or the receiver may be required to make transformations between oral and visual presentations, between words (as in translation), between pictorial forms, and between words and pictorial form; and the same individual may be both a sender and receiver. Although well-trained individuals can make transformations effectively, such transformations are sources of error, perhaps especially so when the same individual must make two transformations as receiver, recorder, and sender.

The quantity of information transmitted by one organization to another, including from one echelon to a higher one, can be reduced by limiting the information to that which enables the receiver to take action; this procedure eliminates information which merely satisfies curiosity.

There are various ways to arrange telephone or radio networks or channels for a team of operators reporting to some receiver or for a group of individuals communicating with each other. Each operator may report to the receiver or each individual to another on a point-to-point basis; or all can be connected in a conference-line hook-up. In the latter arrangement each participant can have continuous access to the party line, or he can have successive access. In the case of successive access, an individual can make a request at any time (and join a queue), or he can make a request only when another speaker has finished. One method may be as effective as another. Larger groups benefit more from successive access than smaller groups.

Without pre-established procedures for selecting a chairman and for making clear what he should do, a group of individuals intercommunicating through a conference-line hook-up finds it difficult to solve problems common to all participants.

Verbal communication between operators may be carried on through an intercom or on a face-to-face basis. Possible distraction effects from face-to-face communication should be considered in making the choice.

Tasks vary according to the operator actions required and the objects of the actions. Tasks can be assigned among members of a team by requiring of each operator (1) the same single action for a large number of objects, or (2) a number of different actions for one or a few objects. When differing actions occur in succession in a system function, the assignment of all to each of the operators along with a particular object produces parallel processing. Series or in-line processing occurs when one operator carries out the same action for each of many objects, and another carries out a subsequent action for them. Relative advantage seems to depend on how smart the operators are and whether they work face to face.

When the objects not only are many but differ in two or more categories, the operator's task becomes more demanding, especially if it also consists of multiple, different actions. Homogeneity of objects for each operator in task distribution can improve team functioning.

Better distribution of tasks can unburden an overloaded operator–for example, by requiring an effector agent (e.g., pilot) to take over control when the regular control agent (e.g., controller) is coping with heavy loads.

Interception and air traffic control tasks can be assigned by giving each operator a certain area, by giving each a particular target or aircraft destination, or by systematically alternating between operators. Among factors to consider in selecting the method are concentration of visual attention, load balancing among operators, transfer of track or controlled object from operator to operator, and flexibility in control.

Task distribution may depend on provision of equipment to carry out the task. When the task might be performed by either of two operators–for example, a ground controller or a pilot–provision of the required display only to the controller can obscure the advantage of giving the task instead to the pilot. (Task distribution may be regarded also as a matter of *Organization*.)

## ORGANIZATION

Co-ordination and integration between team members place a load on the team which may counteract much of the benefit that might otherwise accrue from increasing the size of a small team to handle heavy loads. Another possible reason why reduction in team size need not degrade performance is that individuals work harder to compensate for the reduction.

Effects of structural variables are difficult to ascertain in experiments because self-procedurization and changes in individual effort may obscure them. In addition, it is not clear how much can be generalized about centralized and decentralized structures in system organization, due to differences between systems.

Some systems may profit from combining operational and maintenance activities in the same individuals or at least making the traditional division of responsibilities less severe. For instance, operational exercises might be associated with activities needed for preventive maintenance and trouble-shooting.

In man-machine systems certain linkages between operator positions and certain nodal positions seem to have special significance for system functioning. One such linkage is that between sensor and processor, another between controller and effector. Nodal positions include those through which sensors and inputs are distributed among processor personnel and outputs and effectors among controller personnel. Others deal with threat evaluation and resource management.

In the data processing part of a manual system the effects of turnover seem to be most damaging in the positions close to the system input, in positions overloaded by input, and in positions linked by communications (e.g., telephone) to the turnover position.

(The self-procedurization described under *Procedures* can be regarded as organizational adaptation. The question of team size discussed above can be viewed also as a matter of *Personnel Requirements*.)

## PERSONNEL REQUIREMENTS

The part played by individual skill and personality in man-machine systems is dramatized by the large differences among small teams and individuals who man the same functions and tasks in these systems. Such differences characterize senior decision-makers as well as surveillance operators. The selection of personnel sometimes can have a greater impact on system effectiveness than equipment-related design variables.

To design the personnel part of the system, it is necessary to determine skill requirements for each task, fashion methods of testing individuals for their skill levels, and select those who meet the requirements. Apparently this process is not fully carried out for modern man-machine systems.

Among the operator skills involved in such systems are executing complex sequences of switch-actions rapidly at consoles, distributing tasks among other operators, altering procedures to allow for contingency situations, handling multi-item codes of numeral and pictorial symbols, understanding computer programming and logic, and operating alphanumeric keyboards. (When decision-makers are not adept at operating such keyboards, they should have aides as clerks.)

Individual differences in decision-making include degree of risk-taking, extent of adhering to established procedures, likelihood of perseverating or reacting in a stereotyped fashion in interpreting information, number of alternative organizations of information units used, number of rules used, and number of relationships adduced.

## DECISION-MAKING

Decision-makers sometimes not only reject decisions made by peers but also reject their own prior decisions, making different choices at different times although major factors remain the same.

Much as other system personnel filter out apparently minor items when the input load is heavy, commanders under pressure may disregard some of the more subtle cues given them and thereby forego making fine discriminations. If such cues mean the difference between a feint and a real attack, threat may be misdiagnosed.

In action selection, decision-makers may pay more attention to quantitative than to qualitative criteria, trying to counter every threat at the cost of selecting the best means for doing so.

Decision-makers need to integrate feedback from diverse sources and make use of feedback about their own force's attrition as well as adversary reactions.

Commanders sometimes are prodigal with their weapons, squandering them when the supply is plentiful or appears to exceed the demand. This tendency to use up resources has several implications: (1) A commander's decision criteria should include economic and logistics considerations, even during battle. (2) False cues coming from the adversary could seriously deplete the decision-maker's resources by leading him to overcommitment. (3) The apparent motivation to prevent damage to his own forces could limit the commander's caution in avoiding overcommitment and unjustified risk-taking.

The time span between a decision and the feedback of its consequences may influence the nature of decision-making if the decision-maker is swayed more by short-term than by long-term consequences. The decision-consequence or decision-feedback duration may rank with odds and pay-offs (probabilities and consequences) as a major variable in decision-making.

In conflict or competitive situations, decision-makers have to predict the decisions of their opponents. Sometimes they do this poorly.

Because decision-makers fail to revise their estimates enough given new data, they may be helped by computer-supported Bayesian processing. The extent varies according to the measure used and parameters of system input, such as data volume and degradation. Under some circumstances Bayesian processing fails to produce better decision-making which, incidentally, can improve with practice when humans engage in it. (More detailed treatment of the effects of Bayesian processing can be found in Chapter 21.)

Computer support of decision-making can also be furnished through automated statistical summaries and analysis, such as correlation analysis.

## TRAINING

Systems and subsystems in various configurations can be improved by functioning in exercises during which operators at their regular positions and equipment respond to inputs which are simulations. Such system training involves both the enhancement of interactive and individual skills and the development of interactional procedures, the relative contribution of each to system improvement being yet undetermined. The effectiveness of system training stems from a combination of feedback about performance, including knowledge of results, and discussion of procedures among team members; the relative contribution of each of these to system training is also undetermined.

Untrained system personnel may be grossly unprepared, lacking knowledge, skill, and procedures. System training can bring system performance to a level well above that which the system reaches without it.

A subsystem can be trained by itself if the inputs it would receive in real life can be properly simulated (which is not always feasible). By putting two subsystems into the same training exercise, however, the inputs to one are the outputs of the other, thus obviating the need for special simulation, and the interactions between them can be practiced and procedurized as well.

Remote subsystems and operators need to be included in system training if their performance is important to system functioning and if the rest of the system receives their outputs or they receive the system's outputs. Subsystems at the sensors (front end) of some systems fit this category, as do those among the effectors.

System training is particularly needed when a system first begins operations (or beforehand) and when novel situations arise, although it may be helpful also in maintaining performance at a desired level. For it to counteract the effects of turnover of personnel, apparently turnover must not reach the point where the skills acquired through training are steadily diluted.

System or team training seems to be of special benefit to "invisible" functions, where an operator is unable to determine the consequences of his actions during operations.

When operators process information in series, practice seems to help the first operator in the series in particular, at least at first. Until his error rate drops, the others fail to receive the error-free high rate input which will require better performance. Each gets more training when the one preceding him is trained to a high performance level. The same phenomenon occurs with the training of subsystems which operate in series. One index that the whole team or system is well-trained is the absence of any interaction between serial position and load.

To cope effectively with high input loads, teams and system must practice with such loads. If loads are progressively increased, it is likely that system output will also increase, but this does not necessarily mean that teams should always be trained by progressively raising the input loads. Constant high-load practice, at least under some circumstances, can be even more helpful than practice of equivalent duration in which loads progressively increase.

Team or system training which presents simulation inputs to operators at their system positions is not the only kind of training which improves performance. Initial practice with as highly abstracted simulation of the system can be beneficial. Individual training is also needed, and it may be an even better preparation for tasks which also involve a modest amount of interaction between operators than training which includes the interactions; the interaction requirement may interfere with learning that part of the task centered on individual performance.

## SALIENT MISUNDERSTANDINGS

From the author's viewpoint, man-machine system experiments have demonstrated a number of ways in which man-machine systems have been incompletely understood.

*Feedback.* This is an important factor in design, in training techniques, in procedurization, and in decision-making. Not only has it been widely neglected, its various forms and effects need much more research.

*Individual Differences.* Because most human factors researchers have been interested primarily in human engineering or training, the differences between operators of all types with respect to capability and personality (even after training) have received almost as little attention from experimenters as from system designers and developers. Personnel selection has been underplayed.

*Evaluation.* Some system designers have placed reliance on the opinions and preferences of so-called expert system operators. This is foolhardy. Effectiveness ratings, intuitions, and preference choices may provide suggestive leads but are not reliable guides, as demonstrated by their repeated disagreement with objective data. Experts also frequently disagree among themselves. Subjective data should be focused on specifics.

*Alternatives to Design.* Although the value of better training techniques, personnel selection, organization, and procedure development has been demonstrated by man-machine system experiments, these have not been taken as seriously as they should be in system development. Perhaps the major reason is that, not being hardware, such improvements do not make money for industrial organizations that develop systems. Apparently there is little financial profit in devising ways to make people perform better in systems. No wonder that most man-machine system experiments have been conducted by nonprofit organizations, universities, and government agencies.

*Intervening Variables.* Man-machine systems contain intervening variables. These are dependent variables in relation to system inputs and independent variables in relation to system outputs. For example, in serial information processing the outputs of one subsystem constitute the inputs to the next subsystem in line. Also, teams may develop performance procedures to handle system inputs and these procedures then help determine system outputs. The dual nature of such mediating variables must be recognized by system researchers.

# Glossary of Technical Terms and Abbreviations

(Note: Terms are defined according to their use in this book. Some have other meanings.)

AAW: Anti-air warfare (Navy air defense).

ACTER: Anti-countermeasures trainer developed to simulate ECM with the AN/GPS-T2 in the system training program.

ADDC: Air Defense Direction Center, in the manual air defense system.

Ad hoc: Type of experiemnt concerned with a particular system or situation.

ADIS: Air Defense Integrated System proposed by the University of Michigan Willow Run Research Center.

AEW: Airborne Early Warning.

AEW&C: Airborne Early Warning and Control.

AFB: Air Force Base.

Algorithm: Precise rule or procedure, as in computer instructions.

All-computer simulation: Simulation or experiment entirely within a computer.

Alphanumeric: Consisting of letters and numerals.

AMC: Air Materiel Command.

AMCP: Alternate Mobile Command Post.

Analog computer: A computer whose inputs and outputs are continuously varying amounts such as voltages.

Analog-digital conversion: Change from continuously varying voltages to binary coded pulses.

Analysis of variance: A technique for testing the statistical significance of results in an experiment which contains more than two states of an independent variable or more than one independent variable.

AN/FSG-32V: Advanced computer built for the Air Force.

AN/GPA-23: Air defense analog computing equipment for tracking and interception.

AN/GPS-T2: Transducer for converting markings on film into simulated radar signals.

AN/TSQ-13: Equipment for a proposed tactical air defense system.

API: Airborne Position Indicator for pilots.

APL: Applied Physics Laboratory of The Johns Hopkins University.

APRO: Army Personnel Research Office, later Behavior and Systems Research Laboratory.

ARDS: Aviation Research and Development Service of the FAA.

ARPA: Advanced Research Projects Agency of the Department of Defense.

ARTCC: Air Route Traffic Contol Center.

Assembly: Type of computer program which translates instructions in symbolic terminology into machine-processible form and assigns storage locations.

ASW: Antisubmarine warfare.

ATC: Air traffic control.

ATS: Air Traffic Service of the FAA.

Automatic tracking: Tracking in which the electronic elements which supply position data to a computer are associated automatically with radar signals.

Azimuth: True compass direction to an object.

AZRAN: Conversion of radar position data from azimuth and range descriptors to grid co-ordinates in the SAGE computer rather than in AEW&C aircraft.

Bayes theorem: Probability estimation of a situation or hypothesis by making and aggregating statements about the probabilities that fragmentary data have resulted from particular situations, to modify prior estimations of each situation's probability.

Bayesian processing: Use of the Bayes theorem to derive probability estimations.

Beam interception: Interception in which the angle between the headings of the interceptor and bomber approximates 90 degrees.

Bearing (absolute): Compass direction to an object.

Bearing (relative): Angle between a heading and direction to an object.

Benchmark system: Current system with which a proposed system is being compared.

Blip: Visible radar echo from an object, on a CRT display.

Blip/scan ratio: Ratio between the number of blips and number of composite scans of (looks at) an object, one per antenna rotation of a search radar.

Bomarc: Unmanned antiaircraft missile controlled by SAGE.

Buffer: Storage and scheduling device interposed between input source and processor and/or between processor and display (or transmission) of outputs.

BUIC: Back-Up Interceptor Control System which would take over air defense from a damaged SAGE in wartime.

CAA: Civil Aeronautics Administration.

Cartrac: Proposed analog computing system for tracking radar signals of aircraft.

Category testing: Air Force testing of new systems and equipment.

CDEC: Army's Combat Development Experimentation Center.

CDCEC: Combat Developments Command Experimentation Command.

Charactron display: Special CRT display of alphanumeric and other symbols.

CIC: Combat Information Center, a Navy information and control location.

CINC: Commander-in-chief.

Clock code: Set of symbols composed of the positions of the two clock hands.

Closing angle: In the closing phase of an interception, the angle between the interceptor and bomber headings.

Closing phase: In a beam interception, that portion where the interceptor is headed toward a virtual collision with the bomber.

Clutter: Large amount of data within some area of a geographical-type display.

Code: A set of symbols or other designations, and sometimes also the categories they label.

COIN: Royal Canadian Air Force's Committee on Information Needs.

Combat Center: SAGE location for co-ordinating a number of Direction Centers.

Comcon: Command-Control Simulation Facility at Ohio State University.

Command and control system: Military system for processing information in support of some command or control function.

Compiler: Computer program which changes instructions in a higher-order language into a more detailed, machine-processible language.

Conditional probability: In Bayesian processing, the probability that some fragmentary data resulted from a particular situation.

Confounding: Possible distortion of the effects of an independent variable by another variable.

Console: Combination of switch-action devices, displays, and communication terminal.

Constancy: Counteraction by which a variable maintains the same state, or the state of a variable remains the same, throughout an experiment.

Contamination: Distortion of the actual relationship between an independent and a dependent variable as a result of some aspect of the measurement process.

Contrast: Counteraction by which an experiment includes at least two states of an independent variable, one of which may be a zero state—that is, absence of the variable.

Controller: Operator who guides aircraft or other vehicles at a distance.

Counteraction: Any method of eliminating or minimizing confounding or contamination and thereby increasing the internal validity of results of an experiment.

Counterbalancing: Counteraction whereby effects of the order of presenting the states of independent variables are equalized.

Critical incident: Important event whose apparent cause and consequences can be noted by an observer.

CRT: Cathode ray tube for electronic displays.

Cross-telling: Reporting data from one location to another.

CSM: Command Service Module for the Apollo project.

Cursor: A marker superimposed on a display.

Data link: Communication of binary coded pulses by radio or wire.

DDR&E: Department of Defense Research and Engineering.

Dead reckoning: Advancing a vehicle's track in the absence of new signals, on the basis of the previous path and environmental factors.

Debriefing: Post-exercise review meeting.

Debugging: Trouble-shooting a computer program.

Deck: Observation area in a laboratory; also, a set of IBM cards for a particular function.

Digital-analog conversion: Change from binary coded pulses to continuously varying voltages.

Digitize: Convert into binary code.

Direction Center: A SAGE location for computer and human processing of radar data and guidance of interceptor aircraft.

Discriminative stimulus: Cue for taking (or not taking) some particular reinforced (or unreinforced) action.

DM: Decision-maker.

Dualex: Device for data link transmission from AEW&C aircraft.

Dyad: Pair, two interacting persons.

Dynamic simulation: Simulation which changes—not static.

EAM: Electronic accounting machinery or peripheral devices, such as printers, card-punching, card-sorting, and card-reading devices.

ECCM: Electronic counter-countermeasures.

ECM: Electronic countermeasures.

Effector: Subsystem or associated system for taking action.

Electronic countermeasures: Methods of obstructing an adversary's use of radar or radio.

Electronic warfare: Application of ECM, or conflict between ECM and ECCM.

Embedding: Larger context within which a simulated entity is situated.

Enroute: Air traffic outside of the control areas near airports.

Entry device: Method of manually introducing data, queries, or instructions into a computer.

Equivalence: Counteraction by which repeated instances of the state of a variable are equivalent to each other throughout an experiment.

Error variance: Variations in a measure not attributable to intentional differences between states of independent variables in an experiment.

FAA: Federal Aviation Agency (later, Administration).

Factor analysis: Statistical technique for extracting explanatory factors from multiple correlations.

Factorial design: Experimental design in which every state of each independent variable is combined with each state of every other variable.

Fade: Disappearance of a radar signal.

Feedback: Information concerning prior performance to help guide future performance.

Fidelity: Realism of simulation.

Filter: Remove noise, unwanted signals, or unnecessary data.

Fire control system: Radar and missile-firing equipment in an interceptor with which the pilot guides his aircraft near the target and fires his missiles.

First generation: The initial version of a system or equipment undergoing continuing development.

Fix: An airspace location to or from which aircraft are routed; also, the intersection of two or more absolute bearings from as many different positions.

Focal state: That state of an independent variable of particular interest, usually compared with a zero state—the absence of the variable.

FPO: Filter Plot Officer.

Friendly: One's own forces.

Game: Simulation of organizational behavior by people, usually in competition or conflict.

GCA: Ground Controlled Approach, in air traffic control.

GCI: Ground Control of Interception, in air defense.

GEOREF: A grid co-ordinate arrangement for expressing geographical position.

Graeco-Latin square: Incomplete factorial design (lacking some interactions) for combining four independent variables all of which have the same number of states.

Graphical simulation: Static simulation by diagraming equipment, operations, and environment.

Grease pencil: China-marking pencil for indicating an aircraft's radar-detected positions on a PPI display.

Grid co-ordinates Arrangement of horizontal and vertical lines for defining geographical positions.

Ground support equipment: The equipment which fuels, aims, and fires an ICBM.

Handover: Transfer of responsibility for or control of some object from one operator to another.

Hard copy: Permanent display and record, e.g., on paper, of computer output.

Heading: Compass direction in which a vehicle is pointed.

Heading (command): Compass direction in which a vehicle should be pointed according to instructions.

Heuristic: Serving discovery or problem solving, not precise like an algorithm.

Homolog: Structurally similar.

HumRRO: Human Resources Research Office, subsequently Human Resources Research Organization.

ICBM: Intercontinental Ballistic Missile.

IDA: Institute for Defense Analyses.

IDC: Indoctrination Center, an SDC facility.

Identification: Air defense function of determining whether an aircraft is friendly, hostile, or unknown.

IEC: International Electric Corporation, former subsidiary of International Telephone and Telegraph Corporation (ITT).

IFR: Conditions which require piloting an aircraft by instrument and instrument flight rules.

Illegal action: An operator switch-action rejected by the computer as improper.

ILS: Instrument Landing System for piloting an aircraft to a landing.

In-line: Method of task allocation whereby different operators successively bear responsibility for the same object.

Input: What a system, subsystem, individual, or computer receives for processing.

Intercept Director: SAGE air defense operator who guides interceptor aircraft (or missiles).

Interception: An interceptor aircraft's (or missile's) attack on a hostile bomber.

Interceptor: Air defense aircraft or missile defending against enemy bombers.

Intercom: Internal telephone-type communications.

Interdiction: Tactical air attack against hostile ground elements.

IPAC: Information Processing and Control facility at Ohio State University.

ISO: Intelligence Staff Officer.

Jamming: Electronic countermeasures from a hostile aircraft to keep a ground radar from ascertaining its range, by producing many similar radar signals in azimuth.

KCADS: Kansas City Air Defense Sector of SAGE.

Keyset: Keyboard, set of keys or buttons for actuating switches in communicating with a computer.

Kill: Destroy a hostile aircraft.

Knowledge of results: Feedback of information about performance.

KOR: Knowledge or results.

KR: Knowledge of results.

Latin square: Incomplete factorial design (lacking some interactions) for combining three independent variables all of which have the same number of states.

Launch complex: An interconnected set of ICBMs, one control location, and ground support equipment.

LEM: Lunar Excursion Module for the Apollo project.

Library: Collection of simulation inputs from which to draw in creating a set or sets for an experiment or training exercise.

Light pen: Device for communicating with a computer by pointing it at some element on a CRT display.

Load: The demand made by inputs on performance.

Magnetic tape: Computer storage device for data and instructions.

Manual: Operated by people rather than automatically.

MDC: Master Direction Center in SAGE Mode III.

Miss distance: Distance between two aircraft when one crosses the other's track.

Mock-up: Relatively detailed physical representation of equipment.

Mode II: Early primary back-up arrangement for SAGE, other sectors assuming the disabled sector's coverage.

Mode III: Early secondary back-up arrangement for SAGE, radar sites assuming the disabled sector's coverage.

Molar: Involving large-scale or aggregated units.

Monte Carlo: Technique for randomly establishing occurrences of alternatives and values of variables within distributions in a model and making repeated calculations with them.

Multivariate: More than two correlated variables or more than one independent variable in an experiment.

NAFEC: National Aviation Facilities Experimental Center.

Netting: Intercommunication among a system's operating units.

NMCS: National Military Command System.

Nodal: Type of system position where two or more operators or tasks are interrelated.

Noise: Aspects of an input which make it difficult to discriminate the signal or message.

NORAD COC: North American Air (later Aerospace) Defense Command Combat Operations Center.

NORM: Normative Operations Recording Method for deriving SAGE performance measures.

NRL: Naval Research Laboratory.

NTDS: Navy Tactical Data System.

NUDETS: System for detecting and reporting nuclear detonations.

On-line: Computer operations in almost immediate response to system inputs and directly connected to their sources.

Organization compression: Simulation of an entire organization by one or a few individuals.

Orthogonality: Counteraction whereby each state of each independent variable is combined with each combination of states of other variables; statistically, a condition in which comparisons in a set are all independent.

OTC: Officer in tactical command.

Output: What a system, subsystem, operator, or computer produces for display, control, or communication.

Pattern feeder controller: Air traffic controller responsible for aircraft in an area near an airport.

PDP-1: A computer manufactured by the Digital Equipment Corporation.

PHADS: Phoenix (Ariz.) Air Defense Sector of SAGE.

Philco 2000: A computer manufactured by the Philco-Ford Corporation.

PIP: Probabalistic information processing in which humans make estimations of conditional probabilities in Bayesian processing.

Plan Position Indicator: A geographical-type display of radar signals using a CRT.

Player: Participant in a game or subject in an experiment.

Plot: Inscribe data on a display, generally to show the path of some vehicle.

Polar co-ordinates: Radius vector (range) to a point and angular difference between that vector and 360 degrees (azimuth).

Posterior probability: Probability of a situation or hypothesis after application of the Bayes theorem.

Post-test: Measurement after some state of a variable is introduced.

PPI: Plan Position Indicator.

Preclusion: Counteraction which prevents the intrusion into an experiment of variables that could plausibly account for the results.

Pre-test: Measurement before some state of a variable is introduced.

Printout: A sheet or sheets of paper displaying the results of computer processing.

Prior probability: Probability of a situation or hypothesis before application of the Bayes theorem.

Problem: Integrated simulation inputs for an entire training or evaluation exercise.

Protocol: Initial record of the results of an experiment, including any statements by subjects about it.

Pseudopilot: A quasi subject simulating a pilot.

Quasi subject: Member of an experimental staff or aide who simulates an organization, system, or individual with whom subjects must interact.

Random access: Access to computer storage independent of the location of the preceding access.

Randomizing: Counteraction whereby selection of equivalent states of a variable for association with other states is made on a random or chance basis.

Range: Distance to an object.

RAPCON: Radar Approach Control Center for air traffic control.

Raster: Line-by-line pattern of scanning by the electron beam of a cathode ray tube.

RATCC: Radar Air Traffic Control Center.

Reactive simulation: Simulation in which the nature of the inputs is contingent on the actions of the subjects.

Real-time processing: Computer processing of inputs from events as these occur or would occur in the real world, and timing of outputs to meet real-world requirements.

Refinement: Counteraction whereby a composite state of a qualitatively varying independent variable is fractionated to determine which component is responsible for the results.

Regenerative recording: Replayable registration of inputs and the computer operations which resulted from them.

Regression analysis: Technique for functionally relating two continuous variables each of which has many uncategorized values.

Replication: Counteraction consisting of the repetition of experimental conditions.

RL-101: Buffer equipment for the Philco 2000 in SDC's SSRL.

RTB: Return-to-base—the process of bringing interceptor aircraft back to their airfield.

Run: A sequence of experimental sessions.

Run-through: One complete occurrence of all experimental conditions.

SAC: Strategic Air Command.

SAGE: Semi-Automatic Ground Environment system of air defense supported by digital computer processing.

Sample: A selection from among past recorded flights of real aircraft for air traffic control simulation.

Scenario: Verbal simulation.

Scrub: Eliminate from a display.

SDC: System Development Corporation.

Search radar: Radar which is used to detect objects in range and azimuth.

Sector: Geographical area or portion of airspace subdivided horizontally or vertically; in SAGE, a Direction Center's area of responsibility.

Senior Weapons Director: Co-ordinator of SAGE Weapons Directors.

Sensor: Subsystem or associated system for acquiring data.

SETE: System Exercising for Training and Evaluation in the BUIC system.

Significance test: A mathematical operation to determine the statistical significance of the results of an experiment or correlational analysis

SimCon: Simulation Control Team for 473L exercises.

SimFac: Simulation Facility, an SDC laboratory in New Jersey.

Situation Display: SAGE geographical-type display of processed radar, tracking, and interception data.

SOP: Standing Operating Procedure.

Splash: Term announcing an aircraft has been shot down.

SRDS: System Research and Development Service of the FAA.

SRI: Stanford Research Institute.

SRL: Systems Research Laboratory of the RAND Corporation.

SSRL: Systems Simulation Research Laboratory of SDC.

Stack: A holding location for aircraft at different altitudes.

Statistical significance: Relationship between the dispersion among means and the random dispersion of measurements, expressed as a ratio.

Status board: A display showing availability, characteristics, and allocation of resources.

STL: Space Technology Laboratory.

Stochastic: Random.

STP: System Training Program.

SUBTAG: Submarine Tactics Analysis and Gaming facility of the Electric Boat Division of General Dynamics.

Switch-action: Manual actuation of a button, key, or toggle operating a switch to communicate with a computer.

Symbology: Collective term for many different symbols on displays.

Tabular display: Display of alphanumeric and other symbols in a tabular format.

TACC: Tactical Air Control Center.

Tag: Electronic marker which is superimposed on radar signals on a display and shows their track.

Talker: Individual in a control center who communicates data or commands by intercom, telephone, or radio.

Tally-ho: Term to indicate that the pilot of an interceptor has detected an assigned target by means of his fire control system or visually.

Target: An object which must be detected and tracked and, if hostile, destroyed.

TATCS: Terminal Air Traffic Control System.

TE: Threat Evaluator.

TEAS: Threat Evaluation/Action Selection.

Teleconferencing: Communication over some medium among widely separated members of a group.

Teller: Individual in a control center who communicates data to a plotter at a display.

Terminal area: Airspace near an airport.

Time compression: Representation of events in a shorter time span than they would occupy in the real world.

Time density: Ratio between some unit of time and the number of events requiring operator action during it.

TOR: Training Operations Report, the trainers' record of an exercise in the System Training Program.

TRACE: SDC computer program for reducing and analyzing data in on-line experiments.

Track: An object's path described in terms of successive geographical positions, course, and speed.

Tracking computer: A computer which establishes the tracks of detected objects.

Tracking gates: Electronic elements which become associated with radar signals to furnish position data to a tracking computer.

Transducer: A device which transforms signals in one medium to corresponding signals in another.

TTY: Teletype.

UCLA: University of California at Los Angeles.

USAEPG: U.S. Army Electronic Proving Ground.

Variance: Dispersion of measurements as indicated by the square of their standard deviation.

| | |
|---|---|
| VFR: | Conditions which permit piloting visually and by visual flight rules. |
| VOR: | System of ground radio beacons for navigation of aircraft. |
| Weapons director: | Co-ordinator of SAGE Intercept Directors. |
| Weapons system: | An effector system consisting of weapons and means for delivering them. |
| WEST: | Weapons Evaluation and Subsystem Training for SAGE. |
| WSEG: | Weapons Systems Evaluation Group. |
| 15-J-lc: | Electromechanical device to generate and move simulated radar signals on a PPI display. |
| 425L: | NORAD COC. |
| 465L: | SAC Control System. |
| 473L: | Air Force's Headquarters Command Post. |

# References

Abt, C. 1964. War gaming. *International Science and Technology* 32: 29–37.

Adams, J. A., and Webber, C. E. 1963. Monte Carlo model of tracking behavior. *Human Factors* 5: 81–102.

Adams, O. S., and Chiles, W. D. 1961. *Human performance as a function of the work-rest ratio during prolonged confinement.* ASD Technical Report 67-720. Wright-Patterson Air Force Base, Ohio: Aerospace Medical Laboratory.

Adiletta, J. G., and Chapman, R. L. 1951. *Layout of the combat information center in the PO-2W Aircraft.* Technical Report SPECDEVCEN 279-3-5. New York: New York University.

Aircraft Armaments, Inc. (J. S. Kidd) 1963. *Teleconferencing. An experimental task.* Research Paper P-112. Arlington, Va.: Institute for Defense Analyses.

Alexander, L. T. 1955. *Systems behavior. I. The learning process.* Report P-662. Santa Monica, Calif.: RAND Corporation.

_____. 1962. Terminal air traffic control system. *System Development Corporation Magazine* (July). Santa Monica, Calif.: System Development Corporation.

_____. 1963. *Terminal air traffic control follow-on research.* Report TM-639/007/00. Santa Monica, Calif.: System Development Corporation.

Alexander, L. T., and Ash, M. 1962. *Terminal air traffic control: A laboratory model for man-machine system research.* Report SP-1016. Santa Monica, Calif.: System Development Corporation.

Alexander, L. T., and Cooperband, A. S. 1961. *A laboratory model for systems research: A terminal air traffic control system.* Report TM-639. Santa Monica, Calif.: System Development Corporation.

_____. 1964*a*. Schematic simulation: A technique for the design and development of a complex system. *Human Factors* 6: 87–92.

_____. 1964*b*. The effect of rule flexibility on system adaptation. *Human factors* 6: 209–20.

Alexander, L. T., and Porter, E. H. 1963. *Terminal air traffic control and problems of system design.* Report TM-639/008/00. Santa Monica, Calif.: System Development Corporation.

Alexander, L. T.; Kepner, C. H.; and Tregoe, B. B. 1962. The effectiveness of knowledge of results in a military system-training program. *Journal of Applied Psychology* 46: 202–11.

Alluisi, E. A. 1956. *Human engineering aspects of air traffic control systems.* Final Quarterly Report. Columbus, Ohio: Ohio State University Research Foundation.

Alluisi, E. A.; Chiles, W. D.; and Hall, T. J. 1964. *Combined effects of sleep loss and demanding work-rest schedules on crew performance.* Report TDR-64-63. Wright-Patterson Air Force Base, Ohio: Aerospace Medical Research Laboratories.

Alluisi, E. A.; Chiles, W. D.; and Smith, R. P. 1964. *Human performance in military systems: Some situational factors influencing individual performance.* Report ITR-64-1. Louisville, Ky.: University of Louisville.

Alluisi, E. A.; Chiles, W. D.; Hall, T. J.; and Hawkes, T. R. 1963. *Human group performance during confinement.* Report TDR-63-87. Wright-Patterson Air Force Base, Ohio: Aerospace Medical Research Laboratories.

Altman, I., and Haythorn, W. W. 1967. The ecology of isolated groups. *Behavioral Science* 12: 169-82.

Anderson, C. M., and Dowling, C. E. 1954. *Evaluation by simulation techniques of a proposed traffic control procedure for the New York Metropolitan Area.* Report No. 245. Indianapolis, Ind.: Civil Aeronautics Administration Technical Development and Evaluation Center.

Anderson, C. M., and Vickers, T. K. 1955. *Simulation tests of a tactical airway system.* Report No. 279. Indianapolis, Ind.: Civil Aeronautics Administration Technical Development and Evaluation Center.

Anderson, C. M.; Armour, T. E.; et al. 1957. *Dynamic simulation tests of Baltimore Friendship Airport at increased traffic densities.* Report. Indianapolis, Ind.: Civil Aeronautics Administration Technical Development and Evaluation Center.

Armour, T. E.; Johnson, A. N.; Vickers, T. K.; and Miller, R. S. 1958. *Simulation tests of the factors affecting IFR traffic capacity at Chicago O'Hare Airport.* Report No. 341. Indianapolis, Ind.: Civil Aeronautics Administration Technical Development and Evaluation Center.

Army Research Office Professional Summer Study Group. 1964. *An examination of test and evaluation methodology at the United States Army Electronic Proving Ground.* Final Report (draft). Fort Huachuca, Arizona: Test Plans and Evaluation Department, United States Army Electronic Proving Ground.

Astholz, R. T., and Vickers, T. K. 1958. *A preliminary report on the simulation of proposed ATC procedures for civil jet aircraft.* Report No. 352. Indianapolis, Ind.: Civil Aeronautics Administration Technical Development and Evaluation Center.

Bailey, G. C. 1951. *A study of height-finding with the AN/CPS-6B and the AN/CPS-4 radars.* Report UMM-86. Ann Arbor, Mich.: University of Michigan Engineering Research Institute.

______. 1964. *An experimental comparison of point-to-point and party-line teleconferences.* Technical Note 1. McLean, Va.: Human Sciences Research, Inc.

Bailey, G. C., and Jenny, A. 1965. *A study of teleconference control.* Technical Note 3. McLean, Va.: Human Sciences Research, Inc.

Bailey, G. C.; Nordlie, P. G.; and Sistrunk, F. 1963. *Teleconferencing. Literature review, field studies, and working papers.* Research Paper P-113. Arlington, Va.: Institute for Defense Analyses.

Baker, R. A.; Cook, J. G.; Warnick, W. L.; and Robinson, J. P. 1964. *Development and evaluation of systems for the conduct of tactical training at the tank platoon level.* Technical Report 88. Alexandria, Va.: Human Resources Research Office.

Baker, R. E.; Grant, A. L.; and Vickers, T. K. 1953. *Development of a dynamic air traffic control simulator.* Report No. 191. Indianapolis, Ind.: Civil Aeronautics Administration Technical Development and Evaluation Center.

Balachowski, E. F.; Brown, D. O.; Cassell, R.; Rosi, A. P.; and Miller, R. S. 1960. *Simulation test of feeder fix locations and development of associated procedures for simultaneous approaches to parallel runways.* Bureau of Research and Development Report. Atlantic City, N. J.: Federal Aviation Agency National Aviation Facilities Experimental Center.

Bamford, H. 1964. Systems design laboratory: Air Force and industry both stand to benefit. *Armed Forces Management* (March): 25, 27.

Barrett, G. V.; Kobayashi, M.; and Fox, B. H. 1968. Feasibility of studying driver reaction to sudden pedestrian emergencies in an automobile simulator. *Human Factors* 10: 19–26.

Bayless, J. A. 1962. *Research development plan for the Command Systems Laboratory.* Report FN-6748. Santa Monica, Calif.: System Development Corporation.

_____. 1964. *Multivariate threat analysis: Final experimental results.* Report TM-1818. Santa Monica, Calif.: System Development Corporation.

Bayless, J. A.; Erickson, E. B.; Grant, E. E.; and Horst, D. P. 1963. *Multivariate threat analysis: Experimental results.* Report TM-1328. Santa Monica, Calif.: System Development Corporation.

Becker, G. M.; Reeves, L. M.; Buckley, E. P.; Chauvette, L. M.; and Sinaiko, H. W. 1956. *Dissemination of combat information: Voice links in CIC.* Memo Report 642. Washington, D.C.: Naval Research Laboratory.

Behan, R. A.; Bughman, C. R.; Bumpus, J. N.; and Gilbert, Sally B. 1959. *An experimental investigation of the interaction between problem load and level of training.* Report TM-352. Santa Monica, Calif.: System Development Corporation. Also 1961. *Human Factors* 3: 53–59.

Belden, T. G., and Sinaiko, H. W. 1963. *Coherent language: An experiment in linguistic categorization.* Report R-104. Arlington, Va.: Institute for Defense Analyses.

Bellman, R.; Clark, C.; Craft, C.; Malcolm, D. G.; and Ricciardi, F. 1957. *On the construction of a multistage, multiperson business game.* Report P-1056. Santa Monica, Calif.: RAND Corporation.

Bennett, E.; Haines, E. C.; and Summers, J. K. 1965. *AESOP: A prototype for on-line user control of organizational data storage, retrieval and processing.* Report MTP-23. Bedford, Mass.: MITRE Corporation.

Bennett, E.; Dittman, P. E.; Doughty, J. M.; Miller, I. W.; and Summers, J. K. 1966. *Simulation of computer and operator procedures for tactical force employment planning.* Report MTP-14A. Bedford, Mass.: MITRE Corporation.

Benson, O. 1962. Simulation of international relations and diplomacy. In *Computer applications in the behavioral sciences*, ed. H. Borko. Englewood Cliffs. N. J.: Prentice-Hall, Inc.

Bergum, B. O., and Lehr, D. J. 1962. Vigilance performance as a function of paired monitoring. *Journal of Applied Psychology* 46: 341-43.

Berkowitz, M. I.; Best, H. L.; and Rockett, F. C. 1958. *Feedback through the supervisory loop.* Report TM-197. Santa Monica, Calif.: System Development Corporation.

Berkowitz, S. M., and Doering, Ruth R. 1954. *Analytical and simulation studies of several radar-vectoring procedures in the Washington, D.C. Terminal Area.* The Franklin Institute Laboratories Report No. 222. Indianapolis, Ind.: Civil Aeronautics Administration Technical Development and Evaluation Center.

Berkowitz, S. M., and Fritz, E. L. 1955. *Analytical and simulation studies of terminal-area air traffic control. Summary of joint FIL-TDEC simulation activities in air traffic control.* The Franklin Institute Laboratories Report No. F-2384. Indianapolis, Ind.: Civil Aeronautics Administration Technical Development and Evaluation Center.

Berkowitz, S. M.; Fritz, E. L.; and Miller, R. S. 1957. *Summary of joint FIL-TDC simulation activities in air traffic control.* Report No. 297. Indianapolis, Ind.: Civil Aeronautics Administration Technical Development and Evaluation Center.

Berkun, M. M. 1964. Performance decrement under psychological stress. *Human Factors* 6: 21-30.

Berkun, M. M.; Bialek, H. M.; Kern, R. P.; and Yagi, K. 1962. Experimental studies of psychological stress in man. *Psychological Monographs* 76(15): 1-39.

Beverly, R. S. 1958. *Communications analysis in SP-1.* Report P-1457. Santa Monica, Calif.: RAND Corporation.

Bidwell, J. B. 1967. Driving modeling and driving simulation. In *Mathematical models and simulation of automobile driving,* ed. T. B. Sheridan. Cincinnati, Ohio: U.S. Public Health Service.

Blair, W. C., and Kaufman, H. M. 1959. *Command control I: Multiple display monitoring II. Control display spatial arrangement.* Technical Report. New London, Conn.: General Dynamics Corporation, Electric Boat Division.

Blanchard, H. A. 1961. *Final report: Considerations in the design of a TEAS simulation research facility.* Planning Research Corporation Report R-195. Bedford, Mass.: Air Force Cambridge Research Laboratories.

Bogdanoff, E.; Brooks, H. E.; Jasinski, F. J.; Keys, L. B.; Michael, A. L.; Molnar, A. R.; Proctor, G. L.; Reeves, E. Y.; and Thorsell, B. A. 1960. *Simulation: An introduction to a new technology.* Report TM-499. Santa Monica, Calif.: System Development Corporation.

Boguslaw, R.; Davis, R. H.; and Glick, E. B. 1964. Plans-I. *A vehicle for studying national policy in a less armed world.* Report TM-WD-119. Santa Monica, Calif.: System Development Corporation.

Borah, L. A., Jr. 1963. The effects of threat in bargaining: Critical and experimental analysis. *Journal of Abnormal and Social Psychology* 66: 37–44.

Bottomley, D.; Hansen, R. E.; Johnson, T. R.; Rohland, H. T.; Rossiter, S. B.; and Wright, E. H. 1962. *Dynamic simulation study and evaluation of a proposed air traffic procedural plan and control equipment for the Washington, D.C. area.* Bureau of Research and Development Report No. 101–112V. Atlantic City, N.J.: Federal Aviation Agency National Aviation Facilities Experimental Center.

Bowen, H. M.; Andersen, R.; and Promisel, D. 1966. Studies of divers' performance during the SEALAB II Project. *Human Factors* 8: 183–99.

Brady, F.; DeVoe, R. P.; and Pittsley, J. L. 1959. *A surveillance game.* Report 2144–395–T. Ann Arbor, Mich.: University of Michigan Willow Run Laboratories. A classified report–Confidential.

Bray, C. W. 1948. *Psychology and military proficiency. A history of the Applied Psychology Panel of the National Defense Research Committee.* Princeton, N. J.: Princeton University Press.

_____. 1962. Toward a technology of human behavior for defense use. *American Psychologist* 17: 527–41.

Briggs, G. E., and Johnston, W. A. 1966*a*. Stimulus and response fidelity in team training. *Journal of Applied Psychology* 50: 114–17.

_____. 1966*b*. Influence of a change in system criteria on team performance. *Journal of Applied Psychology* 50: 467–72.

Briggs, G. E., and Naylor, J. C. 1964. *Experiments on team training in a CIC-type task environment.* Technical Report NAVTRADEVCEN 1327–1. Columbus, Ohio: Ohio State University Laboratory of Aviation Psychology.

_____. 1965. Team versus individual training, training task fidelity, and task organization effects on transfer performance by three-man teams. *Journal of Applied Psychology* 49: 387–92.

Briggs, G. E., and Schum, D. A. 1965. Automated Bayesian hypothesis-selection in a simulated threat-diagnosis system. In *Second Congress on the Information System Sciences*, eds. J. Spiegel and D. E. Walker. Washington, D.C.: Spartan Books.

Brinton, J. H., Jr., and Miller, R. S. 1961. *Summary of joint FIL–FAA research in air traffic control during period from September 1958 to January 1961.* Report No. F–A2221. Philadelphia: Franklin Institute Laboratories.

Brown, W. A.; DeVoe, R. P.; Jebe, E. H.; and Pittsley, J. L. 1960. *Combat surveillance experiments in tracking military concentrations.* Memorandum Report 2900–182–R. Ann Arbor, Mich.: University of Michigan Willow Run Laboratories.

Bughman, C. R., and Jaffe, J. 1958. *A research program on specificity of feedback.* Report TM–190. Santa Monica, Calif.: System Development Corporation.

Bumpus, J. N. 1959. *MDC operations in SAGE Mode III: A field research project.* Report TM–332. Santa Monica, Calif.: System Development Corporation.

Burdick, D. S., and Naylor, T. H. 1966. Design of computer simulation experiments for industrial systems. *Communications of the ACM* 9(5): 329–39.

Burnaugh, H. P., and Moore, W. H. 1964. *Variable display: A display-formatting system with selective data retrieval.* Report SP-1757. Santa Monica, Calif.: System Development Corporation.

Burwen, L. W.; Ellis, W. R.; Jensen, B. T.; Leping, L T.; and Terebinski, S. J. 1960. *Effect of debriefing-communications links.* Report TM-529. Santa Monica, Calif.: System Development Corporation.

Campbell, D. T. 1969. Reforms as experiments. *American Psychologist* 24: 409-49.

Campbell, D. T., and Stanley, J. C. 1966. *Experimental and quasi-experimental designs for research.* Chicago: Rand McNally. Also 1963. In *Handbook of research on teaching*, American Educational Research Association. Chicago: Rand McNally.

Cartwright, D., and Zander, A. 1953. *Group dynamics–research and theory.* Evanston, Ill.: Row, Peterson.

Cassell, R.; Conway, R.; Douglass, L., Maurer, J.; and Stephens, R. 1962. *Evaluation of a Honolulu terminal radar service area and the Hawaiian Islands domestic enroute area.* Bureau of Research and Development Final Report. Atlantic City, N. J.: Federal Aviation Agency National Aviation Facilities Experimental Center.

Chapanis, A. 1947. *The relative efficiency of a bearing counter and bearing dial for use with PPI presentations.* Report 166-I-26. Baltimore: The Johns Hopkins University Systems Research Field Laboratory.

_____. 1956. *The design and conduct of human engineering studies.* Technical Report No. 14. San Diego, Calif.: San Diego State College Foundation.

_____. 1959. *Research techniques in human engineering.* Baltimore: The Johns Hopkins Press.

_____. 1961. Men, machines, and models. *American Psychologist* 16: 113-31.

Chapanis, A., and Leyzorek, M. 1950. Accuracy of visual interpolation between scale markers as a function of the number assigned to the scale interval. *Journal of Experimental Psychology* 40: 655-67.

Chapanis, A.; Garner, W. R.; and Morgan, C. T. 1949. *Applied experimental psychology: Human factors in engineering design.* New York: Wiley.

Chapman, R. L. 1951. *Experimental methods of evaluating a system: The airborne CIC.* Technical Report SPECDEVCEN 279-3-7. New York: New York University.

_____. 1955. *Description of the air defense experiments. III. Data collection and processing.* Report P-658. Santa Monica, Calif.: RAND Corporation.

_____. 1956. *A man-machine study. An air defense example: The Cobra experiment.* Report RM-1427. Santa Monica, Calif.: RAND Corporation. A classified report–Confidential.

_____. 1960*a*. Simulation and human behavior. Paper read at the American Management Assocation's First National Simulation Forum, May 1960.

_____. 1960*b*. System simulation. Paper read at the annual meeting of the American Psychological Association, September 1960.

_____. 1960*c*. *Data for testing a model of organizational behavior.* Report RM-1916. Santa Monica, Calif.: RAND Corporation.

_____. 1961*a*. Simulation and human behavior. Some notes on the use of simulation techniques to study organizational problems. In *Simulation and gaming: A symposium.* Report No. 55. New York, N. Y.: American Management Association. Also 1960*a*.

_____. 1961*b*. All-computer vs. man-computer systems in the simulation of group processes. Paper read at the annual meeting of the American Psychological Association, September 1961.

_____. 1962. A method for generating information needed for system designers. In *Proceedings of the Military Operations Research Symposia (MORS)* 2, No. 2, Part 1.

_____. 1965. The case for information system simulation. In *Second Congress on the Information System Sciences*, eds. J. Spiegel and D. E. Walker. Washington, D.C.: Spartan Books.

Chapman, R. L., and Kennedy, J. L. 1955. The background and implications of the Systems Research Laboratory studies. In *Symposium on Air Force human engineering, personnel, and training research*, eds. G. Finch and F. Cameron. Washington, D.C.: National Academy of Sciences–National Research Council.

Chapman, R. L., and Weiner, M. G. 1957. *The history, purpose and script of Cogwheel.* Report P-1105. Santa Monica, Calif.: RAND Corporation.

Chapman, R. L.; Biel, W. C.; Kennedy, J. L.; and Newell, A. 1952. *The Systems Research Laboratory and its program.* Report RM-890. Santa Monica, Calif.: RAND Corporation.

Chapman, R. L.; Kennedy, J. L.; Newell, A.; and Biel, W. C. 1959. The Systems Research Laboratory's air defense experiments. *Management Science* 5: 251-69.

Chauvette, L. M.; Sinaiko, H. W.; and Buckley, E. P. 1957. *Simulation studies of CIC: Air defense performance as a criterion of system effectiveness.* Report 5003. Washington, D.C.: Naval Research Laboratory. A classified report–Confidential.

Clarkson, G. P. E., and Simon, H. A. 1960. Simulation of individual and group behavior. *American Economic Review* 50: 908-19.

Coburn, R. 1960. *An experimental polar transplot system for CIC use.* Report 969. Washington, D.C.: U.S. Navy Electronics Laboratory.

Cockrell, J. T., and Murphy, J. V. 1961. *WEST Field Trial.* Report FN-6006. Santa Monica, Calif.: System Development Corporation.

Cohen, I. K. 1963. *The design and objectives of laboratory problem IV.* Report RM-3354-PR. Santa Monica, Calif.: RAND Corporation.

Cohen, I. K., and Van Horn, R. L. 1964. *A laboratory exercise for information system development.* Report RM-4300-PR. Santa Monica, Calif.: RAND Corporation. Also, 1965. In *Second Congress on the Information System Sciences*, eds. J. Spiegel and D. E. Walker. Washington, D.C.: Spartan Books.

Cohen, K. J., and Rhenman, E. 1961. The role of management games in education and research. *Management Science* 7: 131-66.

Combat Development Experimentation Center. 1961. *Locating battlefield casualties, medical experiment, spring 1961.* Final Report. Fort Ord, Calif.: Combat Development Experimentation Center.

Connolly, D. W. 1958. *Multiple point interdiction tests of AN/TSQ-13 (XD-1).* Report TN 58-3. L. G. Hanscom Field, Bedford, Mass.: Air Force Cambridge Research Center. A classified report–Confidential.

____. 1959. *GCI performance as a function of control load and control method: Manual control vs. air map computer.* Report TN 59-6. L. G. Hanscom Field, Bedford, Mass.: Air Force Cambridge Research Center. A classified report–Confidential.

Connolly, D. W., and Capuano, D. J. 1954. *The design and evaluation of two intercept course computers.* Technical Report SPECDEVCEN 279-3-19. New York: New York University.

Connolly, D. W., and Page, H. J. 1953. *An experimental investigation of the validity of an empirical measure of threat of enemy raids using ACO performance as a criterion.* Technical Report SPECDEVCEN 279-3-17. New York: New York University.

Connolly, D. W.; Fox, W. R.; and McGoldrick, C. C. 1961. *Tactical decision making: II. The effects of threatening weapons performance and uncertainty in information displayed to the decision maker on threat evaluation and action selection.* Report ESD-TR-61-45. L. G. Hanscom Field, Bedford, Mass.: Electronic Systems Division.

Connolly, D. W.; McGoldrick, C. C.; and Fox, W. R. 1961. *Preliminary summary report; tactical decision making: II. The effects of track load on damage, cost and kills.* Report ESD-TR-61-43. L. G. Hanscom Field, Bedford, Mass.: Electronic Systems Division.

Connolly, D. W.; Page, H. J.; and Veniar, F. 1955. *An experimental evaluation of three AEW intercept systems in the airborne CIC.* Technical Report SPECDEVCEN 279-3-21. New York: New York University.

Cooney, S. 1964. SDC/ARPA research in command and control. *System Development Corporation Magazine* (Summer). Santa Monica, Calif.: System Development Corporation.

Cooperband, A. S., and Alexander, L. T. 1965. *The detection of compound motion.* Report SP-1964/001/00. Santa Monica, Calif.: System Development Corporation.

Cooperband, A. S.; Alexander, L. T.; and Schmitz, H. S. 1963. *Test results of the terminal air traffic control laboratory system.* Report TM-639/004/00. Santa Monica, Calif.: System Development Corporation.

Cranston, R.; Holmes, E. H.; and Maatsch, J. L. 1958. *ACW team competition utilizing STP.* Report TM-93. Santa Monica, Calif.: System Development Corporation.

Crawford, M. P. 1966. Dimensions of simulation. *American Psychologist* 21: 788-96.

Cunningham, R. P.; Sheldon, M. S.; and Zagorski, H. W. 1965. *Project NORM pilot study report.* Report TM-2232/000/00. Santa Monica, Calif.: System Development Corporation.

Cusack, B. L. 1963. *Display I: Background, discussion and mission.* Report TM-1504/001/00. Santa Monica, Calif.: System Development Corporation.

____. 1964*a*. *Display I. Experimental design.* Report TM-1504/003/00. Santa Monica, Calif.: System Development Corporation.

_____. 1964*b*. *Display I: Final report*. Report TM-1504/009/00. Santa Monica, Calif.: System Development Corporation.

Cusack, B. L., and Parsons, H. M. 1953. *The effect of number of search operators, reporting procedures and load on the performance of an airborne CIC system*. Technical Report SPECDEVCEN 279-3-13. New York: New York University.

Cusack, B. L.; Flint, Rhea; Gibbons, R. D.; Haney, T. P.; and Swavely, D. C. 1966. *Final report on emergency operations simulation research*. Report TM-2938/001/00. Santa Monica, Calif.: System Development Corporation.

Danaher, J. W.; Eberhard, J. W.; and Coleman, K. W. 1959. *Prediction of operator effectiveness in dynamic air traffic control simulation*. Project N Report 30. Philadelphia: Courtney and Company.

Davage, R. H.; DeVoe, R. P.; and Pittsley, J. L. 1954. *An integrated air defense system–X–Tests of the laboratory model of an advanced weapon assignment section*. Report UMS-153. Ann Arbor, Mich.: University of Michigan Willow Run Research Center.

Davis, C. G.; Danaher, J. W.; et al. 1963. *The influences of selected sector characteristics upon ARTCC controller activities*. Technical Report. Arlington, Va.: Matrix Corporation.

Davis, R. H. 1960. On designing systems to study systems. Paper read at the 6th Western Military Operations Research Symposium. Los Angeles: Thompson Ramo Wooldridge, Inc.

_____. 1963. Arms control simulation: The search for an acceptable method. *Journal of Conflict Resolution* 7:590-602, and *Journal of Arms Control* 1:684-96.

Davis, Ruth H., and Behan, R. A. 1962. Evaluating system performance in simulated environments. In *Psychological principles in system development*, ed. R. M. Gagne. New York: Holt, Rinehart, and Winston.

Davis, R. M. 1965. A history of automated displays. *Datamation* (January): 24-28.

DeCicco, R.; DeVoe, R. P.; Hoagbin, J.; and Pittsley, J. 1962. *An illustration of the use of reference data in processing sensor data*. Report 2900-330-X. Ann Arbor, Mich.: University of Michigan Willow Run Laboratories. A classified report–Secret.

Deutsch, M., and Krauss, R. M. 1960. The effects of threat upon interpersonal bargaining. *Journal of Abnormal and Social Psychology* 61: 181-89.

Devoe, D. B. 1963. Toward an ideal guide for display designers. *Human Factors* 5: 583-91.

Dodson, J. D. 1961. *Simulation system design for a TEAS simulation research facility*. Report AFCRL 1112. L. G. Hanscom Field, Bedford, Mass.: Air Force Cambridge Research Laboratories.

Dodson, J. D.; Fox, R. K.; Garfunkel, I. M.; Nystrom, C. L.; and Schwartz, J. J. 1961. *Concepts and objectives for a threat evaluation/action selection simulation research facility*. Planning Research Corporation Report R-191. Bedford, Mass.: Air Force Cambridge Research Laboratories.

Doten, G. E.; Cockrell, J. T.; and Sadacca, R. 1966. *The use of teams in image interpretation: Information exchange, confidence, and resolving disagreements*. Report 1151. Washington, D.C.: Army Personnel Research Office.

Doughty, J. M. 1960. *A simulation facility for the experimental study of decision making in complex military systems.* Report AFCCD TN 60–32. L. G. Hanscom Field, Bedford, Mass.: Command and Control Development Division.

_____. 1963. *National command post simulation: Game series I.* Report TM-3627. Bedford, Mass.: MITRE Corporation.

_____. 1967. The AESOP testbed: Test series I/2. In *Information system science and technology papers prepared for the Third Congress*, ed. D. E. Walker. Washington, D.C.: Thompson Book Company.

Doughty, J. M.; Schwartz, S. L.; and Cohen, R. I. 1965. *The AESOP Testbed: Test Series I/1.* Report TM-04268/0000/00/0/00. Bedford, Mass.: MITRE Corporation.

Drabek, T. E. 1965. *Laboratory simulation of a police communication system under stress.* Ph.D. dissertation. Columbus, Ohio: Ohio State University.

Drabek, T. E., and Haas, J. E. 1966. Laboratory simulation of a police communication system under stress: Preliminary findings. Paper read at the annual meeting of the American Sociological Association.

Eberhard, J. W. 1966. Sleep requirements and work-rest cycles for long term space missions. Paper read at the annual meeting of the Human Factors Society.

Edwards, J. B., and Morrill, C. S. 1965. *AESOP field study II.* Technical Report 45. Bedford, Mass.: MITRE Corporation.

Edwards, W. D. 1962. Dynamic decision theory and probabilistic information processing. *Human Factors* 4: 59–73.

_____. 1963. *Probabilistic information processing in command and control systems.* Report ESD-TRD-63-345. L. G. Hanscom Field, Bedford, Mass.: Electronic Systems Command.

_____. 1966. Introduction. *IEEE Transactions on Human Factors in Electronics* 7: 1–6.

Edwards, W. D., and Phillips, L. D. 1964. Man as a transducer for probabilities in Bayesian command and control systems. In *Human judgment and optimality*, eds. M. W. Shelly, II, and G. L. Bryan. New York: Wiley. Also, paper read at the 1962 meeting of the American Association for the Advancement of Science.

Egerman, K.; Glaser, R.; and Klaus, D. J. 1963. *Increasing team proficiency through training. 4. A learning-theoretic analysis of the effects of team arrangement on team performance.* Report AIR-B64-9/63-TR. Pittsburgh: American Institute for Research.

Egerman, K.; Klaus, D. J.; and Glaser, R. 1962. *Increasing team proficiency through training. 3. Decremental effects of reinforcement in teams with redundant members.* Report AIR-B64-6/62-TR. Pittsburgh: American Institute for Research.

Eichenlaub, J. H.; Conway, R. C.; et al. 1961. *A simulation study of operational procedures for civil turbojet aircraft.* Bureau of Research and Development Final Report. Atlantic City, N.J.: Federal Aviation Agency National Aviation Facilities Experimental Center.

Ellis, W. R.; Jensen, B. T.; and Terebinski, S. J. 1960. *Studies in crew development: I. Methods of study.* Report TM-502. Santa Monica, Calif.: System Development Corporation.

Ellis, W. R.; Jensen, B. T.; Jordan, N.; and Terebinski, S. J. 1960. *Progress report on the crew development study.* Report FN-2763. Santa Monica, Calif.: System Development Corporation.

Enke, S. 1957. *Logistics Laboratory problem I after two (simulated) years.* Report RM-1993. Santa Monica, Calif.: RAND Corporation.

_____. 1958. *Use of a simulation laboratory to study the organization and effectiveness of Air Force logistics.* Report P-1343. Santa Monica, Calif.: RAND Corporation.

Ernst, A. A. 1959. *Feasibility study for a man-machine systems research facility.* Technical Report 59-51. Wright-Patterson Air Force Base, Ohio: Aerospace Medical Laboratory.

Evans, C.; Wallace, G.; and Sutherland, G. 1967. *Simulation using digital computers.* Englewood Cliffs, N.J.: Prentice-Hall.

Fagan, Pauline J. 1963. *Description of procedures for manual simulation of AAW battle games conducted in the APL ECM battle simulator facility.* Report TG-536/BSF-128. Baltimore: The Johns Hopkins University Applied Physics Laboratory.

Faison, W. E., and Sluka, A. L. 1961. *Dynamic simulation studies of pictorial navigation displays as aids to air traffic control in a high-density terminal area and a medium-density terminal area.* Bureau of Research and Development Report. Atlantic City, N.J.: Federal Aviation Agency National Aviation Facilities Experimental Center.

Faison, W. E.; Olson, W. D.; Slattery, H. F.; Wright, E. H.; and Miller, R. S. 1960. *Dynamic simulation tests and systems study of instrument flight rule operations in the San Diego area.* Bureau of Research and Development Report. Atlantic City, N.J.: Federal Aviation Agency National Aviation Facilities Experimental Center.

Feallock, J. B., and Briggs, G. E. 1963. *A multiman-machine system simulation facility and related research on information processing and decision-making tasks.* Report TRD-63-48. Wright-Patterson Air Force Base, Ohio: Aerospace Medical Laboratory.

Fend, A. V., and Cloutier, R. L. 1958. *Umpire techniques and procedures final report.* Draft Report RR-102. Combat Development Experimentation Center, Fort Ord, Calif.: Technical Operations, Inc.

Fitts, P. M. 1947. Psychological research on equipment design in the AAF. *American Psychologist* 2: 93-98.

Fitts, P. M.; Schipper, L.; Kidd, J. S.; Shelly, M.; and Kraft, C. 1958. Some concepts and methods for the conduct of system research in a laboratory setting. In *Air Force human engineering, personnel and training research,* eds. G. Finch and F. Cameron. Washington, D.C.: National Academy of Sciences-National Research Council.

Fitzpatrick, R. 1962. Toward a theory of simulation. Paper read at the annual meeting of the Human Factors Society.

Flagle, C. D. 1960. Simulation techniques. In *Operations research and systems engineering*, ed. C. D. Flagle, W. H. Huggins, and R. R. Roy. Baltimore: The Johns Hopkins Press.

Flagle, C. D.; Huggins, W. H.; and Roy, R. H. (eds.) 1960. *Operations research and systems engineering.* Baltimore: The Johns Hopkins Press.

Forbes, T. W. 1938. Age performance relationship among accident-repeater automobile drivers. *Journal of Consulting Psychology* 2: 143-48.

Ford, J. D., Jr., and Katter, R. V. 1960*a*. *Weapons director and weapons director technician, SAGE system: An observational study of the performance of operator teams.* Report TM-430. Santa Monica, Calif.: System Development Corporation.

_____. 1960*b*. *The development of cooperation in a two-man team carrying out a decision making function.* Report TM-541. Santa Monica, Calif.: System Development Corporation.

Foster, Mildred. 1963. *Final report on the empirical determination of minimum hold times for 465L MCC forced wall displays.* Report TM-(L)-876/000/00. Santa Monica, Calif.: System Development Corporation.

Fox, B. H. 1960. *Engineering and psychological users of a driving simulator.* Bulletin No. 261: 14-37. Washington, D.C.: Highway Research Board.

Fox, G. J., and Connolly, D. W. 1953. *ACO performance as a function of number of simultaneous raids and time available for interception.* Report SPECDEVCEN 279-3-15. New York: New York University.

Fox, G. J., and Page, H. J. 1954. *The effect of overlapping raid situations on ACO intercept performance.* Report SPECDEVCEN 279-3-22. New York: New York University.

Fox, W. R. 1960. *Utility of blinking new target information on a situation display.* Report TN 60-1. L. G. Hanscom Field, Bedford, Mass.: Air Force Cambridge Research Center. A classified report—Confidential.

Fox, W. R., and Vance, W. H., Jr. 1961. *Tactical decision making: I. Action selection as a function of track load, threat complexity, reliable data presentation and weapon uncertainty.* Report ESD-TDR 61-42. L. G. Hanscom Field, Bedford, Mass.: Electronic Systems Division.

Fraser, T. M. 1966*a*. *Philosophy of simulation in a man-machine space mission system.* National Aeronautics and Space Administration Report NASA SP-102. Washington, D.C.: U.S. Government Printing Office.

_____. 1966*b*. *The effects of confinement as a factor in manned space flight.* Report CR-511. Washington, D.C.: National Aeronautics and Space Administration.

Freed, A. M. 1961*a*. *The AZRAN modification to the AEW & C information processing system.* Report FN-5512. Santa Monica, Calif.: System Development Corporation.

_____. 1961*b*. *Research report. The AZRAN modification in the AEW & C information processing system.* Report FN-5616. Santa Monica, Calif.: System Development Corporation.

Garner, W. R. 1946. *A study of factors affecting operation of the VG remote PPI.* Report 166-I-1. Baltimore: The Johns Hopkins University Systems Research Field Laboratory.

_____. 1950. *The validity of prediction from laboratory experiments to naval operational situations in the area of human engineering and systems research.* Report 166-I-130. Baltimore: The Johns Hopkins University Institute for Cooperative Research.

Garner, W. R.; Saltzman, D. C.; and Saltzman, I. J. 1949. *Some design factors affecting the speed of identification of range rings on polar coordinate displays.* Report 166-I-95. Baltimore: The Johns Hopkins University Psychological Laboratory.

Gebhard, J. W. 1948. *Some experiments with the VF aided tracking equipment.* Report 166-I-53. Baltimore: The Johns Hopkins University Psychological Laboratory.

Gebhard, J. W., and Hanes, R. M. 1963. *Problems of commander interaction with stored logic machines in Navy AAW operations.* Report TG-520. Baltimore: The Johns Hopkins University Applied Physics Laboratory. A classified report–Confidential.

_____. 1964. *Evaluation of the realism achieved in the APL simulation of tactical command-action.* Report TG-581. Baltimore: The Johns Hopkins University Applied Physics Laboratory. A classified report–Confidential.

Gebhard, J. W., and Newton, K. V. 1947. *Brightness of grease pencil marks on a vertical plotting board.* Report 166-I-23. Baltimore: The Johns Hopkins University Systems Research Field Laboratory.

Geisler, M. A. 1957. *A first approach to logistics system simulation.* Report P-1234. Santa Monica, Calif.: RAND Corporation.

_____. 1958. *A first experiment in logistics system simulation.* Report P-1415. Santa Monica, Calif.: RAND Corporation.

_____. 1959. *The use of man-machine simulation in the design of control systems.* Report P-1780. Santa Monica, Calif.: RAND Corporation.

_____. 1960. *Development of man-machine simulation techniques.* Report P-1945. Santa Monica, Calif.: RAND Corporation.

_____. 1962. *The size of simulation samples required to compute certain inventory characteristics with stated precision and confidence.* Report RM-3242-PR. Santa Monica, Calif.: RAND Corporation.

Geisler, M. A., and Ginsberg, A. S. 1965. *Man-machine simulation experience.* Report P-3214. Santa Monica, Calif.: RAND Corporation.

Geisler, M. A.; Haythorn, W. W.; and Steger, W. A. 1962. *Simulation and the Logistics Systems Laboratory.* Report RM-3281-PR. Santa Monica, Calif.: RAND Corporation.

Glanzer, M., and Glaser, R. 1961. Techniques for the study of group structure and behavior: II. Empirical studies of the effects of structure in small groups. *Psychological Bulletin* 58: 1-27.

Glaser, R.; Klaus, D. J.; and Egerman, K. 1962. *Increasing team proficiency through training. 2. The acquisition and extinction of a team response.* Report AIR-B64-5/62-TR. Pittsburgh: American Institute for Research.

Goldhamer, H., and Speier, H. 1959. *Some observations on political gaming.* Report P-1679-RC. Santa Monica, Calif.: RAND Corporation.

Goldstein, I. L.; Southard, J. F.; and Schum, D. A. 1967. Feedback in a complex multiman-machine system. *Journal of Applied Psychology* 51: 346-51.

Good, I. J. 1954. Discussion, in Symposium on Monte Carlo Methods. *Journal of the Royal Statistical Society* B16:68–69.

Goode, H. H. 1951. Simulation: Its place in system design. *Proceedings of the IRE* 39: 1501–6.

Goodwin, W. R. 1957. The System Development Corporation and system training. *American Psychologist* 12: 524–28.

Grant, E. E.; O'Connell, S. L.; and Stoker, K. L. 1960. *The Human Factors Laboratory.* Report TM–561. Santa Monica, Calif.: System Development Corporation.

Grant, G., and Hostetter, R. 1961. *Display problems in aerospace surveillance systems.* Report ESD–TDR–61–57. L. G. Hanscom Field, Bedford, Mass.: Electronic Systems Division.

Green, B. F., Jr. 1963. *Digital computers in research.* New York: McGraw–Hill.

Greyson, M. 1964. *Second war gaming symposium.* Washington, D.C.: Washington Operations Research Council.

Grodsky, M. A. 1966. The use of full scale mission simulation for the assessment of complex operator performance. Paper read at the annual meeting of the American Psychological Association.

Grodsky, M. A., and Bryant, J. P. 1962. *Crew performance during simulated lunar missions.* Report ER 12693. Baltimore: The Martin Company.

Grodsky, M. A.; Glaser, D. L.; and Hopkins, A. R. Jr. 1966. *Analysis of crew performance in the Apollo Command Module, Phase I.* Report ER 14264. Baltimore: The Martin Company.

Grodsky, M. A.; Moore, H. G.; and Flaherty, T. M. 1966. Crew reliability during simulated space flight. *Journal of Spacecraft and Rockets* (July).

Grodsky, M. A.; Roberts, D. L.; and Mandour, J. A. 1966. *Test of pilot retention of simulated lunar mission skills.* Report ER 14139. Baltimore: The Martin Company.

Grodsky, M. A.; Mandour, J. A.; Roberts, D. L.; and Woodward, D. P. 1966. *Crew performance studies for manned space flight.* Summary Technical Report ER 14141–S. Vol. 1, *Description of study and results.* Report ER 14141–1. Baltimore: The Martin Company.

Grubmeyer, R. S. 1956. *Requirements for a universal air traffic control simulator.* Working Paper No. 2. Philadelphia: Franklin Institute Laboratories.

Guetzkow, H. 1959. The use of simulation in the study of inter-nation relations. *Behavioral Science* 4: 183–91.

_____. (ed.) 1962. *Simulation in social science: Readings.* Englewood Cliffs, N.J.: Prentice–Hall.

Guetzkow, H.; Alger, C. F.; Brody, R. A.; Noel, R. C.; and Snyder, R. C. 1963. *Simulation in international relations.* Englewood Cliffs, N.J.: Prentice–Hall.

Gullahorn, J. T., and Gullahorn, Jeanne E. 1962. *Simulating elementary social behavior.* Report SP–938/000/00. Santa Monica, Calif.: System Development Corporation.

_____. 1965. *Computer simulation of role conflict resolution.* Report SP–2261/000/00. Santa Monica, Calif.: System Development Corporation.

Hall, R. L. 1957. Group performance under feedback that confounds responses of group members. *Sociometry* 20: 297–305.

Hanes, R. M., and Gebhard, J. W. 1963. *Studies of tactical command decision. III. Commander acceptance and use of automated command-control procedures.* Report TG-450-3. Baltimore: The Johns Hopkins University Applied Physics Laboratory. A classified report–Secret.

_____. 1964. *Studies of tactical command decisions. IV. Commander selection and override of specific and general automated procedures. Part I.* Report TG-450-4. Baltimore: The Johns Hopkins University Applied Physics Laboratory. A classified report–Secret.

_____. 1965*a*. *Studies of tactical command decision. V. Commander selection and override of automated procedures: Effect of program switching under different levels of uncertainty.* Report TG-450-5. Baltimore: The Johns Hopkins University Applied Physics Laboratory. A classified report–Confidential.

_____. 1965*b*. *Studies of tactical command decision. IV. Commander selection and override of specific and general automated procedures. Part II.* Report TG-450-4A. Baltimore: The Johns Hopkins University Applied Physics Laboratory. A classified report–Confidential.

_____. 1966. The computer's role in command decision. *U.S. Naval Institute Proceedings* (September): 61-68.

Hanes, R. M.; Gebhard, J. W.; and Emch, G. F. 1962. *Studies of tactical command decision. I. Effects of attack tactics on high performance defense in an ECM environment.* Report TG-450-1. Baltimore: The Johns Hopkins University Applied Physics Laboratory. A classified report–Secret.

_____. 1963. *Studies of tactical command decision. II. Effect of quality of target position information on a high performance defense.* Report TG-450-2. Baltimore: The Johns Hopkins University Applied Physics Laboratory. A classified report–Confidential.

Hare, A. P.; Borgatta, E. F.; and Bales, R. F. 1955. *Small groups–studies in social interaction.* New York: Knopf.

Harman, H. H. 1960. *The Systems Simulation Research Laboratory,* Report TM-498. Santa Monica, Calif.: System Development Corporation.

_____. 1961. Simulation: A survey. Paper read at the Joint IRE-AIEE-ACM Computer Conference.

_____. 1963*a*. Comments in panel on application of simulation techniques to tactical and logistical problems. *Proceedings of the U.S. Army Operations Research Symposium* (March).

_____. 1963*b*. A computer based man-machine research laboratory. *Data Processing for Science/Engineering* (December).

Harrison, J. O., Jr. 1964. *Computer-aided information systems for gaming.* Report RAC TP-133. McLean, Va.: Research Analysis Corporation.

Hausrath, A. H. In publication. *Venture simulation in war, business, and politics.* New York: McGraw-Hill.

Havron, M. D., and McGrath, J. E. 1962. The contribution of the leader to the effectiveness of small military groups. In *Leadership and interpersonal behavior,* eds. L. Petrullo and B. M. Bass. New York: Rinehart and Winston.

Havron, M. D.; Fay, R. J.; and Goodacre, D. M. 1951. *Research on the effectiveness of small military units.* Personnel Research Section Report 885. Washington, D.C.: Personnel Bureau, The Adjutant General's Office.

Havron, M. D.; Fay, R. J.; and McGrath, J. E. 1952. *The effectiveness of small military units.* Personnel Research Section Report 980. Washington, D.C.: Personnel Bureau, The Adjutant General's Office.

Havron, M. D.; Lybrand, W. A.; and Cohen, E. 1954. *The assessment and prediction of rifle squad effectiveness.* Report. Arlington, Va.: Psychological Research Associates.

Havron, M. D.; Burdick, H.; Hutchins, E. B.; and Buckley, E. P. 1954. *Experimental investigation of tables of organization and equipment of infantry rifle squad.* Report 54-2, Part I. Arlington, Va.: Psychological Research Associates.

Havron, M. D.; Gorham, W. A.; Nordlie, P. G.; and Bradford, R. G. 1954. *A research study of the tactical training of the infantry rifle squad.* Final Report. Arlington, Va.: Psychological Research Associates.

Havron, M. D.; Whittenburg, J. A.; McGrath, J. E.; Vaughan, W. S.; Kassebaum, R. G.; and Walker, P. G. 1957. *Fire capability of infantry weapons.* Report 57-6. Arlington, Va.: Psychological Research Associates.

Havron, M. D.; Whittenburg, J. A.; Pratt, R. A.; Barry, E. H.; Motsko, M.; Buckley, E. P ; and Hutchins, E. B. 1955. *Experimental investigation of tables of organization and equipment of infantry rifle squad.* Report 55-1, Part II. Arlington, Va.: Psychological Research Associates.

Hayes, J. R. 1962. *Human data processing limits in decision making.* Report ESD-TDR-62-48. L. G. Hanscom Field, Bedford, Mass.: Electronic Systems Division.

Haythorn, W. W. 1957. *Simulation in RAND's Logistics Systems Laboratory.* Report P-1075. Santa Monica, Calif.: RAND Corporation.

_____. 1958. *Simulation in RAND's Logistics Systems Laboratory: Laboratory problem I.* Report P-1456. Santa Monica, Calif.: RAND Corporation.

_____. 1959*a*. *The use of simulation in estimating intrasquadron logistics: A description of LP-II, Phase 1.1.* Report P-1656. Santa Monica, Calif.: RAND Corporation.

_____. 1959*b*. *The use of simulation in logistics policy research.* Report P-1791. Santa Monica, Calif.: RAND Corporation.

_____. 1963*a*. System simulation as a technique in systems research. In *Human factors in technology*, eds. E. Bennett, J. Degan, and J. Spiegel. New York: McGraw-Hill. Also, 1961. *Human factors in systems research.* Report P-2337. Santa Monica, Calif.: RAND Corporation.

_____. 1963*b*. Information systems simulation and modeling. In *Information system simulation and modeling techniques, session 7, First Congress on the Information System Sciences.* Report ESD-TDR-63-474-7. L. G. Hanscom Field, Bedford, Mass.: Electronic Systems Division.

_____. 1967. Project ARGUS—A program of isolation and confinement research. *Naval Research Reviews.*

Haythorn, W. W., and Altman, I. 1967. Personality factors in isolated environments. In *Psychological stress*, eds. M. H. Appley and R. Trumbull. New York: Appleton-Century-Crofts.

Hazle, Marlene, and Lee, W. C. 1964. *Area planning experiments (APEX) final report (Project 920.0).* Report TM-03933/0000/01/11. Bedford, Mass.: MITRE Corporation.

Headquarters, Combat Developments Command Experimentation Center. 1964. *Outline plan of experimentation "Operations at Night."* Report 64-5. Fort Ord, Calif.: Combat Developments Command Experimentation Center.

____. 1965. *Detailed plan of experimentation "Operations at Night."* Report 64-5. Fort Ord, Calif.: Combat Developments Command Experimentation Center.

Held, J. S., and Wolff, W. 1962. *Simulation test environment to evaluate team load (STEEL): Summary analysis report.* Report TM -3263. Bedford, Mass.: MITRE Corporation.

Hett, W. D.; Coulopolos, W. D.; Wolff, W.; and Kowalski, R. A. 1962. *Package D testing with air movements data only (DAMDOT): Final Report.* Report TM-3339. Bedford, Mass.: MITRE Corporation.

Hixson, W. C.; Harter, G. A.; Warren, C. E.; and Cowan, J. D., Jr. 1954. *An electronic radar target simulator for traffic control studies.* Report TR 54-569. Wright-Patterson Air Force Base, Ohio: Wright Air Development Center.

Hollingdale, S. H. (ed.) 1967. *Digital simulation in operational research.* New York: American Elsevier Publishing Company.

Holmen, M. G. 1963. *Applications of simulation in command and control systems.* Report SP-1455. Santa Monica, Calif.: System Development Corporation.

Hovland, C. I. 1960. Computer simulation of thinking. *American Psychologist* 15: 687-93.

Howell, W. C. 1967. *Some principles for the design of decision systems: A review of six years of research on a command-control system simulation.* Report TR-67-136. Wright-Patterson Air Force Base, Ohio: Aerospace Medical Research Laboratories.

Howell, W. C.; Christy, R. T.; and Kinkade, R. G. 1959. *System performance following radar failure in a simulated air traffic control situation.* Report TR 59-573. Wright-Patterson Air Force Base, Ohio: Wright Air Development Center.

Hudson, B. B., and Searle, L. V. 1944. *Description of the Tufts tracking trainer.* OSRD Report 3286. Washington, D.C.: Applied Psychology Panel, NDRC.

Hulbert, S., and Wojcik, C. 1964. *Driving simulator devices and applications.* Report 803A. New York: Society of Automotive Engineers.

Irish, A. S.; Plowman, E. L.; Nichols, J. L.; Wiederspahn, W. H.; and Karroll, J. E. 1955. *Evaluation of the Electronic Data System.* Report 503. Washington, D.C.: Naval Research Laboratory.

Jackson, J. R. 1958. *UCLA executive decision games.* Report 58. Los Angeles, Calif.: University of California Management Sciences Research.

____. 1959. Learning from experience in business decision games. *California Management Review* 1: 92-107.

Jacobs, J. F. 1965. *Practical evaluation of command and control systems.* Report MTP-7. Bedford, Mass.: MITRE Corporation.

Jacoby, Joan E., and Harrison, S. 1960. *Efficient experimentation with simulation models.* Report 60-2. Washington, D.C.: Technical Operations Inc., Omega Operations Evaluation Group Air Force.

Jaffe, J. 1958. *Final report: 85th Division research project.* Report TM-168. Santa Monica, Calif.: System Development Corporation.

_____. 1959. *Preliminary planning for the simulation facility (Simfac).* Report FN-LO-63. Santa Monica, Calif.: System Development Corporation.

Jaffe, J., and Adamson, R. 1959. *Status of Simfac planning; III. Research.* Report FN-LO-179. Santa Monica, Calif.: System Development Corporation.

Jensen, B. T. 1961. *Instruction for "The Railroad Game."* Report TM-608. Santa Monica, Calif.: System Development Corporation.

Jensen, B. T.; Tilton, J. R.; and Anderson, D. N. 1958. *Feedback and debriefing–Report of a study at M-96.* Report TM-404. Santa Monica, Calif.: System Development Corporation.

Johnson, B. E.; Williams, A. C., Jr.; and Roscoe, S. N. 1951. *A simulator for studying human factors in air traffic control systems.* Report No. 11. Urbana, Ill.: University of Illinois National Research Council Committee on Aviation Psychology.

Johnston, W. A. 1966. Transfer of team skills as a function of type of training. *Journal of Applied Psychology* 50: 102–8.

Joint War Games Agency. 1966. *Catalog of war gaming models.* Report JWGA-1-66. Washington, D.C.: Joint Chiefs of Staff.

Jones, J. H.; Madvig, R. M.; Lynch, W. M.; Amara, R. C.; and Elpel, E. A. 1959. *CDEC instrumentation study–phase I.* Project 2900 Report. Menlo Park, Calif.: Stanford Research Institute.

Jordan, N.; Jensen, B. T.; and Terebenski, S. J. 1963. The development of cooperation among three-man crews in a simulated man-machine information-processing system. *Journal of Social Psychology* 59: 175–84.

Kahn, H., and Mann, I. 1957. *War gaming.* Report P-1167. Santa Monica, Calif.: RAND Corporation.

Kaplan, A. 1964. *Conduct of inquiry.* San Francisco: Chandler.

Kaplan, M. A.; Burns, A. L.; and Quandt, R. M. 1960. Theoretical analysis of the "balance of power." *Behavioral Science* 5: 240–52.

Kaplan, R. J., and Newman, J. R. 1963. *A study in probabilistic information processing (PIP).* Report TM-1150/000/00. Santa Monica, Calif.: System Development Corporation.

_____. 1964*a*. *PIP study no. 3: Probability judgments in a strategy decision task.* Report TM-1150/003/00. Santa Monica, Calif.: System Development Corporation.

_____. 1964*b*. *Studies in probabilistic information processing.* Report SP-1743/000/00. Santa Monica, Calif.: System Development Corporation.

_____. 1966. Studies in probabilistic information processing. *IEEE Transactions on Human Factors in Electronics* 7: 49–63.

Kaplan, R. J.; Lichtenstein, Sarah; and Newman, J. R. 1963. *PIP study number 2: Probabilistic information processing under conditions of varying payoff and task difficulty.* Report TM-1150/001/00. Santa Monica, Calif.: System Development Corporation.

Kappauf, W. E. 1947. History of psychological studies of the design and operation of equipment. *American Psychologist* 2: 83–86.

Kaufman, L.; Payne, E. J.; and Bailey, G. C. 1959. *A method and automatic apparatus for simulation of inputs by teletypewriter in experimental command post exercises.* Report 2144-258-R. Ann Arbor, Mich.: University of Michigan Willow Run Laboratories.

Kennedy, J. L. 1952. The uses and limitations of mathematical models, game theory, and systems analysis. In *Psychology in the world emergency.* Pittsburgh: University of Pittsburgh.

_____. 1962*a*. The system approach: Organizational development. *Human Factors* 4: 25-52.

_____. 1962*b*. Environment simulation as a technique for studying human behavior. Paper read at the First Congress on the Information System Sciences.

_____. 1965. Something old, something new. Review of E. H. Porter, *Manpower development. Contemporary Psychology* 10: 260-61.

Kepner, C. H., and Tregoe, B. B. 1959. *On the concept of visible and invisible functioning.* Report TM-409. Santa Monica, Calif.: System Development Corporation.

Kerlinger, F. 1965. *Foundations of behavioral research.* New York: Holt, Rinehart, and Winston.

Kershner, R. B. 1960. A survey of systems engineering tools and techniques. In *Operations research and systems engineering*, eds. C. D. Flagle, W. H. Huggins, and R. R. Roy, Baltimore: The Johns Hopkins Press.

Kibbee, J. M.; Craft, C. J.; and Nanus, B. 1961. *Management games.* New York: Reinhold Publishing Corporation.

Kidd, J. S. 1959*a*. *A comparison of two methods of controller training in a simulated air traffic control task: A study in human engineering aspects of radar air traffic control.* Report TR 58-449. Wright-Patterson Air Force Base, Ohio: Wright Air Development Center. Also 1961*b*.

_____. 1959*b*. *A comparison of one-, two-, and three-man control units under various conditions of traffic input rate.* Report TR 59-104. Wright-Patterson Air Force Base, Ohio: Wright Air Development Center. Also 1961*c*.

_____. 1959*c*. *A summary of research methods, operator characteristics, and system design specifications based on the study of a simulated radar air traffic control system.* Report TR 59-236. Wright-Patterson Air Force Base, Ohio: Wright Air Development Center.

_____. 1961*a*. *Some sources of load and constraints on operator performance in a simulated radar air traffic control task.* Report TR 60-612. Wright-Patterson Air Force Base, Ohio: Wright Air Development Center.

_____. 1961*b*. A comparison of two methods of training in a complex task by means of task simulation. *Journal of Applied Psychology* 45: 165-69.

_____. 1961*c*. A comparison of one-, two-, and three-man work units under various conditions of work load. *Journal of Applied Psychology* 45: 195-200.

_____. 1961*d*. *Work team effectiveness as a function of mechanical degradation of the intrateam communication system.* Report ESD-TN-61-57. L. G. Hanscom Field, Bedford, Mass.: Electronic Systems Division. Also, 1963.

———. 1962. A new look at system research and analysis. *Human Factors* 4: 209–16.

———. 1963. Work team effectiveness as a function of the mechanical degradation of the intrateam communication system. *Journal of Engineering Psychology* 2: 1–14.

———. 1965*a*. *Preliminary investigation of measurement sensitivity.* Technical Note 2. McLean, Va.: Human Sciences Research, Inc.

———. 1965*b*. *An indelicate experiment on telephone conference processes: Some effects of group size under various task conditions and network configurations.* Technical Note 5. McLean, Va.: Human Sciences Research, Inc.

Kidd, J. S., and Christy, R. T. 1961. Supervisory procedures and work–team productivity. *Journal of Applied Psychology* 45: 388–92.

Kidd, J. S., and Hooper, J. J. 1959. *Division of responsibility between two controllers and load balancing flexibility in a radar approach control team. A study in human engineering aspects of radar air traffic control.* Report TR 58–473. Wright–Patterson Air Force Base, Ohio: Wright Air Development Center.

Kidd, J. S., and Kinkade, R. G. 1958. *Air traffic control system effectiveness as a function of division of responsibility between pilots and ground controllers: A study in human engineering aspects of radar air traffic control.* Report TR 58–113. Wright–Patterson Air Force Base, Ohio: Wright Air Development Center.

———. 1959. *Operator change-over effects in a complex task.* Report TR 59–235. Wright–Patterson Air Force Base, Ohio: Wright Air Development Center. Also 1962.

———. 1962. Operator change-over effects in a complex task. *Journal of Engineering Psychology* 1: 83–91.

Kidd, J. S., and Michels, K. M. 1959. *Staff development in systems research techniques.* Report 29. Philadelphia: Courtney & Co.

Kidd, J. S.; Shelly, M. W.; Jeantheau, G.; and Fitts, P. M. 1958. *The effect of enroute flow control on terminal system performance: A study in human engineering aspects of radar air traffic control.* Report TR 57–663. Wright–Patterson Air Force Base, Ohio: Wright Air Development Center.

Kidd, J. S.; Kinkade, R. G.; Ichniowski, F. C.; Widhelm, W. G.; and Urback, S. 1963*a*. *Overview of project No. 102–8X.* Report ER–3238, Vol. I. Cockeysville, Md.: Aircraft Armaments, Inc.

Kidd, J. S.; Widhelm, W. N.; Ichniowski, F. C.; and Kinkade, R. G. 1963*b*. *Method development for ATC system simulation research.* Report ER–3238, Vol. II. Cockeysville, Md.: Aircraft Armaments, Inc.

Kinkade, R. G., and Kidd, J. S. 1959*a*. *The effect of procedural variations in the use of target identification and airborne position information equipment on the performance of a simulated radar approach control system.* Report TR 58–264. Wright–Patterson Air Force Base, Ohio: Wright Air Development Center.

———. 1959*b*. *The effect of different proportions of monitored elements on operator performance in a simulated radar air traffic control system.* Report TR 59–169. Wright–Patterson Air Force Base, Ohio: Wright Air Development Center.

_____. 1959c. *The use of an operational game as a method of task familiarization.* Report TR 59-204., Wright-Patterson Air Force Base, Ohio: Wright Air Development Center. Also 1962.

_____. 1962. The use of an operational game as a method of task familiarization. *Journal of Applied Psychology* 46: 1-5.

Kinkade, R. G.: Kidd, J. S.; Urback, S.; Ichniowski, F. C.; and Widhelm, W. N. 1963. *Handbook of ATC system test methods.* Report ER 3238 (Vol. III). Cockeysville, Md.: Aircraft Armaments, Inc.

Kite, W. R., and Vitz, P. C. 1966. *Teleconferencing: effect of communication medium, network, and distribution of resources.* Study-233. Arlington, Va.: Institute for Defense Analyses.

Klaus, D. J., and Glaser, R. 1960. *Increasing team proficiency through training. I. A. program of research.* Report AIR-B64-60-TR-137. Pittsburgh: American Institute for Research.

Kraft, C. L. 1956. *A broad-band blue lighting system for radar approach control centers: Evaluation and refinements based on three years of operational use.* Report TR 56-71. Wright-Patterson Air Force Base, Ohio: Wright Air Development Center.

Kraft, C. L., and Fitts, P. M. 1954. *A broad-band lighting system for radar air traffic control centers.* Report TR 53-416. Wright-Patterson Air Force Base, Ohio: Wright Air Development Center.

Krumm, R. L., and Farina, A. J. 1962. *Effectiveness of integrated flight simulator training in promoting B-52 crew coordination.* Report MRL-TDR-62-1. Wright-Patterson Air Force Base, Ohio: Aerospace Medical Research Laboratories.

Kurke, M. I. 1963. *Field test methodology; a survey of practices.* Report CORG M-177. Fort Belvoir, Va.: Combat Operations Research Group.

_____. 1965. Operational concept analysis and sources of field data. *Human Factors* 7: 537-44.

_____. 1966. Progress in Army troop test methodology. Paper read at the annual meeting of the Human Factors Society.

Lanzetta, J. T., and Roby, T. B. Effects of work-group structure and certain task variables on group performance. *Journal of Abnormal and Social Psychology* 53: 307-14.

_____. 1957. Group learning and communication as a function of task and structure demands. *Journal of Abnormal and Social Psychology* 55: 121-31.

Lee, W. C. 1963. *Final report, TRICOM and TRICOM extension.* Report TM-3639. Bedford, Mass.: MITRE Corporation.

Lefford, A. 1949. *Unaided visual estimation of bearing under conditions of scope decentration.* Report SPECDEVCEN 279-3-2. New York: New York University.

Lefford, A., and Taubman, R. E. 1950. *Preliminary studies of detection time and other factors involved in AEW performance.* Report SPECDEVCEN 279-3-4. New York: New York University.

Lesiw, W. 1967. Field experiments and system tests in NORAD COC development. In *Information system science and technology papers prepared for the Third Congress*, ed. D. E. Walker. Washington, D.C.: Thompson Book Company.

Licklider, J. C. R. 1962. *Studies in the organization of man-machine systems.* Report 970. Cambridge, Mass.: Bolt, Beranek, and Newman.

Lindquist, E. F. 1953. *Design and analysis of experiments in psychology and education.* Boston: Houghton Mifflin.

Logistics Systems Laboratory. 1957*a*. *First tooling up exercise for the Logistics Systems Laboratory (October to November 1956).* Report RM-1924. Santa Monica, Calif.: RAND Corporation.

_____. 1957*b*. *Second tooling up exercise for the Logistics Systems Laboratory (January to February 1957).* Report RM-1961. Santa Monica, Calif.: RAND Corporation.

McGlothlin, W. H. 1958. *The simulation laboratory as a developmental tool.* Report P-1454. Santa Monica, Calif.: RAND Corporation.

McGrath, J. E.; Nordlie, P. G.; and Vaughan, W. S., Jr. 1960. *A systematic framework for comparison of system research methods.* Report HSR-TN-59/7a-SM. McLean, Va.: Human Sciences Research, Inc.

McGuigan, F. J. 1960. *Experimental psychology: A methodological approach.* Englewood Cliffs, N.J.: Prentice-Hall.

McGuire, J. C.; Kester, S. G.; Parsons, H. M.; and Douglas, R. E. 1966. *System project management procedures for integrated management of the human factors (personnel-related) aspects of Army system development.* Report TM-2908/000/01. Santa Monica, Calif.: System Development Corporation.

McHugh, F. J. 1961. *Fundamentals of war gaming, 2nd edition.* Newport, R.I.: Naval War College.

McKelvey, R. K. 1967. The use of driving simulators at the USPHS Driving Research Laboratory. In *Mathematical models and simulation of automobile driving*, ed. T. B. Sheridan. Cincinnati, Ohio: U.S. Public Health Service.

McKelvey, R. K., and Cohen, J. D. 1954. *The behavior of individuals and personnel systems in the surveillance functions of an Air Force Defense Center. I. Experimental method.* Report AFPTRC TR 54-98. Mather Air Force Base, Calif.: Air Force Personnel and Training Center.

McKelvey, R. K.; Ontiveros, R.; et al. 1961*a*. *Simulator comparison of three runway landing zone lighting patterns.* Bureau of Research and Development Report. Atlantic City, N.J.: Federal Aviation Agency National Aviation Facilities Experimental Center.

_____. 1961*b*. *Simulator comparisons of narrow gauge landing zone lighting patterns in longitudinal vs. lateral arrays.* Bureau of Research and Development Report. Atlantic City, N.J.: Federal Aviation Agency National Aviation Facilities Experimental Center.

McLane, R. C., and Wolf, J. D. 1965. *Final research report on the experimental evaluation of symbolic and pictorial displays for submarine control.* Technical Report. St. Paul, Minn.: Honeywell, Inc.

Mackie, R. R. 1968. Translation and application of psychological research in relation to systems development. Paper read at the Second Naval Material Command Symposium on Human Factors in Systems Effectiveness.

Mackie, R. R., and Harabedian, A. 1964. *A study of simulation requirements for sonar operator trainers.* Report No. 711-1, NAVTRADEVCEN 1320-1. Goleta, Calif.: Human Factors Research, Inc.

March, J. C., and Simon, H. A. 1958. *Organizations.* New York: Wiley.

Martin, E. W. 1961*a*. *Operational description of system simulation research facility (SIMFAC).* Report FN-LO-458. Santa Monica, Calif.: System Development Corporation.

_____. 1961*b*. *Description of SIMFAC data production and processing activity.* Report FN-LO-471. Santa Monica, Calif.: System Development Corporation.

Martin, F. F. 1968. *Computer modeling and simulation.* New York: Wiley.

Martin, H. H. 1959. If an attack comes . . . *Saturday Evening Post* 25: 79-80.

Meeker, R. J.; Shure, G. H.; and Rogers, M. S. 1962. *A review of the group decision-making research projects.* Report TM-731. Santa Monica, Calif.: System Development Corporation.

Meister, D., and Rabideau, G. F. 1965. *Human factors evaluation in system development.* New York: Wiley.

Members of the D16 Staff. 1962. *SATIN Package C test results.* Report SR-59. Bedford, Mass.: MITRE Corporation.

Merrifield, P. R., and Erickson, E. B. 1964*a*. *System support for hypothesis generation.* Report TM-1939. Santa Monica, Calif.: System Development Corporation.

_____. 1964*b*. *Target analysis in the HEMP context—the transitional experiment, part one.* Report TM-2030/000/00. Santa Monica, Calif.: System Development Corporation.

_____. 1964*c*. *Target analysis in the HEMP context—the transitional experiment, part two.* Report TM-2030/001/00. Santa Monica, Calif.: System Development Corporation.

Miller, I. W. 1965. *The AESOP Testbed System description—TACC manual I/2 (model of the fighter section, TACC current plans division).* Report TM-04277/0000/00/0/00. Bedford, Mass.: MITRE Corporation.

Miller, R. B.; Chapman, R. L.; Ely, J. H.; Howland, D.; Kryter, K. D.; and Rath, G. J. 1959. *Report of the task group on design and use of man-machine systems.* Report. Washington, D.C.: Smithsonian Institution.

Miller, R. S. 1958. *Summary of joint FIL-TDS research in air traffic control during period from April 1957 to September 1958.* Report. Philadelphia: Franklin Institute.

Mills, D. L.; Vollmer, H. M.; and Anderson, B. F. 1961. *Personnel factors and unit performance. A study of vehicle commanders, BMU experiment, Spring 1959.* Research Memorandum RO-RM 11. Menlo Park, Calif.: Stanford Research Institute.

Moore, H. G. 1961. *The effects of load and accessibility of information upon performance of small teams.* Report AFCSR-1636. Washington, D.C.: Air Force Office of Scientific Research.

Morey, J. L., and Yntema, D. B. 1964. Experiments on systems. In *Second Congress on the Information System Sciences*, eds. J. Spiegel and D. E. Walker. Washington, D.C.: Spartan Books.

Morgan, C. T. 1950. *Problems of system coordination and plans for dealing with them.* Appraisal Report 166-I-129. Baltimore: The Johns Hopkins University Institute for Cooperative Research.

Morgan, C. T.; Chapanis, A.; Cook, J. S.; and Lund, M. W. 1963. *Human engineering guide to equipment design.* New York: McGraw-Hill.

Morgenthaler, G. W. 1961. The theory and application of simulation in operations research. In *Progress in operations research, Vol. 3*, ed. J. S. Avonofsky New York: Wiley.

Moros, D. 1963. *Final report: Terminal handover testing (Project 910.0).* Report TM-03894/0000/00/0/00. Bedford, Mass.: MITRE Corporation.

Mosimann, J.; LaRoche, F.; and DeVoe, R. P. 1955*a*. *Operation HUSKY Report.* Report 2144-773-M. Ypsilanti, Michigan: University of Michigan Willow Run Research Center.

_____. 1955*b*. *Results of Operation Slowdown.* Report 2144-776-M. Ypsilanti, Michigan: University of Michigan Willow Run Research Center.

_____. 1955*c*. *Results of Operation Slowdown II.* Report 2144-782-M. Ypsilanti, Michigan: University of Michigan Willow Run Research Center.

Murdoch, F. J., and Edmondson, C. C. 1962. Field experimentation as an aid to operations research. Paper read at the 1962 Army Science Conference.

Murphy, J. V.; Katter, D. A.; Wattenbarger, G. H.; and Pool, H. M. 1962. *An investigation of a basic design feature in computerized systems.* Report TM-732/002/00. Santa Monica, Calif.: System Development Corporation.

Naylor, J. C., and Briggs, G. E. 1965. Team-training effectiveness under various conditions. *Journal of Applied Psychology* 49:223-29.

Naylor, T. H.; Balintfy, J. L.; Burdick, D. S.; and Chu, K. 1966. *Computer simulation techniques.* New York: Wiley.

Nelson, H. W., and Peterson, J. W. 1962. *Integrated supply-support policies: the LP-III experience.* Report RM-2839. Santa Monica, Calif.: RAND Corporation.

New York University Cadillac Staff. 1956. *Project Cadillac summary report. Recommendations for operating procedures and personnel allocation in the CIC compartment of the WV-2, Aircraft.* Technical Report SPECDEVCEN 279-3-23. New York: New York University. A classified report—Confidential.

Nichols, J. L., and Karroll, J. E. 1955. *A partial evaluation of the target designation features of the NRL Electronic Data System.* Memo Report 471. Washington, D.C.: Naval Research Laboratory.

Nichols, J. L., and Plowman, E. L. 1956. *Evaluation of the Electronic Data System.* Memo Report 566. Washington, D.C.: Naval Research Laboratory.

Nichols, T. F.; Ward, J. S.; Fooks, N. I.; Brown, F. L.; and Rosenquist, H. S. 1962. *Performance evaluation of light weapons infantrymen.* Technical Report 81. Alexandria, Va.: Human Resources Research Office.

Obermayer, R. W. 1964. Simulation, models, and games: Sources of measurement. *Human Factors* 6: 607-19.

Okanes, M. M. 1962. *SAGE air surveillance subsystem exercising: An example of isolating human performance variables.* Report SP-236. Santa Monica, Calif.: System Development Corporation. Also, Paper read at the 1961 Western Electronic Show and Convention (WESCON).

Overholt, J. (ed.) 1961. *First war gaming symposium.* Washington, D.C.: Washington Operations Research Council.

Page, H. J., and Connolly, D. W. 1954. *The effect of number of air controllers on the air intercept control performance of the airborne CIC.* Technical Report SPECDEVCEN 279-3-20. New York: New York University.

Page, T. L. 1960. A survey of operations research tools and techniques. In *Operations research and systems engineering*, eds. C. D. Flagle, W. H. Huggins, and R. R. Roy. Baltimore: The Johns Hopkins Press.

Panel on Psychology amd Physiology. 1949. *Human factors in undersea warfare.* Washington, D.C.: National Research Council Committee on Undersea Warfare.

Parsons, H. M. 1954*a*. *Experimental laboratory studies of the AN/GPA-23 (XW-1).* Technical Report T-1/A-IV. New York: Columbia University Electronics Research Laboratories.

_____. 1954*b*. *Field tests of the AN/GPA-23 (XW-1).* Technical Report T-2/A-IV. New York: Columbia University Electronics Research Laboratories.

_____. 1957. *Operational study of the AN/GPA-23 system for air defense direction centers.* Report AFCRC TR-57-53. New York: Columbia University Electronics Research Laboratories.

_____. 1960*a*. *Performance of the Temco video correlator with ACTER inputs.* Report FN-4078. Santa Monica, Calif.: System Development Corporation. A classified report—Secret.

_____. 1960*b*. *The development and installation of a system training program: The SAGE ECCM model.* Report SP-265. Santa Monica, Calif.: System Development Corporation.

_____. 1962. System trouble-shooting. In *Proceedings of the Military Operations Research Symposia (MORS)* 2 (1): 333-38.

_____. 1964. *What the Navy's Anti-Air Warfare Training Program can learn from air defense system training experience.* Memorandum Report. Washington, D.C.: Office of Naval Research.

_____. 1966. *STAVE: Stress Avoidance/Escape.* Report SP-2459. Santa Monica, Calif.: System Development Corporation. Also, paper read at the annual meeting of the Human Factors Society.

Parsons, H. M., and Perry, W. E. 1966. *Concepts for command and control systems.* Report TM-WD-227/000/00. Santa Monica, Calif.: System Development Corporation.

Parsons, H. M., and Sciorra, A. F. 1954. *Analysis and implementation of the GCI system—Laboratory testing and training program, tracking computing group, radar AN/GPA-23(XW-2).* Final Report F/A-V. New York: Columbia University Electronics Research Laboratories. A classified report—Confidential.

Parsons, H. M.; Sinaiko, H. W.; and McDonald, K. 1952. *The use of status boards in CICs: A preliminary survey.* Technical report SPECDEVCEN 279-3-14. New York: New York University.

Paul, L. E., and Buckley, E. P. 1967. *Human factors evaluation of a large screen radar display.* Report RD-66-105. Atlantic City, N.J.: Federal Aviation Agency National Aviation Facilities Experimental Center.

Phillips, L. D., and Edwards, W. D. 1966. Conservatism in a simple probability inference task. *Journal of Experimental Psychology* 72: 346-54.

Phillips, L. D.; Hays, W. L.; and Edwards W. D. 1966. Conservatism in complex probabilistic inference. *IEEE Transactions in Human Factors in Electronics* 7: 7–18.

Plowman, E. L.; Nichols, J. L.; Shea, J. D., Jr.; Ridgway, R. H.; Sally, A.; and Chaillet, R. F. 1956. *A comparison of the performance of four air summary plotting systems.* Memo Report 649. Washington, D.C.: Naval Research Laboratory.

Pool, I. de S. 1964. Simulating social systems. *International Science and Technology* (March): 62–70.

Porter, E. H. 1964. *Manpower development.* New York: Harper and Row.

Porter, E. H., and Proctor, J. H. 1962. *Information system performance evaluation.* Report SP-897/000/01. Santa Monica, Calif.: System Development Corporation.

Proctor, J. H. 1963. Normative exercising: An analytical and evaluative aid in system design. *IEEE Transactions in Engineering Management.*

Radloff, R., and Helmreich, R. 1968. *Groups under stress. Psychological research in Sealab II.* New York: Appleton-Century-Crofts.

Rauner, R. M. 1958. *Laboratory evaluation of supply and procurement policies: The first experiment of the Logistics Systems Laboratory.* Report R-323. Santa Monica, Calif.: RAND Corporation.

Rauner, R. M., and Steger, W. A. 1960. *Simulation of ICBM support: The second experiment of the Logistics Systems Laboratory.* Report R-369. Santa Monica, Calif.: RAND Corporation.

_____. 1961*a*. *Simulation and long-range planning for resource allocation.* Report P-2223–1. Santa Monica, Calif.: RAND Corporation. Also 1962.

_____. 1961*b*. *Game simulation and long-range planning.* Report R-2355. Santa Monica, Calif.: RAND Corporation.

_____. 1962. Simulation and long-range planning for resource allocation. *Quarterly Journal of Economics* 66: 219–45.

Redgrave, M. J. 1962. *Some approaches to simulation, modeling, and gaming at System Development Corporation.* Report SP-721. Santa Monica, Calif.: System Development Corporation.

Research Group in Psychology and the Social Sciences. 1960. *The technology of human behavior. Recommendations for defense support of research in psychology and the social sciences.* Washington, D.C.: Smithsonian Institution.

Rhine, R. J. 1960. *SIMFAC briefing.* Report FN-LO-350. Santa Monica, Calif.: System Development Corporation.

Rittenhouse, C. H. 1962*a*. *The use of military experience in the evaluation of tactical performance. A discussion of the measurement techniques used in the squad and platoon experiment, Spring 1961.* Research Memorandum RO-RM 19. Menlo Park, Calif.: Stanford Research Institute.

_____. 1962*b*. Judgemental evaluation of tactical performance. In *Proceedings of the Military Operations Research Symposia (MORS)*, Vol. 2, No. 1, Part 1.

_____. 1966. The development and use of rating techniques in conjunction with objective measures in field research. Paper read at the annual meeting of the Human Factors Society.

Road Research Laboratory. 1965. *Research on road traffic.* London: Her Majesty's Stationery Office.

Roby, T., and Lanzetta, J. 1957. Conflicting principles in man-machine system design. *Journal of Applied Psychology* 41: 170–78.

Rogers, M. S., and Shure, G. H. 1962. *Personality factor stability for three ability levels.* Report SP-652/000/01. Santa Monica, Calif.: System Development Corporation.

Rogers, M. S.; Ford, J. D., Jr.; and Tassone, J. 1959. *A laboratory study of crew turnover. Part I.* Report TM-423. Santa Monica, Calif.: System Development Corporation.

_____. 1960. *A laboratory study of crew turnover. Part II.* Report TM-423. Santa Monica, Calif.: System Development Corporation.

_____. 1961. The effects of personnel replacement on an information-processing crew. *Journal of Applied Psychology* 45: 91–96.

Rogers, M. S.; Shure, G. H.; and Meeker, R. J. 1962. *Decision making styles in a military command setting.* Report SP-709/001/00. Santa Monica, Calif.: System Development Corporation.

Rome, Beatrice K., and Rome, S. C. 1964*a*. *Communication and large organizations.* Report SP-1690. Santa Monica, Calif.: System Development Corporation. Also, two lectures at the Air Force Office of Scientific Research Summer Scientific Seminar on Communication Cybernetics.

_____. 1964*b*. Programming the bureaucratic computer. *IEEE Spectrum* (December): 72–79.

_____. 1965. Leviathan. *System Development Corporation Magazine* (April): 17–25.

_____. 1967. *Leviathan: Hierarchical process in simulated large organizations.* Report TM-3489/000/00. Santa Monica, Calif.: System Development Corporation.

Rome, S. C., and Rome, Beatrice K. 1961. The Leviathan technique for large-group analysis. *Behavioral Science* 6: 148–52.

_____. 1962. Computer simulation toward a theory of large organizations. In *Computer applications in the behavioral sciences*, ed. H. Borko. Englewood Cliffs, N.J.: Prentice-Hall.

Rosenberg, S. 1958. A laboratory approach to interpersonal aspects of team performance. *Ergonomics* 2: 335–48.

_____. 1959. The maintenance of a learned response in controlled interpersonal conditions. *Sociometry* 22: 124–38.

_____. 1960. Cooperative behavior in dyads as a function of reinforcement parameters. *Journal of Abnormal and Social Psychology* 60: 318–33.

Rosenberg, S., and Hall, R. 1958. The effects of different social feedback conditions upon performance in dyadic teams. *Journal of Abnormal and Social Psychology* 57: 271–77.

Rosove, P. E. 1967. *Developing computer-based information systems.* New York: Wiley.

Rowell, J. T. 1962. *Result of system training for SAGE air defense crews.* Report TM-719. Santa Monica, Calif.: System Development Corporation.

Rowell, J. T., and Streich, E. R. 1964. The SAGE system training program for the Air Defense Command. *Human Factors* 6: 537–48.

Rubin, L. S. 1954. *An experimental evaluation of four communication procedures in the airborne CIC.* Technical Report SPECDEVCEN 279-3-11. New York: New York University.

Rubin, L. S., and Connolly, D. W. 1954. *The effect on ACO performance of four airborne CIC height-finding procedures.* Technical Report SPECDEVCEN 279-3-16. New York: New York University.

Ruby, W. J.; Jocoy, E. H.; and Pelton, F. M. 1963. Simulation for experimentation: A position paper. In *AIAA Simulation for Aerospace Flight Conference volume of technical papers.* New York: American Institute of Aeronautics and Astronautics.

Rundquist, E. A. 1963. *Training personnel subsystems at two simulation levels.* Report TM-887. Santa Monica, Calif.: System Development Corporation.

Sackman, H. 1963. *An experimental demonstration of improved man-machine dialogue in the SAGE system.* Report TM-930/000/01. Santa Monica, Calif.: System Development Corporation.

_____. 1964*a*. Regenerative recording in man-machine digital systems. In *Conference proceedings,* 1964 National Winter Convention on Military Electronics.

_____. 1964*b*. A study of operator performance in the SAGE system. In *Proceedings of the Fifth National Symposium on Human Factors in Electronics.* Also, Report SP-1377/000/01. Santa Monica, Calif.: System Development Corporation.

_____. 1967. *Computers, system science, and evolving society.* New York: Wiley.

Sackman, H., and Munson, J. B. 1963. *Investigation of computer operating time and system capacity for man-machine digital systems.* Report SP-1462. Santa Monica, Calif.: System Development Corporation. Also 1964. *Journal of the Association for Computing Machinery* 11: 450–64.

SAE Committee AGE, Training. 1964. *A survey report of simulators used as tools for research, design and development.* Report AIR 779. New York: Society of Automotive Engineers.

Schapiro, H., and Guastella, M. 1955. *An experimental evaluation of integrated airborne early warning systems in the airborne CIC.* Technical Report SPECDEVCEN 279-3-18. New York: New York University. A classified report—Confidential.

Schipper, L. M., and Versace, J. 1956. *Human engineering aspects of radar air traffic control: I. Performance in sequencing aircraft for landing as a function of control time availability.* Report TR 56-57. Wright-Patterson Air Force Base, Ohio: Wright Air Development Center.

Schipper, L. M.; Kidd, J. S.; Shelly, M. W.; and Smode, A. F. 1957. *Terminal system effectiveness as a function of the method used by controllers to obtain altitude information: A study in human engineering aspects of radar air traffic control.* Report TR 57-278. Wright-Patterson Air Force Base, Ohio: Wright Air Development Center.

Schipper, L. M.; Kraft, C. L.; Smode, A. F.; and Fitts, P. M. 1957. *The use of displays showing identity versus no-identity: A study in human engineering aspects of radar air traffic control*. Report TR 57–21. Wright–Patterson Air Force Base, Ohio: Wright Air Development Center.

Schipper, L. M.; Versace, J.; Kraft, C. L.; and McGuire, J. C. 1956*a*. *Human engineering aspects of radar air traffic control: II and III. Experimental evaluations of two improved identification systems under high density traffic conditions*. Report TR 56–58. Wright–Patterson Air Force Base, Ohio: Wright Air Development Center.

_____. 1956*b*. *Human engineering aspects of radar air traffic control: IV. A comparison of sector and in-line control procedures*. Report TR 56–69. Wright–Patterson Air Force Base, Ohio: Wright Air Development Center.

Schlesinger, L. E.; Karmel, B.; and Cohen, S. 1964. System analysis of driving simulation. *Human Factors* 6: 383–92.

Schohan, B.; Rawson, H. E.; and Soliday, S. M. 1965. Pilot and observer performance in simulated low altitude high speed flight. *Human Factors* 7: 257–65.

Schroder, H. M.; Driver, M. J.; and Streufert, S. 1967. *Human information processing: Individuals and groups functioning in complex social situations*. New York: Holt, Rinehart, and Winston.

Schroder, H. M.; Streufert, S.; and Weeden, D. C. 1964. *The effect of structural abstractness in interpersonal stimuli on the leadership role*. Report No. 3 for the Office of Naval Research. Princeton, N.J.: Princeton University.

Schum, D. A. 1965. *Inferences on the basis of conditionally nonindependent data*. Report AMRL–TR–65–161. Wright–Patterson Air Force Base, Ohio: Aerospace Medical Research Laboratories. Also 1966*a*.

_____. 1966*a*. Inferences on the basis of conditionally nonindependent data. *Journal of Experimental Psychology* 72: 401–9.

_____. 1966*b*. Prior uncertainty and amount of diagnostic evidence as variables in a probabilistic inference task. *Organizational Behavior and Human Performance* 1: 31–54.

_____. 1966*c*. *A further evaluation of computer-assisted processing of complex evidence sets in a simulated military threat-diagnosis task*. Report (in preparation). Wright–Patterson Air Force Base, Ohio: Aerospace Medical Research Laboratories.

_____. 1967. Concerning the evaluation and aggregation of probabilistic evidence by man-machine systems. In *Information system science and technology papers prepared for the Third Congress*, ed. D. E. Walker. Washington, D.C.; Thompson Book Company.

Schum, D. A.; Goldstein, I. L.; and Southard, J. F. 1965*a*. *The influence of experience and input information fidelity upon posterior probability estimation in a simulated threat-diagnosis system*. Report AMRL–TR–65–25. Wright–Patterson Air Force Base, Ohio: Aerospace Medical Research Laboratories.

_____. 1965*b*. *Further investigation of the effects of reduced input data fidelity upon the determination of posterior probabilities in a simulated threat-diagnosis system*. Report TR–65–233. Wright–Patterson Air Force Base, Ohio: Aerospace Medical Research Laboratories.

_____. 1966. Research on a simulated Bayesian information-processing system. *IEEE Transactions on Human Factors in Electronics* 7: 37–48.

Schum, D. A.; Southard, J. F.; and Wombolt, L. F. 1966. *Concerning the aggregation of conditionally nonindependent evidence by information-processing systems when the rate of accumulation of the evidence is varied.* Report (in preparation). Wright-Patterson Air Force Base, Ohio: Aerospace Medical Research Laboratories.

Schum, D. A.; Goldstein, I. L.; Howell, W. C.; and Southard, J. F. 1967. Subjective probability revisions under several cost-payoff arrangements. *Organizational Behavior and Human Performance* 2: 84–104.

Schwartz, M., and the Control Simulation Project. 1963. *An investigation of the use of 465L displays within the context of the operations control area during a peacetime force exercise.* Report TM–LO–797/000/01. Santa Monica, Calif.: System Development Corporation.

Schwartz, S. L.; Cohen, R. I.; Green, E. C.; Joel, K. L.; and Winston, J. S. 1965. *The AESOP testbed: Test series I/1 (supplementary technical information).* Report TM–04268/0000/00/1/00. Bedford, Mass.: MITRE Corporation.

Scott, E. L. 1961. *SIMFAC quarterly report.* Report FN–LO–635/000/00. Santa Monica, Calif.: System Development Corporation.

_____. 1962. *SIMFAC quarterly report.* Report FN–LO–635/001/00. Santa Monica, Calif.: System Development Corporation.

Scott, H. T., Jr.; Weiner, M. G.; Sterrett, J. K.; and Erwin, H. W. 1953. *Psychological and technical evaluation of the Polaroid Land projection display system.* Memo Report 169. Washington, D.C.: Naval Research Laboratory.

Seitz, C. P., and Freeberg, N. 1965. Aerospace station simulation for the man, the system and the vehicle. In *Civilian and military uses of aerospace*, ed. H. E. Whipple. New York: Annals of the New York Academy of Sciences.

Shapero, A., and Erickson, C. J. 1961. Human factors testing in weapon system development. Paper read at the ARS Missile and Space Vehicle Testing Conference, Los Angeles.

Sharkey, V. J.; Connolly, D. W.; Doughty, J. M.; Fox, W. R.; and Sulzer, R. I. 1958. *Research and development field test of the AN/TSQ–13 (XD–1) tactical air control system.* Report TR 57–3. Bedford, Mass.: Air Force Cambridge Research Center. A classified report—Confidential.

Shaw, C. J. 1966. *Computer programming and command and control.* Report TM–2857/000/00. Santa Monica, Calif.: System Development Corporation.

Sheldon, M. S., and Zagorski, H. J. 1965. *Project NORM Mission I research report.* Report TM–2232/001/00. Santa Monica, Calif.: System Development Corporation.

Sheridan, T. B. (ed.) 1967. *Mathematical models and simulation of automobile driving.* Cincinnati, Ohio: U.S. Public Health Service.

Sherwin, C. W., and Isenson, R. S. 1967. Project hindsight. *Science* 156: 1571–77.

Shure, G. H. 1967. Behavioral gaming and simulation research. In *Research and Technology Division report for 1966,* ed. C. Baum. Report TM–530/010/00. Santa Monica, Calif.: System Development Corporation.

Shure, G. H., and Meeker, R. J. 1963. *Real-time computer studies of bargaining behavior: The effects of threat upon bargaining.* Report SP-1143/000/01. Santa Monica, Calif.: System Development Corporation.

_____. 1967. *Bargaining and negotiation behavior, quarterly technical progress report.* Report TM-2304/100/00. Santa Monica, Calif.: System Development Corporation.

Shure, G. H.; Meeker, R. J.; and Hansford, E. A. 1965. The effectiveness of pacifist strategies in bargaining games. *Journal of Conflict Resolution* 9: 106-17.

Shure, G. H.; Meeker, R. J.; and Moore, W. H. 1963. *Human bargaining and negotiation behavior: Computer-based empirical studies I. The effects of threat upon bargaining.* Report TM-1330/000/00. Santa Monica, Calif.: System Development Corporation.

Shure, G. H.; Rogers, M. S.; and Meeker, R. J. 1961. *Complexity, realism and simulation in decision research: A study in command staff decision making.* Report FN-5768. Santa Monica, Calif.: System Development Corporation.

_____. 1963. Group decision-making under conditions of realistic complexity. *Human Factors* 5: 49-58. Also, 1961. Report SP-344/000/01. Santa Monica, Calif.: System Development Corporation.

Sidman, M. 1960. *Tactics of scientific research.* New York: Basic Books.

Sidorsky, R. C. 1966. Predicting the decision behavior of a knowledgeable opponent. Paper read at the annual meeting of the Human Factors Society.

Sidorsky, R. C.; Houseman, Joan F.; and Ferguson, D. E. 1964. *Behavioral operational aspects of tactical decision making in AAW and ASW.* Report NAVTRADEVCEN 1329-1. New London, Conn.: General Dynamics Corporation, Electric Boat Division.

Siegel, A. I. 1967. Prediction of individual and crew performance by computer simulation. Paper read at Naval Material Command Symposium on Human Performance Quantification in Systems Effectiveness.

Siegel, A. I., and Wolf, J. J. 1961. A technique for evaluating man-machine system design. *Human Factors* 3: 18-28.

_____. 1962. A model for digital simulation of two-operator man-machine systems. *Ergonomics* 5: 557-72.

_____. 1963. Computer simulation of man-machine systems. In *Unusual environments and human behavior: Psychological problems of man in space*, eds. N. M. Burns, R. M. Chambers, and E. Hendler. London: Free Press of Glencoe.

_____. 1969. *Man-machine simulation models.* New York: Wiley.

Sinaiko, H. W. 1954. Simulation laboratories for CIC systems. *ONR Research Reviews.*

_____. 1958. *Artful: An experimental study of an automatic air defense system under varying conditions of human intervention.* Report R-104. Urbana, Ill.: University of Illinois Coordinated Science Laboratory.

_____. 1962. The fifth revolution. *Proceedings of the Military Operations Research Symposia (MORS) 2*, No. 1.

_____. 1963. *Teleconferencing: Preliminary experiments.* Paper P-108. Arlington, Va.: Institute for Defense Analyses.

_____. 1964*a*. Preliminary experiments in teleconferencing. *Proceedings of the Fifth National Symposium on Human Factors in Electronics.*

_____. 1964*b*. Experiments in international teleconferencing. Paper read at the 15th International Congress of Applied Psychology.

Sinaiko, H. W., and Belden, T. G. 1965. The indelicate experiment. In *Second Congress on the Information System Sciences,* eds. J. Spiegel and D. E. Walker. Washington, D.C.: Spartan Books.

Sinaiko, H. W., and Buckley, E. P. 1961. Human factors in the design of systems. In *Selected papers on human factors in the design and use of control systems*, ed. H. W. Sinaiko. New York: Dover.

Sinaiko, H. W., and Cartwright, G. P. 1959. *Careful: A pilot study of the effects of heavy target load on human and automatic decision makers.* Report R-115. Urbana, Ill.: University of Illinois Coordinated Science Laboratory.

Sinaiko, H. W., and Shpiner, L. 1960. *Experiments on the performance of an automatic air defense system.* Report R-113. Urbana, Ill.: University of Illinois Coordinated Science Laboratory.

Sinaiko, H. W.; Lefford, A.; and Taubman, R. E. 1951. *Functional relationship of time density to the detection of discrete radar target events.* Technical report SPECDEVCEN 279-3-6. New York: New York University.

Sinaiko, H. W.; Buckley, E. P.; Chauvette, L. M.; and Erwin, W. W. 1955. *The transplot system: A method of improving CIC summary air plotting.* Report 4565. Washington, D.C.: Naval Research Laboratory.

Sinaiko, H. W.; Bailey, W. E.; Buckley, E. P.; Chauvette, L. M.; and Erwin, W. W. 1956. *Evaluation of the Sea Mink Mk. III: A clamp-on projection plotter for shipboard use.* Memo Report 587. Washington, D.C.: Naval Research Laboratory.

Sinaiko, H. W.; Bailey, W. E.; Erwin, W. W.; Irish, A. S.; Quinnell, E. H.; and Scott, H. T. 1954. *Performance evaluation of the Miller optical projection system and the Kenyon rapromatic target PPI system.* Memo Report 322. Washington, D.C.: Naval Research Laboratory.

Singleton, J. W. 1964. Simulation in command and control. Paper read at the University of California, Los Angeles.

Slattery, H. F. 1965. *Air traffic control simulation program conference, report of the chairman.* Research and Development Service. Atlantic City, N.J.: Federal Aviation Agency National Aviation Facilities Experimental Center.

Sleight, R. B. 1952. The relative discriminability of several geometric forms. *Journal of Experimental Psychology* 43: 324-28.

Sluka, A. L. 1963. *Dynamic simulation studies of pictorial navigation displays as aids to air traffic control in a low-density terminal area and in an enroute area.* Final Report. Atlantic City, N.J.: Federal Aviation Agency National Aviation Facilities Experimental Center.

Sluka, A. L.; Bradley, J. R.; Yungman, D. W.; and Martin, D. W. 1962. *A simulation study of IFR helicopter operations in the New York area.* Final Report. Atlantic City, N.J.: Federal Aviation Agency National Aviation Facilities Experimental Center.

Smith, S. L., and Duggar, B. C. 1964. *Group vs. individual displays for a search and counting task.* Report MRS-15. Bedford, Mass.: MITRE Corporation.

Smode, A. F.; Hall, E. R.; and Meyer, D. E. 1966. *An assessment of research relevant to pilot training.* Report AMRL-TR-66-196. Wright-Patterson Air Force Base, Ohio: Aerospace Medical Research Laboratories.

Snyder, R. C. 1963. Some perspectives on the use of experimental techniques in the study of international relations. In *Simulation in international relations,* eds. H. Guetzkow; C. F. Alger; R. A. Brody; R. C. Noel; and R. C. Snyder. Englewood Cliffs, N.J.: Prentice-Hall.

Soliday, S. M., and Schohan, B. 1965. Task loading of pilots in simulated low-altitude high-speed flight. *Human Factors* 7: 45-53.

Southard, J. F.; Schum, D. A.; and Briggs, G. E. 1964*a. An application of Bayes Theorem as a hypothesis-selection aid in a complex information-processing system.* Report AMRL-TDR-64-51. Wright-Patterson Air Force Base, Ohio: Aerospace Medical Research Laboratories.

_____. 1964*b. Subject control over a Bayesian hypothesis-selection aid in a complex information-processing system.* Report AMRL-TR-64-95. Wright-Patterson Air Force Base, Ohio: Aerospace Medical Research Laboratories.

Space Technology Laboratories. No date. *Crew procedures research and trainer development program.* Los Angeles, Calif.: Space Technology Laboratories.

Spiegel, J.; Summers, J. K.; and Bennett, E. M. 1966. *AESOP: A general purpose approach to real-time, direct access management information systems.* MITRE Corporation Report MTP-33. Also, Report ESD-TR-66-289. L. G. Hanscom Field, Bedford, Mass.: Electronic Systems Division.

Sprowls, R. C. 1962. Business simulation. In *Computer applications in the behavioral sciences,* ed. H. Borko. Englewood Cliffs, N.J.: Prentice-Hall.

Staff, Systems Research Laboratory. 1953, revised 1954. *Crew learning in an experimental air defense organization.* Report RM-1024-1. Santa Monica, Calif.: RAND Corporation.

Stager, P. 1966. *Conceptual level as a composition variable in small group decision making.* Princeton University Report No. 10. Washington, D.C.: Office of Naval Research.

Stavid Engineering, Inc. 1959. *Research directed toward installation, testing, development, and modification of experimental data processing equipment, AFCRC Electronic Research Directorate.* Final Engineering Report. Plainfield, N.J.: Stavid Engineering, Inc.

Streufert, S.; Graber, F. J.; and Schroder, H. M. 1964. *Performance and perceptual complexity in a tactical decision making task.* Princeton University Report No. 1. Washington, D.C.: Office of Naval Research.

Streufert, S.; Schroder, H. M.; and Grenoble, W. L. 1964. *Some effects of discrepancy from optimal information load and diversity.* Princeton University Report No. 2. Washington, D.C.: Office of Naval Research.

Sulzer, R. L. 1959. *SAINT computer intercept tests.* Report TM 59-3. L. G. Hanscom Field, Bedford, Mass.: Operational Applications Laboratory.

Sulzer, R. L. and Cameron, D. B. 1959. *Tracking studies: Joystick and blanking Cartrac comparison.* Report TM-59-2. L. G. Hanscom Field, Bedford, Mass.: Operational Applications Laboratory.

Summers, J. K., and Bennett, E. 1967. AESOP–A final report: A prototype on-line interactive information control system. In *Information system science and technology papers prepared for the Third Congress*, ed. E. D. Walker. Washington, D.C.: Thompson Book Company.

Summers, J. K., and Hazle, Marlene. 1965. AESOP A/1 system description Part I, data base manipulations. In *System design and implementation subdepartment, information sciences department, AESOP prototype specifications*. Report MTP–13. Bedford, Mass.: MITRE Corporation.

Swain, A. D. 1967. Field calibrated simulation. Paper read at the Symposium on Human Performance Quantification in Systems Effectiveness of the Naval Material Command and the National Academy of Engineering.

Sweetland, A. 1961. *Operations and support scheduling methods derived from laboratory problem II (LP–II): A manned ICBM simulation*. Report RM–2669–PR. Santa Monica, Calif.: RAND Corporation.

Sweetland, A., and Haythorn, W. W. 1961. An analysis of the decision-making functions of a simulated air defense direction center. *Behavioral Science* 6: 105–16.

System Development Corporation. 1959. *Final report: Contract No. AF19(604)–2635*. Report AFCRC–TR–59–56. L. G. Hanscom Field, Bedford, Mass.: Air Force Cambridge Research Center.

Systems Research Laboratory Staff. 1945*a*. *AA target designation studies: Baltimore class heavy cruiser*. Research Report 1. Cambridge, Mass.: Harvard University Systems Research Laboratory Field Station.

_____. 1945*b*. *Time and motion analysis of combat information centers: CL–106 class, light cruiser*. Research Report 2. Cambridge, Mass.: Harvard University Systems Research Laboratory Field Station.

_____. 1945*c*. *Motion and time analysis of A/A coaching and CIC layout, CL–89, USS Miami*. Research Report 3. Cambridge, Mass.: Harvard University Systems Research Laboratory Field Station.

_____. 1945*d*. *Motion and time analysis of combat information centers, Tucson–Nashville–Louisville–Boston*. Research Report 4. Cambridge, Mass.: Harvard University Systems Research Laboratory Field Station.

_____. 1945*e*. *Equipment evaluation studies. The use of remote PPI's: VF and VD*. Research Report 5. Cambridge, Mass.: Harvard University Systems Research Laboratory Field Station.

_____. 1945*f*. *AA target designation studies. Battleship North Carolina class*. Research Report 6. Cambridge, Mass.: Harvard University Systems Research Laboratory Field Station.

_____. 1945*g*. *AA target designation studies: Voice transmission of target information*. Research Report 7. Cambridge, Mass.: Harvard University Systems Research Laboratory Field Station.

_____. 1945*h*. *Experimental methods and special exercises used for shipboard CIC studies*. Research Report 8. Cambridge, Mass.: Harvard University Systems Research Laboratory Field Station.

_____. 1945*i*. *Equipment evaluation studies: Automatic target positioner*. Research Report 9. Cambridge, Mass. Harvard University Systems Research Laboratory Field Station.

_____. 1945*j*. *Effect of plotting load on the quality of plots*. Research Report 10. Cambridge, Mass.: Harvard University Systems Research Laboratory Field Station.

_____. 1945*k*. *A study of plotting techniques*. Research Report 11. Cambridge, Mass.: Harvard University Systems Research Laboratory Field Station.

_____. 1945*l*. *Systems performance studies–value of plotting boards at command stations*. Research Report 12. Cambridge, Mass.: Harvard University Systems Research Laboratory Field Station.

_____. 1945*m*. *Analysis of AA target designation systems*. Research Report 13. Cambridge, Mass.: Harvard University Systems Research Laboratory Field Station.

_____. 1945*n*. *CIC problem generator and display system. General electronic and constructional features*. Research Report 14. Cambridge, Mass.: Harvard University Systems Research Laboratory Field Station.

_____. 1945*o*. *CIC problem generator and display system, operating and maintenance manual*. Research Report 15. Cambridge, Mass.: Harvard University Systems Research Laboratory Field Station.

Taylor, F. V. 1963. Human engineering and psychology. In *Psychology: A study of a science*, ed. S. Koch. New York: McGraw-Hill. Also, 1959. Washington, D.C.: U.S. Naval Research Laboratory.

Teare, R. J. 1965. *Voice degradation and teleconference behavior*. Technical Note 4. McLean, Va.: Human Science Research, Inc.

Teichroew, D., and Lubin, J. F. 1966. Computer simulation–Discussion of the technique and comparison of languages. *Communications of the ACM 723–41.*

Test and Experimentation Division. 1960. *A report on dynamic simulation tests and study of the Bureau of Air Traffic Management plan for the positive control of air traffic on an area basis in the Chicago and Indianapolis air route traffic control center areas*. Atlantic City, N.J.: Federal Aviation Agency National Aviation Facilities Experimental Center.

Thibaut, J. W., and Kelley, H. H. 1959. *The social psychology of groups*. New York: Wiley.

Tompkins, S. S., and Messick, S. (eds.) 1963. *Computer simulation of personality*. New York: Wiley.

Townsend, J. C. 1953. *Introduction to experimental method*. New York: McGraw-Hill.

Truesdell, R. D. 1963. *Final report: DIAL (display alleviation) test series*. Report TM-3635. Bedford, Mass.: MITRE Corporation.

Twery, R. J. 1961. *An evaluation of a casualty assessment system employed in the CDEC TUELE I experiment*. Research Memorandum RO-RM 15. Menlo Park, Calif.: Stanford Research Institute.

Twery, R. J., Barson, S.; and Johnson, J. J. 1963. *Field and computer simulation of tank-antitank combat*. Research Memorandum RO-RM 32. Menlo Park, Calif.: Stanford Research Institute.

Uhlaner, J. E., and Drucker, A. J. 1964. Criteria for human performance research. *Human Factors* 6: 265–78.

Underwood, B. J. 1957. *Psychological research.* New York: Appleton-Century-Crofts.

U.S. Army Combat Developments Command Experimentation Command. 1966. *USACDCEC Experimentation Manual.* Fort Ord, Calif.: U.S. Army Combat Developments Command Experimentation Command.

U.S. Army Strategy and Tactics Analysis Group–STAG. 1962. *Directory of organizations and activities engaged or interested in war games.*

Van Cott, H. P., and Kinkade, R. G. 1968. Human simulation applied to the functional design of information systems. *Human Factors* 10: 211–16.

Vaughan, W. S., and Kassebaum, R. G. 1957. *Target definition, location and selection. Research Study Report V.* Report 57–11. Arlington, Va.: Psychological Research Associates.

Veniar, S. 1953. *The effect of continuous operation on the AEW function of air control officers in the Airborne CIC.* Technical Report SPECDEVCEN 279–3–12. New York: New York University.

Versace, J. 1956. *The effect of emergencies and communications availability with differing entry rates: A study in human engineering aspects of radar air traffic control.* Report TR 56–70. Wright-Patterson Air Force Base, Ohio: Wright Air Development Center.

Vickers, T. K. 1954. *Development of traffic control procedures for tactical airlift operation.* Report No. 235. Indianapolis, Ind.: Civil Aeronautics Administration Technical Development and Evaluation Center.

_____. 1957. *Simulation tests for Army air traffic control.* Report No. 298. Indianapolis, Ind.: Civil Aeronautics Administration Technical Development and Evaluation Center.

_____. 1959. *The use of simulation in ATC systems engineering.* Report No. 410. Indianapolis, Ind.: Civil Aeronautics Administration Technical Development and Evaluation Center.

_____. 1965. *Air traffic control simulation program conference, report of the chairman.* Atlantic City, N.J.: Federal Aviation Agency National Aviation Facilities Experimental Center.

Vickers, T. K., and Miller, R. S. 1956. Recent developments in the simulation of terminal area and enroute area air traffic control problems. *IRE Transactions on Aeronautical and Navigational Electronics* ANE–3: 51–55.

von Buelow, R., 1962. *The SSRL display system.* Report SP–535. Santa Monica, Calif.: System Development Corporation.

von Buelow, R., and The Computer Systems Engineering Group. 1961. *The equipment of the Systems Simulation Research Laboratory.* Report SP–513. Santa Monica, Calif.: System Development Corporation.

Wakeman, R. L. 1962. *Simulation study I: A preliminary report.* Report LO–689/000/00. Santa Monica, Calif.: System Development Corporation.

Walker, S. H. 1962. *Computer simulation and gaming in logistics research.* Report SP–187. Bethesda, Md.: Research Analysis Corporation.

Ward, J. S., and Fooks, N. I. 1965. *Development of improved rifle squad tactical and patrolling programs for the light weapons infantryman.* Technical Report 65–16. Alexandria, Va.: Human Resources Research Office.

Weiner, M. G. 1959*a*. *War gaming methodology.* Report RM-2413. Santa Monica, Calif.: RAND Corporation.

_____. 1959*b*. *An introduction to war games.* Report P-1773. Santa Monica, Calif.: RAND Corporation.

_____. 1960. *Gaming limited war.* Report P-2123. Santa Monica, Calif.: RAND Corporation.

_____. 1961. *The use of war games in command and control analysis.* Report P-2466. Santa Monica, Calif.: RAND Corporation.

Weiner, M. G., and Sinaiko, H. W. 1953. *Study of information filtering activities in a command ship.* Report 4213. Washington, D.C.: Naval Research Laboratory.

Westbrook, C. B. 1961. *Simulation in modern aerospace vehicle design.* AGARD Report 336.

Whittenburg, J. A.; Havron, M. D.; McGrath, J. E.; and Kassebaum, R. G. 1956. *A study of the infantry rifle squad TOE.* Report 56-3. Arlington, Va.: Psychological Research Associates.

Wiechers, J. E. 1963. *The system training program applied to command control systems.* Report SP-1416. Santa Monica, Calif.: System Development Corporation.

Wiener, E. L. 1964. The performance of multi-man monitoring teams. *Human Factors* 6: 179-84.

Williams, S. B., and Hanes, R. M. 1949. Visibility on cathode-ray tube screens: Intensity and color of ambient illumination. *Journal of Psychology* 27: 231-44.

Williges, R. C.; Johnston, W. A.; and Briggs, G. E. 1966. Role of verbal communication in teamwork. *Journal of Applied Psychology* 50: 473-78.

Willis, R. H., and Long, Norma Jean. 1967. An experimental simulation of an international truel. *Behavioral Science* 12: 24-32.

Wolin, B. R. 1959. *Methodology note: On the design and redesign of systems.* Report AFCRC-TN-59-70. L. G. Hanscom Field, Bedford, Mass.: Air Force Cambridge Research Center. Also, Report FN-2600. Santa Monica, Calif.: System Development Corporation.

Wood, R. C., and Friedman, M. P. 1964. *Force allocation experimentation: Report of experiment II.* Report TM-(L)-939/002/00. Santa Monica, Calif.: System Development Corporation.

Young, J. P. 1957. *History and bibliography of war gaming.* Baltimore: The Johns Hopkins University Operations Research Office.

Zajonc, R. B. 1961. *The effects of feedback and group task difficulty on individual and group performance.* Technical Report 15. Washington, D.C.: Office of Naval Research.

Zelditch, M., Jr., and Evans, W. M. 1962. Simulation hierarchies: A methodological analysis. In *Simulation in social science: Readings,* ed. H. Guetzkow. Englewood Cliffs, N.J.: Prentice-Hall.

# Name Index

# Subject Index

THE JOHNS HOPKINS UNIVERSITY PRESS

Composed in Press Roman
by Jones Composition Company

Printed on 50-lb. Sebago
by Universal Lithographers, Inc.

Bound in Interlaken Matte
by The Maple Press Company